# METHODS IN MOLECULAR BIOLOGY™



For other titles published in this series, go to
www.springer.com/series/7651

# Proteomics

## Methods and Protocols

Edited by

**Jörg Reinders* and Albert Sickmann†**

**University of Regensburg, Institute of Functional Genomics, Joseph-Engert Strasse 9 93053 Regensburg, Germany*
*†Institut für Spektrochemie und Angewandte Spektroskopie (ISAS), Bunsen-Kirchoff Str. 11 44139 Dortmund, Germany*

*Editors*
Jörg Reinders
University of Regensburg
Institute of Functional
Genomics
Joseph-Engert-Strasse 9
93053 Regensburg
Germany

Albert Sickmann
Institut für Spektrochemie und
Angewandte Spektroskopie
(ISAS)
Bunsen-Kirchoff Str. 11
44139 Dortmund
Germany

ISSN: 1064-3745 e-ISSN: 1940-6029
ISBN: 978-1-60761-156-1 e-ISBN: 978-1-60761-157-8
DOI: 10.1007/978-1-60761-157-8
Springer Dordrecht Heidelberg London New York

Library of Congress Control Number: 2009927501



Printed on acid-free paper

Springer is part of Springer Science+Business Media (www.springer.com)

# Preface

Proteins are essential players in all cellular processes, facilitating various functions as enzymes and structure-forming or signal-transducing molecules. Their enormous versatility in primary structure, folding, and modification enables a complex, highly dynamic, but nevertheless robust, network carrying out all the necessary tasks to ensure proper function of each cell and concerted activity of cellular associations up to complex organisms. Therefore, proteins have always been, and presumably will always be, the target of all kinds of studies in biological sciences.

Protein purification and separation methods have a longstanding record as they were a prerequisite for enzymological studies and chemical protein identification methods such as Edman-sequencing. Thus, various elaborate and mostly time-consuming techniques for the isolation of distinct proteins have been developed often based on chromatography or electrophoresis, and the identification of the protein's primary structure was accomplished afterwards by no less intricate methods. However, the relatively recent development of MALDI- and ESI-ionization techniques for mass spectrometric analysis of large and fragile biomolecules enabled protein identification in an automated fashion, thereby speeding up protein identification by a multiple. This turned out to be a major breakthrough in protein analysis enabling high-throughput protein identification on a global scale, leading to approaches to study the entirety of all proteins of a cell, tissue, organ, etc.

In 1995, the term "Proteome" was introduced by Marc Wilkins and Keith Williams as the entirety of all proteins encoded in a single genome expressed under distinct conditions representing the turning point in the journey from studying genes to studying proteins, from "Genomics" to "Proteomics." Since then, great efforts have been undertaken to characterize a "healthy" or a "diseased" proteome, but it soon turned out that a proteome is far too complex and dynamic to be defined by such simple terms. The enormous progress that has been accomplished both technically and biologically has not only granted deeper insight into the cellular network but has also raised further questions and set further challenges to proteomic research.

The enormous range of protein abundance, dynamics, and interactions as well as the spatio-temporal distribution of a proteome gave rise to the evolution of several new fields like phospho-, glyco-, subcellular, and membrane proteomics, etc. Many techniques have been developed or significantly increased in these fields and will contribute to the understanding of the cellular networks in the future.

Leading scientists have contributed to this volume, which is intended to give an overview of the contemporary challenges and possibilities in the various areas of proteomics and to offer some detailed protocols as examples for successful analysis in proteomics studies. Therefore, we hope that this book can raise your interest in proteomics and be a valuable reference book for your laboratory work.

# Contents

## Contributors

ROBERT AHRENDS • *Department of Chemistry, Humboldt-Universität zu Berlin, Brook-Taylor Str. 2, 12489 Berlin, Germany*

ROLF APWEILER • *EMBL Outstation – Hinxton, European Bioinformatics Institute, Wellcome Trust Genome Campus, Hinxton, Cambridge CB10 1SD, UK*

GEORG J. ARNOLD • *Laboratory for Functional Genome Analysis LAFUGA, Gene Center, Ludwig-Maximilians-University Munich, Feodor-Lynen-Str. 25, 81377 Munich, Germany*

MOHAN BABU • *Banting and Best Department of Medical Research, University of Toronto, Donnelly Center for Cellular and Biomolecular Research, 160 College Street, Toronto, Ontario, Canada M5S 3E1*

KARSTEN BOLDT • *Department of Protein Science, Helmholtz Zentrum München, Ingolstaedter Landstr. 1, 85764 Neuherberg, Germany; Institute of Human Genetics, Klinikum rechts der Isar, Technical University of Munich, Munich, Germany; Helmholtz Zentrum München – German Research Center for Environmental Health, Department of Protein Science, Ingolstaedter Landstr. 1, 85764 Neuherberg, Germany*

JACQUELINE BURRÉ • *Department of Neuroscience, The University of Texas Southwestern Medical Center, 6000 Harry Hines Boulevard, Dallas, TX, 75390-911, USA*

GARETH BUTLAND • *Banting and Best Department of Medical Research, University of Toronto, Donnelly Center for Cellular and Biomolecular Research, 160 College Street, Toronto, Ontario, Canada M5S 3E1; Life Science Division, Lawrence Berkeley National Lab, 1 Cyclotron Road MS 84R0171, Berkeley, CA 94720*

D. ALLAN BUTTERFIELD • *Department of Chemistry, Center of Membrane Sciences, and Sanders-Brown Center on Aging, University of Kentucky, Lexington, KY 40506-0055, USA*

KO-YI CHIEN • *Department of Molecular and Structural Biochemistry, North Carolina State University, Raleigh, NC 27695-7622, USA*

STEVEN D. CLOUSE • *Department of Horticultural Science, North Carolina State University, Raleigh, NC 27695-7609, USA*

RAINER CRAMER • *The BioCentre and Department of Chemistry, The University of Reading, Whiteknights, Reading, RG6 6AS, UK*

THOMAS DANDEKAR • *Bioinformatik, Biozentrum, Am Hubland, 97074 Universitaet Wuerzburg, Germany*

LUTZ A. EICHACKER • *Universitetet i Stavanger, Centre for Organelle Research, Kristine-Bonnevisvei 22, 4036 Stavanger, Norway*

ANDREW EMILI • *Banting and Best Department of Medical Research, University of Toronto, Donnelly Centre for Cellular and Biomolecular Research, 160 College Street, Toronto, Ontario, Canada M5S 3E1*

THOMAS FRÖHLICH • *Laboratory for Functional Genome Analysis LAFUGA, Gene Center, Ludwig-Maximilians-University Munich, Feodor-Lynen-Str. 25, 81377 Munich, Germany*

CHRISTIAN JOHANNES GLOECKNER • *Department of Protein Science, Helmholtz Zentrum München – German Research Center for Environmental Health, Ingolstaedter Landstrasse 1, 85764 Neuherberg, Germany*

ANGELIKA GÖRG • *Technische Universität München (TUM), Life Science Center Weihenstephan (WZW), Area: Proteomics, Am Forum 2, 85350 Freising-Weihenstephan, Germany*

MICHAEL B. GOSHE • *Department of Molecular and Structural Biochemistry, North Carolina State University, 128 Polk Hall, Campus Box 7622, Raleigh NC 27695-7622, USA*

JACK F. GREENBLATT • *Banting and Best Department of Medical Research, University of Toronto, Donnelly Center for Cellular and Biomolecular Research, 160 College Street, Toronto, Ontario, Canada M5S 3E1; Department of Medical Genetics and Microbiology, University of Toronto, 1 King's College Circle, Toronto, Ontario, Canada M5S 1A8*

MARCUS GROETTRUP • *Division of Immunology, Department of Biology, University of Konstanz, D-78457 Konstanz, Germany*

SAM HANASH • *Fred Hutchinson Cancer Research Center, 1100 Fairview Avenue N., M5-C800, P.O. Box 19024, Seattle, WA 98109, USA*

UMA KOTA • *Department of Molecular and Structural Biochemistry, North Carolina State University, Raleigh, NC 27695-7622, USA*

BEATE KRÜGER • *Bioinformatik, Biozentrum, Am Hubland, 97074 Universitaet Wuerzburg, Germany*

ANDREAS KÜHN • *Department of Chemistry, Humboldt-Universität zu Berlin, Brook-Taylor Str. 2, 12489 Berlin, Germany*

JOYCE LI • *Banting and Best Department of Medical Research, University of Toronto, Donnelly Center for Cellular and Biomolecular Research, 160 College Street, Toronto, Ontario, Canada M5S 3E1*

MICHAEL W. LINSCHEID • *Department of Chemistry, Humboldt-Universität zu Berlin, Brook-Taylor Str. 2, 12489 Berlin, Germany*

KATHARINA LOHRIG • *Department of Analytical Chemistry, Ruhr-University Bochum, Universitaetsstr. 150, 44780 Bochum, Germany*

FRIEDRICH LOTTSPEICH • *Protein Analytics, Max-Planck-Institute of Biochemistry, Martinsried, Germany*

BINGWEN LU • *Department of Chemical Physiology, The Scripps Research Institute, La Jolla, CA, USA*

LENNART MARTENS • *EMBL Outstation – Hinxton, European Bioinformatics Institute, Wellcome Trust Genome Campus, Hinxton, Cambridge, UK*

HARALD MISCHAK • *Mosaiques diagnostics GmbH, Mellendorfer Str. 7-9, 30625 Hannover, Germany*

SRIJEET K. MITRA • *Department of Horticultural Science, North Carolina State University, Raleigh, NC 27695-7609, USA*

HANS GERD NOTHWANG • *Abteilung Neurogenetik, Institut für Biologie und Umweltwissenschaften, Carl von Ossietzky Universität, 21111 Oldenburg, Germany*

ROBIN PARK • *Department of Chemical Physiology, The Scripps Research Institute, La Jolla, CA, USA*

MARZIA PERLUIGI • *Department of Biochemical Sciences, University of Rome "La Sapienza", 00185, Rome, Italy*

STEFAN PIEPER • *Department of Chemistry, Humboldt-Universität zu Berlin, Brook-Taylor Str. 2, 12489 Berlin, Germany*

OXANA POGOUTSE • *Banting and Best Department of Medical Research, University of Toronto, Donnelly Center for Cellular and Biomolecular Research, 160 College Street, Toronto, Ontario, Canada M5S 3E1*

REGINA PREYWISCH • *Division of Immunology, Department of Biology, University of Konstanz, Konstanz, Germany*

MICHAEL PRZYBYLSKI • *Department of Chemistry, Laboratory of Analytical Chemistry and Biopolymer Structure Analysis, University of Konstanz, 78457 Konstanz, Germany*

VERONIKA REISINGER • *Universitetet i Stavanger, Centre for Organelle Research, Kristine-Bonnevisvei 22, 4036 Stavanger, Norway*

HERMANN SCHÄGGER • *Molekulare Bioenergetik, Zentrum der Biologischen Chemie, Fachbereich Medizin, Universität Frankfurt, Theodor-Stern-Kai 7, Haus 26, D-60590 Frankfurt am Main, Germany*

ERIC SCHIFFER • *Mosaiques diagnostics GmbH, Mellendorfer Str. 7-9, 30625 Hannover, Germany*

JENS SCHINDLER • *Abteilung Neurogenetik, Institut für Biologie und Umweltwissenschaften, Carl von Ossietzky Universität, 21111 Oldenburg, Germany*

CARLA SCHMIDT • *Bioanalytical Mass Spectrometry Group, Max Planck Institute for Biophysical Chemistry, Am Fassberg 11, 37077 Göttingen, Germany*

ANNETTE SCHUMACHER • *Department of Protein Science, Helmholtz Zentrum München, Ingolstaedter Landstr. 1, 85764 Neuherberg, Germany*

ALBERT SICKMANN • *Institut für Spektrochemie und Angewandte Spektroskopie (ISAS), Bunsen-Kirchoff Str. 11 44139 Dortmund, Germany*

OLIVER SIMON • *Rudolf-Virchow-Center, DFG-Research Center for Experimental Biomedicine, Wuerzburg, Germany*

RUKHSANA SULTANA • *Department of Chemistry, Sanders-Brown Center on Aging, University of Kentucky, Lexington, KY, USA*

MARIUS UEFFING • *Department of Protein Science, Helmholtz Zentrum München, Ingolstaedter Landstr. 1, 85764 Neuherberg, Germany; Institute of Human Genetics, Klinikum rechts der Isar, Technical University of Munich, Munich, Germany*

HENNING URLAUB • *Bioanalytical Mass Spectrometry Group, Max Planck Institute for Biophysical Chemistry, Am Fassberg 11, 37077 Göttingen, Germany*

HONG WANG • *Fred Hutchinson Cancer Research Center, Seattle, WA 98109, USA*

REINHOLD WEBER • *Laboratory of Analytical Chemistry and Biopolymer Structure Analysis, Department of Chemistry, University of Konstanz, Konstanz, Germany*

FLORIAN WEILAND • *Fachgebiet Proteomik, Technische Universität München, Freising-Weihenstephan, Germany*

WALTER WEISS • *Technische Universität München, Fachgebiet Proteomik, Am Forum 2, D-85350 Freising-Weihenstephan, Germany*

ILKA WITTIG • *Molekulare Bioenergetik, Zentrum der Biologischen Chemie, Centre of Excellence "Macromolecular Complexes", Fachbereich Medizin, Johann Wolfgang Goethe-Universität Frankfurt, Theodor-Stern-Kai 7, D-60590 Frankfurt am Main, Germany*

DIRK WOLTERS • *Department of Analytical Chemistry, Ruhr-University Bochum, Universitaetsstr. 150, 44780 Bochum, Germany*

STEFANIE WORTELKAMP • *Institut für Spektrochemie und Angewandte Spektroskopie (ISAS), Bunsen-Kirchoff Str. 11 44139 Dortmund, Germany*

TAO XU • *Department of Chemical Physiology, The Scripps Research Institute, La Jolla, CA, USA*

JOHN R. YATES III • *Department of Chemical Physiology, The Scripps Research Institute, SR11, 10550 North Torrey Pines Rd., La Jolla, CA 92037, USA*

NIKOLAY YOUHNOVSKI • *Laboratory of Analytical Chemistry and Biopolymer Structure Analysis, Department of Chemistry, University of Konstanz, Konstanz, Germany; Algorithme Pharma Inc., Montreal, Montreal H7V 4B4, Canada*

PETRA ZÜRBIG • *Mosaiques diagnostics GmbH, Mellendorfer-Str. 7-9, 30625 Hannover, Germany*

# Part I

## Introduction

# Chapter 1

## Introduction to Proteomics

**Friedrich Lottspeich**

### Summary

In this chapter, the evolvement of proteomics from classical protein chemistry is depicted. The challenges of complexity and dynamics led to several new approaches and to the firm belief that a valuable proteomics technique has to be quantitative. Protein-based vs. peptide-based techniques, gel-based vs. non-gel-based proteomics, targeted vs. general proteomics, isotopic labeling vs. label-free techniques, and the importance of informatics are summarized and compared. A short outlook into the near future is given at the end of the chapter.

**Key words:** History, Quantitative proteomics, Targeted proteomics, Isotopic labeling, Protein-based proteomics, Peptide-based proteomics

### 1. The History and the Challenge

In the end of the last century, a change of paradigm from the pure function driven biosciences to systematic and holistic approaches has taken place. Following the successful genomics projects, classical protein chemistry has evolved into a high throughput and systematic science, called proteomics. Starting in 1995, the first attempts to deliver a "protein complement of the genome" used the established high-resolving separation techniques like two-dimensional (2D) gel electrophoresis and almost exclusively identified the proteins by the increasingly powerful mass spectrometry. Soon, fundamental and technical challenges were recognized. Unlike the genome, the proteome is dynamic, responding to any change in genetic and environmental parameters. Furthermore, the proteome appears to be orders of magnitude more complex than a genome owing to splicing and

Jörg Reinders and Albert Sickmann (eds.), *Proteomics,* Methods in Molecular Biology, Vol. 564
DOI: 10.1007/978-1-60761-157-8_1, © Humana Press, a part of Springer Science+Business Media, LLC 2009

editing processes at the RNA level and owing to all the post-translational events on the protein level, like limited processing, post-translational modifications, and degradation. The situation is even more difficult, since many important proteins are only present in a few copies/cells and have to be identified and quantified in the presence of a large excess of many other proteins. The dynamic range of the abundant and the minor proteins often exceeds the capabilities of all analytical methods.

So far, only few solutions are available to handle the complexity and dynamic range. One is to reduce the complexity of the proteome and to separate the low abundant proteins from the more abundant ones. This, for example, can be achieved by multidimensional separation steps. But, unpredictable losses of proteins and a large number of resulting fractions make this approach time-consuming and thus also very costly. Alternatively, the proteome to be investigated can be simplified by starting with a specific biological compartment or by reducing the complexity using a suitable sample preparation (e.g. enzyme ligand chips, functionalized surface chips, class-specific antibodies). Successful examples are the analysis of functional complexes or most interaction proteomics approaches. In another approach, a selective detection is performed, which visualizes only a certain number of proteins that exhibit specific common properties. This can be achieved by antibodies, selective staining protocols, protein ligands, or selective mass spectrometry techniques like MRM (multiple reaction monitoring) or SRM (single reaction monitoring) *(1)*. The most straightforward application of this approach is "targeted proteomics," which monitors a small set of well-known proteins/peptides.

However, in the later years of the past century, the main focus of proteomics projects was to decipher the constituents of a proteome. It was realized only slowly that for solving biological problems and realizing the potential of holistic approaches, the changes and the dynamics of changes on the protein level have to be monitored quantitatively.

## 2. Gel-Based Proteomics

Since 1975 by their introduction in by O'Farrel *(2)* and Klose *(3)*, 2D gels have fascinated many scientists owing to their separation power. The combination of a concentrating technique, i.e. isoelectric focusing, with a separation according to molecular mass, i.e. SDS gel electrophoresis, provides a space for resolving more than 10,000 different compounds. Consequently, 2D gels were the method of choice when dealing with very complex protein mixtures like proteomes. Unfortunately, gel-based proteomics had inherent limitations in reproducibility and dynamic range. Standard operating procedures had to be carefully followed to get almost reproducible results even within one lab. Results produced from identical samples in different labs were hardly comparable on a quantitative level. A significant improvement was the introduction of the DIGE technique (GE Healthcare), a multiplexed fluorescent Cy-Dye staining of different proteome states, which eliminated to a large extent the technical irreproducibility *(4)*. With the cysteine-modifying "DIGE saturation labeling," impressive proteome visualization can be achieved with only a few micrograms of starting material *(5)*. A disadvantage is that only two different fluorescent reagents are commercially available for "complete DIGE" and the costs of the reagents are rather prohibitive for larger proteomics projects. Additionally, limitations in load capacity, quantitative reproducibility, difficulties in handling, and interfacing problems to mass spectrometry limited the analysis depth and comprehensiveness of the gel-based proteomics studies.

## 3. Seeking Alternatives

### 3.1. Non-Gel-Based Electrophoresis

How to overcome the limitations of gels and at the same time keep the advantages of a concentrating separation mode like iso-electric focusing? Several instruments were developed that are able to separate proteins in solution but nevertheless use a focusing technique. Probably, the most recognized realizations of these concepts are free-flow electrophoresis instruments like "Octopus" (Becton Dickinson) and the "Off-Gel" system (Agilent). Undoubtedly, when these rather new systems are compared with 2D gels, distinct advantages in recovery and improvements in the amount that can be applied have been realized, but interfacing to a further separation dimension is hampered by rather large volumes and buffer constituents. Thus, the resolution of 2D gels had not been reached so far. In the near future, technical and applicative improvements are to be expected to partly overcome some of the limitations.

### 3.2. Chromatography

In the limited landscape of separation methods, chromatography seemed to have the potential as an alternative tool for in-depth proteome analysis. However, from classical protein chemistry, it was well known that proteins did not give quantitative recovery in many chromatographic modes. So far, only one non-gel multidimensional approach based on chromatographic methods was commercially realized. In the "ProteomeLab™ PF-2D" system

(Beckman), a chromatofocusing column coupled with a reversed phase chromatography fractionates the sample into more than 1,000 fractions. However, here also the advantage to keep the proteins in solution is compromised with the fact that the resolution of the fully chromatographic solution is considerably lower than that with 2D gels.

### 3.3. Peptide-Based Proteomics

Thus, since obviously quantitative multidimensional separations of proteins proved to be notoriously difficult, other alternatives were searched for. One conceptual new idea was to transfer the separation and quantification problem from the protein to the peptide level. If this could be achieved, a new dimension of speed, automation, and reproducibility can be obtained. Thus, new peptide-based strategies, e.g. MudPIT *(6)*, were developed where after cleaving the proteome into peptides, highly automated multidimensional liquid chromatography separations were followed by identification of the peptides using tandem mass spectrometry. Mainly owing to this switch to peptide-based proteomics, chromatography experienced a new boom, and miniaturiztion of peptide separation columns to diameters below 100 μm and introduction of instruments that were capable to deliver nanoliter flow rates became available. Nano-LC with online or off-line mass spectrometric detection became routine. However, in multidimensional mode, nano-LC is still on the border of technical practicability and it still suffers from lack of robustness and ease of handling.

With the application of the peptide-based proteomics strategies, several severe disadvantages became obvious. By cleaving the proteins into peptides, not only the complexity of the proteome was increased by tenfold, but important information concerning the protein identification was also destroyed. Many peptides are identically found in functionally completely different proteins. Thus, from a peptide, the progenitor usually cannot be deduced unequivocally. Furthermore, different isoforms, post-translationally modified proteins, or processing and degradation products of a protein, all produce a large set of identical peptides. As a result, the quantitative information for a certain protein becomes quite uncertain. Amounts of a peptide that are present in more than one protein species do not reflect the quantity of a single protein species, but rather the quantity of the sum of all protein species that contain this peptide.

Due to the complexity and the necessity to analyze and identify each peptide by tandem mass spectrometry, proteome analysis time and costs increased markedly. Strictly speaking, today even the most rapid mass spectrometers are not able to analyze in detail all the masses present in one LC run. Therefore, often especially minor peptides are not analyzed. This so-called "undersampling" is certainly one of the reasons for the usually bad reproducibility of proteome studies, where often a simple repetition of the analysis gives only 20%–30% of overlapping data.

As a consequence of all these aspects, reduction of complexity in quantitative proteomics should be done at protein level. The behavior of a protein during a separation is a characteristic parameter and should also be used for detailed identification and discrimination of single protein species.

## 4. Quantitative Proteomics Using Stable Isotopic Labeling

To improve the quantitative proteomics results, "isotope labeling" techniques were introduced. These "isotopic dilution" strategies were already well known for the analysis of small molecules, drugs, and metabolites. The pioneering work to introduce this technique into the proteomics field was done by the Aebersold group, where the cysteine residues in all proteins of two proteomic states were modified with a biotin-containing either heavy or light version of a reagent (isotope coding affinity tag, ICAT®) *(7)*. Then, the labeled proteomes were combined and cleaved into peptides. Only the cysteine-containing peptides carrying the label are isolated by affinity purification using streptavidin. Peptide separation and mass analysis revealed the identity of the peptides and at the same time determined by the signal intensity of the isotopic peptide pair the quantitative ratio of the peptides in the original proteomes. Improved versions of isotopic reagents were developed, e.g. isotope coding protein label, ICPL®(Serva), small amino group reactive reagents, which gave better reaction yields and increased sequence coverage *(8)*.

Of course, an introduction of the isotopic label as early as possible is desirable, since all the steps performed without the isotopic control may contribute to quantitatively wrong results. Therefore, introducing the isotopic label at an even earlier stage of a proteome analysis was developed. Culture media enriched with $N^{15}$ isotopes or stable isotope labeling of amino acids in cell culture (SILAC) was used in proteomics experiments, especially in cell culture or with microorganisms *(9)*. However, with a remarkable effort, a "SILAC mouse" was also generated and used in proteomics experiments *(10)*. The metabolic labeling approaches are usually restricted to cell culture experiments and are not applicable to samples from higher organisms (e.g. body fluids, tissues, etc.)

Also, for peptide-based approaches, a number of isotopic reagents were proposed. The most popular is iTRAQ (ABI), a family of eight isobaric amino group reactive reagents *(11)*. Because of the identical mass of all variants of the reagent, a certain peptide derived from different proteome states will appear with the identical mass and thus - in contrast to non-isobaric isotopic reagents – the labeling does not increase the complexity in the mass spectrum. However, with a simple, cheap, and rapid MS analysis, no quantitative data can be obtained. Only during MS/MS analysis, specific reporter ions for the different reagents will be liberated and can be quantified. To produce quantitative correct results, the mass selected for MS/MS analysis has to be rather pure. This often is not the case in crowded chromatograms. Consequently, the advantages of high multiplexing with isobaric reagents are somewhat diminished by the limitation to rather low complex peptide mixtures and by the task to analyze each derivatized peptide by MS/MS analysis to disclose quantitative results.

## 5. Informatics and Data Mining

One of the major difficulties in larger proteomics projects is the enormous amount of data that will be produced. Tens of thousands of mass spectra from each proteomic state can be analyzed only by using automated software solutions. Because of demanding peak detection in overcrowded spectra and challenging peptide/protein identification and the mere amount of data to be processed today, data analysis and data evaluation is by far the most time-consuming part of a proteome analysis. Software for automatically detecting the interesting proteins that change from one proteome state to another and filtering such proteins out of the complex proteome data can be expected in the near future.

## 6. State of the Art and Future

However, So far many proteomics experiments published did not really deliver solid and valuable scientific content. This partly is connected with the idea of holistic approaches per se, that the observation of the reactions of a perturbed system does not necessarily provide a simple and clear answer, but rather is a hypothesis generating concept. Unfortunately, the technical ability to cope with proteome complexity is still very limited despite the amazing technical progresses in mass spectrometry and nanoseparations. Consequently, it is often tried to analyze a proteome with significant effort, time, and money, though with today's analytics, most of the existing proteins are out of reach. Only a fraction of the proteome can be explored and to judge the significance and validity of the results, biological and statistical repetitions of the experiments are scientifically required. However, because of the large effort and high costs, this is often ignored. The danger is that in the long run, by ignoring good scientific praxis, the reliability of proteomics as an analytical technique may be queried.

Therefore, we are forced to elaborate intelligent and sophisticated strategies to obtain valid and valuable biological information with the existing technologies in sample preparation, separation sciences, mass spectrometry, and informatics. Closest to this goal is probably "targeted proteomics." Already today, this approach is able to monitor hundreds of known proteins quantitatively and sensitively and it will gain increasing acceptance and eventually enter routine clinical diagnostics.

With general comparative proteomics in attempting the holistic concept, the situation is more complicated with general comparative proteomics. Neither analysis depth nor quantitative accuracy is satisfactory today. Post-translational modifications and analysis of many different protein species originating from the same gene present major difficulties in high throughput approaches and require innovative strategies. Isotopic labeling techniques are in competition with label-free techniques. Although label-free approaches have demonstrated amazingly good results with simple protein mixtures, they have to substantiate this at the proteomics level and after multidimensional separation steps also. Most of the problems and shortcomings are recognized and many scientists are working on their solutions. After one decade of rapid improvements in analysis techniques and only slight improvement in the separation field, the acute pressure is now on the further development in separation sciences. Integrated, well–designed, and highly automated workflows using both chromatography and electrophoresis will be necessary to solve the ambitious proteomics separation problem. Novel separation strategies and interfacing solutions of highly automated multidimensional fractionation schemes are a challenging research area and will, to a large extent, determine the success of proteomics as a holistic approach in the future.

## References


1. Anderson L., Hunter C.L. (2006) Quantitative mass spectrometric multiple reaction monitoring assays for major plasma proteins. *Mol. Cell. Proteomics* 5, 573–588.
2. O'Farrell P.H. (1975) High resolution two-dimensional electrophoresis of proteins. *J. Biol. Chem.* 250, 4007–4021.
3. Klose J. (1975) Protein mapping by combined isoelectric focusing and electrophoresis of mouse tissues A novel approach to testing for induced point mutations in mammals. *Humangenetik* 26, 231–243.
4. Unlue M., Morgan M.E., Minden J.S. (1997) Difference gel electrophoresis: A single gel method for detecting changes in protein extracts. *Electrophoresis* 18, 2071–2077.
5. Sitek B., Luettges J., Marcus K., Kloeppel G., Schmiegel W., Meyer H.E., Hahn S.A., Stuehler K. (2005) Application of fluorescence difference gel electrophoresis saturation labelling for the analysis of microdissected precursor lesions of pancreatic ductal adenocarcinoma. *Proteomics* 5(10), 2665–2679.
6. Washburn M.P., Wolters D., Yates J.R. 3rd (2001) Large-scale analysis of the yeast

proteome by multidimensional protein identification technology. *Nat.Biotechnol. Mar*; 19(3), 242–277.

7. Gygi S.P., Rist B., Gerber S.A., Turecek F., Gelb H.M., Aebersold R. (1999) Quantitative analysis of complex protein mixtures using isotope-coded affinity tags. *Nat.Biotechnol.* 17, 994–999.
8. Schmidt A., Kellermann J., Lottspeich F. (2005) A novel strategy for quantitative proteomics using isotope-coded protein labels. *Proteomics* 5, 4–15.
9. Ong S.E., Blagoev B., Kratchmarova I., Kristensen D.B., Steen H., Pandey A., Mann M. (2002) Stable isotope labeling by amino acids in cell culture, SILAC, as a simple and accurate approach to expression proteomics. *Mol. Cell. Proteomics* 1, 376–386.
10. Krueger M., Moser M., Ussar S., Thievessen I., Luber C., Forner F., Schmidt S., Zaniva S., Fässler R., Mann M. (2008) SILAC-mouse for quantitative proteome analysis uncovers Kindlin-3 as an essential factor for red blood cell function. *Cell.* Jul 25; 134(2), 353–364.
11. Ross P.L., Huang Y.N., Marchese J.N., Williamson B., Parker K., Hattan S., Khainovski N., Pillai S., Dey S., Daniels S., Purkayastha S., Juhasz P., Martin S., Bartlet-Jones M., He F., Jacobson A., Pappin D.J. (2004) Multiplexed protein quantitation in Saccharomyces cerevisiae using amine-reactive isobaric tagging reagents. *Mol. Cell. Proteomics* 3, 1154–1169.

# Part II

## Electrophoretic Separations

# Chapter 2

# High-Resolution Two-Dimensional Electrophoresis

Walter Weiss and Angelika Görg

## Summary

Two-dimensional gel electrophoresis (2-DE) with immobilized pH gradients (IPGs) combined with protein identification by mass spectrometry is currently the workhorse for the majority of ongoing proteome projects. Although alternative/complementary technologies, such as MudPIT, ICAT, or protein arrays, have emerged recently, there is up to now no technology that matches 2-DE in its ability for routine parallel expression profiling of large sets of complex protein mixtures. 2-DE delivers a map of intact proteins, which reflects changes in protein expression level, isoforms, or post-translational modifications. High-resolution 2-DE can resolve up to 5,000 proteins simultaneously (~2,000 proteins routinely), and detect and quantify <1 ng of protein per spot. Today's 2-DE technology with IPGs has largely overcome the former limitations of carrier ampholyte-based 2-DE with respect to reproducibility, handling, resolution, and separation of very acidic or basic proteins. Current research to further advance 2-DE technology has focused on improved solubilization/separation of hydrophobic proteins, display of low abundance proteins, and reliable protein quantitation by fluorescent dye technologies. Here, we provide a comprehensive protocol of the current high-resolution 2-DE technology with IPGs for proteome analysis and describe in detail the individual steps of this technique, i.e., sample preparation and protein solubilization, isoelectric focusing in IPG strips, IPG strip equilibration, and casting and running of multiple SDS gels. Last but not the least, a section on how to circumvent the major pitfalls is included.

**Key words:** Immobilized pH gradient, Proteome, Two-dimensional electrophoresis

## 1. Introduction

Two-dimensional electrophoresis (2-DE) couples isoelectric focusing (IEF) in the first dimension and sodium dodecyl sulfate polyacrylamide gel electrophoresis (SDS-PAGE) in the second dimension to separate proteins according to two independent parameters, i.e., isoelectric point (p*I*) in the first dimension and

Jörg Reinders and Albert Sickmann (eds.), *Proteomics,* Methods in Molecular Biology, Vol. 564
DOI: 10.1007/978-1-60761-157-8_2, © Humana Press, a part of Springer Science+Business Media, LLC 2009

molecular mass ($M_r$) in the second. High-resolution 2-DE can resolve up to 5,000 different proteins simultaneously (~2,000 proteins routinely), and detect and quantify <1 ng of protein per spot. Although 2-DE is technically challenging, and may possibly never enable comprehensive characterization of highly complex proteomes because of the extremely diverse physico-chemical properties of proteins, it is still the most widespread method for the majority of ongoing proteome projects (for review *see* **ref.** *1)* and will probably remain so in the foreseeable future. Alternative/complementary technologies, such as MudPIT, ICAT, or protein arrays, have emerged recently, displaying exciting features (reviewed in **ref.** *2)*, but there is up to now no technology that matches 2-DE in its ability for ***routine parallel expression profiling*** of large sets of complex protein mixtures, delivering a map of intact proteins that reflect changes in protein expression level, isoforms, or post-translational modifications *(1)*.

Although widely applied throughout the 1980s, high-resolution 2-DE technology with *carrier-ampholyte-generated pH gradients* in the first dimension, as introduced by O'Farrell in 1975 *(3)*, suffers from several limitations with respect to reproducibility, resolution, separation of very acidic and/or very basic proteins, and sample loading capacity. These shortcomings have been largely overcome by the introduction of ***immobilized pH gradients*** (*IPGs*) for the first dimension of 2-DE *(4–6)*. IPGs are based on the use of the bifunctional Immobiline™ reagents, a series of chemically well-defined acrylamide derivatives with the general structure $CH_2$=CH–CO–NH–R, where R contains either a carboxyl or an amino group *(6)*. These form a series of buffers with different p*K* values ranging between 1 and 13. Since the reactive end is co-polymerized with the acrylamide matrix, extremely stable pH gradients are generated, allowing true steady-state IEF with increased reproducibility. Narrow-overlapping IPGs *(7, 8)* not only provide higher resolution ($\Delta pI = 0.001$), but also permit detection of lower abundance proteins caused by increased sample loading capacity. Last but not the least, alkaline proteins with isoelectric points up to pH 12 have been separated under truly steady-state conditions using IPG technology *(9–11)*.

2-DE has been considerably simplified by the IPG DryStrip technology *(4)*, the use of semi-automated devices such as the IPGphor *(12, 13)* in the first dimension, and multiple SDS-PAGE apparatus for running up to 20 different samples in parallel in the second dimension. Current research to further advance 2-DE technology has focused on improved solubilization and separation of hydrophobic proteins, display of low abundance proteins, and reliable protein quantitation by fluorescent dyes or isobaric tags. Recent developments permit analyses on a single 2-DE gel containing mixed samples differentially labeled by fluorescent dye molecules using difference gel electrophoresis (DIGE) technology *(14)*.

No matter what kind of protein separation technology may eventually replace 2-D electrophoresis, 2-DE has never been as powerful as today with respect to *reproducibility*, *resolution* by using narrow overlapping IPGs, *coverage* of the pH range from 2.5 to 12, and *quantitation* of differentially expressed proteins by using DIGE with an internal standard.

The major steps of the 2-DE-MS workflow include: (1) sample preparation/prefractionation and protein solubilization; (2) protein separation by 2-DE; (3) protein detection and quantitation; (4) computer-assisted analysis of 2-DE patterns; (5) protein identification and characterization by MS; and (6) database construction. In this chapter, we will provide a comprehensive protocol of high-resolution 2-DE with IPGs. The major emphasis is on 2-DE technology with IPGs, whereas items (3)–(6) will not be discussed here. The reader is referred to the corresponding chapters of this book and review articles *(1, 15, 16)* covering these aspects. The original IPG-Dalt protocol, published in 1988 *(4)*, has since then been constantly refined *(1, 5)* and is regularly updated on our web site (http://www.wzw.tum.de/proteomik).

## 2. Materials

### 2.1. Equipment

IPG DryStrip reswelling tray, IPGphor™ IEF device, universal and/or multiple cup loading strip holders (Manifold™), IPG DryStrips, Ettan DALT™ multiple vertical SDS electrophoresis apparatus, SDS gel casting box, cassette rack, glass plates, water bath, microwave oven, and laboratory shaker have been used. The instructions assume the use of electrophoresis equipment from GE Healthcare/Amersham Biosciences (Uppsala, Sweden), but may be adaptable to similar devices from other manufacturers (e.g., Bio-Rad, Hercules, CA).

### 2.2. Reagent Solutions

1. Modified O'Farrell's *(3)urea lysis buffer* for cell lysis and solubilization of hydrophilic as well as moderately hydrophobic proteins: 9.5 M urea, 1.0% (w/v) dithiothreitol (DTT), 2.0% (w/v) CHAPS, 2.0% (v/v) carrier ampholytes (pH range 3–10), 10 mM Pefabloc® proteinase inhibitor. To prepare 50 mL, dissolve 30.0 g of urea (GE Healthcare/Amersham Biosciences, Freiburg, Germany) in deionized water (*see* **Note 1**) and adjust the volume to 50 mL. Add 0.5 g of Serdolit MB-1 mixed-bed ion exchanger resin (Serva, Heidelberg, Germany), stir for 10–15 min, and filter. Add 1.0 g CHAPS (Roche Diagnostics, Mannheim, Germany), 0.5 g DTT (Sigma-Aldrich, St. Louis, MO), and 1.0 mL of Pharmalyte pH range 3–10

(GE Healthcare/Amersham Biosciences) to 48 mL of the filtered urea solution. Store in aliquots at –70°C (*see* **Notes 2** and **3**). Immediately before use, add 1.0 mg Pefabloc® proteinase inhibitor (Merck, Darmstadt, Germany) per mL of lysis buffer.

For the solubilization of hydrophobic proteins, *thiourea/urea lysis buffer* (2 M thiourea, 7 M urea, 4% (w/v) CHAPS, 1% DTT, 2% (v/v) carrier ampholytes) and 10 mM Pefabloc® proteinase inhibitor in combination with urea/thiourea IPG strip rehydration solution is preferred *(17)*. To prepare 50 mL of thiourea/urea lysis buffer, dissolve 22.0 g of urea (GE Healthcare) in deionized water, add 8.0 g of thiourea (Fluka/Sigma-Aldrich, Buchs, Switzerland), and adjust the volume to 50 mL with deionized water. Add 0.5 g of Serdolit MB-1 mixed bed ion exchange resin (Serva), stir for 10–15 min, and filter. Add 2.0 g CHAPS (Roche), 1.0 mL of Pharmalyte pH 3–10 (GE Healthcare), and 0.5 g DTT (Sigma-Aldrich) to 48 mL of the filtrate and store in aliquots at –70°C (*see* **Notes 2** and **3**). Immediately before use, add 1.0 mg Pefabloc® proteinase inhibitor (Merck) per mL of lysis buffer.

2. *IPG DryStrip rehydration solution*: 8 M urea, 1% (w/v) CHAPS, 15 mM DTT, 0.5% (v/v) Pharmalyte pH 3–10. To prepare 50 mL of the solution, dissolve 25.0 g of urea (GE Healthcare) in deionized water and fill up to 50 mL. Add 0.5 g of Serdolite MB-1 (Serva), stir for 10 min, and filter. To 48 mL of this solution, add 0.5 g of CHAPS (Roche Diagnostics), 0.25 mL Pharmalyte pH 3–10 (40%, w/v) (GE Healthcare), and 100 mg of DTT (Sigma-Aldrich) and fill up to 50 mL with deionized water (*see* **Note 4**). For IEF of very alkaline proteins using narrow IPGs between pH 7 and 10, DTT should be replaced by 200 mM hydroxyethyldisulfide (HED); i.e., instead of DTT, 1.2 mL HED (2,2′-dithio-diethanol; Fluka, or DeStreak™ Reagent; GE Healthcare) is added to 48 mL of the filtrate (*see* **Notes 5**). IPG DryStrip rehydration should either be prepared fresh, or be stored in aliquots at –70°C (*see* **Notes 2** and **3**).
3. *IPG strip equilibration buffer*: 6 M urea, 30% (w/v) glycerol, 2% (w/v) SDS in 0.05 M Tris–HCl buffer, pH 8.6. To make 500 mL, add 180 g of urea (Merck), 150 g of glycerol (Merck), 10 g of SDS (Serva), 16.7 mL of SDS gel buffer (*see* below), and a few grains of bromophenol blue (Serva). Dissolve in deionized water and fill up to 500 mL. The buffer can be stored at room temperature up to 2 weeks.
4. *SDS gel buffer* (4×): 1.5 M Tris-HCl pH 8.6% and 0.4% (w/v) SDS. To make 500 mL, dissolve 90.85 g of Trizma base (Sigma-Aldrich) and 2.0 g of SDS (Serva) in about 400 mL of deionized water. Adjust to pH 8.6 with 4 N HCl (Merck) and

fill up to 500 mL with deionized water. Add 50 mg of sodium azide (Merck) and filter. The buffer can be stored at 4°C up to 2 weeks.

5. *Electrode buffer stock solution* (4×): 96 mM Tris base, 800 mM glycine, 0.4% SDS. To make 5.0 L of electrode buffer stock solution, dissolve 58.0 g of Trizma base (Sigma-Aldrich), 299.6 g of glycine (Sigma-Aldrich), and 20.0 g of SDS (Serva) in deionized water and fill up to 5.0 L.
6. *Acrylamide/Bisacrylamide* solution: (30.8% T, 2.6% C): 30% (w/v) acrylamide and 0.8% (w/v) methylenebisacrylamide in deionized water. To make 500 mL, dissolve 150.0 g acrylamide (2 × cryst.) and 4.0 g methylenebisacrylamide (GE Healthcare) in deionized water and fill up to 500 mL. Add 1–2 g of Serdolit MB-1 (Serva), stir for 10 min, and filter. The solution can be stored up to 2 weeks in a refrigerator. Acrylamide is a neurotoxin when unpolymerized and hence care should be taken to not expose it.
7. *N,N,N,N'*-Tetramethyl-ethylenediamine (**TEMED**) (Bio-Rad, Hercules, CA). Use the undiluted solution (*see* **Note 6**). *Ammonium persulfate* solution: 10% (w/v) of ammonium persulfate in deionized water. To make 10 mL of the solution, dissolve 1.0 g of ammonium persulfate (GE Healthcare) in 10 mL of deionized water. This solution may be kept in a refrigerator for up to 24 h.
8. *Displacing-solution*: 50% (v/v) glycerol in deionized water and 0.01% (w/v) bromophenol blue. To make 500 mL, mix 250 mL of glycerol (100%) (Merck) with 250 mL of deionized water, add 50 mg of bromophenol blue (Serva) and stir for a few minutes.
9. *Overlay buffer*: Buffer-saturated 2-butanol. To make 30 mL, mix 20 mL of SDS gel buffer with 30 mL of 2-butanol (Merck), shake, and wait for a few minutes until the two phases have separated. Use the top layer.
10. *Agarose solution*: Suspend 0.5% (w/v) agarose (GE Healthcare) in electrode buffer and melt it in a microwave oven before use. Then, place the beaker or Erlenmeyer flask in a hot water bath (~75°C) to keep the agarose melted.

## 3. Methods

Following cell lysis and protein solubilization, samples are applied onto individual IPG gel strips cast on GelBond plastic backing *(1, 4, 5)*, either by sample in-gel rehydration, or – more commonly – by cup–loading after IPG DryStrip rehydration. After IEF, followed by equilibration with SDS buffer in the presence of urea, glycerol, DTT, and iodoacetamide *(18)*, the IPG strips are applied onto horizontal or vertical SDS gels. On completion of SDS electrophoresis, the separated proteins are visualized by staining with silver nitrate, organic dyes, autoradiography (or phosphor-imaging) of radiolabeled samples, or – preferably – by staining with fluorescent dye molecules *(1, 15)*, whereas 2-D DIGE gels *(14)* with CyDye-labeled proteins are directly scanned on a fluorescence imager.

### 3.1. Sample Preparation

Sample preparation is the first step and hence a key factor for successful 2-DE-based proteome analysis. Treatment of samples prior to 2-DE typically involves cell disruption, sample clean-up (i.e., removal and/or inactivation of interfering protein and non-protein components), and protein solubilization. Although a universal procedure for sample preparation would be desirable, there is no single method that can be applied to all kinds of samples, and most often the optimal procedure must be found empirically, and/or has to be adapted and further optimized for different types of samples. Although the application of unfractionated samples to 2-DE gels is preferred with regard to simplicity and reproducibility, there are exceptions in which subfractionation methods are required, e.g., to reduce sample complexity, to enrich for low-abundance proteins, or to deplete highly abundant protein species (such as albumin in plasma) that dominate the sample and obscure less abundant proteins *(1, 19)*.

Typically, cells or tissues are disrupted by mechanical or physical (e.g., grinding in a liquid nitrogen-cooled mortar, homogenization, sonication, hypotonic lysis, etc.), chemical (e.g., detergent), and/or enzyme-based methods, which may be used individually or in combination. During – or immediately after – cell lysis, interfering compounds such as proteolytic enzymes must be inactivated and high concentrations of salt ions removed. Proteins are then solubilized in urea– or thiourea/urea-lysis buffer. As an example, solubilization of mammalian tissue proteins will be briefly described later. For further reading, please refer to review articles on this subject (e.g., *[20, 21]*).

1. Collect mammalian tissue samples as soon as possible after the death of the "donor" and immediately freeze in liquid nitrogen at –196°C. Biopsy samples should also be deep-frozen at once to prevent proteolytic attack on proteins.

2. Wrap small tissue specimen (e.g., biopsy samples) in aluminum foil, freeze in liquid nitrogen, and crush with a hammer. Larger tissue pieces (e.g., mouse liver, 50 mg) should be ground under liquid nitrogen using mortar and pestle.

3. Transfer the resulting, deep-frozen powder into a centrifuge tube or Eppendorf vial that already contains an appropriate amount of denaturing lysis buffer (typically 1.0 mL urea or thiourea/urea lysis buffer per 50 mg solid) and protease inhibitor(s). Optimal protein concentration of the extract is between 5 and 10 mg/mL.
4. Leave the suspension for 1 h at room temperature with occasional vortexing or stirring with a spatula. Sonication helps to improve cell disruption and protein solubilization. Sonicate briefly (20 s) while cooling on ice, and repeat sonication twice at 20 min intervals.
5. Centrifuge (40,000 *g*, 15°C, 60 min) the extract to remove any insoluble material (e.g., cell debris). The supernatant may either be used immediately for 2-DE, or stored in small aliquots at –70°C for up to several months.

### 3.2. IPG Strip Rehydration and Isoelectric Focusing in IPG Strips (IPG-IEF)

The first dimension of 2-DE with IPGs is performed in individual, 3-mm wide IPG gel strips cast on a supporting GelBond PAGfilm™ plastic sheet. A multitude of 7–24-cm long ready-made Immobiline DryStrips™ of almost any desired pH range has been made available, e.g., wide pH ranges IPG 3–11 or 6–12, medium pH ranges IPG 4–7 or 6–9, as well as narrow pH ranges (e.g., IPG 4–5 or 4.5–5.5) (e.g., GE Healthcare, Bio-Rad, Sigma-Aldrich, Serva). Alternatively, laboratory-made IPG DryStrips can be used. For details on IPG gel casting, the interested reader is referred to previously published protocols *(22)*.

Prior to IEF, the dried IPG strips must be rehydrated in a reswelling cassette or a reswelling tray to the desired thickness of 0.5 mm. IPG DryStrips may either be rehydrated with sample already dissolved in rehydration buffer ("sample in-gel rehydration") *(23)*, or with rehydration buffer without sample. In the latter case, sample is applied onto the *rehydrated* IPG strip by means of "cup-loading." Although sample application by in-gel rehydration is more convenient, this procedure is discouraged for samples containing high molecular weight (>100 kDa), very alkaline, and/or hydrophobic proteins, since these are taken into the gel with difficulty, owing to hydrophobic interactions between proteins and the wall of the reswelling tray, and/or because of size-exclusion effects of the gel matrix. Cross-contamination may also be a problem, and hence the reswelling tray must be thoroughly cleaned before different experiments. For quantitative analyses, in particular, sample-in-gel rehydration has proven less reliable than cup-loading. Because of these reasons, in-gel sample loading will not be described here, and we will confine ourselves to the description of cup-loading. For more information on sample in-gel rehydration, please refer to our web site (http://www.wzw.tum.de/proteomik).

After rehydration, the IPG strips are applied onto the cooling plate of a horizontal IEF apparatus *(4, 24)*, or an integrated system, the IPGphor™ (GE Healthcare). The IPGphor is programmable and can store up to ten different programs. The central part of this instrument has the so-called strip holders made from an aluminum oxide ceramic, in which IEF is performed. The Peltier-cooled strip holder platform regulates temperature precisely (between 19.5°C and 20.5°C) and serves as the electrical connector for the strip holders. Besides easier handling, a second advantage of the IPGphor is shortened focusing time, since IEF can be performed at rather high voltage (up to 8,000 V). Our instructions assume the use of the IPGphor™ electrofocusing unit with cup-loading strip holders available from GE Healthcare, but may be adaptable to similar devices from other manufacturers.

After the rehydrated IPG strips have been applied to the IPGphor strip holders, samples (20–100 μL) dissolved in lysis buffer are pipetted into disposable plastic cups placed onto the surface of the rehydrated IPG strips (*see* **Note 7**). The best results are obtained when the samples are applied at the pH extremes, i.e., either near the anode or cathode. In most cases, sample application near the anode proved to be superior to cathodic application. When protein separation is performed in alkaline pH ranges such as with IPGs 6–12 or 9–12, anodic application is mandatory for all kinds of samples *(5, 9–11)*.

The amount of protein to be loaded onto a single IPG strip depends on several factors. As a rule-of-thumb, (1) the longer the separation distance (i.e., the bigger the gel size), (2) the narrower the pH gradient, and (3) the less sensitive the protein detection method, the more protein must be applied. Recommended protein amounts range between 50 and 100 μg for analytical, silver-stained 20 × 20 $cm^2$ 2-DE gels, and up to 1 mg (or even more) for micropreparative, Coomassie Blue stained 2-DE gels. In case of very narrow pH range IPGs, it is strongly recommended to apply prefractionated samples only *(19)*].

Typical running conditions for IEF using the IPGphor are given in **Tables 1** and **2**. For improved sample entry, low voltage (150–300 V) is recommended at the beginning of IEF *(5)* prior to gradually increasing voltage up to 8,000 V. If IPG strips shorter than 110 mm are used, the maximum voltage should be limited to 5,000 V. For optimum results, in particular for samples with high salt concentrations, or when (ultra)narrow pH intervals are used, moist filter paper pads (size: 3 × 10 $mm^2$) are inserted between the electrodes and the IPG strip and replaced several times (*see* **Note 8**).

*3.2.1. IPG DryStrip Rehydration*

1. For 180-mm long and 3-mm wide IPG DryStrips, pipette 350–400 μL of IPG DryStrip rehydration solution or DeStreak™ rehydration solution into the grooves of the IPG

**Table 1**
**IPGphor running conditions (for sample in-gel rehydration and for cup loading). Reprinted with permission from ref. *5***

| Gel length | 180 mm | | | | |
|---|---|---|---|---|---|
| Temperature | 20°C | | | | |
| Current maximum | 0.05 mA per IPG strip | | | | |
| Voltage maximum | 8,000 V | | | | |
| **I Analytical IEF** | | | | | |
| IPG DryStrip rehydration | 12–16 h | | | | |
| Initial IEF | 150 V, 1 h | | | | |
| | 300 V, 1 h | | | | |
| | 600 V, 1 h | | | | |
| | 1,000 V, 1 h | | | | |
| IEF to the steady state | Gradient from 1,000 to 8,000 V within 30 min | | | | |
| | 8,000 V to the steady state, depending on the pH used: | | | | |
| **1–1.5 pH units** | | **4 pH units** | | **7 pH units** | |
| e.g. IPG 5–6 | 8 h | IPG 4–8 | 4 h | IPG 3–10 L | 3 h |
| e.g. IPG 4–5.5 | 8 h | | | IPG 3–10 NL | 3 h |
| **3 pH units** | | **5–6 pH units** | | **8–9 pH units** | |
| IPG 4–7 | 4 h | IPG 4–9 | 4 h | IPG 3–12 | 3 h |
| | | | | IPG 4–12 | 3 h |
| **II Micropreparative IEF** | | | | | |
| Initial IEF | As of analytical IEF, but sample entry time prolonged (4 × 2 h each) | | | | |
| IEF to the steady state | Volt hours of analytical IEF + additional 50% (approx.) | | | | |

DryStrip reswelling tray (**Fig. 1a**) (*see* **Note 5** and **9**). For longer or shorter IPG strips, rehydration volume has to be adjusted accordingly (e.g., 450–500 μL for 240-mm long IPG DryStrips).

2. Remove the protective plastic covers from the IPG DryStrips and apply the latter, gel side down, into the grooves of the tray without trapping air bubbles. Always use forceps and wear gloves to avoid protein contaminants. The IPG strips must be moveable and not stick to the reswelling tray. Then, cover each IPG DryStrip with 2 mL IPG DryStrip cover fluid (which prevents desiccation during reswelling), and rehydrate the IPG

**Table 2**
**IPGphor running conditions for very alkaline IPGs (sample cup loading). Reprinted with permission from ref. *7***

| | |
|---|---|
| Gel length | 180 mm |
| Temperature | 20°C |
| Current maximum | 0.07 mA per IPG strip |
| Voltage maximum | 8,000 V |
| IPGs | 6–12, 9–12, 10–12 |
| Sample application | Anodic cup loading |
| IPG DryStrip rehydration | 12–16 h |
| Initial IEF | 150 V, h |
| | 300 V, 1 h |
| | 600 V, 1 h |
| IEF to the steady state | Gradient from 600 to 8,000 V within 30 min |
| | 8,000 V to the steady state |
| Total volt hours | 32,000 Vh |

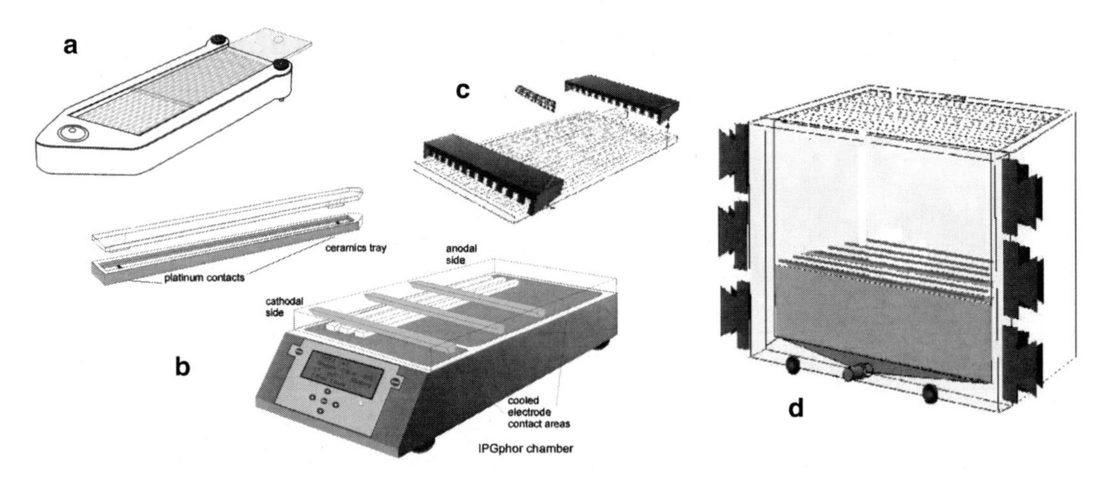


Fig. 1. Procedure of 2-DE with IPGs (IPG-Dalt) based on the protocol of Görg et al. *(4,5)*. (**a**) Rehydration of individual IPG DryStrips in the reswelling tray, (**b**) IEF in individual IPG strip holders on the IPGphor™ instrument, (**c**) IPGphor cup-loading Manifold™ device for improved focusing patterns, and (**d**) multiple vertical SDS gels.

strips overnight at ~20°C. Higher temperatures (>37°C) hold the risk of isocyanate formation with subsequent carbamylation of proteins, whereas lower temperatures (<15°C) should be avoided to prevent urea crystallization on the IPG gel surface.

*3.2.2. IEF on the IPGphor™ Electrofocusing Unit*

1. After the IPG strips have been rehydrated, use clean forceps to remove them from the reswelling tray. Briefly rinse the IPG strips with deionized water and blot them for a few seconds between two sheets of moist filter paper to remove excess liquid to prevent formation of urea crystals on the surface of the IPG gel during IEF.
2. Apply the required number of the cup-loading IPGphor™ strip holders onto the cooling plate/electrode contact area of the IPGphor™ instrument (**Fig. 1b**), and make sure that the pointed (anodic) ends contact the anodic electrode area. Instead of individual strip holders, the Manifold™ device (**Fig. 1c**) may be used, which accommodates up to 12 IPG strips at a time.
3. Transfer the rehydrated IPG strips into the cup-loading strip holders (or into the Manifold™ device), gel side upwards and acidic (pointed) ends facing towards the anode.
4. Moisten two filter paper electrode pads (size: $3 \times 10\ mm^2$) with deionized water, remove excess liquid by blotting with a filter paper, and apply the moistened filter paper pads on the gel surface at the anodic and cathodic ends of the IPG strip, respectively. If necessary (e.g., when the sample contains high amounts of salt), remove and replace these filter paper pads after several hours (*see* **Note 8**).
5. Position the movable electrodes above the electrode filter paper pads. Clip the electrodes firmly onto the electrode paper pads.
6. Position the movable sample cup near the anode (or cathode), and gently press the sample cup onto the surface of the IPG gel strip. The sample cup should form a good seal with the IPG strip but not damage its surface.
7. To confirm that the sample cup does not leak, pipette 100 µL of IPG cover fluid into the cup. If a leak is detected, use a tissue paper to remove the cover fluid and reposition the sample cup. Check again for leakage. Remove the cover fluid before loading the sample.
8. Overlay each IPG strip with 2–4 mL of IPG strip cover fluid (the use of silicone oil or kerosene instead of IPG strip cover fluid should be avoided). In case the cover fluid leaks into the sample cup, re-arrange the cup and use tissue paper to remove the cover fluid from the cup. Check again for leakage, and pipette the sample (20–100 µL) in the sample cup.
9. Program the instrument (desired volt hours, voltage gradient, etc.) and run IEF according to the recommended settings in **Tables 1** and **2**. After the sample has entered the IPG strip, pipette 100 µL of IPG strip cover fluid into the sample cup to prevent desiccation and subsequent formation of urea crystals in the IPG strip at the sample application area.

10. After IEF is complete, proceed with equilibration (*see* **Subheading 3.3**), or store the IPG strips, for up to several months, between two plastic sheets at −70°C.

### *3.3. IPG Strip Equilibration*

Prior to the second-dimension separation (SDS-PAGE), it is essential that the IPG strips are equilibrated to enable the separated proteins to fully interact with SDS. Because of the observation that the focused proteins bind more strongly to the fixed charged groups of the IPG gel matrix than to carrier ampholyte gels, relatively long equilibration times (10–15 min), as well as urea and glycerol to reduce electro-endosmotic effects are required to improve protein transfer from the first to the second dimension. The by far most popular protocol is to incubate the IPG strips for 10–15 min in the buffer originally described by Görg et al. *(18)* (50 mM Tris-HCl pH 8.8, 2% SDS, 1% DTT, 6 M urea, and 30% glycerol) (**Table 3**). This is followed by a further 10–15 min equilibration in the same solution containing 4% (w/v) iodoacetamide instead of DTT. The latter step is used to alkylate any free DTT, as otherwise it migrates through the second-dimension SDS-PAGE gel, resulting in an artifact known as point-streaking that can be observed after silver staining. More importantly, iodoacetamide alkylates sulfhydryl groups and prevents their reoxidation. This two-step reduction/alkylation procedure is highly recommended, since it considerably simplifies downstream sample preparation for spot identification by mass spectrometry. After equilibration, the IPG strips are applied onto the SDS-PAGE gels.

**Table 3**
**IPG strip equilibration protocol (18)**

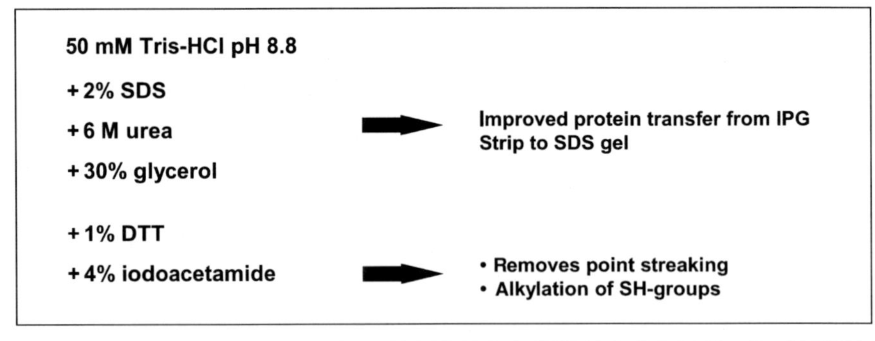


| | DTT | Iodoacetamide |
|---|---|---|
| 1. 15 min (10 min) | + | – |
| 2. 15 min (10 min) | – | + |

1. Dissolve 100 mg of DTT (Sigma-Aldrich) in 10 mL of equilibration buffer to make equilibration buffer I. Make 10 mL per IPG strip. Place the focused IPG strips into individual test tubes (250 mm long; 20 mm internal diameter) and add 10 mL of equilibration buffer I to each tube. Seal the tubes with Parafilm, rock them for 15 min on a shaker, and pour off the equilibration buffer (*see* **Note 10**).
2. Dissolve 0.4 g of iodoacetamide (Sigma-Aldrich) in 10 mL of equilibration buffer to make equilibration buffer II. Make 10 mL per IPG strip. Add this buffer and 50 μL of bromophenol blue (Serva) solution as tracking dye for SDS-PAGE to each tube, and equilibrate for another 15 min with gentle agitation.
3. Pour off equilibration buffer II, and proceed to SDS-PAGE (*see* **Subheading 3.4**). If SDS-PAGE is performed on a horizontal electrophoresis unit (e.g., Multiphor II), briefly rinse the IPG gel strip with deionized water, and place it on a piece of filter paper at one edge for a few minutes to drain off excess equilibration buffer. If SDS-PAGE is performed on a vertical electrophoresis unit (e.g., Ettan DALT II), briefly dip the equilibrated IPG strip in electrode buffer.

### *3.4. SDS-page*

The second dimension (SDS-PAGE) can be carried out on vertical as well as horizontal flat-bed systems *(25)*. Whereas horizontal setups are ideally suited for rapid separations and offer the convenience of ready-to-use SDS gels and buffer strips, vertical systems are preferred for multiple runs in parallel, in particular for large-scale proteome analysis, which usually requires simultaneous electrophoresis of batches of second-dimension SDS gels for higher throughput and maximal reproducibility. For multiple runs, an electrophoresis unit similar to the one originally designed by Anderson and Anderson *(26)*, or modifications thereof, are preferred. In contrast to horizontal SDS-PAGE systems, it is not necessary to use stacking gels with vertical setups, as the nonrestrictive, low polyacrylamide concentration in the IPG gel strip already acts as a stacking zone.

Depending on the total polyacrylamide (%T) concentration (and, to a lesser extent, on the percentage of crosslinker), proteins with $M_r$s ranging between 10,000 and 200,000 can be resolved in "standard" SDS gels. Single percentage polyacrylamide gels offer excellent resolution for a particular $M_r$ window. Widely used second-dimension gels for 2-D electrophoresis are homogeneous gels containing 12.5% total acrylamide (effective separation range: ~15,000–150,000 kDa), whereas 10% T or 15% T gels are employed for higher or lower $M_r$ ranges, respectively. If a polyacrylamide gradient gel is used, the overall separation interval is wider, and sharper spots result because the decreasing pore size of the polyacrylamide gel matrix minimizes diffusion. However, gradient gels require more skill to cast, and hence results may be less reproducible.

The most commonly used buffer system for the second dimension of 2-DE is the discontinuous Tris–glycine buffer system of Laemmli *(27)*, although other buffers have been recommended for special purposes, such as borate buffers for the separation of highly glycosylated proteins, or a Tris–tricine buffer *(28)* for improved resolution of polypeptides in the $M_r$ range between 3,000 and 30,000.

The following instructions assume the use of the Ettan DALT II vertical electrophoresis unit available from GE Healthcare Life Sciences, but are adaptable to similar devices from other manufacturers (e.g., Bio-Rad). For information on second-dimension horizontal SDS slab gel systems, please refer to **refs.** *22, 29*. Vertical second-dimension gels are cast several at a time (up to 14), in a multiple gel caster, and up to 12 SDS gels can be electrophoresed under identical conditions. After equilibration (*see* **Subheading 3.3**), the IPG gel strips are placed on top of the vertical SDS gels, usually with embedding in hot agarose solution (note that the second-dimension gels must be prepared before the equilibration step). For convenience, SDS-PAGE is usually performed overnight, although shorter running times at higher voltages are also applicable. After 2-DE, the separated proteins must be visualized, either by universal (e.g., silver-, Coomassie Brilliant Blue-, or fluorescent staining) or by specific detection methods (e.g., glyco- or phosphoprotein stains) (*see* **Chapter 4**).

#### *3.4.1. SDS Gel Casting*

1. The gel casting cassettes (200 × 260 mm$^2$) are made in the shape of books consisting of two 3-mm thick glass plates connected by a hinge strip, and two 1.0-mm thick spacers in between them (*see* **Note 11**). Stack 14 cassettes vertically into the gel casting box of the Ettan DALT II apparatus with the hinge strips to the right, interspersed with plastic sheets (e.g., 0.05-mm thick polyester sheets) (**Fig. 1d**).
2. Place the front plate of the casting box in place and screw on the nuts.
3. Connect a polyethylene tube (i.d. 5 mm) to a funnel held in a ring-stand at a level of about 30 cm above the top of the casting box. The other end of the tube is placed in the grommet in the casting box side chamber.
4. Fill the side chamber with 100 mL of displacing solution.
5. Prepare the gel solution (*see* **Table 4**) in a flask, omitting the TEMED and ammonium persulfate, and stir for several minutes on a magnetic stirrer. Immediately before gel casting, add

**Table 4**
**Recipes for casting homogeneous (10% T, 12.5% T, or 15% T) vertical SDS gels**

| | 10% T<br>2.6% C | 12.5% T<br>2.6% C | 15% T<br>2.6% C |
|---|---|---|---|
| Acrylamide/bisacrylamide (30.8% T, 2.6% C) | 325 mL | 406 mL | 487 mL |
| SDS Gel buffer | 250 mL | 250 mL | 250 mL |
| Glycerol (100%) | 50.0 g | 50.0 g | 50.0 g |
| Deionized water | 380 mL | 299 mL | 218 mL |
| TEMED (100%) | 50 μL | 50 μL | 50 μL |
| Ammonium persulfate (10%) | 7.0 mL | 7.0 mL | 7.0 mL |
| Final volume | 1,000 mL | 1,000 mL | 1,000 mL |

the TEMED and ammonium persulfate and gently swirl the flask to mix, being careful not to generate air bubbles.

6. To cast the gels, pour the gel solution (830 mL) into the funnel. Avoid introduction of any air bubbles into the tube. Do not fill the cassettes with acrylamide solution completely to leave some space at the top (~10 mm) to later fix the IPG strip to the SDS gel with hot agarose solution. No stacking gel layer is required.
7. Immediately after pouring, remove the tube from the side chamber grommet so that the level of the displacing solution in the side chamber falls.
8. Very carefully pipette about 1 mL of overlay buffer onto the top of each gel to minimize exposure to oxygen and to create a flat gel top surface.
9. Allow the gels to polymerize at ~20°C for a minimum of 3 h, but preferably overnight for better reproducibility.
10. After gel polymerization, remove the front of the casting box and carefully unload the gel cassettes from the box, using a blade to separate the cassettes. Remove the polyester sheets that had been placed between the individual cassettes.
11. Wash each cassette with water to remove any acrylamide adhered to the outer surface, rinse the gel top surface, and then drain excess liquid off the surface. Since only 12 gel cassettes fit into the electrophoresis unit, unsatisfactory gels should be discarded, in particular gels with uneven thickness, i.e., usually those at the outer edges of the gel casting cassette.

12. Gels that are not needed immediately can be wrapped in plastic wrap and stored in a refrigerator (4°C) for up to 3 days.

*3.4.2. Multiple SDS-PAGE Using the Ettan DALT II Vertical Electrophoresis Unit*

1. Add 1,875 mL of electrode buffer stock solution and 5,625 mL of deionized water to the lower electrophoresis buffer tank of the Ettan DALT II unit. Mix and turn on cooling (25°C). Note that precise temperature control improves run-to-run reproducibility.
2. Support the gel cassettes (containing the SDS gels) in a vertical position on the cassette rack to facilitate the application of the IPG gel strips.
3. Dip the equilibrated IPG strip (from **Subheading 3.3**) in the electrode buffer to lubricate it, and place it on top of the gel cassette. Use a spatula or a thin plastic ruler to gently push against the plastic backing of the IPG strip and slide it into the gap between the two glass plates. Add 2 mL of hot (75°C) agarose solution, and continue to slide the IPG strip down so that the entire lower edge of the IPG strip is in contact with the top surface of the slab gel. Ensure that no air bubbles are trapped between the IPG strip and the SDS slab gel surface, or between the plastic backing and the glass plate (*see* **Note 1**).
4. For co-electrophoresis of $M_r$ marker proteins, soak a filter paper pad (2 × 4 mm$^2$) with 5 µL of SDS marker proteins dissolved in electrophoresis buffer, let it dry, and apply it to the left or right of the IPG strip. Dried filter paper pads soaked with molecular weight marker proteins may also be prepared in advance and stored deep-frozen in Eppendorf vials at –70°C.
5. Allow the agarose to solidify for at least 5 min before placing the slab gel into the electrophoresis apparatus (*see* **Step 6**). Repeat the procedure for the remaining IPG strips. Although embedding in agarose is not considered as absolutely necessary, it ensures much better contact between the IPG gel strip and the top of the SDS gel, preventing the IPG strip from moving or floating in the electrophoresis buffer.
6. Wet the outside of the gel cassettes by dipping them into electrode buffer to make them fit more easily into the electrophoresis unit. Insert them in the electrophoresis apparatus. When running fewer than 12 gels, push blank cassette inserts into any unoccupied slots. Seat the upper buffer chamber over the gels and fill it with 2,500 mL of electrode buffer (2×) (1,250 mL of buffer stock solution +1,250 mL of deionized water).
7. Place the safety lid on the electrophoresis unit and start SDS-PAGE. Electrophoresis is performed at constant current in two steps: During the initial migration and stacking period (~2 h), apply 5 mA per SDS gel (100 V maximum setting). Then, continue with 15 mA per SDS gel (200 V maximum setting) for ~16 h overnight, or higher current for faster runs (30 mA per SDS gel) for ~8 h.

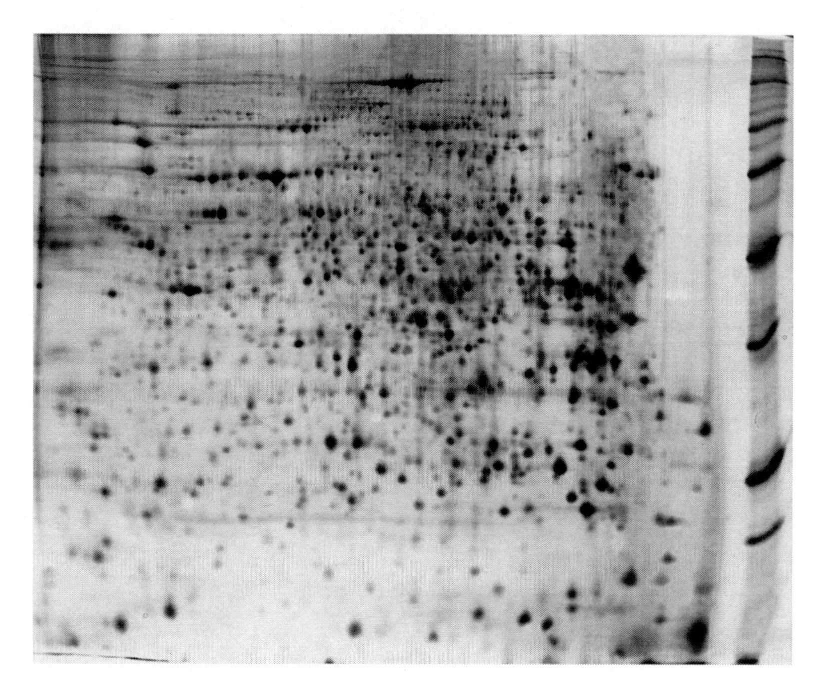

Fig. 2. IPG-Dalt of mouse liver proteins, separated by IEF in a 24-cm long IPG strip containing a wide-range nonlinear pH gradient 3–11, followed by SDS-PAGE in a vertical 12.5% T gel. Sample application by cup-loading near the anode (**left**). Protein detection was by silver-staining. Reprinted with permission from **ref.** *1*.

8. Terminate the run after the bromophenol blue tracking dye front is ~1 mm from the bottom of the gels.
9. After electrophoresis, carefully open the cassettes with a plastic spatula. Use the spatula to separate the agarose overlay from the polyacrylamide gel. Carefully peel the gel from the glass plate, lifting the gel by its lower edge, and place it in a box of fixing or staining solution. An example of results is presented in **Fig. 2**. Note that DIGE gels are directly scanned in the cassette.

## 4. Notes

1. All solutions should be prepared in water having a conductivity of 18.2 MΩ/cm. This standard is referred to as "deionized water" in this text.
2. Urea and thiourea/urea lysis buffer should be prepared fresh. Alternatively, make 1 mL aliquots and store in Eppendorf vials at −70°C for up to several months. Lysis buffer that has been thawed once should not be refrozen.

3. Never heat urea solutions above 37°C. Otherwise, proteins in the sample solution may be carbamylated, resulting in charge artifacts ("spot trains").
4. If proteins have been solubilized in thiourea/urea lysis buffer, thiourea should also be incorporated in the IPG DryStrip-rehydration solution, consisting of a mixture of urea/thiourea (6 M urea, 2 M thiourea), 1% (w/v) CHAPS, 15 mM DTT, and 0.5% (v/v) Pharmalyte pH 3–10).
5. Streaks that distort electrophoresis maps may occur, particularly in narrow pH range IEF gels between pH 7 and 10. Increased sample load worsens the problem. It has been demonstrated that HED (= 2,2′-dithiodiethanol = DeStreak™ reagent) *(30)* reduces non-specific oxidation of proteins and streaking in the alkaline part of these IEF gels. Hence, HED should be incorporated in the IPG DryStrip-rehydration solution of alkaline IPG strips such as IPG 6–10, whereas for IPGs up to pH 12, we strongly recommend to follow the recommendations of our protocol *(1, 10, 11)* displayed in **Table 2**, i.e., IEF with as high voltage as possible and anodic sample application. Alkylation of protein SH-groups prior to IEF may also be helpful.
6. TEMED may be stored in a refrigerator. Buy small bottles as it will gradually decline in quality after opening (i.e., gels may no longer polymerize properly).
7. If sample volumes exceeding 100 μL are to be applied onto IPG strips by cup-loading, pipette 100 μL into the cup, and run IEF with limited voltage (max. 300 V). As soon as the sample has migrated out of the cup into the IPG strip, repeat the procedure until the whole sample has been applied.
8. Theoretically, no further intervention is required after the start of IPG-IEF until IEF has been completed. In practice, however, superior results are obtained if the electrode filter pads between the IPG strip and the electrodes are replaced by new ones after the sample has entered the IPG strip. This is particularly important for samples that contain high amounts of salts and/or protein, because salt contaminants have quickly moved through the gel and have now been collected in the electrode papers, or when very alkaline IPGs (e.g., IPG 6–12 or IPG 9–12) are used. In these cases, the filter paper pads should be replaced every 2 h. For IEF with very alkaline, narrow range IPGs, such as IPG 10–12, this procedure should be repeated once an hour.
9. Instead of a rewelling tray, a reswelling cassette (GE Healthcare) may be used for even and homogeneous IPG strip rehydration. Please refer to **refs.** ***4, 5, 22, 24***.

10. Shorter equilibration times (10 min) may be applied, though at the risk that some proteins may not migrate out of the IPG gel strip during sample entry into the SDS-PAGE. In this case, it is advisable to check, by staining the IPG strip after removal from the SDS gel, whether all proteins have left the IPG strip.
11. Either 1.0 or 1.5 mm-thick spacers may be used for SDS gel casting. Thinner gels stain/destain more quickly, whereas thicker gels have a higher protein loading capacity. Thicker gels are also less fragile and easier to handle.
12. Each SDS gel will require ~2 mL agarose. It takes ~10 min to fully melt the agarose, e.g., in a boiling water bath or a microwave oven. An ideal time to carry out this step is during IPG strip equilibration. After melting, allow the agarose to cool to ~70°C and then slowly pipette (to avoid introducing bubbles) the amount required to seal the IPG strip in place.

## References


1. Görg, A., Weiss, W., and Dunn, M.J. (2004) Current two-dimensional electrophoresis technology for proteomics. *Proteomics* **4**, 3665–3685.
2. Roe, M.R. and Griffin, T.J. (2006) Gel-free mass spectrometry-based high throughput proteomics: tools for studying biological response of proteins and proteomes. *Proteomics* **6**, 4678–4687.
3. O' Farrell, P.H. (1975) High resolution two-dimensional electrophoresis of proteins. *J. Biol. Chem.* **250**, 4007–4021.
4. Görg, A., Postel, W., and Günther, S. (1988) The current state of two-dimensional electrophoresis with immobilized pH gradients. *Electrophoresis* **9**, 531–546.
5. Görg, A., Obermaier, C., Boguth, G., Harder, A., Scheibe, B., Wildgruber, R., and Weiss, W. (2000) The current state of two-dimensional electrophoresis with immobilized pH gradients. *Electrophoresis* **21**, 1037–1053.
6. Bjellqvist, B., Ek, K., Righetti, P.G., Gianazza, E., Görg, A., Westermeier, R., and Postel, W. (1982) Isoelectric focusing in immobilized pH gradients: principle, methodology and some applications. *J. Biochem. Biophys. Meth.* **6**, 317–339.
7. Wildgruber, R., Harder, A., Obermaier, C., Boguth, G., Weiss, W., Fey, S.J., Larsen, P.M., and Görg, A. (2000) Towards higher resolution: two-dimensional electrophoresis of Saccharomyces cerevisiae proteins using overlapping narrow immobilized pH gradients. *Electrophoresis* **21**, 2610–2616.
8. Westbrook, J.A., Yan, J.X., Wait, R., Welson, S.Y., and Dunn, M.J. (2001) Zooming-in on the proteome: Very narrow-range immobilised pH gradients reveal more protein species and isoforms. *Electrophoresis* **22**, 2865–2871.
9. Görg, A., Obermaier, C., Boguth, G., Csordas, A., Diaz, J.J., and Madjar, J.J. (1997) Very alkaline immobilized pH gradients for two-dimensional electrophoresis of ribosomal and nuclear proteins. *Electrophoresis* **18**, 328–337.
10. Wildgruber, R., Reil, G., Drews, O., Parlar, H., and Görg, A. (2002) Web-based two-dimensional database of Saccharomyces cerevisiae proteins using immobilized pH gradients from pH 6 to pH 12 and matrix-assisted laser desorption/ionization-time of flight mass spectrometry. *Proteomics* **2**, 727–732.
11. Drews, O., Reil, G., Parlar, H., and Görg, A. (2004) Setting up standards and a reference map for the alkaline proteome of the Gram-positive bacterium Lactococcus lactis. *Proteomics* **4**, 1293–1304.
12. Islam, R., Ko, C., and Landers, T. (1998) A new approach to rapid immobilised pH gradient IEF for 2-D electrophoresis. *Sci. Tool.* **3**, 14–15.
13. Görg, A., Obermaier, C., Boguth, G., and Weiss, W. (1999) Recent developments in two-dimensional gel electrophoresis with immobilized pH gradients: wide pH gradients up to pH 12, longer separation distances and simplified procedures. *Electrophoresis* **20**, 712–717.

14. Viswanathan, S., Ünlü, M., and Minden J.S. (2006) Two-dimensional difference gel electrophoresis. *Nat. Protoc.* **1**, 1351–1358.
15. Patton, W.F. (2002) Detection technologies in proteome analysis. *J. Chromatogr. B Anal. Technol. Biomed Life Sci.* **771**, 3–31.
16. Dowsey, A.W., Dunn, M.J., and Yang, G.Z. (2003) The role of bioinformatics in two-dimensional gel electrophoresis. *Proteomics* **3**, 1567–1596.
17. Rabilloud, T., Adessi, C., Giraudel, A., and Lunardi, J. (1997) Improvement of the solubilization of proteins in two-dimensional electrophoresis with immobilized pH gradients. *Electrophoresis* **18**, 307–316.
18. Görg, A., Postel, W., Weser, J., Günther, S., Strahler, J.R., Hanash, S.M., and Somerlot, L. (1987) Elimination of point streaking on silver stained two-dimensional gels by addition of iodoacetamide to the equilibration buffer. *Electrophoresis* **8**, 122–124.
19. Görg, A., Boguth, G., Köpf, A., Reil, G., Parlar, H., and Weiss, W. (2002) Sample prefractionation with Sephadex isoelectric focusing prior to narrow pH range two-dimensional gels. *Proteomics* **2**, 1652–1657.
20. Lilley, K.S., Razzaq, A., and Dupree, P. (2002) Two-dimensional gel electrophoresis: recent advances in sample preparation, detection and quantitation. *Curr. Opin. Chem. Biol.* **6**, 46–50.
21. Weiss, W. and Görg, A. (2007) Sample preparation for two-dimensional gel electrophoresis. In: Proteomics Sample Preparation (J.v. Hagen, Ed.), Wiley VCH, pp. 129–143.
22. Görg, A. and Weiss, W. (2000) 2D electrophoresis with immobilized pH gradients. In: Proteome Research: Two Dimensional Electrophoresis and Identification Methods (T. Rabilloud, Ed.), Springer, Berlin, Heidelberg, New York, pp. 57–106.
23. Rabilloud, T., Valette, C., and Lawrence, J.J. (1994) Sample application by in-gel rehydration improves the resolution of two-dimensional electrophoresis with immobilized pH-gradients in the first dimension. *Electrophoresis* **15**, 1552–1558.
24. Görg, A., and Weiss, W. (1999) Analytical IPG-Dalt. *Method. Mol. Biol.* **112**, 189–195.
25. Görg, A., Boguth, G., Obermaier, C., Posch, A., and Weiss, W. (1995) Two-dimensional polyacrylamide gel electrophoresis with immobilized pH gradients in the first dimension (IPG-Dalt): The state of the art and the controversy of vertical versus horizontal systems. *Electrophoresis* **16**, 1079–1086.
26. Anderson, N.L., and Anderson, N.G. (1998) Analytical techniques for cell fractions: multiple gradient-slab gel electrophoresis. *Anal. Biochem.* **85**, 341–354.
27. Laemmli, U.K. (1970) Cleavage of structural proteins during the assembly of the head of bacteriophage T4. *Nature* **227**, 680–685.
28. Schägger, H., and Jagow, G.v. (1987) Tricine-sodium dodecyl sulfate-polyacrylamide gel electrophoresis for the separation of proteins in the range from 1 to 100 kDa. *Anal. Biochem.* **166**, 368–379.
29. Görg, A., and Weiss, W. (1999) Horizontal SDS-PAGE for IPG-Dalt. *Method. Mol. Biol.* **112**, 235–244.
30. Olsson, I., Larsson, K., Palmgren, R., and Bjellqvist, B. (2002) Organic disulfides as a means to generate streak-free two-dimensional maps with narrow range IPG strips as first dimension. *Proteomics* **2**, 1630–1632.

# Chapter 3

## Non-classical 2-D Electrophoresis

**Jacqueline Burré, Ilka Wittig, and Hermann Schägger**

### Summary

Classical 2-D electrophoresis (IEF/SDS 2-DE) using isoelectric focusing (IEF) and SDS-PAGE for the second dimension offers very high resolution for the separation of complex protein mixtures, but hydrophobic proteins can aggregate and are considerably under-represented in these 2-D gels. Non-classical 2-DE, as described here, summarizes several heterogeneous techniques, some of which, like BAC/SDS 2-DE and doubled SDS-polyacrylamide gel electrophoresis (dSDS-PAGE), intend to isolate the difficult hydrophobic proteins that are not accessible by classical 2-DE. Other types of non-classical 2-DE start with 1-D separation of native proteins and complexes, like blue-native electrophoresis (BNE), clear-native electrophoresis (CNE), and high-resolution clear-native electrophoresis (hrCNE). These electrophoretic techniques can substitute for chromatographic isolation of protein complexes, and can even isolate supramolecular physiological assemblies. Subsequent resolution in second dimension can be denaturing to resolve the subunits of complexes, as exemplified with BNE/SDS 2-DE, or native like in BNE/BNE 2-DE (the latter using different cathode buffers for 1-D BNE and 2-D BNE). After isolation of highly pure membrane protein complexes by two native electrophoretic separations, the separation protocol may be finished by denaturing 2-DE like BAC/SDS or doubled SDS-PAGE. Thus, a four-dimensional electrophoretic system with minimal loss of protein results that is useful as an efficient micro-scale protein separation protocol, e.g. for mass spectrometric analyses.

**Key words:** Two-dimensional electrophoresis, Isoelectric focusing, Blue-native, Clear-native, Membrane proteins, Mitochondrial complexes

## 1. Introduction

Membrane proteins are key players in numerous cellular processes, and are therefore in the focus of current biochemical and pharmacological research. Membrane proteins and membrane protein complexes are difficult to isolate in native state, and even separation in denatured state often cannot follow the general protocols for

Jörg Reinders and Albert Sickmann (eds.), *Proteomics*, Methods in Molecular Biology, Vol. 564
DOI: 10.1007/978-1-60761-157-8_3, © Humana Press, a part of Springer Science+Business Media, LLC 2009

water-soluble proteins. In this chapter, we first describe electrophoretic isolation protocols for native membrane protein complexes, and then proceed to denaturing separations. Two-dimensional polyacrylamide gel electrophoresis (2-D PAGE), also abbreviated as 2-D electrophoresis (2-DE), using a native gel for first dimension and a denaturing gel for resolution in second dimension represents one of the simplest types of non-classical 2-DE, the native/denaturing 2-DE. Other types of non-classical 2-DE include native/native or denaturing/denaturing combinations.

Native electrophoretic techniques are micro-scale techniques that often can replace large-scale chromatographic and centrifugation protocols *(1)*. Here, we describe blue-native electrophoresis (BN-PAGE or BNE, **refs.** *2, 3)*, clear-native electrophoresis (CN-PAGE or CNE, **refs.** *4, 5)*, and high-resolution clear-native electrophoresis (hrCNE, **ref.** *6)* that all have advantages and limitations. BNE is the most robust variant with the widest application *(2, 3)*. It is a charge shift technique that uses the binding of many negatively charged Coomassie Blue G-250 molecules (abbreviated here as Coomassie dye) to protein surfaces to impose excess negative charges on the protein and to pull it to the anode. BNE can be applied for determination of native protein masses using minimal amounts of non-purified protein, since proteins are separated according to size in the acrylamide gradient gels used. Excess negative charges repel each other and therefore help to avoid protein aggregation, which is a common feature of membrane proteins. Protein-bound Coomassie dye keeps membrane proteins in solution (in the absence of a detergent) and makes migrating protein bands visible during electrophoresis. Visible blue bands are then excised from the gels and native proteins are easily electroeluted. One disadvantage of BNE is related to the negatively charged Coomassie dye. Combined with neutral detergents, as used for the initial protein solubilization, Coomassie dye forms mixed micelles that are not as aggressive as anionic detergents but can lead to dissociation of labile subunits from a complex or to disassembly of physiological supramolecular structures.

CNE, in contrast to BNE, does not use an anionic component to induce a charge shift on proteins, and relies on the protein intrinsic charge. Therefore, only acidic proteins run to the anode at pH 7.5, the running pH of the gel. Basic proteins migrate towards the cathode and are lost with the cathode buffer. Protein aggregation of membrane proteins is a problem with CNE and smearing bands are often observed. Electroblotting and electroelution following CNE also have not yet found an appropriate solution. However, some membrane proteins, like yeast ATP synthase, migrate as sharp bands. For such proteins, CNE is preferred, since in-gel assays of catalytic activities and detection of fluorescent-labeled proteins are optimal. CNE is the mildest of the three techniques discussed and is especially suited to preserve and isolate physiological supramolecular structures.

hrCNE has been developed recently, to retain the combined advantages of BNE and CNE but to discard their disadvantages *(6)*. This attempt has been partially successful. hrCNE is a charge shift technique like BNE but using non-colored mixed micelles of a mild anionic and a neutral detergent that induce a charge shift on proteins similar to Coomassie dye in BNE. Resolution is as high as in BNE, all membrane proteins migrate to the anode irrespective of their intrinsic isoelectric points, and fluorescence detection is as fine as in CNE. However, hrCNE is not as mild as CNE. Except with special precautions, physiological supramolecular structures are disassembled, mostly to a higher degree than in BNE. It must be noted here that all three native electrophoresis techniques have been developed for the isolation of membrane protein complexes but the techniques can also be used for water-soluble proteins with several restrictions: (1) the isoelectric point (p*I*) of water-soluble proteins must be lower than 7 when using CNE and (2) water-soluble proteins with $pI > 7$ also migrate to the anode in BNE or hrCNE if these proteins bind Coomassie dye (which is observed in BNE with many water-soluble proteins), or if they bind mixed micelles of anionic and neutral detergents (which is rarely observed in hrCNE). Therefore, BNE is clearly preferred over hrCNE with water-soluble proteins. All three native electrophoresis techniques can be combined with SDS-PAGE to two-dimensional native/SDS systems.

A very effective strategy to isolate highly pure proteins immediately from biological membranes is to isolate the largest physiological protein assemblies first using the very mild detergent digitonin for solubilization and BNE, CNE, or hrCNE for 1-D protein separation, followed by another modified BNE for 2-D resolution (using 0.02% dodecylmaltoside (DDM) added to the Coomassie-dye-containing cathode buffer). The mixed DDM/Coomassie dye micelles can dissociate the isolated physiological supramolecular structures into smaller complexes during 2-DE, so that highly pure complexes are accessible by this native/native 2-DE. Protein subunits of the individual membrane protein complexes can optionally be separated by various denaturing electrophoretic techniques, for example, SDS-PAGE, BAC/SDS 2-DE, and dSDS-PAGE, as described in the following. This means that individual subunits finally are isolated by SDS-PAGE in second, third, or even fourth dimension. (One or two native dimensions are used for isolating native protein complexes, followed by one or two denaturing separations for the isolation of the subunits.)

To date, the most powerful gel-based electrophoretic separation method in proteome analysis is classical 2-D PAGE (IEF/SDS 2-DE), a combination of IEF and SDS-PAGE resolving several hundred to 1,000 protein spots on one gel *(7, 8)*. Despite considerable improvement of 2-DE protocols, the technique still has profound limitations regarding hydrophobic proteins because of solubility problems during IEF, and problems with the protein transfer to the second-dimension gel. Highly hydrophobic proteins are still under-represented in 2-DE. In contrast, common 1-D SDS-PAGE is a very robust system, also for hydrophobic proteins, but with low and often non-sufficient resolution. Small gel pieces may contain several different proteins, rendering it impossible to identify all contained components by mass spectrometry. A few water-soluble proteins, and especially proteins with a high number of trypsin cleavage sites, will be identified by peptide mass fingerprints in these mixtures. Low abundance proteins, small proteins, and hydrophobic proteins, however, most likely will be lost.

To circumvent the limitations of IEF/SDS 2-DE and of 1-D SDS-PAGE for proteomic analyses, alternative 2-DE approaches have been developed like BAC/SDS 2-DE using the cationic detergent benzyldimethyl-*n*-hexadecylammonium chloride (BAC, **refs**. *9–11)*, the very similar CTAB/SDS 2-DE using the cationic detergent cetyltrimethylammonium bromide (CTAB, **refs**. *12–14)*, and the recently developed doubled SDS-PAGE using SDS for both dimensions (dSDS-PAGE, **ref**. *15)*. Resolution in general is inferior to classical 2-DE but superior to 1-D SDS-PAGE. The essential benefit of the alternative 2-DE techniques is the separation of hydrophobic proteins as isolated spots, which is a prerequisite to identify such difficult proteins by mass spectrometry. Chromatography-based approaches such as SDS-PAGE coupled to LC MS/MS or multidimensional LC MS/MS, or shot-gun proteomics such as MudPIT *(16–18)* are rapidly gaining popularity. However, these techniques include no positive control for proteins to be identified as does a defined spot on a high-resolution gel.

First-dimension BAC-PAGE uses the cationic detergent BAC in an acidic discontinuous system. Proteins that are positively charged by binding BAC migrate to the cathode but migration distances do not correlate well with the protein mass, presumably because BAC binding and charge/mass ratio are less strict and uniform when compared with SDS-PAGE. However, this missing strict correlation of protein mass and migration distance in BAC-PAGE is beneficial and important for the desired scattering of spots around a diagonal in second-dimensional SDS-PAGE. BAC/SDS 2-DE has been successfully applied to resolve and identify integral membrane proteins with several transmembrane domains from human platelets *(19)*, synaptic vesicles *(20, 21)*, and yeast mitochondria *(22)*.

Doubled SDS-PAGE (dSDS-PAGE) employs SDS for first- and second-dimension gels and thus bypasses the problems with detergent change during transition from first to second dimension. The desired scattering of spots around a diagonal in 2-DE is achieved by strong variation of the acrylamide concentration used for 1-D and 2-D separations and the optional use of urea for one of the two dimensions. Highly hydrophobic proteins can be identified immediately by visual inspection of the 2-D gel. They are all located above the diagonal of more hydrophilic proteins. This has been demonstrated, for example, for membrane protein complexes like bovine respirasomes *(15)*, synaptic vesicles *(21)*, and with the identification of two novel ATP synthase-associated proteins *(23)*.

## 2. Materials

Protean II unit (Bio-Rad) and SE 400 vertical unit (GE Healthcare) are suitable for the electrophoresis types described in this chapter. Power supplies should allow for a voltage around 500 V and a current around 200-500 mA. A peristaltic pump (e.g. P1, GE Healthcare) and a gradient mixer (e.g. Model 385 gradient former, Bio-Rad) are required to cast acrylamide gradient gels.

### 2.1. Native Electrophoresis Techniques (BNE, CNE, and hrCNE)

#### 2.1.1. Acrylamide Gradient Gels for Native Electrophoresis

Unless otherwise indicated, all solutions are stored cold (at 4°C).

1. AB-3, acrylamide/bisacrylamide (Serva) stock solution (0.5 L) with 3% cross-linker, also assigned as 49.5% T, 3% C solution, where T is the total concentration of both monomers (acrylamide and bisacrylamide) and C gives the percentage of the cross-linker bisacrylamide relative to acrylamide: dissolve 240 g acrylamide and 7.5 g bisacrylamide in water to a total volume of 500 mL and then store at 7°C (bisacrylamide can crystallize from the solution at lower temperature). (Caution, wear gloves and use a pipetting aid when handling highly neurotoxic acrylamide/bisacrylamide.)
2. Two molar 6-aminohexanoic acid (Sigma) (1 L).
3. Two molar imidazole (Sigma) (1 L, no pH adjusted).
4. One molar imidazole/HCl (1 L): titrate 68.08 g imidazole with 114 mL 5 N HCl to reach pH 7.0 at 4°C and fill up with water to 1 L.
5. One molar Tris (Sigma) (1 L).
6. One molar Tricine (Sigma) (1 L).
7. GB-nat (3×), triple concentrated gel buffer for native gels: mix 375 mL of the 2 M 6-aminohexanoic acid solution, 37.5 mL 1 M imidazole/HCl, pH 7.0, and add water to a volume of 500 mL. The actual pH is around 7.0-7.3. It must not be corrected except with imidazole or HCl.

8. GB-nat (1×), overlay buffer for native gels, prepared by dilution of GB-nat (3×).
9. APS (10%), ammonium persulfate (Sigma) 10% (w/v) freshly prepared.
10. TEMED, tetramethyl-ethylenediamine (Sigma).

*2.1.2. Isolation and Storage of Biological Membranes*

The preferred buffer for isolation and storage of biological membranes contains 250 mM sucrose, 50 mM imidazole/HCl, pH 7.0 (1 mM EDTA and 5 mM 6-aminohexanoic acid is added for protease inhibition).

*2.1.3. Solubilization and Sample Preparation*

1. Solubilization buffer A: 50 mM NaCl, 50 mM imidazole, 5 mM 6-aminohexanoic, 1 mM EDTA, pH 7.0 at 4°C.
2. Solubilization buffer B: 50 mM imidazole, 500 mM 6-aminohexanoic, 1 mM EDTA, pH 7.0 at 4°C.
3. DDM (10%), DDM (10%, w/v) (Glycon, Luckenwalde, Germany) dissolved in water and stored frozen as 1 mL aliquots.
4. Triton X-100 (10%, w/v) (Fluka), dissolved in water and stored frozen as 1 mL aliquots.
5. Digitonin (20%, w/v), (Fluka, no. 737006, >50% purity, used without recrystallization) dissolved in water with warming to 50-90°C and stored frozen as 1 mL aliquots. Redissolved by warming to 50°C or up to 90°C if necessary.
6. DOC (10%), sodium deoxycholate (10%, w/v) (Merck) dissolved in water and stored frozen as 1 mL aliquots.
7. Fifty percent glycerol (w/v) (Sigma).
8. Fifty percent glycerol (w/v), 0.1% Ponceau S dye (Sigma).
9. Fife percent Coomassie dye, 5% (w/v) Coomassie blue G-250 (Serva Blue G; Serva) suspended in 500 mM 6-aminohexanoic.

*2.1.4. Electrophoresis Buffers and Conditions*

1. Anode buffer (5 L) for BNE, CNE, and hrCNE: 25 mM imidazole/HCl, pH 7.0. Store cold.
2. Cathode buffer C for CNE (5 L): 50 mM Tricine, 7.5 mM imidazole, pH around 7.0. No pH correction except with Tricine or imidazole. Store cold.
3. Cathode buffer B for BNE (2 L): add 0.4 g Coomassie blue G-250 (0.02% Coomassie dye) to 2 L of cathode buffer C for CNE. Stir at room temperature overnight, and keep this buffer at room temperature to avoid formation of large Coomassie micelles that can block the gel pores.
4. Cathode buffer B/10 for BNE (2 L): dilute 200 mL cathode buffer B with 1.8 L of cathode buffer C. Store cold.

5. Cathode buffer for hrCNE-1 (containing 0.02% DDM and 0.05% DOC): add 0.4 mL DDM (10%) and 1.0 mL DOC (10%) to 200 mL cathode buffer C (sufficient for one gel, i.e. to fill one cathode chamber). Add DDM and DOC to cathode buffer C (room temperature) shortly before use.
6. Cathode buffer for hrCNE-2 (containing 0.05% Triton X-100 and 0.05% DOC): add 1.0 mL Triton X-100 (10%) and 1.0 mL DOC (10%) to 200 mL cathode buffer C. Add Triton X-100 and DOC to cathode buffer C (room temperature) shortly before use.
7. Cathode buffer for hrCNE-3 (containing 0.01% DDM and 0.05% DOC): add 0.2 mL DDM (10%) and 1.0 mL DOC (10%) to 200 mL cathode buffer C. Add DDM and DOC to cathode buffer C (room temperature) shortly before use.

### *2.2. Two-Dimensional Native 2-DE*

#### *2.2.1. Electrophoresis of 2-D Native Gels*

1. Cathode buffer B + DDM for modified BNE containing 0.02% DDM: add 0.4 mL DDM (10%) to 200 mL cathode buffer B shortly before use.
2. Cathode buffer B + TX for modified BNE containing 0.03% Triton X-100: add 0.6 mL Triton X-100 (10%) to 200 mL cathode buffer B shortly before use.

### *2.3. Native/SDS 2-DE*

#### *2.3.1. Tricine-SDS Gels*

1. GB-Tric (3×), triple concentrated gel buffer for Tricine-SDS-PAGE (1 L): 3 M Tris, 1 M HCl, 0.3% SDS, pH 8.45.
2. SDS 20% (w/v) (Sigma).
3. SDS 1%, 1% mercaptoethanol (Sigma) (prepared shortly before use to keep mercaptoethanol in reduced state).

#### *2.3.2. Tricine-SDS-PAGE Buffers and Conditions*

1. Anode buffer (10×) (1 L): 1 M Tris, 225 mM HCl, pH 8.9.
2. Cathode buffer (10×) (1 L): 1 M Tris, 1 M Tricine, 1% SDS, pH around 8.25.

### *2.4. BAC/SDS 2-DE*

#### *2.4.1. Gel Preparation for BAC-PAGE*

1. Acrylamide/bisacrylamide solution (40% (w/v), 29:1, or 40% T, 3.4% C) (1 L, Roth) – toxic and possibly carcinogenic and teratogenic. Store cold.
2. Three hundred millimolar potassium dihydrogen phosphate adjusted with phosphoric acid to pH 2.1. Abbreviated as "300 mM potassium phosphate, pH 2.1" (1 L, Sigma).
3. A "75-mM potassium phosphate, pH 2.1" (100 mL, Sigma).
4. Five hundred millimolar potassium dihydrogen phosphate adjusted with phosphoric acid to pH 4.1. Abbreviated as "500-mM potassium phosphate, pH 4.1" (250 mL, Sigma).
5. One percent bisacrylamide (w/v) solution (250 mL, Sigma) – toxic and possibly carcinogenic and teratogenic. Dissolve 2.5-g bisacrylamide in water to a total volume of 250 mL,

stir at room temperature until bisacrylamide is dissolved, and store at 4°C.

6. Eighty milllimolar ascorbic acid (5 mL, AppliChem). Prepare always fresh.
7. Five millimolar ferrous sulfate (1 mL, Merck). Prepare fresh as tenfold solution, dissolve at 37°C for 5 min and subsequently dilute to 5 mM with water.
8. Two hundred and fifty millimolar BAC (1 mL, Roth). Prepare always fresh and dissolve at 37°C until the solution turns clear.
9. A 0.025% and 0.04% $H_2O_2$ (v/v) (30% solution, Roth). Prepare dilutions always fresh. Store 30% hydrogen peroxide stock solution and dilutions at 4°C in the dark.

*2.4.2. Sample Preparation and 1-D BAC-PAGE*

1. BAC sample buffer (1×): 2.25-g urea, 0.5-g BAC, 630-mg glycerol, 4-mL water. Warm up to 37°C with shaking. Upon solubilization of urea and BAC (solution turns clear) add 0.25-mL 1.5-M DTT, 50-μL 5% pyronin Y (w/v) (Sigma) and fill up with water to 10 mL. The buffer can be stored at –20°C (e.g. as 500 μL aliquots). The buffer is unstable at room temperature and the solution should not be used for more than 5 days. Frozen samples can be warmed up to 37°C for solubilization.
2. BAC electrophoresis buffer (1×, 1 L) contains 2.5-mM BAC, 150-mM glycine, and 50-mM *o*-phosphoric acid (pH around 1.8). The solution is stirred at room temperature for at least 1 h to completely dissolve BAC. The buffer is stable at room temperature.

*2.4.3. Fixation, Staining, and Equilibration of BAC-SDS Gels*

1. Fixation solution: 35% isopropanol (v/v), 10% acetic acid (v/v).
2. Coomassie-staining solution: 0.1% Coomassie Brilliant Blue-R250 (w/v), 7.5% acetic acid (v/v), 50% ethanol (v/v).
3. Destaining solution I: 7.5% acetic acid (v/v), 50% ethanol (v/v), 5% glycerol (w/v).
4. Destaining solution II: 7.5% acetic acid (v/v), 5% ethanol (v/v), 5% glycerol (w/v).
5. Storage solution: 5% acetic acid (v/v).
6. Equilibration solution: 100-mM Tris/HCl, pH 6.8.

*2.4.4. 2-D Laemmli SDS-PAGE*

1. SDS sample buffer: 12% glycerol (w/v), 2% SDS (w/v), 0.08-M Tris/HCl, pH 6.8, 0.0012% bromophenol blue (w/v), 100 mM DTT.
2. Electrophoresis buffer (for anode and cathode in the Laemmli system): 25 mM Tris, 192 mM glycine, 0.1% SDS (w/v). The resulting pH is around 8.4.

3. Thirty percent acrylamide (w/v) (1 L, Roth) – toxic and possibly carcinogenic and teratogenic.
4. One molar Tris/HCl pH 8.7 (1 L).
5. A 0.25 M Tris/HCl pH 6.8 (250 mL).
6. Twenty percent SDS (w/v) (100 mL).
7. TEMED, tetramethyl-ethylenediamine (Roth).
8. APS (10%) (Biorad), ammonium persulfate 10% (w/v), freshly prepared, 1 mL stocks can be frozen at –20°C.

### 2.5. Doubled SDS Polyacrylamide Gel Electrophoresis

SDS sample buffer for Tricine-SDS-gels: 12% SDS (w/v), 6% mercaptoethanol (v/v), 30% glycerol (w/v), 0.05% Coomassie dye, 150 mM Tris/HCl, pH 7.0.

#### 2.5.1. First-Dimensional Tricine-SDS-PAGE (see Subheadings 2.3.1 and 2.3.2)

#### 2.5.2. Acidic Incubation

Solution for acidic incubation (0.5 L): 100 mM Tris, 150 mM HCl, pH 1.6 (50-100 times the gel volume is commonly used).

#### 2.5.3. Second-Dimensional Laemmli-SDS PAGE (see Subheading 2.4.4)

1. GB-Lae (4×): gel buffer stock solution for Laemmli gels containing 0.4% SDS and 1.5 M Tris adjusted with HCl to pH 8.6 at room temperature.
2. SGB (10×): Side gel buffer stock solution (100 mL) contains 1.0 M Tris, 1 M HCl, pH 7.4.

## 3. Methods

Proteins and complexes are first isolated here by electrophoretic techniques preserving the native state and followed by 2-D SDS-PAGE. Other non-classical 2-DE variants include two native separations or, like BAC/SDS 2-DE and dSDS-PAGE, are denaturing in both dimensions.

### 3.1. Native Electrophoresis Techniques (BNE, CNE, and hrCNE)

All three native electrophoresis variants described here use the same gels (*see* **Note 1**), essentially the same protein solubilization protocols (with minor modifications as indicated in the protocols below; *see* **Notes 2** and **3**), and the same electrophoresis conditions but different cathode buffers (*see* **Subheading 2.1.4**) that specify BNE, CNE, and three variants of hrCNE (hrCNE-1, -**2**, and -**3**). The resolution of identical samples by BNE, CNE, and hrCNE-1 using purified bovine mitochondria solubilized by DDM, is exemplified in **Fig. 1a–c**. Using milder solubilization by digitonin (**Fig. 1d**, **e**), supercomplexes of respiratory complexes

($S_0$–$S_2$) and oligomeric ATP synthase (M, D, T, and H) can be preserved by BNE and hrCNE-3. The advantage of hrCNE-1 for detection of specifically fluorescent-*tagged* proteins (mitochondrial complex I from *Yarrowia lipolytica*, **Fig. 1f**) compared with BNE (**Fig. 1g**) is immediately apparent.

#### *3.1.1. Acrylamide Gradient Gels for Native Electrophoresis*

Casting one 4–13% acrylamide gel (0.16 × 14 × 14 cm) is exemplified.

1. Prepare the 3.5% acrylamide sample gel mixture (6 mL): Mix 0.44 mL acrylamide-bisacrylamide solution (AB-3), 2 mL gel buffer for native gels (GB-nat 3×), 3.4 mL cold water, and store cold (4°C or on ice).
2. Prepare the 4% acrylamide separation gel mixture (18 mL): Mix 1.5 mL AB-3, 6 mL GB-nat (3×), 10.4 mL cold water, and store cold.
3. Prepare the 13% acrylamide separation gel mixture (15 mL): Mix 3.9 mL AB-3, 5 mL GB-nat (3×), 3 g glycerol, 3 mL cold water, and store cold.
4. Mount the gel cassette using 2 glass plates (18 × 18 cm) and spacers (0.16 × 2 × 18 cm), add 1 mL water to the bottom, mount a long cannula that reaches the bottom of the gel cassette, and store cold.
5. Add the 4% and 13% acrylamide mixtures to the two arms of a gradient mixing vessel with the connecting channel closed, and store cold.
6. Add 10% APS (100 and 75 μL), and TEMED (10 and 7.5 μL), to the 4% and 13% acrylamide mixtures, respectively, in the gradient mixer and agitate using a magnetic stirrer. Then pump in a few milliliter of the 4% solution, open the connecting channel of the gradient mixer and continue with pumping until liquid level reaches ~1 cm from top. (Alternatively, if the gel-casting device includes a hole in the spacers between the two glass plates, the gradient gel can also be pumped from the bottom.)
7. Remove cannula and let the gel polymerize at room temperature (15–30 min).
8. Remove water from the top of the gel after polymerization, add 50 μL 10% APS and 5 μL TEMED to the 3.5% acrylamide sample gel mixture (*see* **step 1**), pour this solution on top of the polymerized gradient separation gel, and insert the comb for 0.5 cm or 1 cm sample wells.
9. The sample gel polymerizes within 10–15 min.
10. Remove the comb and overlay with overlay buffer (GB-nat 1×).
11. Store the gel at 4°C, and renew the overlay buffer at 3-day intervals.

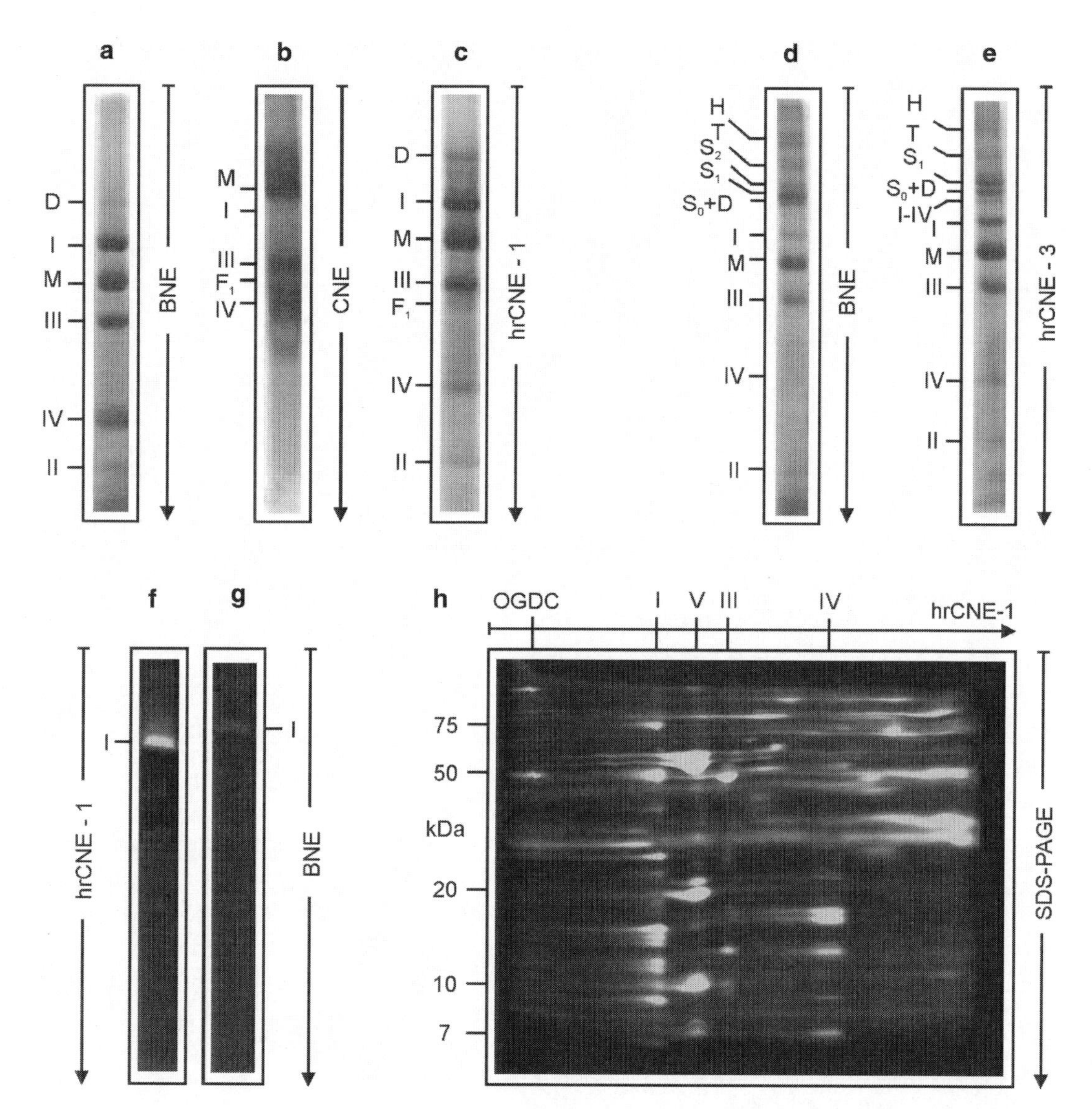


Fig. 1. Performance of various types of 1-D native electrophoresis techniques, and one example for 2-D SDS-PAGE. (**a-e**) 1-D native gels were Coomassie-stained after electrophoresis. (**a-c**) Bovine mitochondria were solubilized by dodecylmaltoside and identical samples were separated by (**a**) BNE, (**b**) CNE, and (**c**) hrCNE-1 to separate mitochondrial complexes I-IV, monomeric and dimeric complex V (M and D), and the $F_1$-part of complex V ($F_1$). The complexes were resolved according to native mass (e.g. complex II, 130 kDa; dimeric complex V, around 1.5 MDa). (**d** and **e**) Bovine mitochondria were solubilized by digitonin and identical samples were separated by (**d**) BNE, and (**e**) hrCNE-3 to separate in addition physiological associates of complexes I-IV ($S_0$-$S_2$) and the tetrameric (T, ~3 MDa) and hexameric (H, ~4.5 MDa) structures of complex V. (**f** and **g**) Mitochondria from the yeast *Yarrowia lipolytica* expressing a complex I-subunit tagged with the yellow fluorescent protein was solubilized by digitonin, and identical samples were resolved by (**f**) hrCNE-1, and (**g**) BNE to demonstrate fluorescence quenching in BNE and the advantage of hrCNE. (**h**) Two-dimensional resolution of a 1-D native gel strip comparable with (**c**) using a 10% acrylamide gel for Tricine-SDS-PAGE. Mitochondrial proteins were generally labeled here by NHS-Fluorescein dye (Pierce, *see* **ref.** *6)* prior to electrophoresis, and detected immediately in the non-stained 2-D gel using a Typhoon scanner (GE Healthcare). OGDC, oxoglutarate dehydrogenase complex (2.5 MDa).

#### 3.1.2. Isolation and Storage of Biological Membranes

1. Suspend biological membranes preferentially in carbohydrate or glycerol containing buffers (*see* **Subheading 2.1.2, Item 1**) to avoid freezing of water within multiprotein complexes that can dissociate the complexes (*see* **Note 4**).
2. Shock-freeze aliquots and store at −80°C.

#### 3.1.3. Solubilization and Sample Preparation

General solubilization principles are exemplified with mitochondrial and bacterial membranes. Mitochondrial complexes are also used as markers for native mass determination (*see* **Note 5**). Solubilization and sample preparation are routinely performed at 4°C or on ice.

1. Pellet aliquots of isolated mitochondria (400 μg protein) by 10 min centrifugation at 20,000 × *g*, or bacterial membranes (400 μg protein) by 30 min ultracentrifugation at 100,000 × *g*.
2. Add 40 μL solubilization buffer A (*see* **Subheading 2.1.3, Item 1**) or B (*see* **Subheading 2.1.3, Item 2**, *see* **Note 6**).
3. Homogenize the membranes by pipetting or with a tiny spatula.
4. Add detergent from 10% stock solutions, e.g. 4–12 μL DDM (DDM 10%) or Triton X-100 (TX 10%), if isolation of individual complexes is desired. These detergent amounts correspond to detergent/protein ratios of 1.0–3.0 g/g (*see* **Note 7**).
5. Leave samples for 5–10 min.
6. Centrifuge for 15 min at 100,000 × *g*, or 20 min at 20,000 × *g* to retain complexes larger than 5 MDa in the supernatant.
7. Add 5 μL 50% glycerol/0.1% Ponceau S (*see* **Note 8**) to the supernatants for CNE and hrCNE; for BNE add 5 μL glycerol (50%) and then Coomassie dye from a 5% Coomassie stock to set a detergent/Coomassie ratio of 8 (g/g), e.g. 3 μL from the 5% Coomassie stock when 12 μL 10% DDM was used for solubilization.
8. For sample loading see below.

#### 3.1.4. Electrophoresis Buffers and Conditions

Electrophoresis is generally performed at 4–7°C, since broadening of bands has been observed at room temperature.

1. Mount the gel into the electrophoresis chamber and fill in the anode buffer and the specific cathode buffer selected for CNE, BNE, or hrCNE-1, -2, and -3 (**Subheading 2.1.4**). Apply samples to native gel wells by underlaying the sample under the cathode buffer. Set the power supply to 500 V and 15 mA per gel (this is the maximum current for gels with dimensions 0.16 × 14 × 14 cm) and start electrophoresis. The initial voltage (around 200 V) gradually rises during electrophoresis until 500 V and 15 mA is reached, and electrophoresis continues with constant voltage (500 V) and decreasing current.

2. Run CNE and hrCNE for about 3–4 h until the red dye Ponceau S reaches the gel front. Similarly, BNE can be run for about 3–4 h until Coomassie dye approaches the gel front. Commonly, cathode buffer B is removed by a suction pump after about one-third of the run and replaced by cathode buffer B/10 with tenfold lower dye concentration (*see* **Notes 9** and **10**).

### 3.2. Two-Dimensional Native 2-DE

Multiple variations of BNE, CNE, and hrCNE for two-dimensional native electrophoresis (native in both dimensions) seem conceivable. However, one specific combination is clearly preferred: the combination of 1-D BNE (using digitonin for protein solubilization, **Fig. 2a**) and modified 2-D BNE (with 0.02% DDM added to the cathode buffer) as exemplified in **Fig. 2b** (*see also* **Note 11**).

#### 3.2.1. Handling of 1-D Native Gels for 2-D Native Separation

1. Prepare a 3.5% acrylamide native gel mixture (9 mL): Mix 0.63 mL AB-3 (*see* **Subheading 2.1.1**), 3.0 mL GB-nat (3x) (*see* **Subheading 2.1.1**), 5.28 mL water, and store at room temperature.

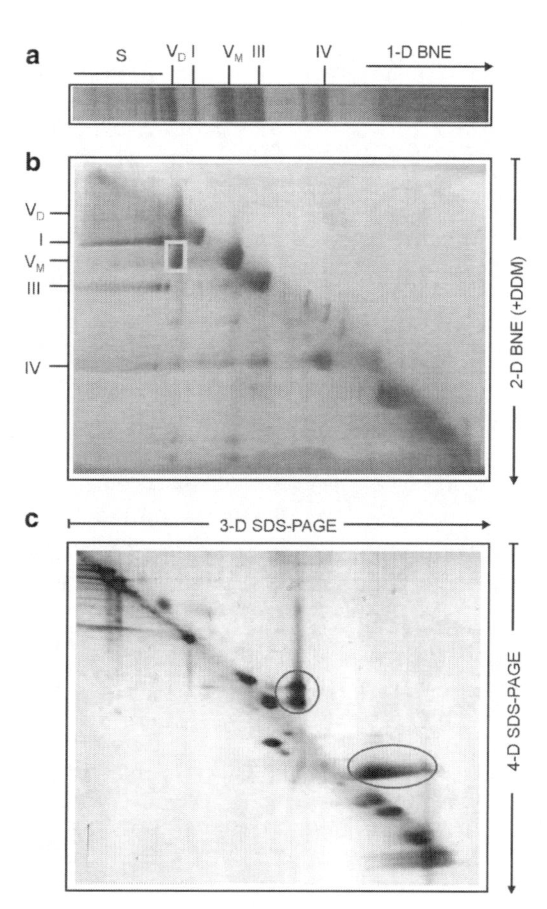


Fig. 2. Four-dimensional separation of the subunits of mitochondrial ATP synthase. (**a**) Crude rat heart mitochondria were solubilized by digitonin and the mitochondrial complexes I, III, IV, monomeric and dimeric complex V ($V_M$ and $V_D$), and several respiratory chain supercomplexes (summarized as S) were separated by 1-D BNE. (**b**) A gel strip from 1-D BNE was further resolved by 2-D BNE but using a modified cathode buffer for BNE that was supplemented with 0.02% dodecylmaltoside (DDM). This modified cathode buffer dissociated supercomplexes into the individual complexes, and most of the dimeric complex V ($V_D$) into monomeric complex V ($V_M$, *boxed yellow*). (**c**) Monomeric complex $V_M$ (*boxed yellow*) was then cut out from the BNE/BNE 2-D gel and subjected to 3-D Tricine-SDS-PAGE using a 10% acrylamide gel supplemented with 6 M urea. The gel strip from 3-D SDS-PAGE was further resolved by 4-D Tricine-SDS-PAGE using a 16% acrylamide gel, according to the common protocol for doubled SDS-PAGE (dSDS-PAGE, *see* **ref.** *15*). Highly hydrophobic subunits (*circled red*) are separated from the diagonal of common subunits.

2. Prepare a 5% acrylamide mixture for the separation gel (18 mL): Mix 1.8 mL AB-3, 6 mL GB-nat (3×), 10.2 mL cold water, and store cold.
3. Prepare a 16% acrylamide mixture for the separation gel (15 mL): Mix 5.0 mL AB-3, 5 mL GB-nat (3×), 3 g glycerol, around 3.2 mL cold water, and store cold.
4. Cut strips out of 1-D native gels (1 cm width, extending from sample gel to the running front that is defined by Ponceau S and Coomassie dyes).
5. Wet the strips with water and place them on glass plates at the positions commonly occupied by sample gels.
6. Mount the gel cassette using 2 glass plates (18 × 18 cm) and spacers (0.16 × 2 × 18 cm), add 1 mL water to the bottom, and mount a long cannula that reaches the bottom of the gel cassette.
7. Add the 5% and 16% acrylamide mixtures (*see* **steps 2** and **3**) to the two arms of a gradient mixing vessel with the connecting channel closed, and store cold.
8. Add 10% APS (75 and 50 μL), and TEMED (7.5 and 5 μL), to the 5% and 16% acrylamide mixtures, respectively, in the gradient mixer. Then pump in a few milliliter of the 5% acrylamide solution, open the connecting channel and continue with pumping. (Alternatively, if the gel-casting device includes a hole in the spacers between the two glass plates, the gradient gel can also be pumped from the bottom.) Stop pumping when water on top of the gradient reaches the 1-D native gel strip.
9. Let the gel polymerize at room temperature (around 30 min).
10. Remove water from the top of the gel after polymerization. Then push the native gel strip down to the gradient gel using appropriate plastic cards. Add 100 μL APS (10%) and 10 μL TEMED to the 3.5% acrylamide sample gel mixture (*see* **step 1**), pour this solution on top of the polymerized gradient separation gel besides the native gel strip but do not cover the gel strip.
11. The sample gel polymerizes within 10–15 min.
12. A 2-D native electrophoresis can be started shortly afterwards. It can also be started the next day if the gel is overlaid with water or with GB-nat (1×).

*3.2.2. Electrophoresis of 2-D Native Gels*

1. Install the gel in the electrophoresis chamber and fill in the anode buffer, which is the same as used before for 1-D BNE, CNE, and hrCNE (*see* **Subheading 2.1.4, Item 1**).

2. Independent of the 1-D native gel type used, select cathode buffer B + DDM or cathode buffer B + TX (*see* **Subheading 2.2.1**) containing 0.02% DDM or 0.03% Triton X-100, respectively, for the following modified 2-D BNE. Start modified 2-D BNE under the running conditions of 1-D BNE (setting 500 V and 15 mA for gels with dimensions 0.16 × 14 × 14 cm).
3. Stop electrophoresis late, in contrast to 1-D native gels, e.g. after 5 h, until Coomassie dye begins to elute from the gel, to focus the 1 cm broad band from the 1-D native gel into sharp spots. Vertical gel strips (from low to high acrylamide) or horizontal gel strips (focusing on a specific protein and its associations in various supramolecular structures) can then be analyzed by SDS-PAGE in third dimension.

### 3.3. Native/SDS 2-DE

All 1-D native and 2-D native gels are commonly processed by generally applicable protocols for final SDS-PAGE (2-D, 3-D, or 4-D SDS-PAGE depending on the number of preceding electrophoretic separation steps). The performance of such native/SDS 2-D gels is exemplified by **Fig. 1h** using hrCNE-1 for separation of fluorescent-labeled mitochondrial complexes, and Tricine-SDS-PAGE for the 2-D resolution of the subunits (*see also* **Note 12**).

#### 3.3.1. Tricine-SDS Gels

Casting of a 10% acrylamide 2-D Tricine-SDS gel (0.07 × 14 × 14 cm) is exemplified.

1. Prepare a 10% acrylamide native gel mixture (2.5 mL): Mix 0.5 mL AB-3, 0.83 mL GB-nat (3×) (*see* **Subheading 2.1.1**), and 1.17 mL water.
2. Prepare a 10% acrylamide mixture for the separation SDS-gel (15 mL): Mix 3 mL AB-3, 5 mL GB-SDS (3×) (**Subheading 2.3.1**), 1.5 g glycerol, and add water to a volume of 15 mL.
3. Cut a gel strip out of the 1-D native gel (0.5 cm width, extending from sample gel to the running front defined by Ponceau S and Coomassie dye) and place the strip on a glass plate at the common location of sample gels.
4. Wet the strip with 1% SDS for 10-30 min or with 1% SDS, 1% mercaptoethanol for 1-2 h if cleavage of disulfide bonds is essential. However, in the latter case, remove mercaptoethanol carefully before mounting the gel in the gel cassette, since mercaptoethanol is an effective inhibitor of polymerization.
5. Mount the gel cassette using spacers (0.07 × 2 × 18 cm) and 2 glass plates (18 × 18 cm). The 0.5 cm strip from the native gel (0.16 cm) is thereby squeezed to a width around 1.0 cm.
6. Add 75 µL APS (10%) and 7.5 µL TEMED to the 10% acrylamide mixture (**step 2**), and pour in the solution through the gaps between native gel strip and spacers. The separating gel solution should end 2–3 mm beneath the native gel strip.

7. Overlay with water that may contact the native gel strip and let the gel polymerize at room temperature (around 10–15 min).
8. Remove water from the top of the gel after polymerization, and push the native gel strip down to the separating gel using a plastic card that is slightly thinner than gel spacers.
9. Add 15 μL APS (10%) and 1.5 μL TEMED to the 10% acrylamide native gel mixture (*see* **step 1**), and pipette this solution besides the native gel strip but do not cover the gel strip.
10. The native gel polymerizes within 5 min.
11. A 2-D SDS-PAGE is commonly started shortly afterwards.

#### *3.3.2. Tricine-SDS-PAGE Buffers and Conditions*

1. Install the gel in the electrophoresis chamber and fill in the anode and cathode buffers (10× stock solutions diluted with 9 volumes of water) into the lower and upper chambers, respectively.
2. Start electrophoresis at room temperature with voltage and current limited to 200 V and 50 mA (for gel dimensions 0.07 × 14 × 14 cm).
3. When the current falls below 50 mA, the voltage can be increased up to 300 V with the current still limited to 50 mA. The run time is around 3–4 h.

### *3.4. BAC/SDS 2-DE*

Good resolution of proteins in the mass range 10-180 kDa is achieved using 8% acrylamide gels for 1-D BAC-PAGE followed by 15% acrylamide gels for 2-D Laemmli SDS-PAGE, as exemplified below (*see* **Note 13**). Typical performance of BAC/SDS 2-DE is shown in **Fig. 3b** and compared with dSDS-PAGE in **Fig. 3c**.

#### *3.4.1. Gel Preparation for BAC-PAGE*

Mount minigels (0.8 mm, 5 × 8 cm)

1. Prepare the separation gel for an 8% acrylamide BAC gel (40 mL): Mix 7.5 g urea, 8 mL acrylamide/bisacrylamide solution (40%; *see* **Subheading 2.4.1, Item 1**), 10 mL 300 mM potassium phosphate, pH 2.1, 1.4 mL 1% bisacrylamide solution, and 11 mL water. Add 2 mL 80 mM ascorbic acid, 64 μL 5 mM ferrous sulfate, 400 μL 250 mM BAC, and stir until all components are dissolved.
2. For polymerization, add 1.6 mL 0.025% $H_2O_2$, mix and wait until all air bubbles have disappeared. Pour the gel between the glass plates and overlay ~1 mL 75 mM potassium phosphate pH 2.1 on top of the separating gel solution to achieve a smooth surface. Polymerization is completed after ~15 min.

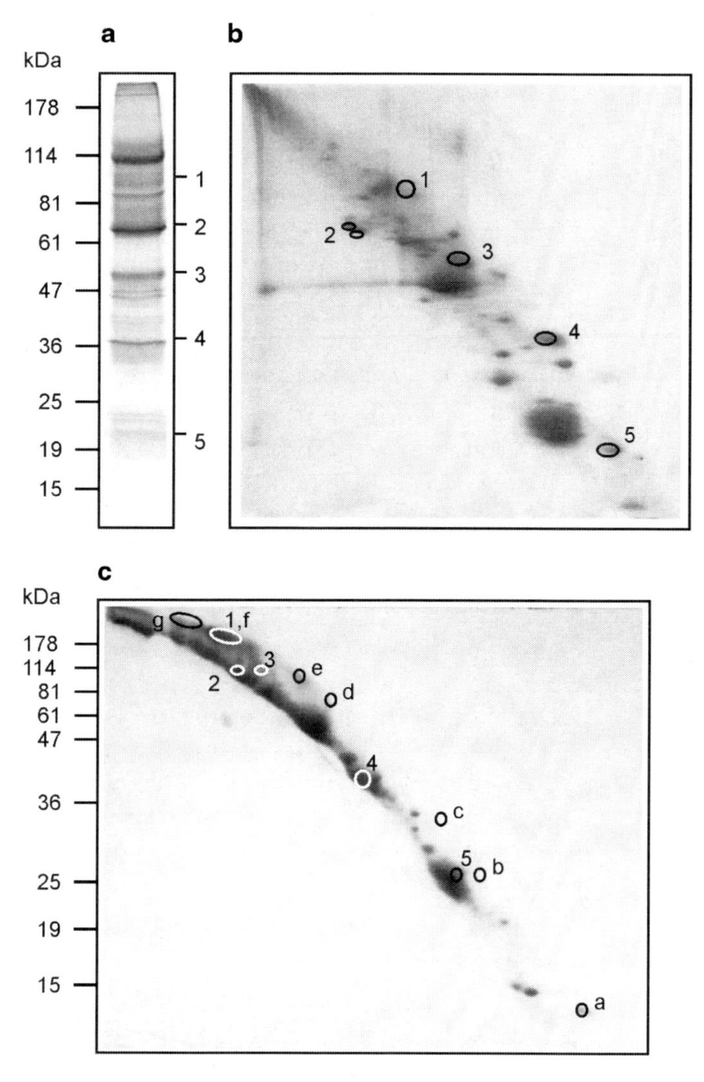


Fig. 3. Performance of 1-D SDS-PAGE, BAC/SDS 2-DE, and dSDS-PAGE exemplified with the separation of 150 µg immunoisolated synaptic vesicle proteins. (**a**) 1-D SDS-PAGE using a 15% acrylamide gel for Laemmli SDS-PAGE, (**b**) BAC/SDS 2-DE using an 8% acrylamide gel for 1-D BAC-PAGE and a 15% acrylamide gel for 2-D Laemmli SDS-PAGE, (**c**) dSDS-PAGE using a 9% acrylamide gel (without urea) for 1-D Tricine-SDS-PAGE and a 15% acrylamide gel for 2-D Laemmli SDS-PAGE (using doubled gel buffer concentration when compared with standard Laemmli SDS-PAGE). Highly hydrophobic proteins (marked a-g) are located above the diagonal of common proteins. Some proteins that were identified in all three gel types (proteins 1-5) were marked for comparison. Reproduced with permission from **ref.** *21*.

3. Remove the potassium phosphate solution and prepare 10 mL of the 4% stacking gel solution: 1 g urea, 1 mL of the 40% acrylamide/bisacrylamide solution (*see* **Subheading 2.4.1, Item 1**), 2.5 mL 500 mM potassium phosphate pH 4.1, 2.4 mL 1% bisacrylamide solution, and 3 mL water. Add 500 µL 80 mM ascorbic acid, 8.5 µL 5 mM ferrous sulfate, 70 µL 250 mM BAC and start polymerization by addition of 500 µL 0.04% $H_2O_2$.

4. The gel can be used immediately or stored at 4°C for up to one week.

#### 3.4.2. Sample Preparation and 1-D BAC-PAGE

1. Dissolve the sample in BAC sample buffer (e.g. 15 μg protein for Western blotting or 200 μg for a preparative gel in 30 μL) and incubate for 5 min at 60°C (*see* **Note 14**). Spin the sample with a table centrifuge at maximum speed for 10 s, and load 20 μL of the supernatant on the gel (for minigels).
2. Install the gel in the electrophoresis chamber, fill in the electrophoresis buffer into the lower and upper chamber, respectively, and load the sample wells (*see* **Note 15**).
3. For minigels (0.8 mm, 5 × 8 cm), start electrophoresis at 10 mA towards the cathode for 30 min until the pyronine dye has entered the separating gel. Then, hold the current at 20 mA for 2.0-2.5 h until the pyronine dye is around 1 cm apart from gel front. For larger gels (14 × 16 cm), the current is initially set to 25 mA and finally raised to 80 mA for 8–9 h.

#### 3.4.3. Fixation, Staining, and Equilibration

To avoid protein losses and to assess the quality of the actual protein separation by BAC-PAGE, proteins are stained in the 1-D gel strips using silver staining protocols before these strips are equilibrated for SDS PAGE.

1. Remove the gel from the glass plates, mark one edge of the gel, since the pyronine dye will be washed away, and shake four times for 15 min in fixation solution.
2. Stain the gel for 15 min with Coomassie staining solution and destain three times for 15 min with destaining solution I, and overnight with destaining solution II (*see* **Note 16**).
3. Store the gel in 4% acetic acid at 4°C for several days or use it immediately for the second dimension SDS gel. Gel documentation can be done at this stage.
4. To prepare the gel strip for the 2-D separation, cut the gel strips from BAC-PAGE using sharp and long blades rather than short scalpels and shake the strips three times for 10 min in equilibration solution (*see* **Notes 17** and **18**). For minigels, up to 0.6 cm wide lanes can be used.

#### 3.4.4. 2-D Laemmli SDS-PAGE

1. Prepare 15 mL 15% acrylamide separating gel (with low cross-linker concentration; *see* **Note 19**) by mixing 7.5 mL 30% acrylamide solution (*see* **Subheading 2.4.1**, **Item 10**), 1.3 mL 1% bisacrylamide solution (*see* **Subheading 2.4.1**, **Item 5**), 5.6 mL 1 M Tris/HCl, pH 8.7, 500 μL water, 75 μL 20% SDS, 15 μL TEMED, and 75 μL 10% APS.
2. Pour the gel and overlay with isopropanol until polymerization is complete.

3. Remove isopropanol and prepare the stacking gel (5 mL) mixing 835 μL 30% acrylamide solution, 650 μL 1% bisacrylamide solution, 2.5 mL 0.25 M Tris-HCl pH 6.8, 975 μL water, 25 μL 20% SDS, 5 μL TEMED, and 25 μL 10% APS. Pour the gel on top of the separating gel.
4. Then mount a fixed, stained, and equilibrated gel strip from 1-D BAC-PAGE on top of the SDS gel with a large well to accommodate the BAC gel strip and an extra well for the molecular mass marker. Avoid any air bubbles between BAC and SDS gels by filling the well with water prior to pushing the BAC gel strip down to the SDS gel. The gel strip should fit exactly into the well. No additional agarose or polyacrylamide seal is necessary at this stage.
5. Install the gel into the electrophoresis chamber and overlay the BAC gel strip with 5× SDS sample buffer including DTT for 5 min. Then, fill both buffer tanks with SDS electrophoresis buffer, add the protein marker to the precast well, and start electrophoresis with a current of 5 mA until bromophenol blue dye (from the sample buffer) enters the separation gel.
6. Continue electrophoresis at 20 mA until the dye leaves the gel (*see* **Note 20**). For larger gels, set 25 mA and 50 mA, respectively. Stain the second-dimensional gel with silver or Coomassie dye or transfer onto nitrocellulose membranes for immunoblotting (*see* **Note 21**).

### 3.5. Doubled SDS Polyacrylamide Electrophoresis

dSDS-PAGE combines two orthogonal SDS gels, as exemplified with **Fig. 2c**. Mitochondrial ATP synthase had been prepared by two native electrophoretic separations (**Fig. 2a** and **b**). dSDS-PAGE then followed as denaturing separations in the third and fourth dimension. dSDS-PAGE is also compared with BAC/SDS 2-DE in **Fig. 3** (*see also* **Note 22**).

#### 3.5.1. First-Dimensional SDS-PAGE

Nine percent acrylamide gels for 1-D Tricine-SDS-PAGE, as described here, cover the 5-100 kDa mass range. Glycerol in **step 1** can optionally be replaced by 6 M urea to increase resolution (*see also* **Notes 22** and **23**).

1. For one 9% acrylamide 1-D Tricine-SDS gel with dimensions (0.07 × 14 × 14 cm), mix 2.7 mL acrylamide/bisacrylamide solution (AB-3; *see* **Subheading 2.1.1, Item 1**), 2 mL gel buffer (GB-Tric (3×); *see* **Subheading 2.1.1, Item 7**), 1.5 g glycerol, and fill up with water to a final volume of 15 mL.
2. Add 7.5 μL TEMED and 75 μL 10% APS for polymerization, pour the mixture into the gel cassette, and overlay with water.
3. After polymerization, remove water, and add the 4% acrylamide stacking gel mixture (5 mL): 400 μL acrylamide/bisacrylamide solution (AB-3; *see* **Subheading 2.1.1, Item 1**), 1.67 mL gel

buffer (GB-Tric (3×); *see* **Subheading 2.1.1**, **Item** 7) and 3 mL water. Polymerization is initiated by 3 µL TEMED and 30 µL 10% APS.

4. Insert the gel comb. After polymerization, the gel can be used immediately or stored at 4°C for several days (overlayed with GB-Tric (1×); *see* **Subheading 2.1.1**, **Item 8**).
5. Dissolve the samples in SDS sample buffer (*see* **Subheading 2.5.1**) and incubate at 40°C for 10 min. Load appropriate volumes (20-30 µL on 0.07 × 1.0 cm gel wells).
6. Start electrophoresis towards the anode at low voltage (<50 V) for about 10 min until the Coomassie dye has entered the separating gel. Then raise the voltage to 180 V (current must be limited to 80 mA). The run time is around 3 h.

*3.5.2. Acidic Incubation*

1. Cut out gel strips (0.7 × 10 × 120 mm) from 1-D SDS gels (*see* **Note 24**).
2. Incubate 1-D gel strips for 5-20 min in acidic incubation buffer (*see* **Subheading 2.5.2**; 50-100 times the volume of the gel strip; *see* **Note 25**).

*3.5.3. Second-Dimensional Laemmli-SDS-PAGE*

1. Place the acidic incubated 1-D gel strip between two glass plates using 0.7 mm (or slightly thinner) spacers for 0.7 mm 1-D gel strips, and remove the acidic solution carefully. Wait for a few minutes before putting the gel upright to avoid gliding down of the gel strip.
2. Prepare the solution for a 15% Laemmli gel (15 mL for gel dimensions 0.07 × 14 × 14 cm) mixing 4.5 mL acrylamide/bisacrylamide solution (AB-3; *see* **Subheading 2.1.1**, **Item 1**), 7.5 mL Laemmli separation gel buffer (GB-Lae (4×); *see* **Subheading 2.5.3**, **Item 1**; the volume is twice the normally used volume; *see* Note 26) and add water to a final volume of 15 mL.
3. Start polymerization adding 7.5 µL TEMED and 75 µL 10% APS.
4. Pour in the acrylamide mixture for the 2-D gel (leaving a 1–2 mm gap to the 1-D strip) and overlay with water (may contact the 1-D strip).
5. After polymerization, the 1-D gel strip is slowly pushed down to the separating gel for the second dimension using a plastic card slightly thinner than the spacers used. Air bubbles between 1-D and 2-D gels thereby are squeezed away.
6. Fill the gaps on both sides (between 1-D strip and side spacers) with a special 9% acrylamide side gel (*see* **Note 27**): Mix 0.9 mL acrylamide/bisacrylamide solution (AB-3; *see* **Subheading 2.1.1**, **Item 1**), 0.5 mL side gel buffer (SGB (10×); *see* **Subheading 2.5.3**, **Item 2**), add water to a volume of 5 mL, and start polymerization by adding 3 µL TEMED and 30 µL 10% APS.

7. Set 200 V initially, and gradually increase the voltage to 350 V. The run time is around 4 h.
8. Stain the 2-D gel with silver or Coomassie dye or transfer onto nitrocellulose or PVDF membranes for immunoblotting *(15, 24, 25)*.

## 4. Notes

Native electrophoresis techniques (BNE, CNE and hrCNE)

1. Gradient gels are commonly used. 4–13% acrylamide gels, overlaid with a 3.5% acrylamide sample gel, are appropriate for the resolution of proteins by BNE and hrCNE in the total mass range 10 kDa to 3 MDa, and optimal for the resolution of proteins larger than 100 kDa and smaller than 1 MDa. Starting with 3% acrylamide in the sample gel and on top of the gradient expands the mass range to 10 MDa.
2. Try to remove nuclei or nuclear DNA from the biological membrane of interest preferentially by differential centrifugation, since DNA blocks the sample wells and prevents Coomassie dye and proteins from entering the gel (DNase treatment was not effective in our hands).
3. Try to remove mitochondria as completely as possible from the biological membranes of interest, since mitochondrial complexes often dominate over other complexes with lower abundance. Presence of low amounts of mitochondrial contaminations, however, might be favorable, since mitochondrial complexes are useful as internal markers in native gels and in 2-D SDS gels.
4. The preferred buffer for storage and the following native electrophoresis is 250 mM sucrose, 50 mM imidazole/HCl, pH 7.0. Other buffers like 5–20 mM Na-Mops or Na-phosphate are tolerated. Avoid potassium and divalent cations, which tend to precipitate Coomassie dye together with bound protein, and avoid salt concentrations higher than 50 mM, which lead to very strong protein stacking in the sample well and often to membrane protein aggregation.
5. Mitochondrial complexes are useful marker complexes for native mass determination in the 100 kDa to 1 MDa range (Complex I, 1 MDa; Complex II, 130 kDa; Complex III, 500 kDa; Complex IV, 200 kDa; Complex V, 700–750 kDa). Several commercially available marker proteins can also be used for determination of native masses, e.g. tyroglobulin (bovine thyreoidea), 669 kDa; apoferritin (equine spleen), 443 kDa;

β-amylase (sweet potato), ~200 kDa; albumin (bovine serum), 66 kDa.

6. Solubilization buffer A is normally used. Solubilization buffer B has lower ionic strength, and is therefore suited to reduce protein stacking in the sample well and is used especially when membrane proteins aggregate using solubilization buffer A.
7. For initial experiments with a novel biological membrane, it seems advisable to test a series of detergent/protein ratios, e.g. 0.5, 1.0, 2.0, and 4.0 (g/g) to find the optimal solubilization conditions. If isolation of physiological supramolecular complexes is desired, the milder detergent digitonin should be used (20% stock solutions, warmed up for solubilization shortly before use to 50-90°C). Doubled detergent/protein ratios (2.0-6.0 g digitonin per gram protein) are commonly required, since the mass of digitonin is roughly twice the mass of DDM or Triton X-100.
8. Ponceau S facilitates sample application, and is also useful as a marker for the electrophoresis front.
9. The change of buffer B to B/10 improves detection of faint bands (e.g. for native electroelution), improves the transition to 2-D SDS-PAGE, and improves protein retention on native blots (since Coomassie dye competes with proteins for binding sites on PVDF membranes).
10. Stop electrophoresis early (for example after 2-h run) if native extraction of a protein band is desired or another native dimension, e.g. in BNE/BNE 2-DE, is to follow. In short runs, proteins have not yet reached their specific pore size limits. They are still mobile and therefore can be efficiently extracted *(3)*.

Two-dimensional native 2-DE

11. First-dimension native gels are intended to separate large physiological complexes, which is achieved using specifically the detergent digitonin and three native electrophoresis techniques, CNE, BNE, and also by hrCNE-3 (ordered according to decreasing suitability). Second dimension must be less mild to separate substructures. The preferred variant for 2-DE is modified BNE with 0.02% DDM added to cathode buffer B (0.02% DDM may be replaced by 0.03% Triton X-100).

Native/SDS 2-DE

12. Uniform 16% acrylamide gels for Tricine-SDS-PAGE are commonly used for the 2-100 kDa range *(24, 25)*. Speed of separation or subsequent Western blotting may direct to lower acrylamide concentration, e.g. to 10% acrylamide gels.

BAC/SDS PAGE

13. Depending on the sample composition, higher or lower percentage acrylamide gels might be useful.

14. Do not incubate protein samples for more than 5 min or at temperatures higher than 60°C in BAC sample buffer, since this can lead to protein aggregation. Samples can be stored in BAC sample buffer at –20°C. If heating of frozen samples for 5 min at 60°C is necessary for complete solubilization of the buffer chemicals, spin samples and use the supernatant to avoid protein aggregates clogging the gel.
15. Fill all empty gel wells with sample buffer to avoid field inhomogeneity during electrophoresis.
16. Stained BAC gels often seem rather diffuse. In spite of that, proceed with the protocol. Two-dimensional gels commonly will be fine.
17. Add 10% glycerol to the equilibration solution if you encounter problems with pushing the BAC gel strip down to the SDS gel. This will shrink the gel but resolution will decrease.
18. Be careful when handling the BAC gel. It is considerably more flexible than an SDS gel.
19. Acrylamide/bisacrylamide mixtures are often defined by the total percentage (% T) of both monomers (acrylamide and bisacrylamide), and by the percentage of the cross-linker bisacrylamide (% C) relative to the total amount of both monomers. The cross-linker concentration (% C) usually is varied in the range 0.6% C (as described in **Subheading 3.4.4**) to 6% C. Low percentage of cross-linker makes gels soft, and shifts the useful mass range. The resolution of 15% T, 0.6% gels, for example, resembles the resolution of 10% T, 3% C gels. Gels with low cross-linker concentration swell considerably during staining/destaining, thereby improving extractability of tryptic fragments for MS analysis. With the Laemmli system 3.4% C gels are mostly used. This is achieved using the 40% T, 3.4% C acrylamide/bisacrylamide stock-solution (*see* **Subheading 2.4.1**, **Item 1**) instead of using acrylamide- and bisacrylamide-solutions separately.
20. Two dyes are visible in the second-dimension gel. Coomassie dye is the slower migrating component; bromophenol blue is leading.

    For analysis of transfer efficiency, stain the BAC gel strip also using silver staining protocols.
21. Up to 300 μg protein can be efficiently separated using mini-gels. The minimum amount for immunoblotting was around 15 μg, even for detection of low copy number proteins.

Doubled SDS-PAGE

22. Separation of highly hydrophobic proteins from the bulk of other proteins relies on the use of strongly differing acrylamide concentrations for the 1-D and 2-D SDS gels in dSDS-PAGE. Highly hydrophobic proteins show anomalous

(too fast) migration in low acrylamide 1-D gels and therefore appear above a diagonal of other common proteins in the 2-D gel with high acrylamide concentration. Further spreading of spots around the diagonal, i.e. higher resolution and detection of more separate spots, optionally is achieved by introducing urea in one of the two dimensions, preferentially in the first-dimension gel.

23. Combining 10% acrylamide gels for 1-D Tricine-SDS-PAGE and 16% acrylamide gels for 2-D Tricine-SDS-PAGE is optimal for the mass range 6–75 kDa (total range 2-100 kDa). For an improved resolution of proteins in the 75–100 kDa range, 9% acrylamide gels had been used for 1-D Tricine-SDS-PAGE and 15% acrylamide gels for 2-D Laemmli SDS-PAGE (but using doubled ionic strength for the Laemmli gel, as described in **Subheading 3.5.3**).
24. The strips must not be fixed and stained. This can lead to complete loss of highly hydrophobic proteins.
25. Using sufficient volumes of the acidic solution, the color of Coomassie dye changes from blue to green. Incubation can be stopped at that point or after 10 min. The incubation solution is acidic to immobilize SDS and SDS-bound proteins. Ionic strength is high to increase the SDS concentration (originating from the 0.1% SDS cathode buffer) in the 1-D gel strip during transfer from 1-D to 2-D separation. The acidic solution contains 100 mM Tris base so that the pH can shift from an acidic to a stabilized alkaline pH during electrophoresis.
26. The volume of Laemmli gel buffer was doubled when compared with the conventional Laemmli SDS-PAGE to increase the ionic strength in the gel by a factor of 2, which in turn concentrates protein spots in the 2-D gel by a factor of 2.
27. A special side gel buffer was introduced to mimic the ionic strength in the acidified 1-D gel strip and thus to guarantee homogenous field over the whole gel width. However, the buffer was neutralized to pH 7.4, since acrylamide gels would not polymerize at acidic pH.

## References


1. Hunte, C., von Jagow, G., and Schägger, H. (eds.) (2003) *Membrane Protein Purification and Crystallization. A Practical Guide*. Academic Press, Orlando, FL.
2. Schägger, H. (2006) Blue native electrophoresis, in *Membrane Protein Purification and Crystallization. A Practical Guide* (Hunte, C., von Jagow, G., and Schägger, H., eds.). Academic Press, Orlando, FL, pp. 105–130.
3. Wittig, I., Braun, H. P., and Schägger, H. (2006) Blue-native PAGE. *Nature Protocols* **1**, 416–428.
4. Schägger, H., Cramer, W. A., and von Jagow, G. (1994) Analysis of molecular masses and oligomeric states of protein complexes by blue native electrophoresis and isolation of membrane protein complexes by two-dimensional native electrophoresis. *Anal. Biochem.* **217**, 220–230.

5. Wittig, I. and Schägger, H. (2005) Advantages and limitations of clear native polyacrylamide gel electrophoresis. *Proteomics* **5**, 4338–4346.
6. Wittig, I., Karas, M., and Schägger, H. (2007) High resolution clear-native electrophoresis for in-gel functional assays and fluorescence studies of membrane protein complexes. *Mol. Cell. Proteomics* **6**, 1215–1225.
7. Klose, J. (1975) Protein mapping by combined isoelectric focusing and electrophoresis of mouse tissues. A novel approach to testing for induced point mutations in mammals. *Humangenetik* **26**, 231–243.
8. O'Farrell, P. H. (1975) High resolution two-dimensional electrophoresis of proteins. *J. Biol. Chem.* **250**, 4007–4021.
9. Macfarlane, D. E. (1983) Use of benzyldimethyl-n-hexadecylammonium chloride (16-BAC), a cationic detergent, in an acidic polyacrylamide gel electrophoresis system to detect base labile protein methylation in intact cells. *Anal. Biochem.* **132**, 231–235.
10. Macfarlane, D. E. (1989) Two dimensional benzyldimethyl-n-hexadecylammonium chloride - sodium dodecyl sulfate preparative polyacrylamide gel electrophoresis: a high capacity high resolution technique for the purification of proteins from complex mixtures. *Anal. Biochem.* **176**, 457–463.
11. Hartinger, J., Stenius, K., Hogemann, D., and Jahn, R. (1996) 16-BAC/SDS PAGE: a two-dimensional gel electrophoresis system suitable for the separation of integral membrane proteins. *Anal. Biochem.* **240**, 126–133.
12. Helling, S., Schmitt, E., Joppich, C., Schulenborg, T., Mullner, S., Felske-Muller, S., Wiebringhaus, T., Becker, G., Linsenmann, G., Sitek, B., Lutter, P., Meyer, H. E., and Marcus, K. (2006) 2-D differential membrane proteome analysis of scarce protein samples. *Proteomics* **6**, 4506–4513.
13. Navarre, C., Degand, H., Bennett, K. L., Crawford, J. S., Mortz, E., and Boutry, M. (2002) Subproteomics: Identification of plasma membrane proteins from the yeast *Saccharomyces cerevisiae*. *Proteomics* **2**, 1706–1714.
14. Marnell, L. L. and Summers, D. F. (1984) Characterization of the phosphorylated small enzyme subunit, NS, of the vesicular stomatitis virus RNA polymerase. *J. Biol. Chem.* **259**, 13518–13524.
15. Rais, I., Karas, M., and Schägger, H. (2004) Two-dimensional electrophoresis for the isolation of integral membrane proteins and mass spectrometric identification. *Proteomics* **4**, 2567–2571.
16. Nesvizhshii, A. I. (2007) Protein identification by tandem mass spectrometry and sequence database searching. *Methods Mol. Biol.* **367**, 87–119.
17. Wu, L. and Han, D. K. (2006) Overcoming the dynamic range problem in mass spectrometry-based shotgun proteomics. *Expert Rev. Proteomics* **3**, 611–619.
18. Florens, L. and Washburn, M. P. (2006) Proteomic analysis by multidimensional protein identification technology. *Methods Mol. Biol.* **328**, 159–175.
19. Moebius, J., Zahedi, R. P., Lewandrowski, U., Berger, C., Walter, U., and Sickmann, A. (2005) The human platelet membrane proteome reveals several new potential membrane proteins. *Mol. Cell. Proteomics* **4**, 1754–1761.
20. Coughenour, H. D., Spaulding, R. S., and Thompson, C. M. (2004) The synaptic vesicle proteome: a comparative study in membrane protein identification. *Proteomics* **4**, 3141–3155.
21. Burre, J., Beckhaus, T., Schägger, H., Corvey, C., Karas, M., Zimmermann, H., and Volknandt, W. (2006) Analysis of the synaptic vesicle proteome using three gel-based protein separation techniques. *Proteomics* **6**, 6250–6262.
22. Zahedi, R. P., Meisinger, C., and Sickmann, A. (2005) Two-dimensional BAC/SDS PAGE for membrane proteomics. *Proteomics* **5**, 3581–3588.
23. Meyer, B., Wittig, I., Trifilieff, E., Karas, M., and Schägger, H. (2007) Identification of two proteins associated with mammalian ATP synthase. *Mol. Cell. Proteomics* published online June 17, as M700097-MCP200.
24. Schägger, H. (2006) SDS electrophoresis techniques, in *Membrane Protein Purification and Crystallization. A Practical Guide* (Hunte, C., von Jagow, G., and Schägger, H., eds.). Academic Press, Orlando, FL, pp. **85**–103.
25. Schägger, H. (2006) Tricine-SDS-PAGE. *Nature Protocols* **1**, 16–22.

# Chapter 4

# Protein Detection and Quantitation Technologies for Gel-Based Proteome Analysis

**Walter Weiss, Florian Weiland, and Angelika Görg**

## Summary

Numerous protein detection and quantitation methods for gel-based proteomics have been devised that can be classified in three major categories: (1) *Universal* (or "*general*") *detection techniques*, which include staining with anionic dyes (e.g., Coomassie brilliant blue), reverse (or "negative") staining with metal cations (e.g., imidazole-zinc), silver staining, fluorescent staining or labeling, and radiolabeling, (2) *specific staining methods* for the detection of post-translational modifications (e.g., glycosylation or phosphorylation), and (3) *differential display techniques* for the separation of multiple, covalently tagged samples in a single two-dimensional electrophoresis (2-DE) gel, followed by consecutive and independent visualization of these proteins to minimize methodical variations in spot positions and in protein abundance, to simplify image analysis, as well as to improve protein quantitation by including an internal standard.

The most important properties of protein detection methods applied in proteome analysis include high sensitivity (i.e., low detection limit), wide linear dynamic range for quantitative accuracy, reproducibility, cost-efficiency, ease of use, and compatibility with downstream protein identification or characterization technologies, such as mass spectrometry (MS). Regrettably, no single detection method meets all these requirements, albeit fluorescence-based technologies are currently favored for most applications; hence, the major focus of this chapter is on fluorescent-dye-based protein detection and quantitation techniques. Although satisfying results with respect to sensitivity and reproducibility are also obtained by methods based on *radioactive labeling* of proteins (which is still unsurpassed in terms of sensitivity), radiolabeling is, however, largely impractical for *routine* proteomic profiling because of the costs and the health and safety concerns associated with handling radioactive compounds.

**Key words:** Protein detection, Quantitation, Proteome, Two-dimensional electrophoresis

Jörg Reinders and Albert Sickmann (eds.), *Proteomics,* Methods in Molecular Biology, Vol. 564
DOI: 10.1007/978-1-60761-157-8_4, © Humana Press, a part of Springer Science+Business Media, LLC 2009

## 1. Introduction

The availability of highly sensitive, reliable protein detection and quantitation technologies is a prerequisite for taking full profit of the high-resolution capabilities of the two-dimensional gel electrophoresis (2-DE) *(1)*. During the past decades, numerous protein visualization methods have been devised, which can be divided into three major categories: (1) *universal detection techniques*, which include staining with anionic dyes (e.g., Coomassie Blue), negative staining with metal cations (e.g., imidazole-zinc), silver staining, fluorescence staining or labeling, as well as incorporation of radioactive isotopes, followed by autoradiography or electronic detection methods; (2) *specific staining methods* for the detection of post-translational modifications (PTMs), such as glycosylation or phosphorylation; and (3) *differential display techniques* for the separation of multiple, covalently tagged samples in a single 2-DE gel, followed by the consecutive and independent visualization of these proteins.

The most important properties of protein detection methods applied in proteome analysis include high sensitivity (i.e., low detection limit), wide linear dynamic range (for quantitative accuracy), reproducibility, cost-efficiency, ease of use, and compatibility with downstream protein identification and characterization technologies, such as MS. Unfortunately, no single detection method meets all these requirements (cf. **Table 1**), albeit fluorescence-based technologies are currently favored for most applications. Satisfying results with respect to sensitivity and reproducibility are also obtained by *radiolabeling* of proteins. Radioactive labeling can be accomplished by in vivo metabolic incorporating of radioactive amino acids such as $^{35}$S-methionine or $^{14}$C-leucine into proteins or by in vitro radiolabeling of proteins using iodination with $^{131}$I or $^{125}$I. The radiolabeled proteins can be detected after 2-D electrophoresis by direct autoradiography or fluorography using x-ray films, which are exposed to the dried gels and then quantified by densitometry or by electronic detection methods based on storage phosphor imaging screen technology. The latter procedure offers higher sensitivity (the detection limit is in the low pg range) and a much wider dynamic range of detection (up to five orders of magnitude) than the film-based approaches. Although radiolabeling is still unsurpassed in terms of sensitivity, it is largely impractical for *routine* proteomic applications because of the high costs and the safety and health concerns associated with handling radioactive compounds. Hence, in this chapter, we will focus on the most popular nonradioactive protein visualization and quantitation methods for gel-based proteome analysis. For more comprehensive information, the reader is referred to review articles *(1–5)*. Typical results of general and PTM-specific stains, as well as of differential display techniques, are displayed in **Fig. 1**.

**Table 1**
**Protein detection and quantification methods for gel-based proteome analysis**

| Detection method | Detection limit[a] | Linear dynamic range[b] | Quantification[b] | Reproducibility | Cost efficiency | Ease of use | Speed | MALDI-MS compatibility | Comments |
|---|---|---|---|---|---|---|---|---|---|
| **Coomassie Blue anionic dye** | | | | | | | | | |
| "Classical" stain (soluble dye) | ~100 ng | ~10–30 | ++ | ++ | ++++ | +++ | ++–+++ | ++++ | Favored for downstream protein identification (MS, Edman sequencing) |
| Colloidal dye particles | ~10 ng | ~30 | +++ | +++ | ++++ | ++++ | ++–+++ | ++++ | |
| **Reverse (negative) stain** | | | | | | | | | |
| Imidazole-zinc | ~10 ng | <10 | + | ++ | ++++ | ++++ | ++++ | ++++ | Favored for downstream protein identification (MS) |
| **Silver stain** | | | | | | | | | |
| Sliver nitrate (acidic) | <1 ng | ~10 | ++ | +(+) | +++ | + | ++–+++ | +–++ | The most sensitive silver stains are MS incompatible (→ protein cross-linkage with aldehydes) |
| Silver ammonia (alkaline) | ~0.1 ng | ~10 | ++ | ++ | +++ | + | ++ | (+) | |
| **Post-electrophoretic fluorescent stain** | | | | | | | | | |
| SYPRO Ruby | ~2 ng | $10^2$–$10^3$ | ++++ | ++++ | ++ | ++++ | +++ | +++ | Sophisticated instrumentation (fluoroimager) is required for high sensitivity |
| Epicoccone (Deep Purple) | <1 ng | ~$10^3$ | ++++ | ++++ | ++ | +++ | +++ | +++ | |

(continued)

**Table 1**
**(continued)**

| | | | | | | | | | |
|---|---|---|---|---|---|---|---|---|---|
| **Pre-electrophoretic fluorescent label** | | | | | | | | | |
| CyDye minimal (Lys) label | <1 ng | ~$10^4$ | ++++ | ++++ | ++ | ++ | +++ | ++ | Sophisticated instrumentation (fluoro-imager) is required for high sensitivity |
| CyDye saturation (Cys) label | <0.1 ng | ~$10^4$ | ++++ | ++++ | ++ | ++ | +++ | ++ | |
| **Radioactive labeling** | | | | | | | | | |
| Multiple x-ray film exposure | <0.1 ng | $10–10^2$ | ++ | ++ | ++ | ++ | + | c | Safety concerns associated with handling radioactivity. |
| Phosphor-imager plates | <0.01 ng | $10^4–10^5$ | ++++ | +++ | + | ++ | +++ | c | Expensive hardware (phosphor-imager) is required |
| **Fluorescent PTM stains** | | | | | | | | | |
| Pro-Q Diamond Phospho-protein stain | 1–20 ng | $10^2–10^3$ | ++–+++ | +++ | ++ | +++ | +++ | +++ | Sophisticated instrumentation (fluoro-imager) is required for high sensitivity |
| Pro-Q Emerald Glycoprotein stain | 1–20 ng | $10^2–10^3$ | ++–+++ | ++ | ++ | ++ | ++ | +++ | |

+ = low; ++ = moderate; +++ = good; ++++ = excellent

[a] Detection limit depends on the amino acid composition of the protein, and/or the number of phosphate groups or carbohydrate moieties

[b] Sophisticated instrumentation (densitometer; fluoro- or phosphorimager) is required for high sensitivity and wide linear dynamic range

[c] MS-analysis of radiolabeled proteins is only rarely performed

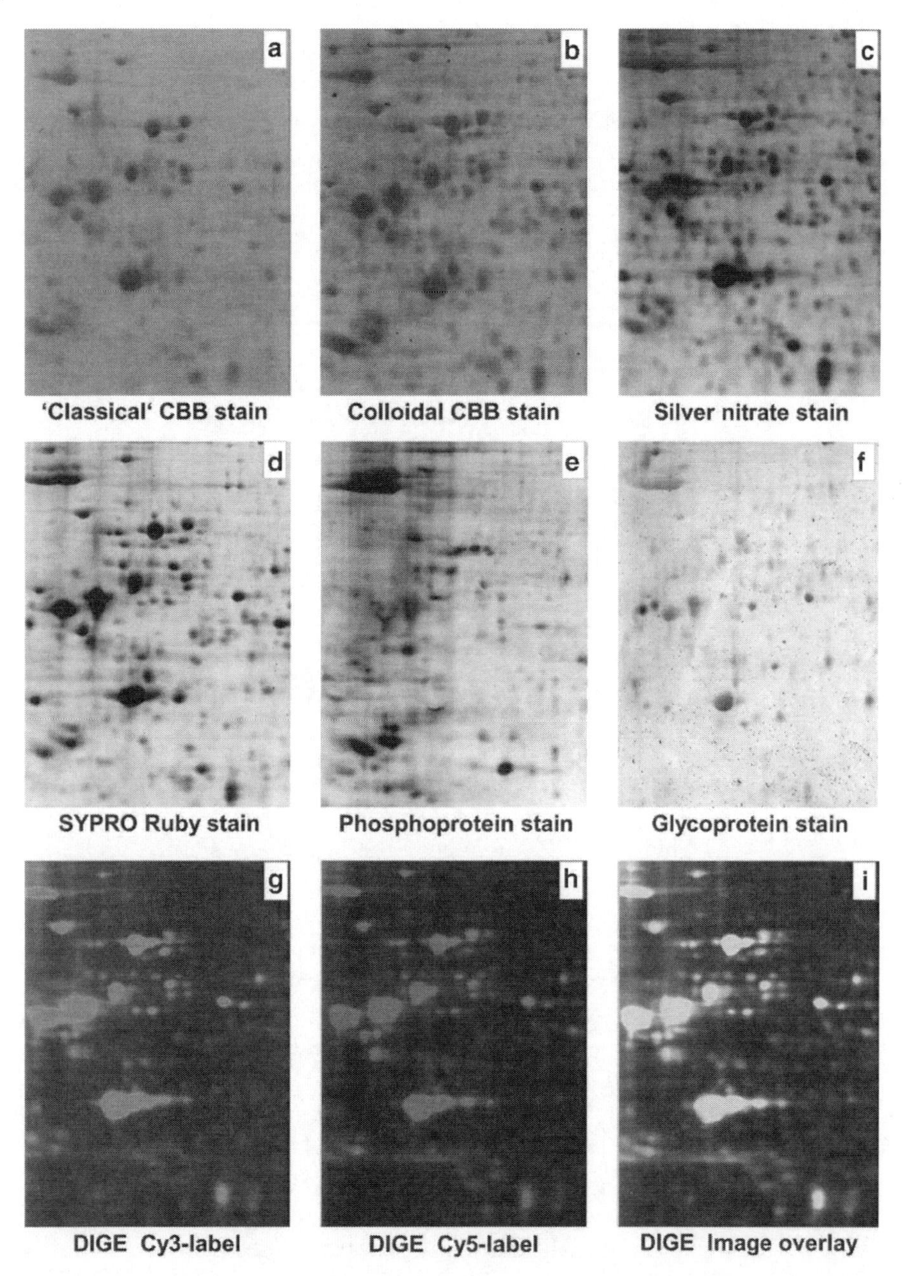


Fig. 1. Visualization of total proteins, phosphoproteins, or glycoproteins in 2-DE gels, and differentially expressed proteins in DIGE gels, respectively. Proteins from tumor cell line lysates were separated by 2-D electrophoresis (**a–f**) and 2-D-DIGE (**g–i**) (detail enlargement). Resulting gels were stained with (**a**) CBB ("classical stain"), (**b**) colloidal CBB, (**c**) silver nitrate, or (**d**) SYPRO Ruby for total protein, or with (**e**) Pro-Q Diamond phosphoprotein stain or (**f**) Pro-Q Emerald 488 glycoprotein stain. Proteins from two different two cell lines were labeled in vitro with two Cyanine dyes (Cy3 and Cy5), differing in their excitation and emission wavelengths. The samples were then mixed and separated on a single 2-DE gel. After consecutive excitation with both wavelengths, the resultant gel images (**g** and **h**) were overlayed (**i**) to visualize differences (i.e., up- or downregulated proteins) between both samples. The CBB- and silver-stained gels were imaged on an Epson 1680 Pro 48 bit scanner, the other gels on an Ettan DIGE imager (GE Healthcare) with 200 dpi resolution. The amount of protein loaded onto the 2-DE gels (size 200 × 250 × 1 $mm^3$) was 500 μg for Coomassie Blue staining, 150 μg for glyco- and phosphoprotein, and 100 μg for all other protein detection methods.

## 2. Materials

### 2.1. Equipment

*Staining trays*: made of glass, plastic, or stainless steel may be used, depending on the type of application. Plastics must be used for fluorescent stains because glass will interfere with the staining procedure, whereas glass trays are best suited for Coomassie Brilliant Blue (CBB) and silver staining. Among plastics, we recommend polypropylene food boxes, whereas for glass, we prefer Pyrex® (http://www.arc-international-cookware.com/en_Home.html) glass ovenware roasters. The bottom of all trays should be flat and at least 20% larger than the area of the gels to be stained, i.e., 25 × 30 $cm^2$ for "standard" (i.e., 20 × 25 $cm^2$) 2-DE gels.

*Laboratory shaker*: Either a slow speed (30–60 rpm, adjustable) *orbital shaker* (i.e., the platform operates on a horizontal plane in a circular motion) or a slow speed (30-100 strokes per minute, adjustable) *reciprocal shaker* (i.e., the platform operates on a horizontal plane in a forward and backward motion) may be used. The shaker (e.g., GFL 3016) (GFL, Burgwedel, Germany) should be robust and sufficiently large (>45 × 45 $cm^2$) to accommodate at least two piles of staining trays and ensure a smooth working motion.

High-resolution *flatbed scanner* (e.g., Epson Express 1680 Pro, Epson Europe, Amsterdam, the Netherlands) is used for digitizing Coomassie Blue or silver stained gels. *Fluorescent imager* (equipped with either a cooled CCD camera, or a multicolor confocal laser scanner) is used for digitizing gels stained/labeled with fluorescent compounds, as well as appropriate software for image analysis. These instructions assume the use of an Ettan DIGE Imager or a Typhoon 9,400 from GE Healthcare Life Sciences (Freiburg, Germany), but similar devices from other manufacturers, such as a Fuji-FLA-5100 (Fujifilm Europe, Düsseldorf, Germany) or a Bio-Rad Molecular Imager FX (Bio-Rad, Richmond, CA, USA) may work equally well. For difference gel electrophoresis (DIGE) analysis, low-fluorescence glass plates (3 mm thick), as well as image analysis software such as DeCyder Differential Analysis Software (GE Healthcare Life Sciences), which is specifically designed for analysis of multiplexed fluorescent images, is required.

### 2.2. Reagent Solutions

All solutions should be prepared in water having a resistivity of 18 MΩ/cm. This standard is referred to as "deionized water" or "Milli Q water" in this text. All chemicals for buffers or reagent solutions should be of analytical or biochemical grade.

*2.2.1. Reagent Solutions for Coomassie Brilliant Blue Stain*

"Classical" Coomassie Brilliant Blue R-250 Stain

1. *CBB R-250 destaining solution*: 40% (v/v) ethanol and 10% (v/v) acetic acid. To prepare 1,000 mL of the solution, mix 500 mL of deionized water, 400 mL ethanol (Merck, Darmstadt, Germany), and 100 mL glacial acetic acid (Merck).
2. *CBB R-250 staining solution*: 0.1% (w/v) CBB R-250 dye in 40% ethanol and 10% acetic acid. To prepare 500 mL (which is sufficient for staining one 20 × 25 × 0.1 $cm^3$ 2-DE gel) of the solution, dissolve 500 mg CBB R-250 (e.g., GE Healthcare, Bio-Rad, or Serva, Heidelberg, Germany) in 200 mL ethanol (96%; Merck) and stir for ~60 min. Then, add 250 mL deionized water and 50 mL glacial acetic acid (Merck) under stirring, and filter.

Colloidal Coomassie Brilliant Blue G-250 Stain *(7, 8)*

1. *Fixing solution*: 40% (v/v) ethanol and 10% (v/v) acetic acid. To prepare 500 mL of the solution, mix 250 mL of deionized water, 200 mL ethanol (Merck, Darmstadt, Germany), and 50 mL glacial acetic acid (Merck).
2. *CBB G-250 staining solution*: 0.12% (w/v) CBB G-250 dye, 10% ammonium sulfate, 10% phosphoric acid, and 20% methanol. To prepare 500 mL of the solution, which is sufficient for staining one 2-DE gel (20 × 25 × 0.1 $cm^3$), dilute 50 mL phosphoric acid (85%) (Merck) in 100 mL of deionized water. Add 50 g of ammonium sulfate (Merck) and stir until the ammonium sulfate has dissolved. Add 600 mg CBB G-250 (e.g., GE Healthcare, Bio-Rad, or Serva) and stir for ~30 min. Then complete to 400 mL with deionized water and add 100 mL methanol (Merck) under stirring. This dye suspension is stable at room temperature for up to 6 months. Stir before use, but do *not* filter the dye suspension.
3. *CBB-G-250 destaining solution*: 5% (w/v) ammonium sulfate and 10% (v/v) methanol. To prepare 1,000 mL of the solution, dissolve 50 g of ammonium sulfate (Merck) in 800 mL deionized water and stir until the ammonium sulfate has dissolved. Then add 100 mL methanol and complete to 1,000 mL with deionized water.

*2.2.2. Reagent Solutions for Imidazole-Zinc Reverse Stain (10, 11)*

1. *Imidazole solution*: 0.2 M imidazole containing 0.1% SDS. To prepare 500 mL, dissolve 6.81 g imidazole (Merck), and 500 mg sodium dodecyl sulfate (SDS) (Serva) in 400 mL of freshly prepared Milli Q water and complete to 500 mL. *Caution*: Imidazole is harmful. Wear suitable protective clothing!
2. *Zinc sulfate solution*: 0.2 M zinc sulfate. To prepare 500 mL of the solution, dissolve 28.76 g zinc sulfate heptahydrate (Merck) in 400 mL of freshly prepared Milli Q water and complete to 500 mL. It is also possible to prepare concentrated (10×) imidazole and zinc sulfate stock solutions in advance. If stored in tightly closed amber bottles at room temperature,

they are stable for months. Dilute 1:10 with deionized water before use.

*2.2.3. Reagent Solutions for Silver Nitrate Stain (14)*

1. *Fixing solution*: 40% (w/v) ethanol and 10% (w/v) acetic acid. To prepare 1,000 mL, mix 500 mL of deionized water, 400 mL ethanol (Merck), and 100 mL glacial acetic acid (Merck).
2. *Sensitizer*: 0.02% sodium thiosulfate-pentahydrate. To prepare 500 mL of the solution, dissolve 100 mg of sodium thiosulfate-pentahydrate in 500 mL deionized water. The solution must be prepared the day of use.
3. *Silver nitrate solution*: 0.2% silver nitrate and 0.02% formaldehyde (37%). To prepare 500 mL of the solution, dissolve 1.0 g of silver nitrate (Merck) in 500 mL deionized water, and add 0.1 mL formaldehyde solution (37%) (Merck). The solution should be prepared just before use (*see* **Note 1**). *Caution*: Formaldehyde is toxic and should be handled with care.
4. *Developer*: 3% sodium carbonate, 0.05% formaldehyde (37%), and 0.0005% sodium thiosulfate pentahydrate. To prepare 1,000 mL of the solution, dissolve 30 g of sodium carbonate (Merck), and 5 mg of sodium thiosulfate pentahydrate in 1,000 mL deionized water and add 0.5 mL of formaldehyde solution (37%) (Merck). Developer should be prepared immediately before use (*see* **Note 2**).
5. *Stop solution*: 0.5% glycine. To prepare 500 mL of the solution, dissolve 2.5 g of glycine (Merck) in 500 mL deionized water.

*2.2.4. Reagent Solutions for SYPRO Ruby Total Protein Stain*

1. *Fixing solution*: 40% (v/v) ethanol and 10% (v/v) acetic acid. To prepare 1,000 mL of the solution, mix 500 mL of deionized water, 400 mL ethanol (Merck), and 100 mL glacial acetic acid (Merck).
2. *SYPRO Ruby staining solution*: Ready-to-use solution (Molecular Probes, Eugene, OR, USA).
3. *Destaining solution*: 10% (v/v) ethanol and 7% (v/v) acetic acid. To prepare 1,000 mL of the solution, mix 830 mL of deionized water, 100 mL ethanol (Merck), and 70 mL glacial acetic acid (Merck).

*2.2.5. Reagent Solutions for Pro-Q Diamond® Phosphoprotein Stain*

1. *Fixing solution*: 40% (v/v) ethanol and 10% (v/v) acetic acid. To prepare 1,000 mL of the solution, mix 500 mL of deionized water, 400 mL ethanol (Merck), and 100 mL glacial acetic acid (Merck).
2. *Pro-Q Diamond phosphoprotein stain*: Ready-to-use solution (Molecular Probes, Eugene, OR, USA).
3. *Pro-Q Diamond phosphoprotein destaining solution*: 25% acetonitrile and 50 mM sodium acetate/acetic acid, pH 4.0.

To prepare 1,000 mL of the solution, dissolve 4.1 g sodium acetate (Merck) in ~700 mL deionized water and titrate to pH 4.0 with glacial acetic acid (Merck). Add 250 mL of acetonitrile (Merck) and complete to 1,000 mL with deionized water. *Caution*: Acetonitrile is toxic and should be handled with care.

#### 2.2.6. Reagent Solutions for Pro-Q Emerald® 488 Glycoprotein Stain

1. *Fixing solution*: 40% (w/v) ethanol and 5% (w/v) acetic acid. To prepare 1,000 mL of the solution, mix 550 mL of deionized water, 400 mL ethanol (Merck), and 50 mL glacial acetic acid (Merck).
2. *Wash solution*: 3% (v/v) acetic acid. To prepare 1,000 mL of the solution, mix 970 mL of deionized water and 30 mL glacial acetic acid (Merck).
3. *Oxidizer*: Periodic acid in wash solution. To prepare 500 mL of the solution, add 250 mL wash solution to the bottle containing the periodic acid (Component C′; Molecular Probes) and mix until completely dissolved. Add another 250 mL wash solution before use.
4. *Pro-Q Emerald glycoprotein staining solution*: Add 0.5 mL of dimethylsulfoxide (DMSO, Merck) per vial of Pro-Q Emerald 488 reagent (Molecular Probes), mix thoroughly, and add 25 mL of staining buffer (Molecular Probes) just before use. The staining solution of 250 mL is sufficient for staining one 2-DE gel ($20 \times 25 \times 0.1$ cm$^3$).

#### 2.2.7. Reagent Solutions for Labeling Protein Samples for DIGE

1. *SDS sample buffer*: 1% (w/v) SDS, 100 mM Tris-HCl (pH 8.5). To prepare 50 mL of this buffer, dissolve 0.5 g of SDS (Serva) and 0.6 g of Tris base (Sigma) in ~40 mL of deionized water. Titrate to pH 8.5 with 4 N HCl, filter, and adjust the final volume to 50 mL with deionized water.
2. *Thiourea/urea/CHAPS buffer A*: 2 M thiourea, 7 M urea, 4% (w/v) CHAPS, 30 mM Tris-HCl (pH 8.5). To prepare 50 mL of this buffer, dissolve 22.0 g of urea and 8.0 g of thiourea (Fluka) in ~25 mL of deionized water. Add 0.5 g of Serdolit MB-1 ion exchange resin (Serva), stir for 30 min, and filter. Add 2.0 g CHAPS (GE Healthcare) and 180 mg Tris base to the filtrate. Titrate to pH 8.5 with 4 N HCl and adjust the final volume to 50 mL with deionized water.
3. *Thiourea/urea/CHAPS buffer B*: 2 M thiourea, 7 M urea, 4% (w/v) CHAPS. To prepare 50 mL of this buffer, dissolve 22.0 g of urea and 8.0 g of thiourea (Fluka) in ~25 mL of deionized water. Add 0.5 g of Serdolit MB-1 ion exchange resin (Serva), stir for 10 min, and filter. Add 2.0 g CHAPS (GE Healthcare) and adjust the final volume to 50 mL with deionized water.

4. *Dithiothreitol (DTT) solution*: DTT 50% (w/v). To prepare 1.0 mL of the solution, dissolve 500 mg of dithiothreitol in 0.7 mL of deionized water. Prepare fresh before use.
5. *CyDye stock solution*: Reconstitute the CyDyes (CyDye DIGE Fluors Cy2, Cy3 and Cy5 minimal dye are available from GE Healthcare) in dimethylformamide (DMF, 99.8% anhydrous) (Sigma-Aldrich) to a stock solution of 1 mM (*see* the CyDyes manual for a detailed protocol). The reconstituted dyes are stable at –20°C for ~6 months (*see* **Note 3**).
6. *Stop solution*: 10 mM lysine. To prepare 10mL of the solution, dissolve 14.6 mg of lysine in 10 mL of deionized water.

## 3. Methods

For most procedures applied for the detection of proteins after electrophoretic separation, the resolved polypeptides have to be fixed in solutions such as ethanol or acetic acid or water for at least several hours (but usually overnight) to remove any compounds (e.g., carrier ampholytes, detergents) that might interfere with the detection, followed by universal or PTM-specific staining. Because of the high sensitivity of most staining procedures, gels should only be manipulated with powder-free nitril or vinyl gloves.

### 3.1. Universal Protein Detection and Quantitation Methods

#### 3.1.1 Coomassie Brilliant Blue Staining

*Coomassie Brilliant Blue* CBB staining methods *(6)* have found widespread use for the postelectrophoretic detection of proteins because of their low cost, simplicity, and compatibility with most downstream protein analysis technologies. However, in terms of the requirements for proteome analysis, the principal limitations of CBB stains lie in their narrow dynamic range that does not significantly exceed one order of magnitude, and in their limited sensitivity, which precludes the detection of low abundance proteins. As the detection limit of the "classical" CBB stain is in the range of ~100 ng of protein per spot, typically no more than a few hundred protein spots can be visualized on a 2-DE gel, even if the protein load is in the milligram range. In contrast, CBB dye in colloidal dispersions *(7)* is capable of detecting down to 10 ng of protein per spot *(8)*. Colloidal CBB stains are based on the observation that CBB G-250 can form microprecipitates in an acidic medium, resulting in a very low concentration of soluble free dye. Since the colloidal dye particles cannot penetrate the gel matrix, the low concentrated free dye molecules preferably bind to the proteins with minimal gel matrix staining, and, therefore, destaining to obtain low background can be considerably shortened, or even omitted. Sensitivity may be additionally enhanced by near-infrared fluorescence imaging for detection *(9)*, but even the best colloidal CBB staining procedures are still less sensitive than the majority of silver and fluorescent stains (cf. **Fig. 1**).

In the classical (i.e., regressive) approach, the gel is first saturated with the dye (usually CBB R-250) dissolved in an alcohol or acetic acid or water mixture, and then destained with the same solvent, however devoid of dye, until a clear background is obtained. Since the affinity of the dye molecules (which preferably bind to basic amino acid residues) is higher to the proteins than to the gel matrix, the latter is destained more rapidly. Unfortunately, the destaining procedure is difficult to standardize, thereby limiting the reproducibility of quantitative results, whereas colloidal CBB stains are progressive end-point stains, and thus more reproducible.

"Classical" Coomassie Brilliant Blue R-250 Staining Protocol (According to Ref. *6*, Modified)

1. Place the gel into a staining dish (*see* **Subheading 2.1**) on a laboratory shaker and fix the proteins for >30 min in 250 mL trichloroacetic acid (TCA, 20% (w/v), Merck) with gentle shaking.
2. Decant the fixing solution and rinse the gel briefly (1–2 min) with CBB R-250 destaining solution (40% ethanol and 10% acetic acid) to remove excess TCA.
3. Immerse the gel in 500 mL CBB R-250 staining solution and incubate at room temperature on the laboratory shaker for at least 3 h. Staining can continue overnight if more convenient.
4. Discard the CBB stain and briefly rinse the gel with deionized water (*see* **Note 4**).
5. Destain the gel (however not completely) two times 30 min each with CBB R-250 destaining solution with shaking until the background is no longer stained dark blue.
6. For higher sensitivity and decreased background staining, immerse the gel in 500 mL of 1% (v/v) acetic acid for up to 24 h with gentle shaking.
7. Digitize the gel with the help of a CCD camera or a flatbed scanner at 300 dpi resolution.

Colloidal Coomassie Brilliant Blue G-250 Staining Protocol (According to Refs. *7*, *8*, Modified)

1. Place the gel into a staining dish (*see* **Subheading 2.1**) on a laboratory shaker and incubate for >3 h in 500 mL fixing solution (40% (w/v) ethanol and 10% (w/v) acetic acid).
2. Wash the gel briefly (2 × 30 s) with deionized water.
3. Immerse the gel in 500 mL colloidal CBB G-250 staining solution and incubate at room temperature with gentle shaking for at least 3-4 h. Staining can continue overnight if more convenient.
4. Decant the staining solution and rinse the gel briefly (2 × 30 s) with deionized water. Then incubate the gel on the laboratory

shaker with 500 mL of CBB-G-250 destaining solution (5% ammonium sulfate and 10% (v/v) methanol) for 20 min, followed by a further 20-min incubation in 250 mL CBB G-250 destaining solution diluted with 250 mL deionized water (*see* **Note 5**).

5. Acquire image with the help of a CCD camera or a flatbed scanner.

#### *3.1.2. Imidazole-Zinc Staining*

*Negative (or reverse) staining* procedures exploit the fact that protein-bound metal cations (*e.g.*, copper, zinc) are less reactive than the free salt in the gel. Thus, the speed of precipitation of metal cations to form an insoluble salt is slower on the sites occupied by proteins than in the protein-free background. This generates transparent protein zones or spots, while the gel background becomes opaque because of the precipitated, insoluble salt. Imidazole-zinc staining *(10, 11)* is currently the most sensitive "negative" postelectrophoretic staining method. The staining procedure is rapid, simple, and sensitive (detection limit: 10-50 ng of protein per spot) and allows easy recovery of proteins from the gel by chelating the metal cations with EDTA. Moreover, it is compatible with subsequent protein identification by MALDI-MS or Edman sequencing, making the stain rather popular for detection of proteins separated on micropreparative 2-D gels. Consecutive staining of CBB-stained gels by imidazole-zinc is also possible, resulting in increased sensitivity. A major disadvantage of imidazole-zinc-staining is its rather restricted linear dynamic range (less than one order of magnitude), and therefore, this staining procedure is of only limited value for detecting quantitative differences on 2-DE gels.

##### Imidazole-Zinc Reverse Staining Protocol (According to Refs. *10, 11)*

1. After electrophoresis, place the gel into a transparent plastic tray and briefly rinse the gel (2 × 30 s) in *freshly prepared* Milli Q water (*see* **Note 6**).
2. Incubate the gel with gentle agitation on a laboratory shaker for 15 min in 500 mL 0.2 M imidazole containing 0.1% SDS.
3. Discard the solution, and rinse the gel briefly (30 s) in 500 mL Milli Q water to remove excess imidazole solution.
4. Add 0.2 M zinc sulfate solution while manually shaking the gel for about 1 min. Observe the gel over a black surface. The gel background will become white because of the precipitation of zinc-imidazole complex on the gel surface, leaving protein spots transparent. Note that prolonged incubation in the $Zn^{2+}$ solution may result in diminished sensitivity or even complete loss of the image.
5. As soon as an adequate image has been obtained, immediately discard the zinc sulfate solution and replace with Milli Q water

to stop development. Wash the gel briefly (three times, 60 s each). Keep the negatively stained gel in water at 4°C until use (*see* **Note 7**).

*3.1.3. Silver Staining*

Silver staining methods *(12–16)* are far more sensitive than CBB or reverse stains, the most sensitive ones exhibiting a detection limit of less than 0.1 ng protein per spot. Regrettably, they usually provide only a poor linear response with more than an approximately 10-fold range in protein concentration. Moreover, silver staining methods are laborious and complex multistep procedures, which are far from stoichiometric and exhibit significant gel-to-gel variability (typically >20%) due to the highly subjective assessment of the end-point of the staining procedure, thus making silver staining less suitable for quantitative studies. In addition, saturation or negative ("doughnut") staining of the higher abundant proteins, which is often encountered, is a challenge for computerized image analysis, although the problem may be alleviated by prestaining with CBB *(17)* or by time-based analysis *(18)*.

Silver staining is based on the reduction of ionic silver to metallic silver on the *protein surface*. During impregnation of the gel, silver ions complex with protein functional groups such as amino groups, sulfur residues, and amide bonds. The less tightly bound silver ions are washed off the gel matrix before image development, whereas the protein-complexed silver ions are reduced with formaldehyde to metallic silver, followed by further autocatalytic reduction and consecutive image formation, which is finally stopped by immersing the gel in an appropriate solution (e.g., diluted acetic acid or glycine).

Silver stains can be classified into two major categories: acidic silver nitrate *(12–14)* and basic silver diamine *(15)* stains. The latter methods, using aldehyde-based fixatives or sensitizers, are the more sensitive ones, but prevent subsequent protein identification and characterization due to protein crosslinkage. If aldehydes are omitted in the fixative and in the subsequent gel impregnating buffers (except in the developer), microchemical characterization by MALDI-MS peptide mass fingerprinting is possible *(19, 20)*, though at the expense of sensitivity (cf. *(16)*).

Fast Silver Nitrate Staining Protocol (According to Ref. 14, Modified)

1. On completion of electrophoresis, place the gel into a glass tray on a laboratory shaker and fix the proteins in 500 mL 40% ethanol and 10% acetic acid at least for 3 h. Replacement of the fixing solution after 1 h is recommended (*see* **Notes 8–13**).
2. Wash the gel in 500 mL of 30% ethanol for 20 min with gentle shaking, followed by two consecutive washes (20 min each) with 15% ethanol and Milli Q water, respectively.
3. Sensitize with 500 mL sodium thiosulfate solution for 1 min with gentle shaking.

4. Discard the sensitizer solution and rinse the gel in Milli Q water for 3 × 20 s while manually shaking.
5. Add 500 mL of silver nitrate solution and gently agitate for ~30 min. Protect the tray from bright light.
6. Decant the silver nitrate solution and rinse the gel in Milli Q water for 3 × 20 s while manually shaking.
7. Quickly add 500 mL of developing solution (3% sodium carbonate, 250 µL 37% formaldehyde). Sometimes, a gray or brown precipitate may form, which normally dissolves within a few seconds by vigorous shaking. The developer may be replaced after 1 min for less background staining.
8. Develop until protein spots are clearly visible (5–15 min). The most intense spots will appear within a few minutes.
9. Stop development as soon as an adequate degree of staining has been achieved to avoid excessive background formation. Discard the development, rinse the gel briefly in deionized water, and immerse it in 500 mL stop solution (0.5% glycine or, alternatively, 4% Tris and 2% acetic acid) for 20 min with gentle shaking. Then, wash the gel (3 × 10 min) with Milli Q water before further processing.
10. Digitize the gel with the help of a CCD camera or a flatbed scanner at 300 dpi resolution.

#### *3.1.4. Fluorescence-Based Protein Detection Methods for Total Protein Staining*

Because of the shortcomings of organic dyes, silver staining or radiolabeling for visualization and quantitation of proteins, methods based on the *fluorescent detection of proteins* have increasingly gained popularity *(2, 21)*. Two major approaches are currently favored: (1) covalent derivatization of proteins with fluorophores prior to IEF, and (2) postelectrophoretic protein staining either by the intercalation of fluorophores into the SDS micelles coating the proteins or by the direct electrostatic interaction with the proteins. The best-known examples for *pre-electrophoretic fluorescent labeling* are the cyanine-based dyes that react with lysyl or cysteinyl residues, respectively. These dyes are commercially available as CyDyes, and their properties will be discussed in more detail in **Subheading 3.3** on fluorescent DIGE *(22)*. The major problem with pre-electrophoretic labeling is the occurrence of protein size or protein charge modifications, which may result in altered protein mobilities alongside the $M_r$ or p$I$ axis. To circumvent this drawback, proteins can be stained with fluorescent dye molecules *after* the electrophoretic separation. The most prominent examples are the ruthenium-based dye SYPRO Ruby *(23)* and Deep Purple stain *(24)*, the latter containing the compound *epicocconone*, which becomes fluorescent on interaction with proteins. Staining is accomplished within a few hours and may be easily adapted for robotic staining devices. A cost-efficient alternative to SYPRO Ruby staining has been devised by Rabilloud et al. *(25)* and further improved by Lamanda et al. *(26)*. The detection limit of staining with SYPRO Ruby is 1–2 ng protein per spot (Deep Purple is even more sensitive), with a linear dynamic range of quantitation exceeding three orders of magnitude. Besides their wide linear dynamic range and their ease of use, another advantage of most postelectrophoretic fluorescent staining procedures is their compatibility with downstream protein identification methods. The major limitation is their lower sensitivity when compared with electronic detection methods of radiolabeled proteins (cf. **Table 1**).

SYPRO Ruby® Staining Protocol

1. After the completion of electrophoresis, place the gel into a high-density polypropylene tray (e.g., a food box). Do **not** use a glass vessel! Fix the proteins in 500 mL 40% ethanol and 10% acetic acid for at least 3 h at room temperature on a reciprocal or orbital shaker (*see* **Note 14**).
2. Discard the fixing solution, and add 500 mL SYPRO Ruby staining solution for a “standard” 2-DE gel with the dimensions $25 \times 20 \times 0.1$ $cm^3$. Up to three gels may be simultaneously stained in one polypropylene box.
3. Cover the box with a tight-fitting lid and with aluminum foil to protect the reagent from bright light. (SYPRO Ruby dye is light-sensitive and thus gel staining must be carried out in an opaque container or in the dark.)
4. Shake the box gently, preferably with a circular action on an orbital shaker, for at least 3 h at room temperature or overnight.
5. Discard the excess stain solution. Wash the gel ($3 \times 10$ min) in deionized water or, preferably, with 10% ethanol and 7% acetic acid to reduce speckling, which is sometimes a problem (*see* **Note 15**).
6. Capture and save the image using an appropriate fluorescence imager. Use the filter sets that match the excitation (maximum, 450 nm) and emission (maximum, 610 nm) wavelength for SYPRO Ruby. Images may also be obtained by using a simple UV transilluminator, but high sensitivity will only be obtained with dedicated fluorescence imagers, e.g., Molecular Dynamics Fluorimager (Sunnyvale, CA, USA), Molecular Imager FX from Bio-Rad, FLA-5100 image analyzer (Fuji), or similar devices. For comprehensive information, *see* http://probes.invitrogen.com/handbook/tables/0388.html).

### 3.2. Methods for the Analysis of Protein Post-translational Modifications

Up to now, numerous PTMs of proteins, including phosphorylation, glycosylation, acetylation, lipidation, sulfation, ubiquination, or limited proteolysis have been reported. Since no information can be gained on PTMs through genome sequencing, the possibility to detect and characterize PTMs is a major strength of proteomics. One of the particular advantages of 2-DE is its capability to readily locate post-translationally modified proteins, as they frequently appear as distinct rows of spots in the horizontal or vertical axis of the 2-DE gel. The most important PTMs are glycosylation, phosphorylation, and limited proteolysis. Protein *phosphorylation* is a key PTM, crucial in the control of numerous regulatory pathways, signal transduction, enzyme activities, and the degradation of proteins, whereas *glycosylation* is associated with biochemical alterations, developmental changes, and pathogenesis. PTM specific stains are employed either directly in the 2-DE gel or after transferring (blotting) onto an immobilizing membrane, which is then probed with specific antibodies (e.g., against phosphotyrosine residues) or with lectins (against carbohydrate moieties).

#### 3.2.1. Phosphoprotein Detection

*Phosphoprotein detection* on 2-DE gels may be achieved by autoradiography after in vivo incorporation of $^{32}P$ or $^{33}P$ orthophosphate into proteins. However, this method is restricted to cell cultures and cannot be applied for clinical samples obtained from patients. An alternative method is immunostaining with phosphoamino acid-specific antibodies after blotting the 2-DE separated proteins onto an immobilizing membrane. However, whereas antiphosphotyrosine antibodies are quite specific, antibodies directed against phosphosryl and phosphothreonyl residues are less and often sensitive to the context of a larger epitope. A technique that is particularly useful for the *characterization* of phosphorylation sites is MS of 2-DE separated proteins after phosphoprotein detection (reviewed by **ref. 27**).

Recently, a fluorescent detection and quantitation method for gel-separated phosphoproteins using Pro-Q Diamond phosphoprotein stain has been introduced *(28)*. The procedure is simple, MS compatible, and also suitable for multiplexed proteomics. It has been reported that the sensitivity limit ranges from ~1 to 16 ng/spot, depending on the phosphorylation state of the protein. The signal is linear over three orders of magnitude, and the strength of the signal correlates with the number of phosphate groups in the protein. Regrettably, the stain is not absolutely specific for phosphoproteins, in particular, with complex protein samples such as in 2-DE applications, since highly abundant, nonphosporylated proteins may also be stained (albeit considerably less intense than the phosphorylated ones), and hence there is no perfect correlation between the phosphoprotein patterns detected by autoradiography and Pro-Q Diamond stain. More reliable results are obtained by determining the ratio of the Pro-Q Diamond signal versus the SYPRO Ruby (i.e., total protein) signal intensity (cf. **Fig. 1d, e**). Note that phosphatase inhibitors should always be included during cell lysis and sample preparation to avoid dephosphorylation of proteins.

Pro-Q Diamond® Phosphoprotein Staining Protocol

1. After completion of electrophoresis, place the gel into a high-density polypropylene tray (e.g., food box). Do neither use a glass vessel nor a staining container that has previously been used for SYPRO® Ruby protein gel stain (as residual SYPRO® Ruby stain may interfere with Pro-Q® Diamond phosphoprotein gel staining).
2. Fix the proteins in 500 mL 40% ethanol and 10% acetic acid (or, alternatively, 50% methanol and 10% acetic acid) overnight at room temperature while rotating the tray on an orbital shaker. Replacement of the fixing solution after 1 h is highly recommended. Fixation is critical to achieving specific staining as incomplete removal of SDS would otherwise result in very poor or no staining of phosphoproteins.
3. Discard the solution and wash the gel with gentle agitation with Milli Q water (500 mL, three changes, 20 min per wash).
4. Incubate the gel with 500 mL Pro-Q Diamond phosphoprotein stain (supplied as a ready-to-use solution) for 2–3 h (however, not overnight). Cover the box with tight-fitting lid and with aluminum foil to protect the reagent from light.
5. Discard the staining solution and destain in the dark with 500 mL destaining solution (25% acetonitrile and 50 mM sodium acetate/acetic acid, pH 4.0; three changes, 30 min per wash).
6. Wash the gel 2 × 5 min with Milli Q water.
7. Continue with image acquisition. Pro-Q® Diamond stain has excitation/emission maxima of ~555/580 nm and can be detected by use of a visible-light scanning instrument, a visible-light transilluminator, or (with reduced sensitivity) a 300 nm transilluminator (*see* **Note 16**).

*3.2.2. Glycoprotein Detection*

Apart from autoradiography after incorporation of $^{14}C$ labeled sugars (a method that is only rarely applied because of the safety considerations associated with radiolabeling), two major principles for the detection of glycoproteins separated by 2-DE prevail: One is based on sugar binding proteins, so-called *lectins*. A wide range of lectins with different carbohydrate specificities (preferably labeled with a reporter group, such as fluorescent dye molecules, enzymes, or colloidal gold) is available and can be applied either singly (for glycoprotein differentiation) or as a mixture (for total glycoprotein staining). This method is unsuitable for glycoprotein detection in electrophoresis gels, however, and must be applied after blotting. The second principle for glycoprotein detection is based on coupling a *carbonyl reactive group* (usually a substituted hydrazine) to the aldehyde groups generated in the carbohydrate part of the glycoproteins after periodate oxidation of vicinal hydroxyls. Visualization of the glycoproteins depends on the kind of reporter group attached to the hydrazine and is achieved by UV illumination in case of fluorescent molecules (e.g., dansyl hydrazine) or through the reaction product (insoluble colour, chemiluminescence, etc.) in case of hydrazine-conjugated enzymes. Recently, the glycoprotein stain Pro-Q Emerald has been devised that reacts with periodic acid oxidized carbohydrate groups, and which generates a green fluorescent signal on glycoproteins. The stain permits detection of ~1–20 ng of glycoprotein per band/spot, depending upon the nature and the degree of protein glycosylation *(29)*. Two variants of the stain ("488" and "300", respectively) are available, depending on their excitation wavelengths. Pro-Q Emerald 300 dye-labeled proteins may be poststained with SYPRO Ruby, allowing sequential two-colour detection of glycosylated and total proteins. Functional characterization of 2-DE gel-separated proteins using a sequential staining procedure with Pro-Q Diamond (for phosphoproteins), followed by Pro-Q Emerald 488 (for glycoproteins) and SYPRO Ruby stain (for total proteins) is also applicable *(28, 30)*.

Pro-Q Emerald® 488 Glycoprotein Staining Protocol

1. After completion of electrophoresis, place the gel into a high-density polypropylene tray (e.g., food box). Do not use a glass vessel! Fix the proteins in 1,000 mL 40% ethanol and 10% acetic acid (or, alternatively, 50% methanol and 10% acetic acid) at room temperature while rotating the tray on an orbital rotator. Replace the fixing solution after 1 h and continue with fixing overnight.
2. Discard the solution, and wash the gel with gentle agitation with Milli Q water (1,000 mL, two changes, 20 min per wash).
3. Incubate the gel in 500 mL oxidizing solution in the dark for 1 h with gentle agitation.
4. Wash the gel with Milli Q water (1,000 mL, six changes, 15–20 min per wash).
5. Incubate the gel with gentle agitation in 250 mL Pro-Q Emerald 488 staining solution for ~3 h (or overnight). Cover the box with tight-fitting lid and with aluminum foil to protect the reagent from light.
6. Incubate the gel with 1,000 mL of wash solution (three changes, 30–45 min per wash).
7. Visualize the glycoproteins by, e.g., utilizing 488 nm laser excitation and 530 nm bandpass emission filter (the Pro-Q Emerald 488 stain has an excitation maximum at ~510 nm

and an emission maximum at ~520 nm). After image acquisition, the gel may be additionally stained with SYPRO Ruby for total proteins (cf. **Note 16**).

### 3.3. Difference Gel Electrophoresis

Proteomics is typically based on the comparison of different protein profiles, e.g., from diseased versus healthy or stressed versus untreated samples. In the conventional 2-DE methodology, protein samples are separated on individual gels, stained, and quantified, followed by image comparison with computer-aided image analysis programs. Because this multistep technology often prohibits different images from being perfectly superimposable, image analysis is frequently very time-consuming. To simplify this laborious procedure, Ünlü et al. *(22)* have developed a method called fluorescent DIGE, in which two samples are labeled in vitro using two different fluorescent cyanine dyes (CyDyes) differing in their excitation and emission wavelengths, then mixed before IEF, and separated on a single 2-D gel. After consecutive excitation with both wavelengths, the images are overlaid and "subtracted," whereby only differences (e.g., up- or downregulated, or post-translationally modified proteins) between the two samples are visualized (cf. **Fig. 1g–i**). Because of the comigration of both samples, methodological variations in spot positions and protein abundance are avoided, resulting in improved reproducibility and more reliable protein quantitation, in particular, if an internal standard has been included. This internal standard, typically a pooled mixture of all the samples in the experiment labeled with a third CyDye, is used for the normalization of data between different gels *(31)*.

Because of the increase in molecular mass ($M_r$) by ~450 Da per labeled lysyl residue, labeling of more than one lysine residue per protein molecule must be avoided; otherwise, it would result in multiple spots in the vertical axis of the 2-DE gel. In practice, approximately only 3–5% of protein is labeled (hence, this procedure is also referred to as "minimal labeling"). Since the bulk of the protein remains unlabeled, the slight increase in $M_r$ sometimes presents a problem for spot excision for MS analysis, particularly with lower $M_r$ proteins. This offset has to be taken into account when automatic spot pickers are used. An alternative is to stain the separated proteins additionally with SYPRO Ruby before spot picking from micropreparative gels. Instead of "minimal labeling" of lysyl-residues, so-called "saturation labeling" with similar cyanine dyes, which label, however, cysteine residues to saturation *(32)* is sometimes employed. The advantage of saturation labeling is the greater sensitivity of the stain (detection limit is <0.1 ng of protein), compared with ~1 ng with minimal labeling, but saturation labeling has also several drawbacks, such as precipitation and loss of proteins during the labeling reaction due to the introduction of the hydrophobic dye molecules. For more comprehensive information on DIGE technology please *see* **Refs. 3, 33**.

*3.3.1. Minimal (Lys) Labeling the Protein Samples*

1. Extract proteins, preferably with SDS-buffer. Determine the protein concentration of the sample with a protein assay (e.g., 2DQuant kit, GE Healthcare). Dilute the SDS extract with a fourfold excess of thiourea/urea/CHAPS buffer B (final SDS concentration ≤ 0.2%). The protein concentration should be ≥1 and ≤20 mg/mL. Best results are obtained with final protein concentrations between 5 and 10 mg/mL. Instead of using SDS buffer, proteins may be directly solubilized in thiourea/urea/CHAPS buffer A.
2. Directly before labeling, dilute the CyDye stock solution to a final concentration of 0.4 mM (400 pmol/μL), e.g., mix 3 μL of DMF with 2 μL of reconstituted CyDye.
3. Prepare an aliquot of 50 μg of the protein sample and add 400 pmol (1 μL) of the CyDye. Cool on ice. *Important*: (1) Make sure the pH is 8.5 by checking with a pH indicator strip. Adjust the pH if necessary. (2) Keep the dyes and the samples in the dark and on ice!
4. Vortex and centrifuge the mixture in a microfuge at 10,000 g for 5 s.
5. Incubate the sample in the dark for 30 min on ice.
6. Add 1 μL of 10 mM lysine to stop the labeling reaction.
7. Mix, centrifuge briefly, and incubate in the dark for 10 min on ice.
8. Add 1 μL of DTT solution and 1 μL of Pharmalyte 3–10 (GE Healthcare) per 50 μL of sample solution.
9. Pool the protein samples that are going to be separated on the same first- and second-dimension gel, and immediately apply the mixture to the IPG strip for IEF. Alternatively, labeled samples can be stored for at least 3 months at –70°C.
10. Run IEF and SDS-PAGE according to the Protocols described in Chapter **2** on high-resolution 2-D electrophoresis. Protect the IPGphor during the IEF and the buffer tank during SDS-PAGE from light to minimize photobleaching of the fluorescence dyes.

*3.3.2. Image Acquisition and Analysis*

1. For best results during image acquisition, use low-fluorescence glass plates. Directly scan the gels in the glass cassettes after completion of SDS-PAGE to ensure that all gels have the same dimensions; doing so simplifies spot matching of different gels. Carefully clean the exterior of the glass plates with deionized water and dry with a lint-free laboratory wipe before the gel cassette is positioned on the scanner.
2. Excite each fluorescent dye consecutively to avoid fluorescence crosstalk and scan at a resolution of at least 200 μm with a proper filter (*see* **Table 2**).

**Table 2**
**Excitation wavelengths and filters for proper scanning of CyDye labeled proteins**

| Dye | Absorption maximum | Fluorescence maximum | Laserlight wavelength | Emission Filter |
|---|---|---|---|---|
| Cy2 | 491 nm | 509 nm | 488 nm | 520 nm, band pass 40 |
| Cy3 | 553 nm | 569 nm | 532 nm | 580 nm, band pass 40 |
| Cy5 | 645 nm | 664 nm | 633 nm | 670 nm, band pass 40 |

3. After the scan, software such as ImageQuant (GE Healthcare) easily provides an initial image overlay of the scanned channels of one gel, thus giving a quick overview of differences between the labeled extracts is given.

## 4. Notes

1. Because formaldehyde polymerizes in the cold, the formaldehyde solution (37%) should be stored at room temperature and not used longer than 6 months as the amount of active formaldehyde will gradually decrease.
2. Optimal temperature of the silver nitrate staining reagent solutions is 20–25°C. More intense background staining will be observed at temperatures >30°C.
3. DMF should be fresh (less than 3 months old after first opening), because it degrades to amine compounds, which interfere with the labeling reaction.
4. Do not reuse Coomassie Blue staining solutions as abundant proteins may partially leach out and contaminate subsequent gels. This is a particular problem for protein identification by MS.
5. Colloidal CBB staining kits based on the protocol of Neuhoff et al. *(7)* are available from different manufacturers, e.g., from Carl Roth (Karlsruhe, Germany). Roti®-Blue staining is quick (2–3 h), and destaining is usually not required.
6. Water is a critical reagent in the imidazole-zinc method, and only *freshly* produced (to avoid incorporation of $CO_2$ from the air) high-quality water should be used in all steps.

Alternatively, it has been recommended to briefly rinse the gel in 1% sodium carbonate solution before incubation in the imidazole solution. However, this procedure seems unnecessary if high-quality imidazole and freshly prepared Milli Q water are used *(11)*.

7. To recover the protein(s) for downstream protein identification, excise the spot of interest and complex the zinc ions by incubating the gel plug in 100 mM EDTA disodium salt in, e.g., PBS, pH 7.4 (or a similar buffer) for 1–2 min.
8. To ensure even staining, proper and constant agitation by means of a laboratory shaker is required for all incubations lasting longer than 1 min. For shorter steps, manual agitation is more convenient.
9. To remove carrier ampholytes, SDS and other detergents, which otherwise produce a strong background ("clouds") in the lower $M_r$ region, the gel should be fixed for 3 h at least, but preferably overnight. Fixation can be prolonged for up to 72 h, with at least one change of fixing solution after 1 h.
10. Up to four gels per tray may be processed simultaneously. In this case, 1,000 mL of solution are required. Batch processing can be used for every step longer than 5 min, except for image development, where one gel per tray is mandatory.
11. Handling of gels not cast on a supporting plastic sheet is sometimes tricky, in particular, if the polyacrylamide concentration is not very high (<10%) and the gels are, thus, rather fragile. Rabilloud and coworkers *(16)* recommend that the best way to change solutions is to press a rigid plastic sheet (e.g., for separating gel cassettes in multigel casting chambers) on the surface of the gel(s) with the aid of a gloved hand and then to incline the whole setup to allow the emptying of the box while keeping the gel(s) in it.
12. Always use powder-free nonlatex (e.g., nitril or vinyl) gloves when handling gels, as keratin and latex proteins are potential sources of contamination.
13. Silver staining kits are available from different suppliers (e.g., GE Healthcare or Bio-Rad).
14. The use of polypropylene food boxes or polyvinyl chloride photographic staining trays is recommended as they adsorb the least amount of dye. The staining trays must be thoroughly cleaned, and it is best to rinse with ethanol before use. Staining must be performed with trays that have never been in contact with BSA, nonfat dry milk, or any other protein blocking agent to prevent carryover contamination.
15. When using gels that have been cast on a plastic backing, the plastic sheet should be removed before scanning (e.g., with the help of a film remover available from GE Healthcare).

Alternatively, use gels cast on low-fluorescence plastic backings.

16. Subsequent staining with SYPRO® Ruby is recommended, as determining the ratio of Pro-Q® Diamond dye to SYPRO® Ruby dye signal intensities for each spot provides a measure of the phosphorylation level normalized to the total amount of protein.

## References


1. Görg, A., Weiss, W., and Dunn, M.J. (2004) Current two-dimensional electrophoresis technology for proteomics. *Proteomics* **4**, 3665–3685. Review.
2. Patton, W.F. (2002) Detection technologies in proteome analysis. *J. Chromatogr. B* **771**, 3–31. Review.
3. Van den Bergh, G., and Arckens, L. (2004) Fluorescent two-dimensional difference gel electrophoresis unveils the potential of gel-based proteomics. *Curr. Opin. Biotech.* **15**, 38–43. Review.
4. Miller, I., Crawford, J., and Gianazza, E. (2006) Protein stains for proteomic applications: which, when, why? *Proteomics* **6**, 5385–5408. Review.
5. Harris, L.R., Churchward, M.A., Butt, R.H., and Coorssen, J.R. (2007) Assessing detection methods for gel-based proteomic analyses. *J. Proteome Res.* **6**, 1418–1425.
6. Fazekas de St.Groth, S., Webster, R.G., and Datyner, A. (1963) Two new staining procedures for quantitative estimation of proteins on electrophoresis strips. *Biochim. Biophys. Acta* **71**, 377–391.
7. Neuhoff, V., Arold, N., Taube, D., and Ehrhardt, W. (1988) Improved staining of proteins in polyacrylamide gels including isoelectric focusing gels with clear background at nanogram sensitivity using Coomassie Brilliant Blue G-250 and R-250. *Electrophoresis* **9**, 255–262.
8. Candiano, G., Bruschi, M., Musante, L., Santucci, L., Ghiggeri, G.M., Carnemolla, B., Orecchia, P., Zardi, L., and Righetti, P.G. (2004) Blue silver: a very sensitive colloidal Coomassie G-250 staining for proteome analysis. *Electrophoresis* **25**, 1327–1333.
9. Luo, S., Wehr, N.B., and Levine, R.L. (2006) Quantitation of protein on gels and blots by infrared fluorescence of Coomassie Blue and Fast Green. *Anal. Biochem.* 350, 233–238.
10. Fernandez-Patron, C., Castellanos-Serra, L., and Rodriguez, P. (1992) Reverse staining of sodium dodecyl sulfate polyacrylamide gels by imidazole-zinc salts: sensitive detection of unmodified proteins. *Biotechniques* **12**, 564–573.
11. Castellanos-Serra, L., and Hardy, E. (2006) Negative detection of biomolecules separated in polyacrylamide electrophoresis gels. *Nat. Protoc.* **1**, 1544–1551.
12. Switzer, R.C., Merril, C.R., and Shifrin, S. (1979) A highly sensitive silver stain for detecting proteins and peptides in polyacrylamide gels. *Anal. Biochem.* **98**, 231–237.
13. Merril, C.R., Goldman, D., Sedman, S.A., and Ebert, M.H. (1981) Ultrasensitive stain for proteins in polyacrylamide gels shows regional variation in cerebrospinal fluid proteins. *Science* **211**, 1437–1438.
14. Blum, H., Beier, H., and Gross, H.J. (1987) Improved silver staining of plant proteins, RNA and DNA in polyacrylamide gels. *Electrophoresis* **8**, 93–99.
15. Oakley, B.R., Kirsch, D.R., and Morris, N.R. (1980) A simplified ultrasensitive silver stain for detecting proteins in polyacrylamide gels. *Anal. Biochem.* **105**, 361–363.
16. Chevallet, M., Luche, S., and Rabilloud, T. (2006) Silver staining of proteins in polyacrylamide gels. *Nat. Protoc.* **1**, 1852–1858.
17. De Moreno, M.R., Smith, J.F., and Smith, R.V. (1985) Silver staining of proteins in polyacrylamide gels: increased sensitivity through a combined Coomassie Blue-silver stain procedure. *Anal. Biochem.* **151**, 466–470.
18. Becher, B., Knofel, A.K., and Peters, J. (2006) Time-based analysis of silver-stained proteins in acrylamide gels. *Electrophoresis* **27**, 1867–1873.
19. Shevchenko, A., Wilm, M., Vorm, O., and Mann, M. (1996) Mass spectrometric sequencing of proteins from silver-stained polyacrylamide gels. *Anal. Chem.* **68**, 850–858.

20. Mortz, E., Krogh, T.N., Vorum, H., and Görg, A. (2001) Improved silver staining protocols for high sensitivity protein identification using matrix-assisted laser desorption/ionization-time of flight analysis. *Proteomics* **1**, 1359–1363.
21. Harris, L.R., Churchward, M.A., Butt, R.H., and Coorssen, J.R. (2007) Assessing detection methods for gel-based proteomic analyses. *J. Proteome Res.* **6**, 1418–1425.
22. Ünlü, M., Morgan, M.E., and Minden, J.S. (1997) Difference gel electrophoresis: a single gel method for detecting changes in protein extracts. *Electrophoresis* **18**, 2071–2077.
23. Berggren, K.N., Schulenberg, B., Lopez, M.F., Steinberg, T.H., Bogdanova, A., Smejkal, G., Wang, A., and Patton, W.F. (2002) An improved formulation of SYPRO Ruby protein gel stain: comparison with the original formulation and with a ruthenium II tris (bathophenanthroline disulfonate) formulation. *Proteomics* **2**, 486–498.
24. Mackintosh, J.A., Choi, H.Y., Bae, S.H., Veal, D.A., Bell, P.J., Ferrari, B.C., Van Dyk, D.D., Verrills, N.M., Paik, Y.K., and Karuso, P. (2003) A fluorescent natural product for ultra sensitive detection of proteins in one-dimensional and two-dimensional gel electrophoresis. *Proteomics* **3**, 2273–2288.
25. Rabilloud, T., Strub, J.M., Luche, S., van Dorsselaer, A., and Lunardi, J. (2001) A comparison between Sypro Ruby and ruthenium II tris (bathophenanthroline disulfonate) as fluorescent stains for protein detection in gels. *Proteomics* **1**, 699–704.
26. Lamanda, A., Zahn, A., Röder, D., and Langen, H. (2004) Improved Ruthenium II tris (bathophenantroline disulfonate) staining and destaining protocol for a better signal-to-background ratio and improved baseline resolution. *Proteomics* **4**, 599–608.
27. Reinders, J., and Sickmann, A. (2005) State-of-the-art in phosphoproteomics. *Proteomics* **5**, 4052–4061. Review.
28. Steinberg, T.H., Agnew, B.J., Gee, K.R., Leung, W.Y., Goodman, T., Schulenberg, B., Hendrickson, J., Beechem, J.M., Haugland, R.P., and Patton, W.F. (2003) Global quantitative phosphoprotein analysis using multiplexed proteomics technology. *Proteomics* **3**, 1128–1244.
29. Hart, C., Schulenberg, B., Steinberg, T.H., Leung, W.Y., and Patton, W.F. (2003) Detection of glycoproteins in polyacrylamide gels and on electroblots using Pro-Q Emerald 488 dye, a fluorescent periodate Schiff-base stain. *Electrophoresis* **24**, 588–598.
30. Wu, J., Lenchik, N.J., Pabst, M.J., Solomon, S.S., Shull, J., and Gerling, I.C. (2005) Functional characterization of two-dimensional gel-separated proteins using sequential staining. *Electrophoresis* **26**, 225–237.
31. Alban, A., David, S.O., Bjorkesten, L., Andersson, C., Sloge, E., Lewis, S., and Currie, I. (2003) A novel experimental design for comparative two-dimensional gel analysis: two-dimensional difference gel electrophoresis incorporating a pooled internal standard. *Proteomics* **3**, 36–44.
32. Shaw, J., Rowlinson, R., Nickson, J., Stone, T., Sweet, A., Williams, K., and Tonge, R. (2003) Evaluation of saturation labelling two-dimensional difference gel electrophoresis fluorescent dyes. *Proteomics* **3**, 1181–1195.
33. Lilley, K.S., and Friedman, D.B. (2004) All about DIGE: quantification technology for differential-display 2D-gel proteomics. *Expert Rev. Proteomics* **1**, 401–409.

# Part III

## Mass Spectrometry and Tandem Mass Spectrometry Applications

# Chapter 5

# MALDI MS


## Rainer Cramer



## Summary

Matrix-assisted laser desorption/ionization (MALDI) is a key technique in mass spectrometry (MS)-based proteomics. MALDI MS is extremely sensitive, easy-to-apply, and relatively tolerant to contaminants. Its high-speed data acquisition and large-scale, off-line sample preparation has made it once again the focus for high-throughput proteomic analyses. These and other unique properties of MALDI offer new possibilities in applications such as rapid molecular profiling and imaging by MS. Proteomics and its employment in Systems Biology and other areas that require sensitive and high-throughput bioanalytical techniques greatly depend on these methodologies. This chapter provides a basic introduction to the MALDI methodology and its general application in proteomic research. It describes the basic MALDI sample preparation steps and two easy-to-follow examples for protein identification including extensive notes on these topics with practical tips that are often not available in the **Subheadings 2** and **3** of research articles.



**Key words:** Mass spectrometry, MALDI, MS/MS, Proteomics, Sample preparation, Sequencing, Peptide mass fingerprinting, Peptide mass mapping


## 1. Introduction

It is arguable that proteomics would not have thrived without the earlier development of the so-called “soft” ionization techniques in mass spectrometry (MS), namely electrospray ionization (ESI) *(1, 2)* and MALDI *(3–5)*. These two ionization techniques were introduced a decade before the term “proteomics” was used for the first time. During this decade, their further development and application in the field of protein analysis ultimately enabled the transition from classical to high-sensitivity and large-scale protein analysis, some of the key features proteomics is well known for today *(6, 7)*.


Jörg Reinders and Albert Sickmann (eds.), *Proteomics,* Methods in Molecular Biology, Vol. 564
DOI: 10.1007/978-1-60761-157-8_5, © Humana Press, a part of Springer Science+Business Media, LLC 2009

MALDI and ESI MS provide similar levels of analytical sensitivity and comparable mass ranges for most analytes. In proteomics, both ionization techniques are usually employed for mass spectrometric analyses at the peptide level because their practical sensitivity is best in the peptide mass range, particularly once the technical (instrumental, preparative, and data mining) limitations have been taken into account. However, there are distinct differences between MALDI and ESI, which make one or the other more suitable for specific types of proteomic analyses. For instance, ESI produces predominately multiply charged ions. For tryptic peptides, these are mainly doubly and triply charged ion species that typically lead to easy-to-interpret fragment ion spectra using common dissociation techniques such as collision-induced dissociation (CID). MALDI, on the other hand, predominately produces singly charged peptide ions that fragment into a more complex array of fragment ions, making it more difficult to interpret the fragment ion spectrum and elucidate the identification or sequence of the peptide.

However, MALDI has also some distinct advantages such as high-speed data acquisition *(8)*, off-line analysis capabilities in connection with LC separation *(9–11)*, relatively easy sample preparation, highly controllable sample consumption, postanalytical sample archiving and recovery, and molecular imaging at high spatial resolution (down to 1 μm) *(12–15)*, to name but a few. Most of these are in contrast to ESI and have recently become the focus of a revived interest in MALDI MS in proteomics.

## 2. Materials

1. 2,5-dihydroxybenzoic acid (DHB) and α-cyano-4-hydroxycinnamic acid (CHCA) (Bruker Daltonics, Bremen, Germany).
2. HPLC grade water and acetonitrile (ACN) (Rathburn, Walkerburn, UK).
3. Monobasic ammonium phosphate (AP) (Sigma, Poole, UK).
4. Formic acid (FA), trifluoroacetic acid (TFA), ethanol, and acetone (Fisher Scientific, Loughborough, UK).
5. Natural siliconized tubes, 0.5 mL (BioQuote, York, UK).
6. Tips, 0.1–10 μL and 2–200 μL (Eppendorf, Hamburg, Germany).
7. KIMCARE medical wipes (Kimberley-Clark Europe, Reigate, UK).

## 3. Methods

The MALDI process itself is highly convoluted, consisting of several processes (mainly laser energy absorption, sample desorption, and matrix/analyte ionization) with various different subprocesses contributing to the formation of the overall end-products, the intact analyte ions *(4, 16–18)*. Because of this complexity, MALDI is relatively poorly understood making a rational design of the perfect MALDI experiment virtually impossible. However, from a practical point of view, MALDI is easy to apply and learn. In the easiest case, often called the "dried-droplet" preparation, the analyte and matrix is dissolved in appropriate solvents such as water or aqueous organic solvents. These two solutions are then mixed by applying individual solution droplets (~1 μL or less) to the ion source target or by mixing the two solutions off-target and then applying a small droplet to the target's surface. Subsequent evaporation of the solvents leads to the MALDI sample with the analyte well distributed and embedded in the matrix. Concentrations for matrix solutions are typically close to the saturation limits, whereas analyte concentrations down to 1 fmol/μL are usually unproblematic for successful peptide analysis.

Although the success of MALDI MS analyses depends highly on the matrix, only little is known about the criteria that are crucial for a compound to work as a good matrix in MALDI MS. Essentially, all successful matrices have been found empirically owing to the lack of sufficient guidelines for choosing and designing the optimal matrix system for a given analyte. Interestingly, two of the earlier discovered matrices, DHB and CHCA, are still the most commonly used MALDI matrices for the analysis of peptides. For intact protein analyses, several other matrices have been found to be advantageous such as sinapinic acid or a mixture of DHB and methoxy-hydroxybenzoic acid in a ratio of 9:1. In general, for the more labile polypeptide analytes, such as large proteins or post-translationally modified peptides, one usually chooses the so-called "soft" matrices whereas the "hard" (also called "hot") matrix CHCA is typically chosen for the more generic task of analyzing the "common" peptides or peptide mixtures. The advantage of CHCA as a matrix is that it provides high analytical sensitivity and is amenable to various sample preparation techniques, resulting in relatively homogeneous MALDI samples that can reduce some of the problems associated with sample morphology and topology (e.g., fluctuating ion signal and decreased mass resolution and accuracy). **Table 1** provides an overview of some of the MALDI matrices used in proteomic analyses (*see* also **ref.** *19)*.

As can be seen from **Subheading 2** the MALDI sample preparation can be achieved with relatively few chemicals, in the extreme case just with the matrix compound and the analyte.

**Table 1**
**Commonly used MALDI matrices and their applications in proteomics**

| Name | Application | Wavelength (nm) | References |
|---|---|---|---|
| 2,5-Dihydroxybenzoic acid | General protein and peptide analysis; post-translationally modified peptides and proteins; in-source decay | 337 | *(11, 22, 23, 30)* |
| α-Cyano-4-hydroxycinnamic acid | General peptide analysis; post-source decay | 337, 355 | *(22, 31, 32)* |
| 2,6-Dihydroxyacetophenone | Post-translationally modified peptides and proteins | 337 | *(30, 33, 34)* |
| 2,4,6-Trihydroxyacetophenone | Post-translationally modified peptides and proteins | 337 | *(35, 36)* |
| Sinapinic acid | Proteins | 337 | *(37–39)* |
| Succinic acid | General protein and peptide analysis; post-translationally modified peptides and proteins | 2940 | *(30, 40, 41)* |

However, there have been many studies and publications on the various types of sample preparation techniques including the introduction of matrix additives. In general, these techniques have been developed with the objective of circumventing the disadvantages of MALDI. Despite these developments, many MALDI analyses in proteomics can easily be undertaken using one basic sample preparation technique, the dried-droplet technique. This has therefore been chosen for the following examples of MALDI MS analysis in proteomics: the identification of proteins by peptide mass mapping and peptide sequencing.

Although MALDI sample preparation and analysis are typically not considered as high-risk work, health and safety and waste management requirements and regulations should be considered and adhered to and may require a change of the protocols described within this chapter, in particular, when hazardous compounds are analyzed.

### *3.1. Preparation of MALDI Samples*

1. Clean the MALDI target according to the manufacturer's recommendations. For the commonly used stainless steel targets, initial cleaning and removal of previous MALDI samples can be achieved by wiping the surface with wetted lint-free wipes, but the final washing steps should be rinsing of the target surface. Optionally, ultrasonication in cleaning solvents for some 10–20 min can also be employed before the final rinsing steps. In general, alternating washing (e.g., 3–5 times) with organic solvents, such as methanol, and water, respectively, will remove most of the substances from the previous analyses. From time to

time, or if previously used substances are known to be strongly adsorbed on stainless steel surfaces, it is advisable to clean the target with a stronger acid solution such as 5–10% FA using cotton buds. Drying of any residual liquid after the final rinsing step should be achieved by evaporation. For prestructured targets such as AnchorChip™ targets *(20)*, less abrasive cleaning and mild solutions are often used to avoid any damage to the targets (*see* **Note 1**).

2. After the cleaning procedure, the target should be kept in a clean environment.
3. Prepare a saturated aqueous DHB solution and a CHCA solution with a concentration of 5 g/l CHCA in ACN:0.1% TFA (1:1 by volume) in microcentrifuge tubes (*see* **Note 2**). When using hydrophobic target surfaces, with or without hydrophilic anchor spots, such as AnchorChip™ targets, dilute these two matrix solutions with their respective solvents by a factor of 5–10 depending on the surface or size of the anchors used. The above-mentioned solutions are basic matrix solutions that work well in most applications. Nonetheless, optimization of the solvent and matrix compositions depending on the application and target surface is recommended. For instance, the addition of AP (e.g., 10–50 mM) typically reduces the cation adduct and matrix cluster ion signals *(21, 22)*. There are other known additives that have been found for the analysis of some specific analytes such as phosphoric acid that has been reported to improve the analysis of phosphorylated peptides *(23)*. Various different solvent compositions have also been reported (*see* **Note 3**).
4. Mix the analyte solution with the matrix solution in a ratio of 1:1 to 1:10 depending on the abundance level of the analyte and impurities in the analyte solution (*see* **Note 4**).
5. Spot 1–2 μL of the mixed analyte/matrix solution onto the cleaned MALDI sample target, and let the sample droplet dry in a clean lab environment under standard room conditions (*see* **Note 5**). Optionally, warm or cold streams of air can be applied to speed up the solvent evaporation process (*see* **Note 6**).

   **Figure 1** shows microscope images of typical MALDI samples prepared using DHB as matrix according to the above protocol steps.
6. Store the prepared MALDI target in a clean environment under standard room conditions before mass spectrometric analysis.

### *3.2. Peptide Mass Mapping by Axial MALDI TOF-MS*

1. Prepare MALDI samples of the analyte digest mixture and, for external calibration and the optimization of instrumental parameters, a standard peptide mixture as described under **Subheading 3.1** (*see* **Note 7**).
2. Optimize the laser energy and other instrumental parameters by acquiring mass spectra of the standard peptide mixture.

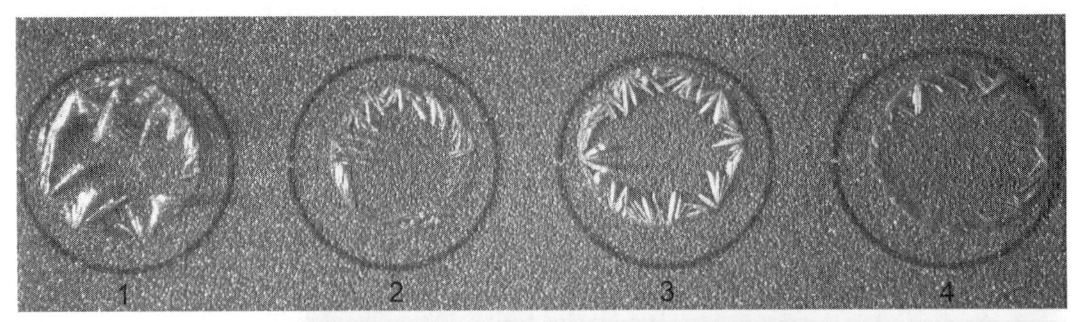


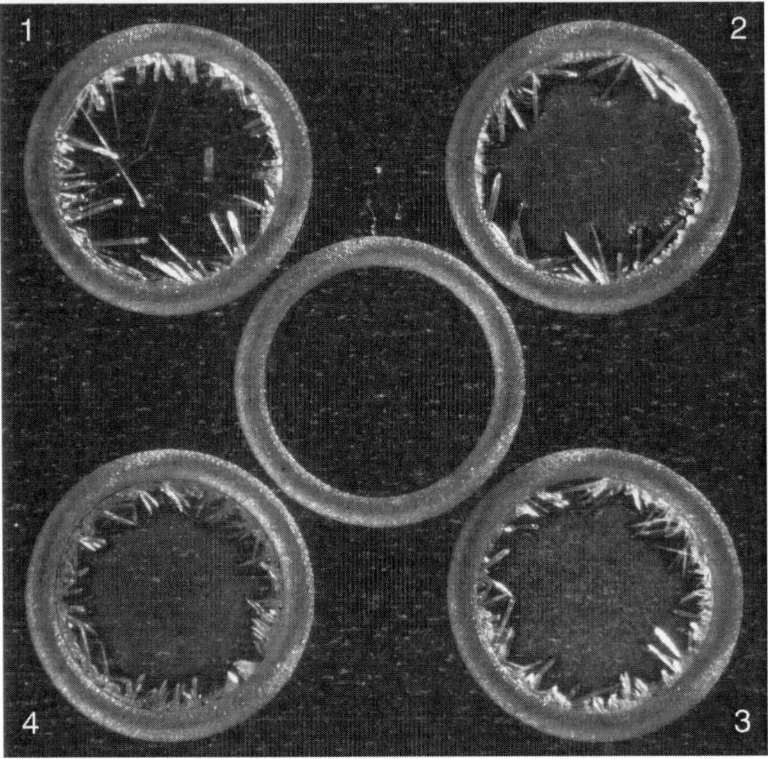


Fig. 1. Microscope images of four MALDI samples on two different stainless steel targets prepared by the dried-droplet technique with DHB as matrix according to steps in **Subheading 3.1**. One pure matrix solution and three different analyte solutions mixed with matrix solutions were used: (1) 2 μL saturated DHB solution, (2) 1 μL saturated DHB solution mixed with 1 μL standard peptide mixture containing several peptides at a concentration of ~1 pmol/μL, (3) 1 μL saturated DHB solution mixed with 1 μL standard peptide mixture containing several peptides at a concentration of ~1 pmol/μL with 50 mM AP, and (4) 1 μL saturated DHB solution mixed with 1 μL in-solution digest (10 pmol protein/μL) with residual digest buffer compounds. The addition of AP reduces the detection of alkali adduct ions and leads to increased crystallization. On hydrophobic targets, this enables the automated analysis of the MALDI spectrum by circumventing the "sweet spot" searching problem due to one crystalline layer or a multitude of virtually nonseparated analyte-embedded crystals. In contrast, the residual digest buffer of the in-solution digest impedes the acquisition of high-quality MALDI analysis owing to adduct ion formation, additional low-mass ion signal intensities, and often inferior analyte/matrix crystallization. However, in this case, these two samples only show slight differences in crystallization. All analyte samples resulted in satisfactory spectra, respectively, displaying the above-mentioned spectral features (data not shown).

Although MALDI analysis is known to be relatively straightforward, the acquisition of MALDI mass spectra itself requires some experience, particularly if one tries to achieve good shot-to-shot stability with regard to ion yields. The "sweet spot" phenomenon, i.e., the recording of good ion yields at some (mainly crystalline) sample positions and very low ion yields at others, has been described extensively in the literature. This potentially highly fluctuating ion signal is one of the main reasons why many researchers still prefer to acquire MALDI mass spectra manually rather than automatically, giving greater control in finding and using the "sweet spots." In general, optimizing the instrumental parameters is therefore easier with matrices and matrix preparations that produce more homogeneous samples. However, one should always optimize the instrumental settings with standard MALDI samples of similar properties as the ones to be analyzed, i.e., for samples prepared with DHB as matrix, the instrument should be optimized and externally calibrated with standards also prepared with DHB. After optimization, acquire a mass spectrum for external calibration and save this spectrum. Calibrate the instrument (*see* manufacturer's instructions and **Note 8**).

3. Acquire a mass spectrum of the analyte sample and, if possible, calibrate it with internal standards that might be present and known in the analyte sample (*see* **Note 9**).

   **Figure 2** shows an example of an internally calibrated mass spectrum of the differentially expressed mouse protein phosphoglycerate mutase 1 separated by two-dimensional gel electrophoresis. Internal calibration was achieved by using two autolysis products of porcine trypsin ($m/z$ 842.5100 and $m/z$ 2211.1046).

4. Create a peak mass list. Most of the instrument manufacturers supply data processing software with peak picking routines that are perfectly adequate for this task. Most of these routines can be fine-tuned by the user (*see* **Note 10**).

   **Table 2** shows the automatically determined peak mass list of the mass spectrum in **Fig. 2** using the manufacturer's peak picking routines as provided, using its default parameters.

5. Submit the peak mass list to database search programs such as MASCOT (http://www.matrixscience.com/search_form_select.html), ProteinProspector (http://prospector.ucsf.edu), or PROWL (http://prowl.rockefeller.edu). Select the correct search parameters including the enzyme employed (e.g., trypsin), fixed modifications (e.g., Cys-carboxymethylation), possible modification (e.g., Met-oxidation), and the mass tolerance. For internal calibration, mass tolerances of less than 50 ppm should be easily achievable with modern axial TOF mass spectrometers (less than 100 ppm for external calibration).

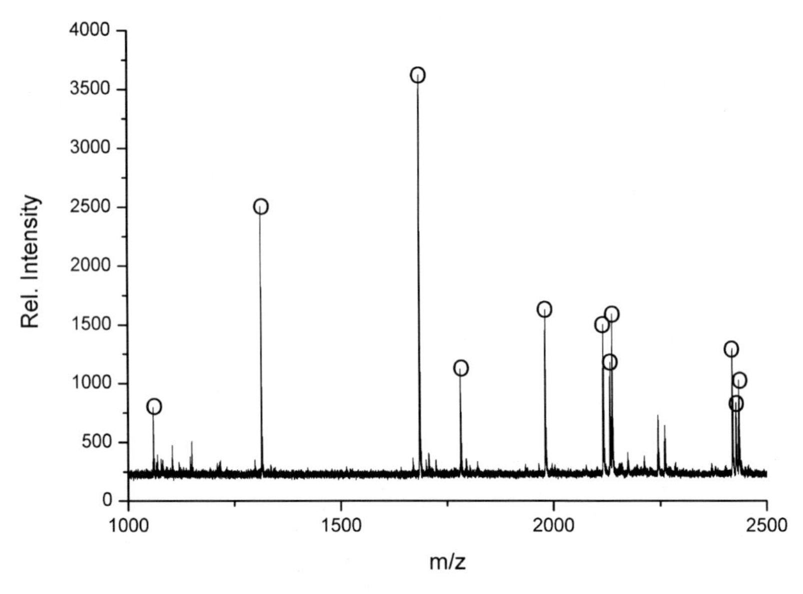


Fig. 2. Axial TOF MALDI mass spectrum of an in-gel digested protein separated by two-dimensional gel electrophoresis. The spectrum is internally calibrated with trypsin autolysis peptide ions. Peaks labeled with circles were automatically picked by the instrument manufacturer's software using default settings.

**Table 2**
**Peak mass list obtained from the mass spectrum shown in Fig. 2 using the MS instrument manufacturer's peak picking software with its default parameters**

| *m/z* |
|---|
| 1059.559 |
| 1312.592 |
| 1683.885 |
| 1780.741 |
| 1979.855 |
| 2115.101 |
| 2131.099 |
| 2135.975 |
| 2417.118 |
| 2426.143 |
| 2433.126 |

**Figure 3** shows the search parameter page (**a**) and results page (**b**) for searching the peak list of **Table 2**.

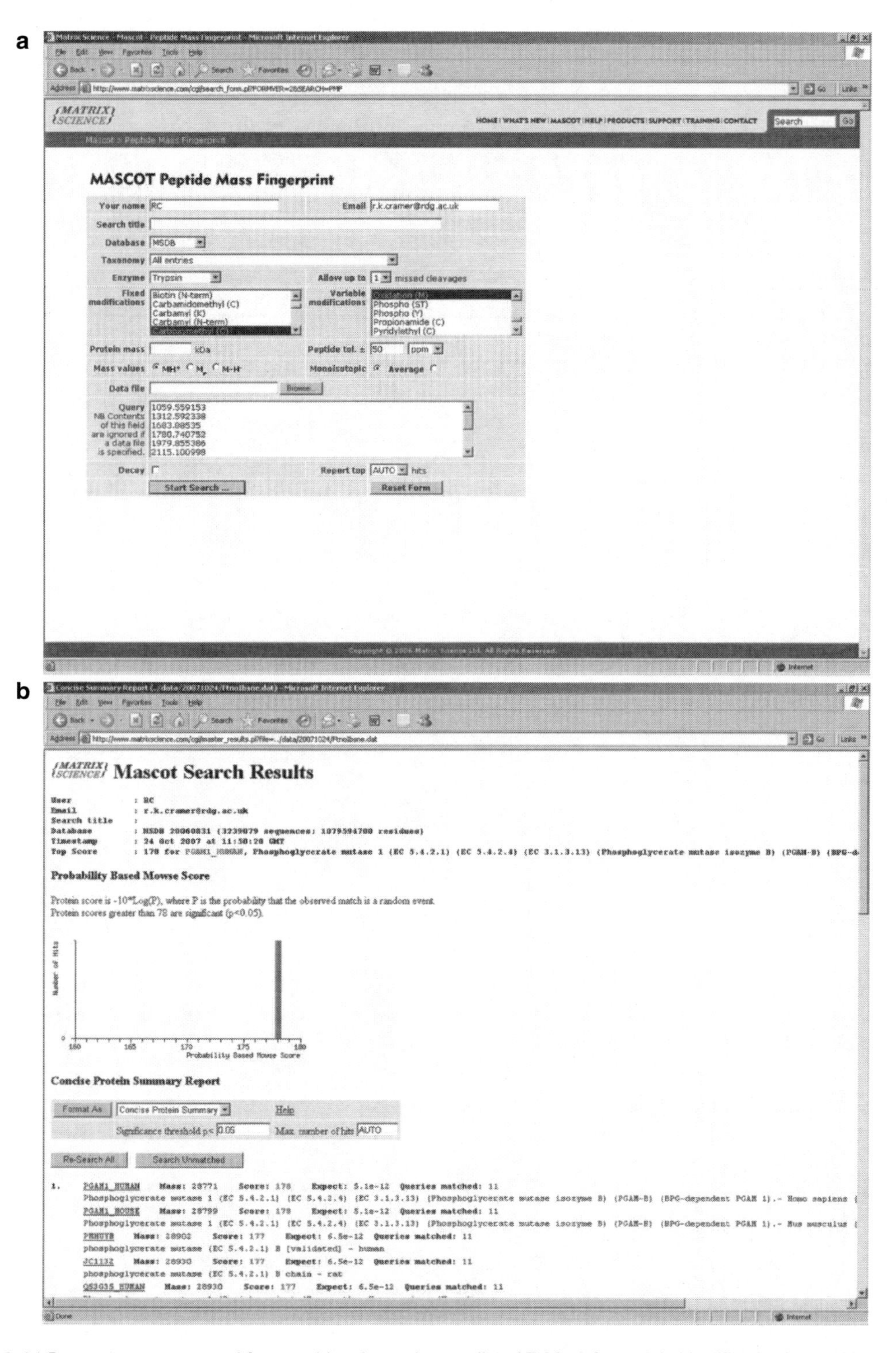


Fig. 3. (**a**) Parameter page as used for searching the peak mass list of **Table 2** for protein identification by peptide mass mapping employing MASCOT Peptide Mass Fingerprint and (**b**) search results page as obtained after searching the peak mass list of **Table 2** using the search engine and the parameters shown in (**a**). The top hit of human phosphoglycerate mutase 1 is highly significant but still ambiguous because of its high homology with its mouse homologue. The species the sample was derived from was in fact mouse demonstrating the need for further data evaluation.

6. As an optional final step, the peak mass list can be searched again using a decoy database to evaluate protein identification owing to pure chance (*see* **refs.** *24, 25* for more information on this topic). Further data mining such as the exclusion of protein isoforms or homologues and data searching such as second searches of the nonmatching ion masses are also recommended (*see* **Note 11**).

#### *3.3. Peptide Sequencing by Orthogonal MALDI Q-TOF-MS/MS Using CID*

1. Follow **steps 1–3** as described under **Subheading 3.2** (*see* **Note 12**). These steps will provide the acquisition of a mass spectrum of the components in the sample's mixture, often called MS survey scan. From this mass spectrum, precursor ions can be selected for subsequent MS/MS analysis, i.e., peptide sequencing in the case of proteomic analyses.
2. Once a precursor ion has been chosen from the MS survey scan switch to the MS/MS mode of the instrument and collect a fragment ion mass spectrum (*see* **Note 13**).
   **Figure 4** shows an example of an externally calibrated fragment ion mass spectrum of a peptide from blood serum that was purified according to **ref.** *26.*
3. Create a peak mass list that consists of the precursor ion and fragment ions (*see* **Note 14**). Most of the instrument manufacturers supply data processing software with peak picking routines that can be employed for this task. However, in many

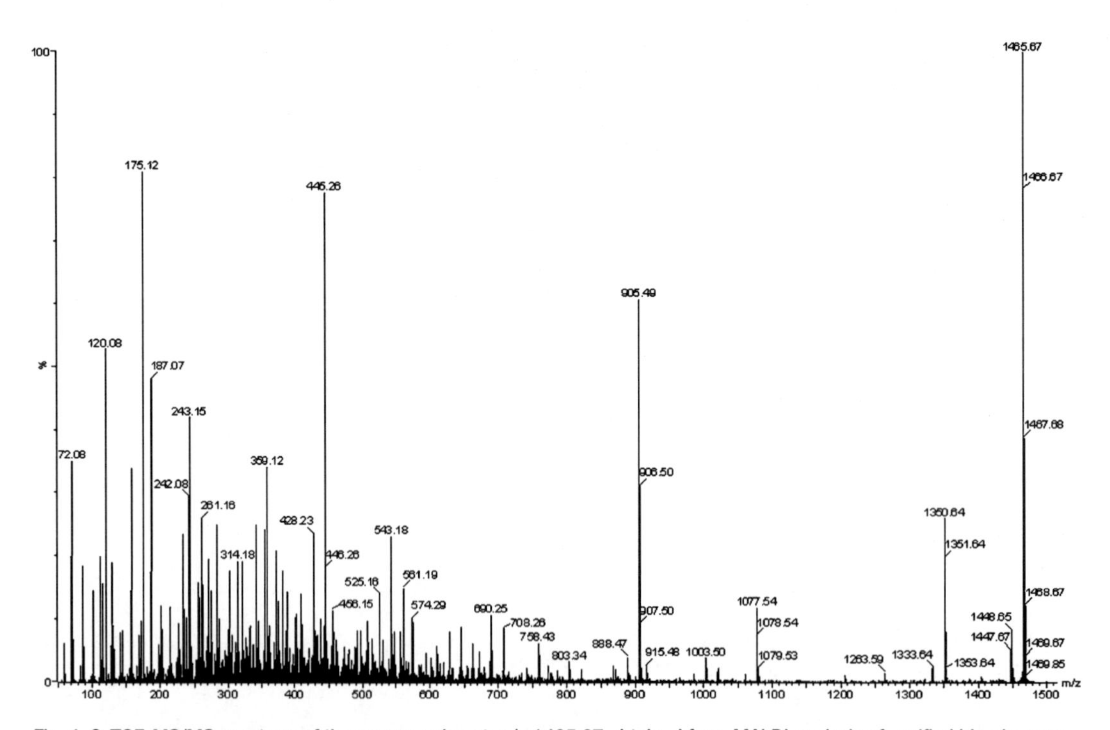


Fig. 4. Q-TOF-MS/MS spectrum of the precursor ion at *m/z* 1465.67 obtained from MALDI analysis of purified blood serum.

cases, this software has been written for ESI data acquisition and processing, typically based on automated LC-MS/MS with set time frames. For MALDI MS/MS, this software might be rather restrictive or complicated to use if one would like to acquire MS/MS spectra manually from various different samples to optimize data quality according to the acquisition needs for the various precursor ions.

Alternatively, after automated peak labeling, one can quickly hand-pick the precursor ion mass and 20 fragment ion masses manually by selecting the 10 most abundant fragment ions in each of the two halves of the mass range that are covered by the fragment ions, i.e., from 0 Da to the mass of the precursor ion. This selection usually works well with most database search engines and often provides a quicker way to undertake the data processing for database searching, particularly if only one or a few searches have to be performed (*see* **Note 15**).

**Table 3** shows the manually determined peak mass list from the fragment ion mass spectrum in **Fig. 4** according to **step 3**.

4. Submit the peak mass list to database search programs such as MASCOT (http://www.matrixscience.com/search_form_select.html) or ProteinProspector (http://prospector.ucsf.edu). Select the correct search parameters including the enzyme employed (e.g., "no enzyme" if no enzyme was used as in the example of **Fig. 4**), fixed modifications (e.g., Cys-carboxymethylation), possible modification (e.g., Met-oxidation), and the mass tolerance. Because of the high specificity of MS/MS data, mass tolerances are not as critical as in peptide mass mapping and can therefore be chosen far more loosely despite the fact that Q-TOF instruments typically perform better in accurate mass measurements than axial TOF instruments (*see* **Note 16**).

   **Figure 5** shows the search parameter page (**a**) and results page (**b**) for searching the peak mass list of **Table 3**.
5. As an optional final step, the peak mass list can be searched again using a decoy database to evaluate protein identification due to pure chance (*see* **refs.** *24, 25* for more information on this topic). Further data mining such as the exclusion of protein isoforms or homologues and a manual comparison of the experimental fragment ion data with a theoretical fragment ion mass spectrum are also recommended (*see* **Note 11**).

### *3.4. Automation and Batch Analysis*

The above-mentioned examples describe just a single MALDI MS and MS/MS measurement for protein identification. They have been chosen to provide an easy-to-follow introduction into the application of MALDI in proteomics. However, proteomics is characterized by large-scale protein analyses and, therefore, automation and batch analysis are important aspects in this field. There are a multitude of articles discussing these aspects showing

**Table 3**
**Manually determined peak mass list from the fragment ion mass spectrum in Fig. 4 (*see* step 3 in Subheading 3.2)**

| *m/z* |
|---|
| 1465.67 |
| 72.08 |
| 120.08 |
| 158.09 |
| 175.12 |
| 187.07 |
| 242.08 |
| 243.15 |
| 261.16 |
| 359.12 |
| 445.26 |
| 758.43 |
| 803.34 |
| 888.47 |
| 905.49 |
| 915.48 |
| 1003.50 |
| 1077.54 |
| 1350.64 |
| 1447.67 |
| 1448.65 |

NB: The first entry of the table is the mass value for the precursor ion.

the enormous variety of approaches and specific setups that highly depend on the exact experimental layout. It is therefore impossible to provide protocols for the general setup of batch analyses, particularly since automation and the associated software can be extremely specific.

In general, MS instrument manufacturers usually supply scripts or programs for adequate data preprocessing resulting in batch files that can then be submitted to automated database searching by at least one of the common search engines. Particularly, in the case

a

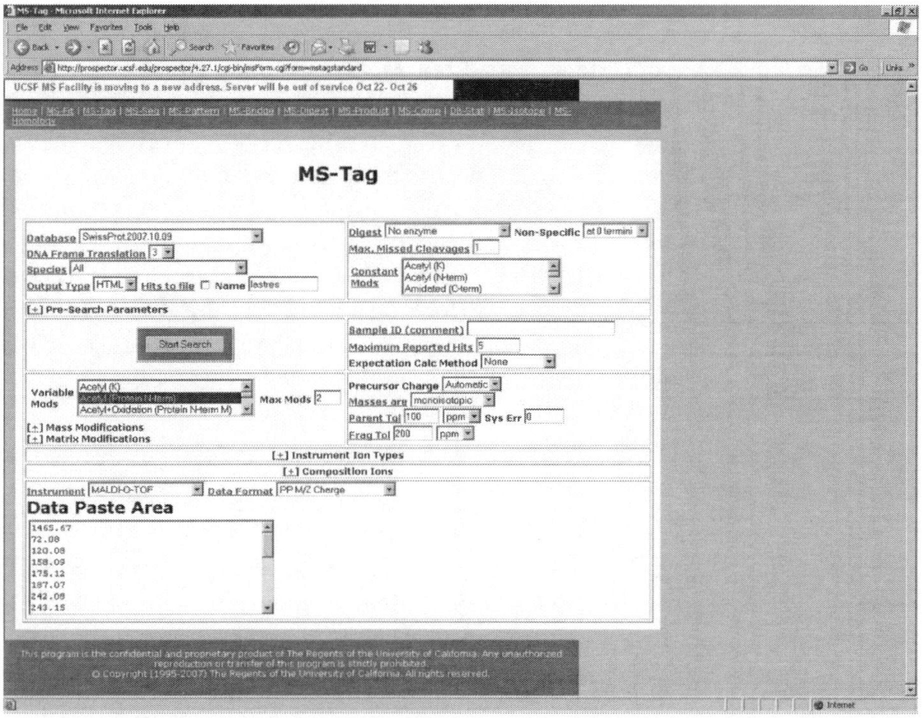

MS-Tag - Microsoft Internet Explorer

File Edit View Favorites Tools Help

Address http://prospector.ucsf.edu/prospector/4.27.1/cgi-bin/msform.cgi?form=mstagstandard

UCSF MS Facility is moving to a new address. Server will be out of service Oct 22- Oct 26

Home | MS-Fit | MS-Tag | MS-Seq | MS-Pattern | MS-Bridge | MS-Digest | MS-Product | MS-Comp | DB-Stat | MS-Isotope | MS-Homology

MS-Tag

Database SwissProt.2007.10.09
DNA Frame Translation 3
Species All
Output Type HTML Hits to file Name lestres

Digest No enzyme Non-Specific at 0 termini
Max. Missed Cleavages 1
Constant Mods Acetyl (K) Acetyl (N-term) Amidated (C-term)

[±] Pre-Search Parameters

Start Search

Sample ID (comment)
Maximum Reported Hits 5
Expectation Calc Method None

Variable Mods Acetyl (K) Acetyl (Protein N-term) Acetyl+Oxidation (Protein N-term M) Max Mods 2

[±] Mass Modifications
[±] Matrix Modifications

Precursor Charge Automatic
Masses are monoisotopic
Parent Tol 100 ppm Sys Err 0
Frag Tol 200 ppm

[±] Instrument Ion Types

[±] Composition Ions

Instrument MALDI-Q-TOF Data Format PP M/Z Charge

Data Paste Area

1465.67
72.08
120.08
158.09
175.12
187.07
242.08
243.15

This program is the confidential and proprietary product of The Regents of the University of California. Any unauthorized reproduction or transfer of this program is strictly prohibited.
© Copyright (1995-2007) The Regents of the University of California. All rights reserved.

Internet

b

MS-Tag Search Results - Microsoft Internet Explorer

File Edit View Favorites Tools Help

Address http://prospector.ucsf.edu/prospector/4.27.1/cgi-bin/mssearch.cgi

Home | MS-Fit | MS-Tag | MS-Seq | MS-Pattern | MS-Bridge | MS-Digest | MS-Product | MS-Comp | DB-Stat | MS-Isotope | MS-Homology

MS-Tag Search Results

Search completed. 1 min 6 sec elapsed. 0 sec remaining.

[±] Parameters

[±] Pre Search Results

Results

MS-Tag search selects 6 entries (results displayed for top 5 matches).

Parent mass: 1465.6700 (+/- 0.147 Da)
20 Ions used in search: 72.0800, 120.0800, 158.0900, 175.1200, 187.0700, 242.0800, 243.1500, 261.1600, 359.1200, 445.2600, 758.4300, 803.3400, 888.4700, 905.4900, 915.4800, 1003.5000, 1077.5400, 1350.6400, 1447.6700, 1448.6500 (+/- 200.00 ppm)

| Rank | # Unmatched Ions | Sequence | Score | m/z Submitted | MH+ Calculated (Da) | Error (ppm) | MS-Digest Index # | Protein MW (Da)/pI | Accession # | Species | Protein Name |
|---|---|---|---|---|---|---|---|---|---|---|---|
| 1 | 1 | (A)DSGEGDFLAEGGGVR(G) | 24.9 | 1465.6700 | 1465.6554 | 9.98 | 67756 | 94974/5.7 | P02671 | HUMAN | Fibrinogen alpha chain precursor [Contains: Fibrinopeptide A] |
| 2 | 4 | (R)DAFISNGVTFANLP(Q) | 19.4 | 1465.6700 | 1465.7322 | -42.4 | 88719 | 34707/5.4 | Q74711 | GEOSL | Porphobilinogen deaminase |
| 3 | 5 | (D)TKESDFLAEGGGVR(-) | 18.8 | 1465.6700 | 1465.7281 | -39.7 | 67754 | 1682/4.3 | Q7M3I6 | HALGR | Fibrinogen alpha chain [Contains: Fibrinopeptide A] |
| 4 | 9 | (S)DENGSVRFATLRT(Y) | 18.5 | 1465.6700 | 1465.7394 | -47.3 | 92485 | 44601/5.6 | Q60Y65 | CAEBR | 4-hydroxyphenylpyruvate dioxygenase |
| 5 | 6 | (D)DANFINSSSIKGNV(I) | 18.2 | 1465.6700 | 1465.7281 | -39.7 | 61876 | 28827/9.7 | Q9M8M5 | ARATH | Ethylene-responsive transcription factor ERF084 |

MS-Tag in ProteinProspector 4.27.1
© Copyright (1995-2007) The Regents of the University of California.

Internet

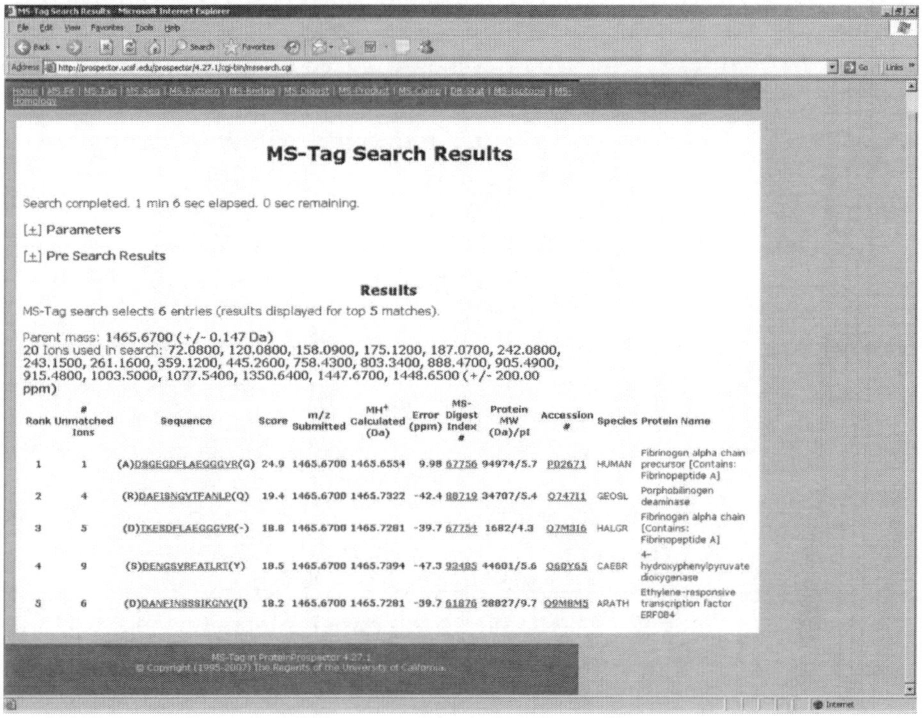

Fig. 5. (**a**) Parameter page as used for searching the peak mass list of **Table 3** for protein identification by peptide sequencing employing ProteinProspector MS-Tag and (**b**) search results page as obtained after searching the peak mass list of **Table 3** using the search engine and the parameters shown in (**a**). Although the ProteinProspector top hit and the corresponding MASCOT MS/MS ion search top hit using the same peak mass list (data not shown) are identical at the peptide level, correctly identifying the peptide's sequence and its occurrence in human fibrinogen $\alpha$ chain precursor, the low probability score in the MASCOT MS/MS ion search only resulted in significant homology at the peptide level, whereas no significant hit was obtained at the protein level. This clearly demonstrates the need for more than one peptide identification per protein using MS/MS.

of MS/MS of several different peptides, it is advisable to exploit batch analysis because the identification of several peptides originating from the same protein will significantly increase the probability of correct protein identification. As mentioned in **Note 11**, there is also a move to stricter requirements for the acceptance of MS/MS data and search results with respect to protein identification. As a consequence, protein identification just based on peptide mass mapping with low sequence coverage and mass accuracies or on a single MS/MS spectrum is becoming less acceptable. Thus, today's proteomic analyses are in most cases not based on single MS/MS analyses, but require expert knowledge in batch analysis and data mining, a subject that is unfortunately beyond the scope of this chapter.

## 4. Notes

1. Ideally, the last washing step uses an aqueous solution that can dissolve the remaining small traces of sodium and potassium salts, which are ubiquitous and therefore often not completely removed or even introduced by the earlier clean-up steps.
2. At a first glance, the preparation of a saturated solution seems to be quite straightforward. However, in the case of DHB, precipitated material is often not only found at the bottom of the solution, but also floating on its surface. Therefore, it is advisable to transfer the saturated solution without any precipitate into a new container and check that no precipitate is present on the solution's surface. This step will avoid any potential pick-up of precipitate in subsequent steps.
3. In most proteomic studies, only one MALDI sample per sample is analyzed. However, if the sample amount allows it, the analysis of two or more MALDI samples per sample, each prepared with a different matrix, often results in substantially more complementary data. For AnchorChip™ targets, it is recommended to use an ethanol:acetone (2:1) mixture as solvent for the CHCA matrix solution.
4. If the sample is somewhat contaminated (e.g., with buffer compounds or salts), a higher dilution factor might facilitate adequate MALDI sample crystallization and, therefore, satisfactory MALDI analysis, despite the lower analyte concentration. If the available volume of the analyte solution is relatively low, e.g., <1 μL, it can be advantageous to spot 1 μL of the matrix solution on the target and directly mix the small volume of the analyte solution with the matrix droplet on the

target, thus avoiding any losses through additional exposure to the mixing container (e.g., analyte adhesion to walls).

5. There is anecdotal evidence that sample preparations with CHCA as matrix in rooms with a relative humidity below 30% can result in MALDI samples of inferior quality, leading, in the extreme case, to no analyte ion signal *(26, 27)*.

6. There is still no consensus on the optimal solvent evaporation method. Many of the suggested methodologies in the literature are most likely optimized for some specific analytes/analyte solutions, but should be suitable for general applications. Therefore, depending on the particulars of the application, some other drying procedures might provide better results in some cases.

7. For the MALDI sample preparation of the standard peptide mixture, one should consider the following points. First, the standard peptide mixture should cover the expected mass range of the analyte peptide mixture for optimal calibration results. Second, to improve the calibration further, the standard peptide sample should be spotted close to the analyte peptide sample. Finally, the peptide concentration should be slightly above the peptide concentration of the analyte sample. This ensures that the concentrations of the analyte and standard sample are similar and, therefore, the adjustment of the laser energy that is optimal for the analysis will be typically in a similar range enabling the optimization of instrumental parameters with the peptide standard and keeping conditions the same between the calibration and the analyte measurements.

8. This step and some of the following steps depend highly on the instrument employed and, therefore, cannot be covered in detail. However, it is usually beneficial to deflect low molecular ions or lower the detector response for these ions, because many impurity and matrix signals appear in this mass range and could saturate the detector. Essentially, these ions are of no analytical value. On the contrary, their recording might negatively affect the quality of the peptide mass list that needs to be created subsequently (*see* **step 4**) by adding nonpeptidic ion signals to it. Typically, the mass range below $m/z$ 500–750 is excluded from the analysis. Nonetheless, for the initial optimization of the instrument, it is often useful to have a look at this low mass range to get a better idea of fundamental parameters that might influence the quality of analysis such as the applied laser energy and the degree of sample contamination.

9. As before, calibration standards, whether internal or external, should cover the mass range of interest. For protein digests, autolysis products of the enzyme employed are often detected and can be used for internal calibration.

10. A few checks on this list are often useful. In general, the list should contain the monoisotopic ion masses only. Obviously, the most abundant ion signals should be present, and ideally, all nonprotonated ion masses (e.g., Na- and K-adduct ion masses) are excluded. Furthermore, peak mass lists with many entries (typically more than 100) often lead to low probability scores and should therefore be avoided.
11. Besides comprehensively described and well-documented results and experimental protocols, some of these additional steps are often required for the publication of research articles by some proteomics journals *(28, 29)*.
12. In general, these initial steps can be identical. However, because of the differences of Q-TOF and axial TOF instruments, there are a few points that are noteworthy. The decoupling of the TOF analysis and the ionization event results in extremely robust external calibrations, practically, eliminating the need for internal calibration and the close positioning of the external calibrant to the analyte sample. It also allows the use of MS/MS calibrations for MS measurements and vice versa, as long as the mass ranges are adequately covered by the calibrant ion signals. Furthermore, it allows the application of much higher laser energy, i.e., well above the energy threshold for ion production, without obtaining the detrimental effects on mass accuracy and resolution as observed in axial TOF mass spectrometers. Nonetheless, one should aim to optimize the laser energy with regard to both maximal analyte ion production and minimal low-mass, nonspecific ion production.
13. This step and some of the following steps depend highly on the instrument employed and, therefore, cannot be covered in detail. In general, the manufacturer's recommendations should be followed. Most instruments have default routines and parameters for the acquisition of MS/MS fragment ion spectra. However, it is noteworthy that the typical ionization technique on these instruments is ESI and not MALDI. Thus, one should check that for the precursor ion selection, the different isotope distributions of singly charged MALDI ions in comparison with the typically doubly (or triply) charged ESI ions at the same $m/z$ value is taken into account. Further, collision energies also need to be adjusted for the singly charged (MALDI) precursor ions.
14. In general, the list should contain the monoisotopic ion masses only.
15. Most search engines require mass list files of a specific format. One format that is commonly accepted and easy to generate is the .pkl format. Files using this format can include several MS/MS mass lists (separated by at least one blank line), ena-

bling batch searching (*see* also **Subheading 3.4**). The individual MS/MS mass lists consist of a first line with a triplet of space-separated values for the measured *m/z*, intensity, and charge state value, respectively, and subsequent lines with a pair of space-separated values for the measured *m/z* and intensity of the fragment ions. Since many search engines make only little use of intensity values, these values can often be set arbitrarily, saving time if intensity values are required but need to be individually typed in.

16. Mass tolerance settings of several hundred parts per million are perfectly acceptable and sometimes advisable. Mass tolerances that have been chosen too strictly or too loosely and their various limitations, from the achievable scoring to search speed, are discussed by some software manufacturers of MS data search software (e.g., http://www.matrixscience.com/pdf/2006WKSHP3.pdf).

## Acknowledgments

The author thanks Dr. Soren Naaby-Hansen for providing the 2DE gel sample, Dr. Ali Tiss for acquiring the data of the human serum sample, and Dr. Eckhard Nordhoff for critical reading of the manuscript. This work was partially supported by the MRC through grant G0301107.

### References

1. Fenn, J. B., Mann, M., Meng, C. K., Wong, S. F., and Whitehouse, C. M. (1989) Electrospray ionization for mass spectrometry of large biomolecules. *Science* **246**, 64–71.
2. Bruins, A. P., and Cook, K. D. (2007) in The Encyclopedia of Mass Spectrometry – Molecular Ionization (Gross, M. L., and Caprioli, R. M., Eds.), Vol. 6, pp. 415–21, Elsevier Ltd, Oxford, UK.
3. Karas, M. and Hillenkamp, F. (1988) Laser desorption ionization of proteins with molecular masses exceeding 10,000 daltons. *Anal Chem* **60**, 2299–301.
4. Cramer, R. and Dreisewerd, K. (2007) in The Encyclopedia of Mass Spectrometry – Molecular Ionization (Gross, M. L., and Caprioli, R. M., Eds.), Vol. 6, pp. 646–61, Elsevier Ltd, Oxford, UK.
5. Hillenkamp, F. and Peter-Katalinic, J. (2007) MALDI-MS, Wiley VCH, Weinheim, Germany.
6. Mann, M., Hendrickson, R. C., and Pandey, A. (2001) Analysis of proteins and proteomes by mass spectrometry. *Annu Rev Biochem* **70**, 437–73.
7. Aebersold, R., and Mann, M. (2003) Mass spectrometry-based proteomics. *Nature* **422**, 198–207.
8. Corr, J. J., Kovarik, P., Schneider, B. B., Hendrikse, J., Loboda, A., and Covey, T. R. (2006) Design considerations for high speed quantitative mass spectrometry with MALDI ionization. *J Am Soc Mass Spectrom* **17**, 1129–41.
9. Mirgorodskaya, E., Braeuer, C., Fucini, P., Lehrach, H., and Gobom, J. (2005) Nanoflow liquid chromatography coupled to matrix-assisted laser desorption/ionization mass spectrometry: sample preparation, data analysis, and application to the analysis of complex peptide mixtures. *Proteomics* **5**, 399–408.

10. Zhang, N., Li, N., and Li, L. (2004) Liquid chromatography MALDI MS/MS for membrane proteome analysis. *J Proteome Res* **3**, 719–27.
11. Brancia, F. L., Bereszczak, J. Z., Lapolla, A., Fedele, D., Baccarin, L., Seraglia, R., and Traldi, P. (2006) Comprehensive analysis of glycated human serum albumin tryptic peptides by off-line liquid chromatography followed by MALDI analysis on a time-of-flight/curved field reflectron tandem mass spectrometer. *J Mass Spectrom* **41**, 1179–85.
12. Caprioli, R. M., Farmer, T. B., and Gile, J. (1997) Molecular imaging of biological samples: localization of peptides and proteins using MALDI-TOF MS. *Anal Chem* **69**, 4751–60.
13. Chaurand, P., Schwartz, S. A., and Caprioli, R. M. (2004) Assessing protein patterns in disease using imaging mass spectrometry. *J Proteome Res* **3**, 245–52.
14. McDonnell, L. A. and Heeren, R. M. (2007) Imaging mass spectrometry. *Mass Spectrom Rev* **26**, 606–43.
15. Luxembourg, S. L., Mize, T. H., McDonnell, L. A., and Heeren, R. M. (2004) High-spatial resolution mass spectrometric imaging of peptide and protein distributions on a surface. *Anal Chem* **76**, 5339–44.
16. Dreisewerd, K. (2003) The desorption process in MALDI. *Chem Rev* **103**, 395–426.
17. Karas, M. and Kruger, R. (2003) Ion formation in MALDI: the cluster ionization mechanism. *Chem Rev* **103**, 427–40.
18. Knochenmuss, R. and Zenobi, R. (2003) MALDI ionization: the role of in-plume processes. *Chem Rev* **103**, 441–52.
19. Blum, H., Beier, H., and Gross, H. J. (1987) Improved silver staining of PLANT-proteins, Rna and DNA in polyacrylamide gels. *Electrophoresis* **8**, 93–9.
20. Schuerenberg, M., Luebbert, C., Eickhoff, H., Kalkum, M., Lehrach, H., and Nordhoff, E. (2000) Prestructured MALDI-MS sample supports. *Anal Chem* **72**, 3436–42.
21. Smirnov, I. P., Zhu, X., Taylor, T., Huang, Y., Ross, P., Papayanopoulos, I. A., Martin, S. A., and Pappin, D. J. (2004) Suppression of alpha-cyano-4-hydroxycinnamic acid matrix clusters and reduction of chemical noise in MALDI-TOF mass spectrometry. *Anal Chem* **76**, 2958–65.
22. Cramer, R. and Corless, S. (2005) Liquid ultraviolet matrix-assisted laser desorption/ionization – mass spectrometry for automated proteomic analysis. *Proteomics* 5, 360–70.
23. Kjellstrom, S. and Jensen, O. N. (2004) Phosphoric acid as a matrix additive for MALDI MS analysis of phosphopeptides and phosphoproteins. *Anal Chem* **76**, 5109–17.
24. Balgley, B. M., Laudeman, T., Yang, L., Song, T., and Lee, C. S. (2007) Comparative evaluation of tandem MS search algorithms using a target-decoy search strategy. *Mol Cell Proteomics* **6**, 1599–608.
25. Peng, J., Elias, J. E., Thoreen, C. C., Licklider, L. J., and Gygi, S. P. (2003) Evaluation of multidimensional chromatography coupled with tandem mass spectrometry (LC/LC-MS/MS) for large-scale protein analysis: the yeast proteome. *J Proteome Res* **2**, 43–50.
26. Tiss, A., Smith, C., Camuzeaux, S., Kabir, M., Gayther, S., Menon, U., Waterfield, M., Timms, J., Jacobs, I., and Cramer, R. (2007) Serum Peptide Profiling using MALDI mass spectrometry. *Proteomics* 7 (S1), 77–89.
27. West-Norager, M., Kelstrup, C. D., Schou, C., Hogdall, E. V., Hogdall, C. K., and Heegaard, N. H. (2007) Unravelling in vitro variables of major importance for the outcome of mass spectrometry-based serum proteomics. *J Chromatogr B Analyt Technol Biomed Life Sci* **847**, 30–7.
28. Carr, S., Aebersold, R., Baldwin, M., Burlingame, A., Clauser, K., and Nesvizhskii, A. (2004) The need for guidelines in publication of peptide and protein identification data: Working Group on Publication Guidelines for Peptide and Protein Identification Data. *Mol Cell Proteomics* **3**, 531–3.
29. Wilkins, M. R., Appel, R. D., Van Eyk, J. E., Chung, M. C., Gorg, A., Hecker, M., Huber, L. A., Langen, H., Link, A. J., Paik, Y. K., Patterson, S. D., Pennington, S. R., Rabilloud, T., Simpson, R. J., Weiss, W., and Dunn, M. J. (2006) Guidelines for the next 10 years of proteomics. *Proteomics* **6**, 4–8.
30. Cramer, R., Richter, W. J., Stimson, E., and Burlingame, A. L. (1998) Analysis of phospho- and glycopolypeptides with infrared matrix-assisted laser desorption and ionization. *Anal Chem* **70**, 4939–44.
31. Gobom, J., Schuerenberg, M., Mueller, M., Theiss, D., Lehrach, H., and Nordhoff, E. (2001) Alpha-cyano-4-hydroxycinnamic acid affinity sample preparation. A protocol for MALDI-MS peptide analysis in proteomics. *Anal Chem* **73**, 434–8.
32. Chaurand, P., Luetzenkirchen, F., and Spengler, B. (1999) Peptide and protein identification by matrix-assisted laser desorption ionization (MALDI) and MALDI-post-source decay time-of-flight mass spectrometry. *J Am Soc Mass Spectrom* **10**, 91–103.

33. Zehl, M. and Allmaier, G. (2004) Ultraviolet matrix-assisted laser desorption/ionization time-of-flight mass spectrometry of intact hemoglobin complex from whole human blood. *Rapid Commun Mass Spectrom* **18**, 1932–8.
34. Gorman, J. J., Ferguson, B. L., and Nguyen, T. B. (1996) Use of 2,6-dihydroxyacetophenone for analysis of fragile peptides, disulphide bonding and small proteins by matrix-assisted laser desorption/ionization. *Rapid Commun Mass Spectrom* **10**, 529–36.
35. Fenaille, F., Le Mignon, M., Groseil, C., Siret, L., and Bihoreau, N. (2007) Combined use of 2,4,6-trihydroxyacetophenone as matrix and enzymatic deglycosylation in organic-aqueous solvent systems for the simultaneous characterization of complex glycoproteins and N-glycans by matrix-assisted laser desorption/ionization time-of-flight mass spectrometry. *Rapid Commun Mass Spectrom* **21**, 812–6.
36. Lowenstein, E. J., Daly, R. J., Batzer, A. G., Li, W., Margolis, B., Lammers, R., Ullrich, A., Skolnik, E. Y., Bar-Sagi, D., and Schlessinger, J. (1992) The SH2 and SH3 domain-containing protein GRB2 links receptor tyrosine kinases to ras signaling. *Cell* **70**, 431–42.
37. Lei, Z., Anand, A., Mysore, K. S., and Sumner, L. W. (2007) Electroelution of intact proteins from SDS-PAGE gels and their subsequent MALDI-TOF MS analysis. *Methods Mol Biol* **355**, 353–63.
38. Loo, R. R. and Loo, J. A. (2007) Matrix-assisted laser desorption/ionization-mass spectrometry of hydrophobic proteins in mixtures using formic acid, perfluorooctanoic acid, and sorbitol. *Anal Chem* **79**, 1115–25.
39. Signor, L., Varesio, E., Staack, R. F., Starke, V., Richter, W. F., and Hopfgartner, G. (2007) Analysis of erlotinib and its metabolites in rat tissue sections by MALDI quadrupole time-of-flight mass spectrometry. *J Mass Spectrom* **42**, 900–9.
40. Cramer, R., Hillenkamp, F., and Haglund, Jr., R.F. (1996) Infrared matrix-assisted laser desorption and ionization by using a tunable mid-infrared free-electron laser. *J Am Soc Mass Spectrom* 7, 1187–93.
41. Overberg, A., Karas, M., Bahr, U., Kaufmann, R., and Hillenkamp, F. (1990) Matrix-assisted infrared-laser (2.94 µm) desorption/ionization mass spectrometry of large biomolecules. *Rapid Commun Mass Spectrom* **4**, 293–96.

# Chapter 6

# Capillary Electrophoresis Coupled to Mass Spectrometry for Proteomic Profiling of Human Urine and Biomarker Discovery

**Petra Zürbig, Eric Schiffer, and Harald Mischak**

## Summary

Currently, the main focus of clinical proteome analysis is on detection and identification of polypeptides that significantly change owing to pathological changes. Capillary electrophoresis coupled online to an electrospray ionization time of flight mass spectrometer (CE-MS) allows the differential display of a large number of polypeptides in a single, reproducible, and time-limited step and enables the comparison of different protein profiles for biomarker discovery. In addition to the reproducibility of the CE-MS setup, many further steps including data processing and mining, usage of biomarkers for diagnosis, and biomarker sequencing are necessary to answer the demands of biomarker discovery of clinical significance. In this chapter, we discuss materials and methods for CE-MS-based clinical proteomics allowing the reproducible profiling of urine.

**Key words:** Capillary electrophoresis, Mass spectrometry, Urine, Polypeptides, Biomarker discovery

## 1. Introduction

In the postgenomics era, literature has been flooded with manuscripts dealing with clinical proteomics. Unfortunately, many of these reports are of questionable value and cannot be interpreted adequately or reproduced for a number of reasons. These include the lack of comparability, use of sample sizes of sufficient statistical power, sample processing, other preanalytical aspects, appropriate quality control, statistical evaluation, and independent validation, and in some cases all of the above. Ideally, proteomic analysis

Jörg Reinders and Albert Sickmann (eds.), *Proteomics,* Methods in Molecular Biology, Vol. 564
DOI: 10.1007/978-1-60761-157-8_6, © Humana Press, a part of Springer Science+Business Media, LLC 2009

should be accomplished with high comparability and resolution, enabling profiling of an adequate number of polypeptides of a sufficient number of samples. A major goal for clinical application is the differential display of a large number of polypeptides in a single, reproducible, and time-limited step enabling comparison of different protein profiles.

Capillary electrophoresis coupled to mass spectrometry (CE-MS) is a powerful alternative to commonly used proteomic technologies *(1–4)*. The approach was successfully applied to answer the demands of biomarker discovery of clinical significance *(5–9)*. The high reproducibility of CE-MS is achieved by the careful selection of the analyzed body fluid and by profiling all samples under highly standardized preparation/analysis conditions and stringent quality control. However, beyond the reproducibility of a proteomic setup, many further steps are necessary to successfully process and analyze the obtained data and to validate the conclusions made.

In this chapter, we give an outline of methods making CE-MS a powerful platform for clinical proteomics. We discuss in detail system configurations, data processing, data mining, usage of biomarkers for diagnosis, and biomarker sequencing without omitting the possible pitfalls and possibilities to avoid them.

## 2. Materials

### 2.1. Sample Preparation

1. Urine samples, immediately frozen and stored at –20°C.
2. Buffer for sample dilution: 2 M urea, 100 mM NaCl, 0.0125% $NH_3$, 0.01% SDS, pH 10–12.
3. Amicon ultracentrifuge filter device (30 kDa; Millipore, Bedford, USA).
4. PD-10 column (GE-Healthcare Biosciences AB, Uppsala, Sweden).
5. Elution buffer: Ammonia buffer: 0.04% $NH_3$,pH 10.5–11.5.
6. Lyophilization Christ Speed-Vac RVC 2–18/Alpha 1–2 (Christ, Osterode am Harz, Germany).
7. BC Assay: protein quantitation kit (Uptima, Interchim, Montlucon, France).

### 2.2. CE-MS Analysis

1. CE P/ACE MDQ system (Beckman Coulter, Fullerton, USA).
2. Capillary: 90 cm, 50 μm I.D. fused silica capillary.
3. Capillary equilibration: 5 mol/L NaOH (Meck KGaA, Darmstadt, Germany).

**Table 1**
**Composition of the standard solution for CE-MS calibration**

| Protein/peptide | Solvent | Molecular weight (Da) |
|---|---|---|
| Lysozyme | $H_2O$ | 14,303 |
| Ribonuclease | $H_2O$ | 13,681 |
| Aprotinin | $H_2O$ | 6,513 |
| REVQSKIGYGRQIIS | DMSO | 1,733 |
| TGSLPYSHIGSRDQIIFMVGR | DMSO | 2,333 |
| ELMTGELPYSHINNRDQIIFMVGR | DMSO | 2,832 |
| GIVLYELMTGELPYSHIN | DMSO | 2,048 |

4. Rinse between runs: 0.1 M NaOH (Meck KGaA, Darmstadt, Germany).
5. CE-Running buffer: 20% acetonitrile (ACN) in 1 M formic acid (98–100%).
6. Sheath liquid 30% (v/v) *iso*-propanol (Sigma-Aldrich, Germany) and 0.4% (v/v) formic acid in HPLC-grade water (flow rate: 2 μL/min).
7. ESI-TOF (electrospray ionization–time of flight) sprayer-kit (Agilent technologies, Palo Alto, CA, USA).
8. Micro-TOF-MS (ESI-TOF; Bruker Daltonik, Bremen, Germany).
9. Standard protein/peptide solution (0.5 pmol/μL) for CE-MS analysis: Lysozyme (Sigma-Aldrich L7651, Germany), Ribonuclease (Sigma-Aldrich, R4875, Germany), Aprotinin (Sigma-Aldrich, A1153, Germany), and four synthetic peptides: GIVLYELMTGELPYSHIN, ELMTGELPYSHINNRDQIIFMVGR, TGSLPYSHIGSRDQIIFMVGR, REVQSKIGYGRQIIS (*see* **Table 1**).

### *2.3. Data Processing and Statistical Analysis*

1. Software for peak detection: MosaiquesVisu: (biomosaiques software, Hanover, Germany).
2. Commercial statistical package: SAS (www.sas.com) and MedCalc (www.medcalc.be).
3. Multitest R-package by Dudoit et al. (*see* e.g., **ref.** ***10*** and references therein) available at www.bioconductor.org.
4. SVM-based program for multiparametric data classification: MosaCluster (biomosaiques software, Hanover, Germany).

## 3. Methods

The methods described here describe *(1)* urine sample preparation, *(2)* CE-MS analysis, *(3)* data processing, *(4)* statistical data mining for feature selection, *(5)* multiparametric modeling and data classification, and *(6)* sequencing of potential biomarkers.

### 3.1. Sample Preparation

Body fluids are highly complex mixtures of proteins and peptides with a wide range of polarity, hydrophobicity, and size. When analyzing such complex samples (e.g., urine), major concerns are the irreproducible loss of analytes (here polypeptides) during sample preparation, consequently compromising reproducibility of the generated data. Ideally, a crude unprocessed sample should be analyzed. This would avoid all artifacts, losses, or biases arising from sample preparation. To date, the presence of large molecules, e.g., albumin, immunoglobulin, and others, hamper this direct approach, because these molecules bare little information in the clinical setting but interfere with the detection of smaller, less abundant proteins and peptides. Thus, a preparation of the samples is advisable, but should be limited to a few steps.

#### 3.1.1. Collection and Storage of Samples

Urine samples were obtained cross-sectional from patients with a defined disease and healthy individuals or longitudinal from patients at different time points of the disease. The samples (at least 1 mL) are typically taken as spontaneous midstream urine from the second urine of the day. Proper handling of the samples is of outmost importance; the urine should be aliquoted, immediately frozen after collection, and should be stored at –20°C until analysis (*see* **Note 1**).

#### 3.1.2. Sample Preparation

Shortly before analysis, a 0.7-mL aliquot is thawed and diluted with 0.7 mL 4 M urea, 20 mM $NH_4OH$ containing 0.2% SDS. This is followed by an ultrafiltration step using Amicon ultracentrifugal filter device (30 kDa; Millipore, Bedford, USA). Samples are spun at 3,000 × *g* until 1.5 mL of filtrate has passed through the filter, thus removing high-molecular weight proteins such as albumin, transferrin, and others (*see* **Note 2**). Subsequently, the filtrate is desalted using PD-10 column, which was equilibrated with 25 mL of ammonia buffer, to remove urea, SDS, salts, and urinary matrix molecules. Polypeptides are eluted with 2 mL ammonia buffer, divided into three aliquots (A, B 900 μL each, and C 200 μL), and lyophilized overnight. Total protein content is determined using the BC assay and aliquot C resuspended in 50 μL of HPLC-grade water. Aliquot A is resuspended in 10 μL of HPLC-grade water shortly before CE-MS analysis. Aliquot B is stored as reserve sample at –20°C.

### 3.2. CE-MS Analysis of Urine

#### 3.2.1. Capillary Electrophoresis

Samples are transferred to appropriate vials and stored in the CE-auto sampler at 5°C. For CE a P/ACE MDQ (Beckman Coulter, Fullerton, USA) system is used equipped with a 90-cm, 50-μm inner diameter, and bare-fused silica capillary. The use of coated capillaries appears not to be beneficial (*see* **Note 3**). The capillary is first rinsed with running buffer (20% ACN, 0.5% formic acid in HPLC-grade water) for 3 min. The sample is injected for 99 s with 1–4 psi, resulting in the injection of 130–525 nL sample volume. Separation is performed with + 30 kV at the injection side and the capillary temperature is set to 35°C for the entire length of the capillary up to the ESI interface. After each run, the capillary is rinsed for 5 min with 0.1 M NaOH, to remove unspecifically adsorbed material. This is followed by 5 water flushing and re-equilibration with running buffer. The CE-MS setup is depicted in **Fig. 1**.

#### 3.2.2. CE-MS-Interface and Analysis

The MS analysis is performed in positive electrospray mode with an ESI-TOF sprayer-kit (Agilent Technologies, Palo Alto, CA, USA) using a Micro-TOF-MS (Bruker Daltonik, Bremen, Germany). The ESI sprayer is grounded, and the ionspray interface potential is set between –3,700 and –4,100 V. Sheath liquid is applied coaxially, consisting of 30% (v/v) *iso*-propanol and 0.4% (v/v) formic acid in HPLC-grade water, at a flow rate of 1–2 μL/min. These conditions result in a detection limit of about 1 fmol of different standard proteins/peptides *(11)*. MS-spectra are accumulated every 3 s, over a mass-to-charge range from 400 to 2,500 or 3,000 *m/z* for about 60 min.

### 3.3. Data Processing and Statistical Analysis

Mass per charge values of MS peaks are deconvoluted into mass and combined, if they represent identical molecules at different charge states using MosaiquesVisu software *(12)*. MosaiquesVisu (www.proteomiques.com) employs a probabilistic clustering algorithm and uses both isotopic distribution and conjugated masses for charge-state determination of polypeptides.

#### 3.3.1. Data Processing: Peak Annotation

In a first step, MosaiquesVisu identifies all peaks within each single spectrum. This usually results in more than 1,00,000 peaks from a single sample. Only signals observed in a minimum of three consecutive spectra with a minimum signal-to-noise ratio (usually of >4) are considered. Single charged peaks are discarded. This reduces the list to 3,000–7,000 so-called "CE/MS" peaks (**Fig. 2a**). Mass spectral ion peaks representing identical molecules at different charge states are deconvoluted into single masses using either the distance between resolved isotope peaks of the ion or according to conjugated signals for unresolved isotope peaks. The resulting list of approximately 1,000 to 3,000 proteins can be converted into a three-dimensional plot, *m/z* on

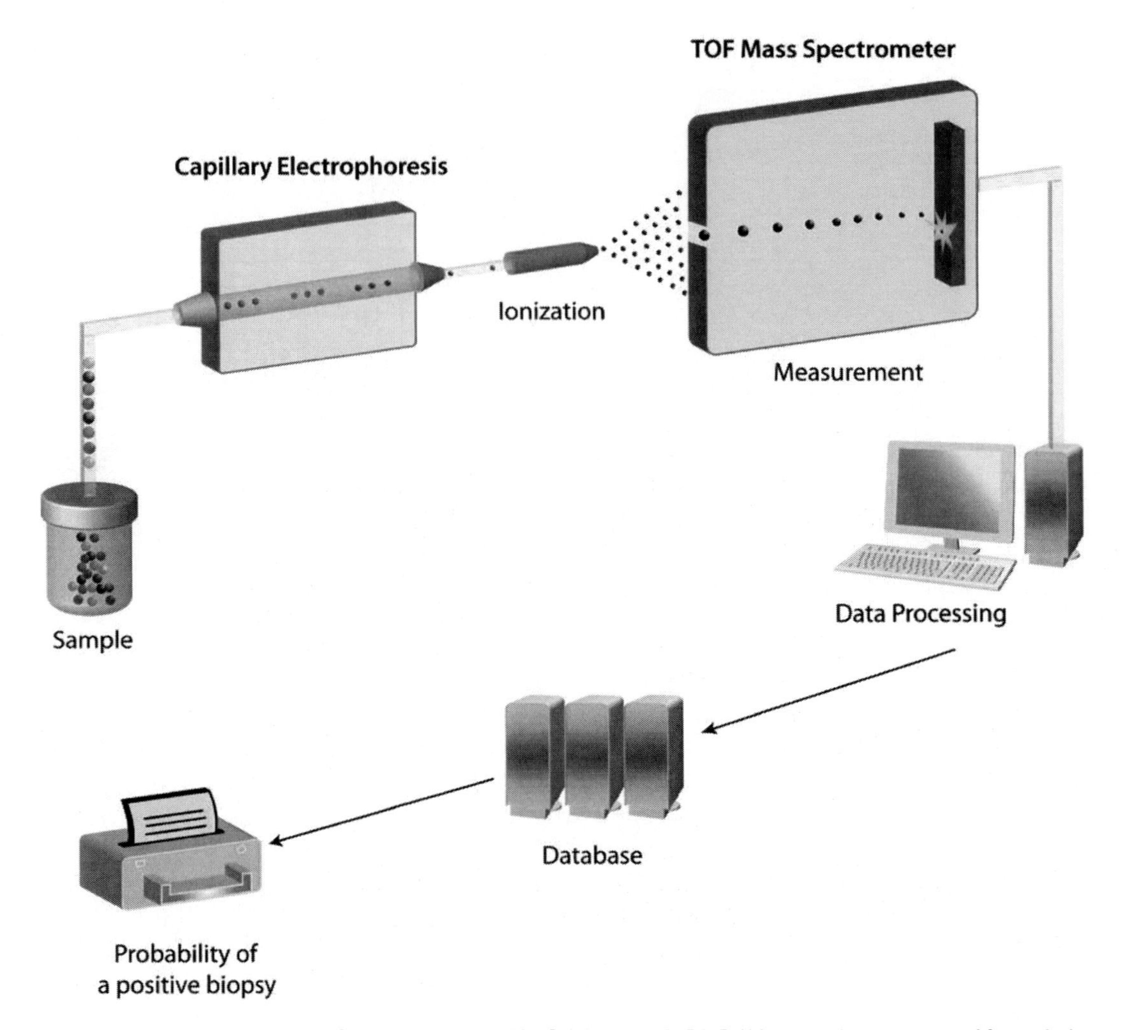


Fig. 1. Schematic drawing of CE-MS protocol (published by Sniehotta et al. *[51]*). Urine samples are prepared for analysis, and polypeptides are separated by CE and directly sprayed into ESITOF-MS. Data are evaluated using specific software solutions. Each polypeptide is defined by its accurate mass and normalized CE migration time. Signal intensity serves as a measure of the relative abundance. The data are stored as peak lists summarizing the information in a database.

the *y*-axis plotted against the migration time on the *x*-axis and the signal intensity in *z*-axis (**Fig. 2b**).

*3.3.2. Data Processing: Calibration*

To obtain high mass accuracy, TOF-MS derived masses are calibrated using 80 precisely (mass deviations <0.5 ppm) FT-ICR characterized reference masses. High FT-ICR MS resolution enables an accurate analysis of the first isotope signal ($z > 6$), which is crucial for the exact mass determination high-molecular-weight peptides. Both migration time and ion-counts as a measure of relative abundance show high variability, mostly due to different amounts of salt and peptides in the sample. Consequently, CE-migration time and ion signal intensity are normalized based on

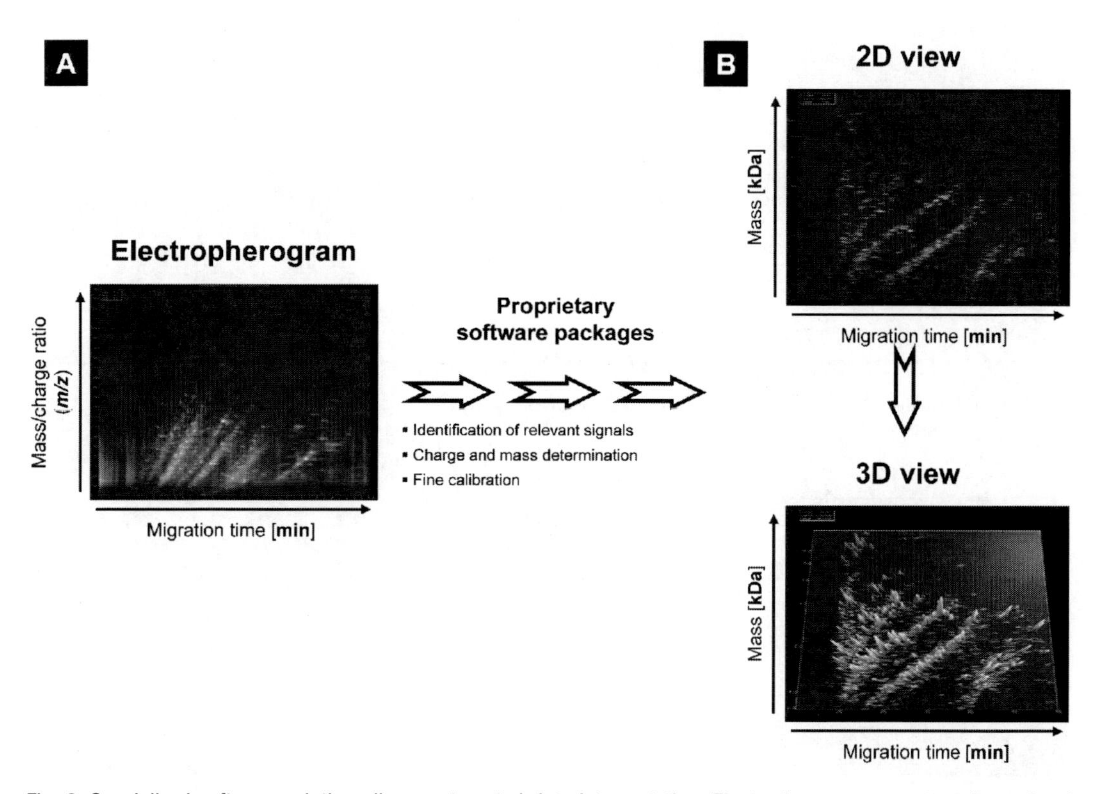


Fig. 2. Specialized software solution allows automated data interpretation. Electropherogram as a two-dimensional contour plot, mass per charge on the *y*-axis against the migration time in minutes (*x*-axis), signal intensity is color-coded. Signals that fit the criteria of "real polypeptide" signals are considered. Charge is assigned to each of these signals, conjugated peaks are combined, and as a result a spectrum based on mass and normalized migration time is obtained. The software enables the compilation of individual data sets to disease specific polypeptide patterns and the depiction of spectra in a 2-D and 3-D view.

reference signals by 200 abundant "housekeeping" peptides generally present in urine, which serve as internal standards ***(13, 14)***. These "internal standards" are present in at least 90% of all urine samples analyzed so far with a relative standard deviation less than 100%. For calibration, a weighted local regression algorithm is performed. The resulting peak list characterizes each protein and peptide by its exact molecular mass (mass deviation 3 ± 9 ppm), calibrated CE-migration time, and normalized signal intensity. All detected peptides are deposited, matched, and annotated in a Microsoft SQL database, enabling digital data compilation (*see* **Fig. 3**). Proteins and peptides within different samples are considered identical, if the absolute mass deviation was linearly increased from 50 ppm (0.8 kDa) to 100 ppm (20 kDa). The CE migration time deviation was linearly increased over the entire electropherogram from 2% to 5%. These clustering parameters show minimal error rates and consider increased peak widths at higher migration times. To eliminate peptides of apparently low significance,

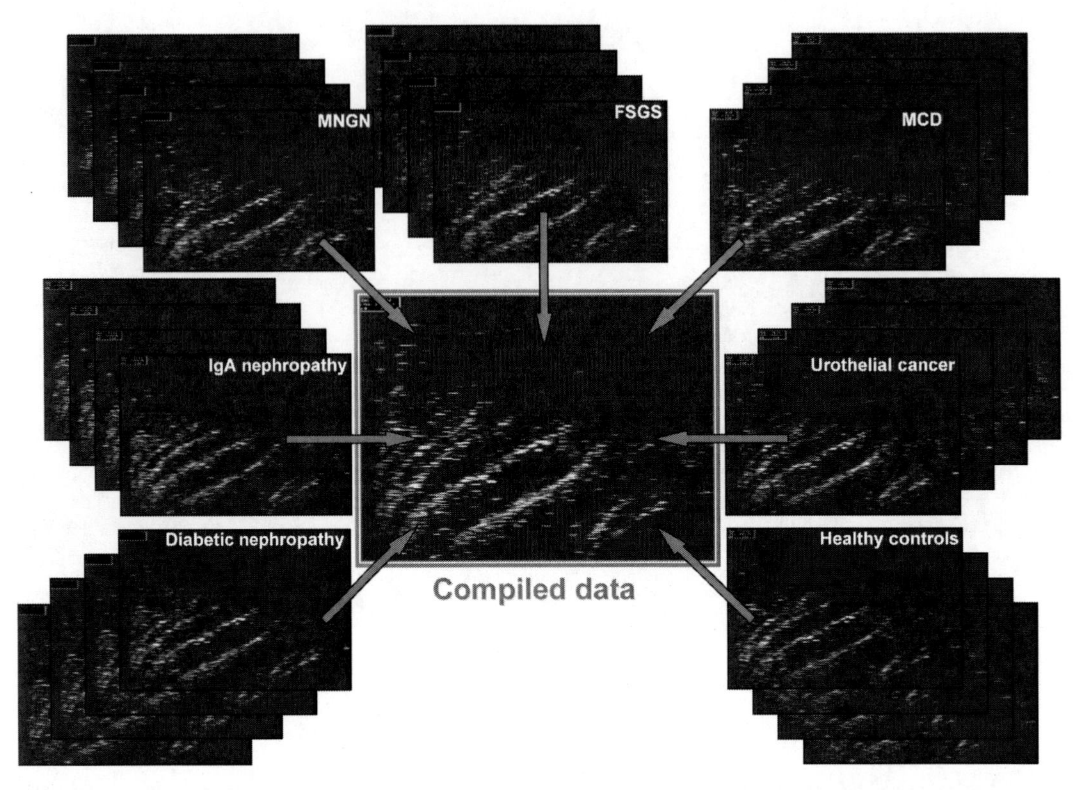


Fig. 3. Digital data compilation of individual data sets from healthy volunteers and from patients with different diseases (diabetic nephropathy, IgA nephropathy, FSGS, MNGN, MCD, urothelial cancer, etc.) into group-specific polypeptide panels. Comparison of the different specific panels enables to pinpoint polypeptides with significant differences in the groups.

which appear sporadically, only those peptides present in more than 20% of the urine samples in at least one group (samples from patients with same disease) are further investigated.

*3.3.3. Statistical Analysis*

The main purposes of proteomic profiling of urine are biomarker identification and validation. Individual data sets of patients from selected pathological conditions can be matched to the SQL database. This process allows the comparison of a specific "disease group" to any desired "control group," enabling the identification of statistically significant changes and the definition of potential biomarkers. The significance of a biomarker correlates with its occurrence in each sample. The significance of a result is also called its $p$-value; the smaller the $p$-value, the more significant the result is said to be. In statistical hypothesis testing, the $p$-value is the probability of obtaining a result at least as extreme as a given data point, assuming the data point was the result of chance alone. The fact that $p$-values are based on this assumption is crucial to their correct interpretation. The $p$-values are calculated using the natural logarithm transformed intensities for variance stabilization and the Gaussian approximation to the *t*-distribution. As a safeguard against multiple tests of statistical significance on the same data, multiple testing adjustments of *p*-values have to be applied to the subset of markers that passed a certain frequency threshold (usually 70%). For example, *p*-values can be calculated using the Westfall and Young maxT-procedure *(15)*. This function computes permutation-based step-down adjusted *p*-values. To ensure stability of the results, the *p*-values by the minP procedure of Westfall and Young are verified to be of similar magnitude *(16)*.

#### 3.3.4. Model Generation

The generation of models for the discrimination of case and control groups can be performed using different classification algorithms, such as logistic regression or support vector machines (SVMs) (*see* **Note 4**). Both approaches are dealing with learning methods, which require a so-called "training set" to generate a polypeptide pattern consisting of significant biomarkers (*see* **Fig. 4**). In the first step, adequate classification parameters (biomarkers) are defined characterizing particular properties (diseases) of a training set. After training, the model should have "learned" the properties of the trainings set, and should be able to generalize to unknown data sets allowing their correct classification. Quality criteria for classification are the diagnostic sensitivity and specificity or even more appreciate positive and negative predictive values considering disease prevalence (*see* **Note 5**).

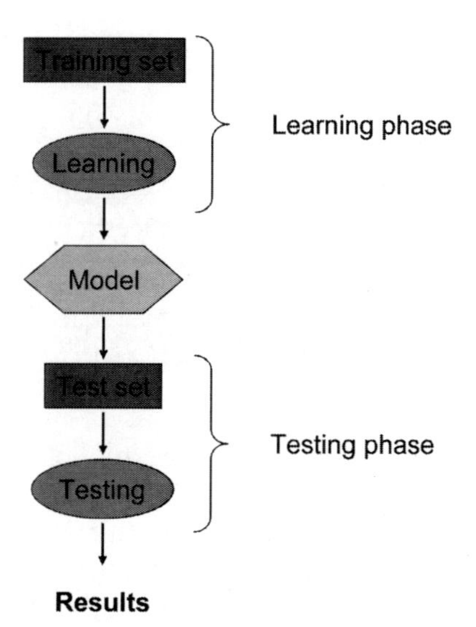


Fig. 4. Procedure of model generation. Workflow for the model generation contains learning and test phase; for this propose, different independent data sets are required.

### Logistic Regression

In statistics, logistic regression is a regression model for binomially distributed response/dependent variables. It is useful for modeling the probability of an event (diseased or not diseased) occurring as a function of other factors (biomarkers). It is a generalized linear model that uses the logit as its link function.

Logistic regression analyzes binomially distributed data of the form

$$Y_i \sim B(p_i, n_i) \text{ for } i = 1, \ldots, m$$

where the numbers of Bernoulli trials $n_i$ (analyzed urine samples) are known and the probabilities of success $p_i$ (probability of disease) are unknown. An example of this distribution is the fraction of urine samples from diseased individuals ($p_i$) after $n_i$ samples are analyzed.

The model is then that for each trial (value of $i$) there is a set of explanatory/independent variables (biomarkers) that might inform the final probability. These variables can be expressed in a $k$ vector $X_i$ and the model then takes the form

$$p_i = E\left(\left.\frac{Y_i}{n_i}\right| X_i\right)$$

The logits of the unknown binomial probabilities (the logarithms of the odds) are modeled as a linear function of the variable $X_i$

$$\operatorname{logit}(p_i) = \ln\left(\frac{p_i}{1-p_i}\right) = \beta_0 + \beta_1 x_{1,i} + \cdots + \beta_k x_{k,i}$$

### Support Vector Machines

SVMs are a set of related supervised learning methods used for classification and regression. SVMs nonlinearly map their $n$-dimensional input space into a high-dimensional feature space. The support vectors are then used to construct an optimal hyperplane in this feature space (*see* **Fig. 5**). This hyperplane is used for classification; the features are the selected polypeptides, each of which represents one dimension. An excellent tutorial has been produced by Burges *(17)*.

After the model is optimized, the accuracy and quality (sensitivity and specificity) can by assessed based on the training set by total cross-validation: In each run, one data set remains unconsidered, and the model is trained with the remaining data sets *(18)*. Subsequently, the excluded patient is classified with this model. The total correct classification rate (after each patient has once been excluded) serves as a measure for the accuracy of the model and allows the calculation of the sensitivity and specificity.

As an example for the classification of two classes (nine members each), an SVM simplified hyperplane in a 3-D feature space based on three polypeptides named *x*, *y*, and *z* is shown in **Fig. 5**. The position of each patient ("control" as cubes and

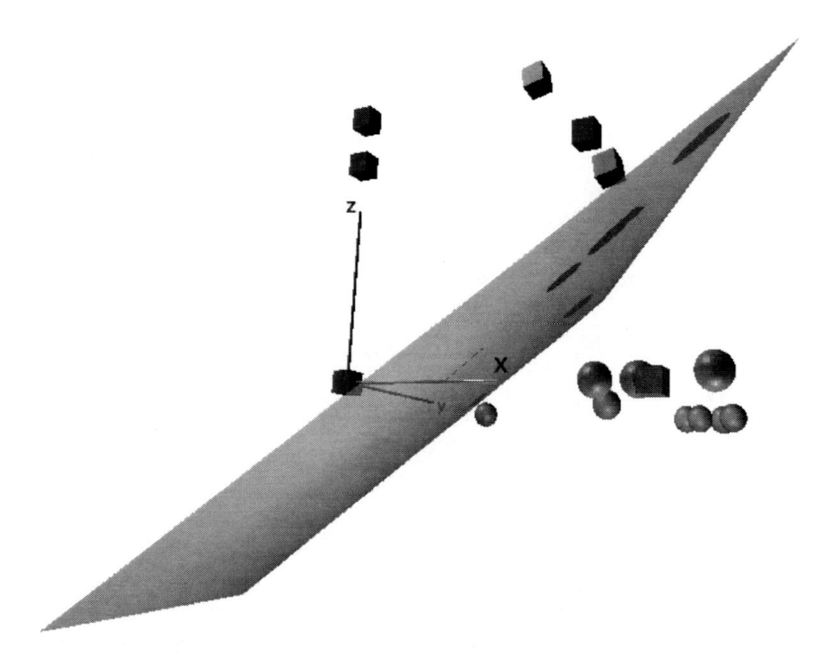


Fig. 5. Simplified SVM model based on three polypeptides (published by Theodorescu et al. *[52]*). The position of each patient ("control" as cubes and "case" as globes) is defined by the data (logarithmic amplitude) of these three polypeptides. Three "control" patients are 0 for all three polypeptides, and hence are indistinguishable from each other. A simplified linear plane serves as separator to distinguish between the two groups. When using this limited number of biomarkers, not all patients are classified correctly and consequently one "control" patient is found in the "case" group. It should be noted that this represents only a very simplified model because it is impossible to graphically depict more than three dimensions, as would be necessary here.

"case" as globes) is defined by the logarithmic amplitude of the three polypeptides. In 3/9 "control" data sets, none of the three selected polypeptides occurred; these are indistinguishable from each other (dark gray squares in the hyperplane origin). Evidently, when using this limited number of biomarkers, not all patients are classified correctly. One "control" patient is found in the space of the "case" group.

#### *3.3.5. Receiver-Operating Characteristic Curve*

The clinical performance of a laboratory test depends on its diagnostic accuracy or the ability to correctly classify subjects into clinically relevant subgroups (e.g., "healthy" versus "diseased"). The receiver-operating characteristic (ROC) curve depicts the overlap between the two populations by plotting sensitivity (true-positive fraction in percent) versus 100-specificity (false-positive fraction in percent) for all available thresholds. The area under the ROC curve (AUC) provides a single measure of the overall accuracy independent of a particular threshold *(19)*. If the value is 0.5, i.e., a diagonal line in the plot, the given parameter does not discriminate between two defined sample groups. An AUC of 1.0 indicates no overlap in the distributions of the parameter under investigation, corresponding to perfect separation of both groups.

### 3.4. Mass Spectrometry for Peptide Identification

Peptide sequencing using tandem MS is essential for identifying the defined biomarkers, helps to clarify pathophysiology of different diseases, and enables the search for therapeutic targets. The so-called "bottom-up" approach (2-D-gel electrophoresis and multidimensional protein identification technology) of analytical protein MS utilizes an initial treatment with a protease (usually trypsin) for protein identification to cut proteins into relatively small peptides of which the mass per charge ratio can be accurately determined by MS. This so-called "peptide mass fingerprinting" allows reliable identification of proteins *(see* e.g., **refs.** *20–25)*.

In biomarker discovery experiments, a different approach has become popular: the so-called "top-down" approach, in which polypeptides are delivered to the mass spectrometer without prior digestion in its naturally occurring form. Depending on the mass accuracy of the applied mass spectrometer, peptides and small proteins up to 10–20 kDa can be recognized. Top-down strategies have been exploited using CE-MS *(26–31)*, LC-MS (liquid chromatography-mass spectrometry) *(32–34)*, and SELDI-TOF-MS (surface–enhanced laser desorption/ionization-time of flight–mass spectrometry) *(35–37)* systems. Absolute identification of proteins is more demanding with top-down approaches than with bottom-up approaches ($m/z$ ratios are often reported rather than protein identifications). For the bottom-up analysis, the first provision is the mass fingerprint for the correct identification of a protein. To confirm the identity of the protein, sequence analysis of a few tryptic peptides can be performed. The sequence analysis of the peptides resulting from the top-down approach is often a make-or-break outcome (*see* **Note 6**). Major obstacles are frequently occurring post-translational modifications (PTMs) that change the observed masses, which then are different from the expected masses, complicating database searches by higher degrees of freedom. Furthermore, search algorithms are generally adapted to the needs of tryptic digests, which differ greatly from the requirements for de novo sequencing of naturally occurring peptides or proteins (*see* also **refs.** *38, 39)*.

From chemical point of view, the identification of the full sequence becomes uncertain, because fragmentation of peptides may result in water elimination (asparagines, aspartic acid, glutamine, glutamic acid, or proline *[40, 41]* residues) and consequently in partial fragment spectra. In this case, only $MS^n$ approaches with the use of an ion trap device may lead to satisfactory results.

## 4. Notes

1. Urine samples must be collected under proper conditions, especially when high-sensitive methods are applied, such as MS. Minor degradation of proteins generally does not cause problems for Western blotting, whereas ELISA or even 2-D-gel electrophoresis, CE-MS, or any other MS-based proteomic analysis will be severely hampered. To prevent degradation and resulting additional fragments of proteins that are irrelevant to the underlying disease, samples require storage at least at –20°C immediately after collection. Immediate cooling of the samples on ice is another way to preserve stability. Compared with serum or plasma, urine is a very stable body fluid *(42)*, making it preferable over the other fluids.
2. High-abundant high-molecular-weight proteins of urine, such as albumin, transferrin, or others, generally appear not to contain significant diagnostic information *(43)*, but cause severe problems during CE-MS analysis, such as precipitation, clogging of the capillary, and overloading. Additionally, the analysis of low-abundant molecules of interest is compromised by large amounts of high-abundant molecules. Therefore, an optimized protocol to remove these large molecules using ultrafiltration has been developed. The most abundant higher molecular weight protein in urine is albumin, a carrier protein that binds a substantial fraction of other proteins and peptides. To prevent binding of small fragments to albumin, we use chaotropic urea in combination with the denaturing detergent SDS for sample dilution. Consistently, reproducible high recovery rates of >80% of added standard polypeptides were observed.
3. Under neutral or basic conditions, the silanoyl groups of the capillary form a negatively charged surface, which unspecifically interacts with positive charges of the proteins. This leads to peak broadening, especially of highly charged molecules. To circumvent these problems, several different coatings and coating protocols have been described *(44–46)*. Hence, we have examined several mostly silyl-based coatings *(42)*. Unfortunately, most of the coating procedures described in the literature are tedious and time-consuming. Coated capillaries gave good results, when using standard polypeptides. However, all coatings were not satisfactory when analyzing proteins of highly complex clinical samples, because these appear to partially deposit on the coating material leading to peak broadening and increased migration times. Consequently, our experience using different types of coating so far were quite unsatisfactory. The optimal approach appears to be the use of

uncoated capillaries. Decreasing the pH of the background electrolyte reduces the negative charge of the surface, thus reducing capillary–protein interaction. Additionally, hydrophobic interactions can be reduced by adding organic solvent to the running buffer. Therefore, best results are obtained at low pH (~2) in the presence of acetonitrile.

Last, but not the least, it is important to point out that most coatings result in a positively charged capillary wall and cause an inverse electro-osmotic flow; consequently, the electrical field has to be reversed.

4. No matter which mathematical approach is used, two basic considerations apply: *(1)* The number of independent variables should be kept to a minimum and should certainly be below the number of samples investigated and *(2)* an approach is valid only when it is tested with a blinded validation set. It should be imperative to include such a blinded data set in any biomarker discovery process *(47)*.

5. Sensitivity is the proportion of true positives of all diseased cases in the population. Specificity is the proportion of true negatives of all negative cases in the population. Positive predictive value is the proportion of true positives of all positives in the test. Negative predictive value is the proportion of true negatives of all negatives in the test.

6. Classical protein sequencing analysis by Edman degradation or peptide mass fingerprint requires nearly purified polypeptides. Because of the high complexity of body fluids, to date, it is impossible to purify a single polypeptide of interest for further examination. Consequently, these polypeptides have to be uniquely defined by highly accurate mass and, if possible, additional identifying parameters (e.g., CE migration time) for subsequent sequencing using MS/MS. CE- or LC-separated sample can either be directly interfaced with an MS/MS instrument or the entire run can be spotted onto a MALDI target plate for subsequent analysis *(28, 48, 49)*. Sequencing using MALDI-TOF-TOF avoids repetitive analysis by optimal fragmentation conditions of the peptides adhered on the target. However, this approach is limited by the signal intensity of peptides found by MALDI-MS technique. In contrast to this, the usage of online-ESI-ion trap-MS methods allows characterizing peptides by the application of $MS^n$. ESI-Q-TOF allows successful sequencing in respect due to its high mass accuracy and the high resolution. Best results are obtained with FT-ICR devices; however, their purchase and operation are cost-intensive. It has been well documented that the dynamic range and coverage obtained from MS/MS proteome analysis is a function of both the quality of the separation applied and the MS device used *(50)*. The

greater challenge of "top-down" approaches in comparison with classical "bottom-up" techniques is the missing availability of peptide mass fingerprint information. Therefore, sequencing of naturally occurring peptides is dependent on the successful fragmentation and interpretation of MS/MS data of each single peptide. In consequence, high-performance mass spectrometers with features such as good fragment ion recovery and high mass accuracy are preferred for peptide identification.

## References


1. Schmitt-Kopplin, P., Frommberger, M. (2003) Capillary electrophoresis–mass spectrometry: 15 years of developments and applications. *Electrophoresis* **24**, 3837–3867.
2. Schmitt-Kopplin, P., Englmann, M. (2005) Capillary electrophoresis–mass spectrometry: Survey on developments and applications 2003–2004. *Electrophoresis* **26**, 1209–1220.
3. Stutz, H. (2005) Advances in the analysis of proteins and peptides by capillary electrophoresis with matrix-assisted laser desorption/ionization and electrospray–mass spectrometry detection. *Electrophoresis* **26**, 1254–1290.
4. Neususs, C., Pelzing, M., and Macht, M. (2002) A robust approach for the analysis of peptides in the low femtomole range by capillary electrophoresis–tandem mass spectrometry. *Electrophoresis* **23**, 3149–3159.
5. Oda, R. P., Clark, R., Katzmann, J. A., and Landers, J. P. (1997) Capillary electrophoresis as a clinical tool for the analysis of protein in serum and other body fluids. *Electrophoresis* **18**, 1715–1723.
6. Hernandez-Borges, J., Neususs, C., Cifuentes, A., and Pelzing, M. (2004) On-line capillary electrophoresis–mass spectrometry for the analysis of biomolecules. *Electrophoresis* **25**, 2257–2281.
7. Guzman, N. A., Park, S. S., Schaufelberger, D., Hernandez, L., Paez, X., Rada, P., Tomlinson, A. J., and Naylor, S. (1997) New approaches in clinical chemistry: On-line analyte concentration and microreaction capillary electrophoresis for the determination of drugs, metabolic intermediates, and biopolymers in biological fluids. *J. Chromatogr. B Biomed. Sci. Appl.* **697**, 37–66.
8. Kolch, W., Neususs, C., Pelzing, M., and Mischak, H. (2005) Capillary electrophoresis-mass spectrometry as a powerful tool in clinical diagnosis and biomarker discovery. *Mass Spectrom. Rev.* **24**, 959–977.
9. Kaiser, T., Wittke, S., Just, I., Krebs, R., Bartel, S., Fliser, D., Mischak, H., and Weissinger, E. M. (2004) Capillary electrophoresis coupled to mass spectrometer for automated and robust polypeptide determination in body fluids for clinical use. *Electrophoresis* **25**, 2044–2055.
10. Dudoit, S., van der Laan, M. J. (2007) *Multiple Testing Procedures and Applications to Genomics*. Berlin: Springer.
11. Kaiser, T., Wittke, S., Just, I., Krebs, R., Bartel, S., Fliser, D., Mischak, H., and Weissinger, E. M. (2004) Capillary electrophoresis coupled to mass spectrometer for automated and robust polypeptide determination in body fluids for clinical use. *Electrophoresis* **25**, 2044–2055.
12. Neuhoff, N., Kaiser, T., Wittke, S., Krebs, R., Pitt, A., Burchard, A., Sundmacher, A., Schlegelberger, B., Kolch, W., and Mischak, H. (2004) Mass spectrometry for the detection of differentially expressed proteins: A comparison of surface-enhanced laser desorption/ionization and capillary electrophoresis/mass spectrometry. *Rapid Commun. Mass Spectrom.* **18**, 149–156.
13. Theodorescu, D., Fliser, D., Wittke, S., Mischak, H., Krebs, R., Walden, M., Ross, M., Eltze, E., Bettendorf, O., Wulfing, C., and Semjonow, A. (2005) Pilot study of capillary electrophoresis coupled to mass spectrometry as a tool to define potential prostate cancer biomarkers in urine. *Electrophoresis* **26**, 2797–2808.
14. Theodorescu, D., Wittke, S., Ross, M. M., Walden, M., Conaway, M., Just, I., Mischak, H., and Frierson, H. F. (2006) Discovery and validation of new protein biomarkers for urothelial cancer: A prospective analysis. *Lancet Oncol.* **7**, 230–240.
15. DeLeo, J. M. (1993) Receiver operating characteristic laboratory (ROCLAB): Software for

developing decision strategies that account for uncertainty. 318–325.

16. Westfall, P. H., Young, S. S. (1993) *Resampling-based Multiple Testing: Examples and Methods for P-Value Adjustment.* New York: Wiley .
17. Burges, C. J. C. (1998) A tutorial on support vector machines for pattern recognition. *Knowledge Discovery and Data Mining* **2**, 121–167.
18. Levner, I. (2005) Feature selection and nearest centroid classification for protein mass spectrometry. *BMC Bioinformatics* **6**, 68.
19. Biron, D. G., Joly, C., Marche, L., Galeotti, N., Calcagno, V., Schmidt-Rhaesa, A., Renault, L., and Thomas, F. (2005) First analysis of the proteome in two nematomorph species, Paragordius tricuspidatus (Chordodidae) and Spinochordodes tellinii (Spinochordodidae). *Infect. Genet. Evol.* **5**, 167–175.
20. Gagnaire, V., Piot, M., Camier, B., Vissers, J. P., Jan, G., and Leonil, J. (2004) Survey of bacterial proteins released in cheese: A proteomic approach. *Int. J. Food Microbiol.* **94**, 185–201.
21. Gras, R., Muller, M. (2001) Computational aspects of protein identification by mass spectrometry. *Curr. Opin. Mol. Ther.* 3, 526–532.
22. Pang, J. X., Ginanni, N., Dongre, A. R., Hefta, S. A., and Opitek, G. J. (2002) Biomarker discovery in urine by proteomics. *J. Proteome Res.* **1**, 161–169.
23. Raharjo, T. J., Widjaja, I., Roytrakul, S., and Verpoorte, R. (2004) Comparative proteomics of Cannabis sativa plant tissues. *J. Biomol. Techol.* **15**, 97–106.
24. Thongboonkerd, V., McLeish, K. R., Arthur, J. M., and Klein, J. B. (2002) Proteomic analysis of normal human urinary proteins isolated by acetone precipitation or ultracentrifugation. *Kidney Int.* **62**, 1461–1469.
25. Kaiser, T., Kamal, H., Rank, A., Kolb, H. J., Holler, E., Ganser, A., Hertenstein, B., Mischak, H., and Weissinger, E. M. (2004) Proteomics applied to the clinical follow-up of patients after allogeneic hematopoietic stem cell transplantation. *Blood* **104**, 340–349.
26. Meier, M., Kaiser, T., Herrmann, A., Knueppel, S., Hillmann, M., Koester, P., Danne, T., Haller, H., Fliser, D., and Mischak, H. (2005) Identification of urinary protein pattern in type 1 diabetic adolescents with early diabetic nephropathy by a novel combined proteome analysis. *J. Diabetes Complicat.* **19**, 223–232.
27. Mischak, H., Kaiser, T., Walden, M., Hillmann, M., Wittke, S., Herrmann, A., Knueppel, S., Haller, H., and Fliser, D. (2004) Proteomic analysis for the assessment of diabetic renal damage in humans. *Clin. Sci. (Lond)* **107**, 485–495.
28. Rossing, K., Mischak, H., Parving, H. H., Christensen, P. K., Walden, M., Hillmann, M., and Kaiser, T. (2005) Impact of diabetic nephropathy and angiotensin II receptor blockade on urinary polypeptide patterns. *Kidney Int.* **68**, 193–205.
29. Weissinger, E. M., Wittke, S., Kaiser, T., Haller, H., Bartel, S., Krebs, R., Golovko, I., Rupprecht, H. D., Haubitz, M., Hecker, H., Mischak, H., and Fliser, D. (2004) Proteomic patterns established with capillary electrophoresis and mass spectrometry for diagnostic purposes. *Kidney Int.* **65**, 2426–2434.
30. Wittke, S., Haubitz, M., Walden, M., Rohde, F., Schwarz, A., Mengel, M., Mischak, H., Haller, H., and Gwinner, W. (2005) Detection of acute tubulointerstitial rejection by proteomic analysis of urinary samples in renal transplant recipients. *Am. J. Transpl.* **5**, 2479–2488.
31. Jurgens, M., Appel, A., Heine, G., Neitz, S., Menzel, C., Tammen, H., and Zucht, H. D. (2005) Towards characterization of the human urinary peptidome. *Comb. Chem. High Throughput Screen* **8**, 757–765.
32. Schrader, M., Schulz-Knappe, P. (2001) Peptidomics technologies for human body fluids. *Trend. Biotechnol.* **19**, S55–S60.
33. Svensson, M., Skold, K., Svenningsson, P., and Andren, P. E. (2003) Peptidomics-based discovery of novel neuropeptides. *J. Proteome Res.* **2**, 213–219.
34. Tan, C. S., Ploner, A., Quandt, A., Lehtio, J., and Pawitan, Y. (2006) Finding regions of significance in SELDI measurements for identifying protein biomarkers. *Bioinformatics* **22**, 1515–1523.
35. Tang, N., Tornatore, P., and Weinberger, S. R. (2004) Current developments in SELDI affinity technology. *Mass Spectrom. Rev.* **23**, 34–44.
36. Yip, T. T., Lomas, L. (2002) SELDI ProteinChip array in oncoproteomic research. *Technol. Cancer Res. Treat.* **1**, 273–280.
37. Fliser, D., Novak, J., Thongboonkerd, V., Argiles, A., Jankowski, V., Girolami, M. A., Jankowski, J., and Mischak, H. (2007) Advances in urinary proteome analysis and biomarker discovery. *J. Am. Soc. Nephrol.* **18**, 1057–1071.
38. Mischak, H., Julian, B. A., and Novak, J. (2007) High-resolution proteome/peptidome analysis of peptides and low-molecular-weight proteins in urine. *Proteomics Clin. Appl.* **1**, 792–804.
39. Geiger, T., Clarke, S. (1987) Deamidation, isomerization, and racemization at asparaginyl and aspartyl residues in peptides. Succinimide-linked reactions that contribute to protein degradation. *J. Biol. Chem.* **262**, 785–794.

40. Stephenson, R. C., Clarke, S. (1989) Succinimide formation from aspartyl and asparaginyl peptides as a model for the spontaneous degradation of proteins. *J. Biol. Chem.* **264**, 6164–6170.
41. Kolch, W., Neususs, C., Pelzing, M., and Mischak, H. (2005) Capillary electrophoresis-mass spectrometry as a powerful tool in clinical diagnosis and biomarker discovery. *Mass Spectrom. Rev.* 24, 959–977.
42. Wittke, S., Fliser, D., Haubitz, M., Bartel, S., Krebs, R., Hausadel, F., Hillmann, M., Golovko, I., Koester, P., Haller, H., Kaiser, T., Mischak, H., and Weissinger, E. M. (2003) Determination of peptides and proteins in human urine with capillary electrophoresis-mass spectrometry, a suitable tool for the establishment of new diagnostic markers. *J. Chromatogr. A* **1013**, 173–181.
43. Belder, D., Deege, A., Husmann, H., Kohler, F., and Ludwig, M. (2001) Cross-linked poly(vinyl alcohol) as permanent hydrophilic column coating for capillary electrophoresis. *Electrophoresis* **22**, 3813–3818.
44. Johannesson, N., Wetterhall, M., Markides, K. E., and Bergquist, J. (2004) Monomer surface modifications for rapid peptide analysis by capillary electrophoresis and capillary electrochromatography coupled to electrospray ionization-mass spectrometry. *Electrophoresis* **25**, 809–816.
45. Liu, C. Y. (2001) Stationary phases for capillary electrophoresis and capillary electrochromatography. *Electrophoresis* **22**, 612–628.
46. Mischak, H., Apweiler, R., Banks, R. E., Conaway, M., Coon, J. J., Dominizak, A., Ehrich, J. H., Fliser, D., Girolami, M., Hermjakob, H., Hochstrasser, D. F., Jankowski, V., Julian, B. A., Kolch, W., Massy, Z., Neususs, C., Novak, J., Peter, K., Rossing, K., Schanstra, J. P., Semmes, O. J., Theodorescu, D., Thongboonkerd, V., Weissinger, E. M., Van Eyk, J. E., and Yamamoto, T. (2007) Clinical Proteomics: A need to define the field and to begin to set adequate standards. *Proteomics Clin. Appl.* **1**, 148–156.
47. Musyimi, H. K., Narcisse, D. A., Zhang, X., Stryjewski, W., Soper, S. A., and Murray, K. K. (2004) Online CE-MALDI-TOF MS using a rotating ball interface. *Anal. Chem.* **76**, 5968–5973.
48. Thongboonkerd, V., McLeish, K. R., Arthur, J. M., and Klein, J. B. (2002) Proteomic analysis of normal human urinary proteins isolated by acetone precipitation or ultracentrifugation. *Kidney Int.* **62**, 1461–1469.
49. Shen, Y., Jacobs, J. M., Camp, D. G., Fang, R., Moore, R. J., Smith, R. D., Xiao, W., Davis, R. W., and Tompkins, R. G. (2004) Ultra-high-efficiency strong cation exchange LC/RPLC/MS/MS for high dynamic range characterization of the human plasma proteome. *Anal. Chem.* **76**, 1134–1144.
50. Sniehotta, M., Schiffer, E., Zürbig, P., Novak, J., and Mischak, H. (2007) Capillary electrophoresis – a multifunctional application for clinical diagnosis. *Electrophoresis* **28**, 1407–1417.
51. Theodorescu, D., Fliser, D., Wittke, S., Mischak, H., Krebs, R., Walden, M., Ross, M., Eltze, E., Bettendorf, O., Wulfing, C., and Semjonow, A. (2005) Pilot study of capillary electrophoresis coupled to MS as a tool to define potential prostate cancer biomarkers in urine. *Electrophoresis* **26**, 2797–2808.

# Chapter 7

# A Newcomer's Guide to Nano-Liquid-Chromatography of Peptides

**Thomas Fröhlich and Georg J. Arnold**

## Summary

LC-MS/MS is one of the most powerful techniques in the field of proteomics allowing high throughput identification of proteins out of complex protein mixtures. Besides high sample throughput, the analytical sensitivity is one of the major benefits of this technology. A prerequisite for sensitive LC-MS/MS approaches is chromatography with very low flow rates in the nanoliter per minute range, usually referred to as nano-liquid chromatography (nano-LC). However, to perform this separation technology, an appropriate instrumental setup as well experienced operators are a prerequisite. The aim of this chapter is to help nano-LC newcomers to get introduced to this fascinating technology. Technical components of nano-LC systems like solvent delivery systems, sample injection systems, and nano-chromatography columns are described. Detailed procedures to mount, test, and operate the system are outlined, and advices for an effective troubleshooting are provided.

**Key words:** Nano-HPLC, Reversed phase chromatography, Proteomics, Proteins, Peptides

## 1. Introduction

In the field of proteomics, the combination of liquid chromatography as a separation tool with tandem mass spectrometry as an identification tool is referred to as LC-MS/MS. This method has become a broadly used key technology to analyze complex protein or peptide mixtures *(1–3)*. For chromatographic separation, usually reversed-phase liquid chromatography (RP-LC) is used, because of the high peak capacity (maximum number of components that can be resolved) of this method *(4)*. Furthermore, the compatibility of RP-LC with mass spectroscopy (MS) allows a direct MS analysis of the column eluate without further

Jörg Reinders and Albert Sickmann (eds.), *Proteomics,* Methods in Molecular Biology, Vol. 564
DOI: 10.1007/978-1-60761-157-8_7, 

treatment. However, especially in more comprehensive holistic proteome approaches, the separation power of RP chromatography is often insufficient. In those cases, usually two (2D-LC) or even more chromatography steps based on different separation mechanisms, the so-called orthogonal separations, have to be combined. Further details of this method referred to as multidimensional liquid chromatography (MDLC) are described in a special chapter of this book. For proteomics purposes, the chromatography directly preceding MS analysis is usually performed with very low flow rates combined with columns of very small inner diameters (IDs) *(5)*. Low flow rates increase the analyte concentration, which leads to higher sensitivity and enables the detection of low amounts of peptides. As an example, suppose a given amount of peptide elutes in a peak volume of 500 μL using a conventional analytical chromatography setup (column diameter 4.6 mm/L mL/min flow rate). Under maintenance of the linear flow velocity, a peak volume of only 135 nL is obtained in a typical nano-flow setup (column diameter 75 μm/270 nL/min flow rate), and the sensitivity is increased by more than 3,700-fold. In case of the broadly used electrospray mass spectrometers, the increase in ionization efficiency is a further benefit, since nano-flow rates lead to smaller droplets with higher surface-to-volume ratios and therefore to an improved ion desorption *(6, 7)*. However, flow rates in the nanoliter per minute range and the usage of capillary columns require appropriate LC-systems and experienced system operators to obtain sophisticated and reproducible results. This practical guide is especially dedicated to nano-LC newcomers and provides technical explanations as well as practical guidelines for the operation of nano-LC systems. Furthermore, some common problems of nano-LC are described, and several tips for troubleshooting are given.

## 2. Materials

### *2.1. Nano-LC System*

To get sophisticated and reproducible results over a long period of time, it is unavoidable to go deeper into technical details. Therefore, a brief description of the most important components of typical nano-LC systems is given in the following chapters.

#### *2.1.1. Solvent Delivery System*

For the precise generation of nanoliter per minute flow rates, several strategies exist. The most common one is to use micro-capillary HPLC pumps delivering microliter per minute flows and to split the mobile phase using the so-called flow-splitters. The simplest flow-splitter consists of a low-volume flow-splitting tee and a long restriction capillary (empty capillary with defined diameter and length). A photo of a splitter used in our lab is shown in **Fig. 1**. The mobile phase enters the tee and is divided into two streams. The major stream flows through the restriction capillary and is connected to the waste, whereas the minor flow (nL/min) is connected to the chromatography system. The split ratio is dependent on both the length and ID of the restriction capillary, as well as the backpressure of the remaining flow path. Therefore, alteration of the backpressure (e.g., by plugging or column aging) results in changes of the split ratio, and as a consequence to altered nano-flow rates that may ultimately impair the reproducibility of LC runs (*see* **Note 1**). To undergo this problem, a variety of more sophisticated splitters were made available, e.g., adjustable flow-splitters, which facilitate manual adjustment of the split ratio, or "active" flow-splitters constantly monitoring the flow and automatically adjusting it by high-speed switching valves. An alternative to flow-splitters are systems directly delivering nano-flow rates, promising even higher reproducibility and faster system response.

*2.1.2. Tubing*

To minimize the void volumes, capillary tubing instead of standard HPLC tubing is used in the nano-flow part of the system. Usually, the capillaries have IDs from 50 μm down to 10 μm. Although smaller IDs help to reduce the overall void volume of a system, it has to be considered that the risk of plugging is enhanced and the systems backpressure increases by the reduction

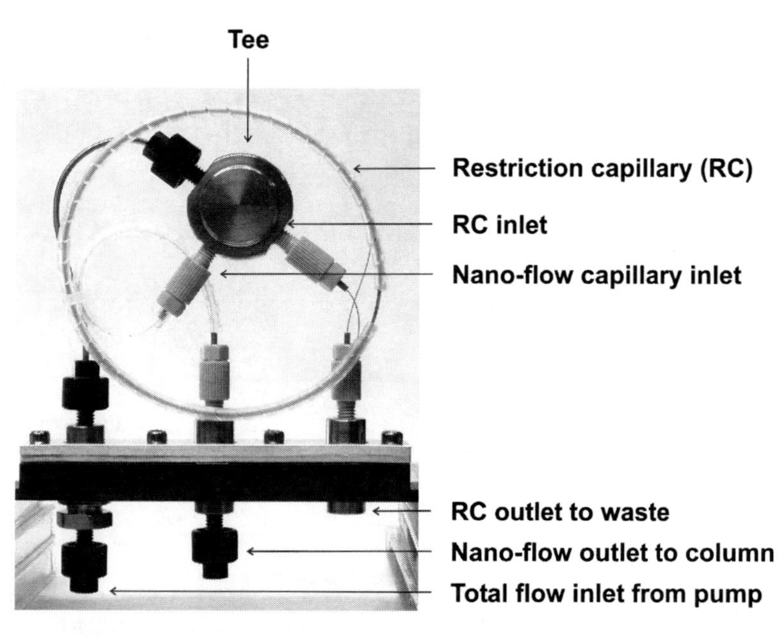


Fig. 1. Photograph of a flow-splitter used in our lab consisting of a tee and a restriction capillary.

of capillary IDs (*see* **Note 2**). It should further be kept in mind that the latter may affect the overall nano-flow rate in systems using passive flow-splitters. Capillary tubing is available in several materials such as polyetheretherketones (PEEK), Teflon, or fused silica. Most commonly used are fused silica tubings coated with a protective thin polyimide coating. A schematic view of fused silica tubing is shown in **Fig. 2**. Fused silica tubing is available in a broad variety of IDs. Beside its ID, the outer diameter (OD) of a capillary is important, especially if they should be connected to other components of the nano-LC system. Most common is fused silica tubing with an OD of 360 μm. A brief overview of how fused silica capillaries can be cut and connected is given in **Subheading 3.1.1**.

#### *2.1.3. Autosampler – Sample Traps*

For nanoliter per minute flow rates, the sample injection system has to fulfill special recommendations. Besides the capability to inject small volumes accurately, internal volumes between the needle and sample loop, as well as between the loop and separation column, must be minimized to avoid long transfer times. However, the loop injection of larger volumes is mostly not feasible. For instance, the transfer of a volume of 10 μL from a loop to the column would take at least 50 min at a standard nano-flow rate of 200 nL/min. In cases where larger volumes in the range of several microliters have to be injected, it is common praxis to use the so-called sample RP trap columns, the geometry of which allows the sample to be transferred with a high flow rate in the microliter per minute range. After sample loading, this trapping column is switched into the RP chromatography system *(8, 9)*. By the application of a gradient with increasing organic solvent concentration, the sample is subsequently eluted from the trap column and transferred to the separation column. Besides the capability of this method to quickly inject large sample volumes, it provides a convenient way to desalt the sample and wash it extensively before the trap column is switched over to the analytical column. However, it has to be considered that hydrophilic peptides may not fully or partially bind to the trap column and may therefore get lost (*see* **Note 3**). Furthermore, the binding

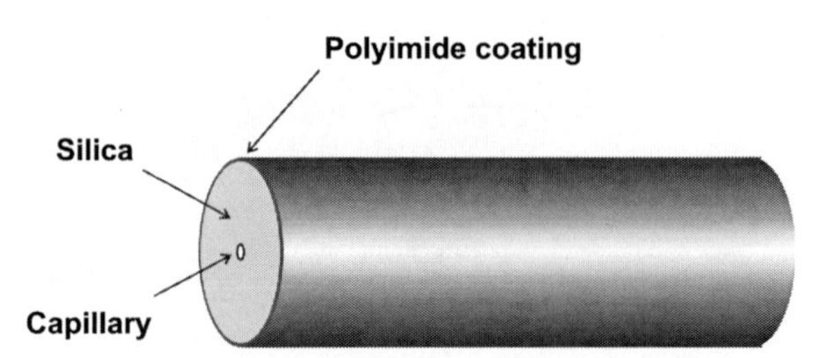


Fig. 2. Schematic representation of a fused silica capillary.

properties of the trap and the analytical column have to be properly adjusted to each other. **Figure 3** shows the flow path for a sample pre-concentration as used in our laboratory. This setup is using two trap columns, allowing a sample loading onto one trap column, whereas the sample previously applied to the other trap column undergoes RP separation. Another common injection technique is the so-called "vented column" strategy *(10, 11)*, a flow-path of which is shown in **Fig. 4**.

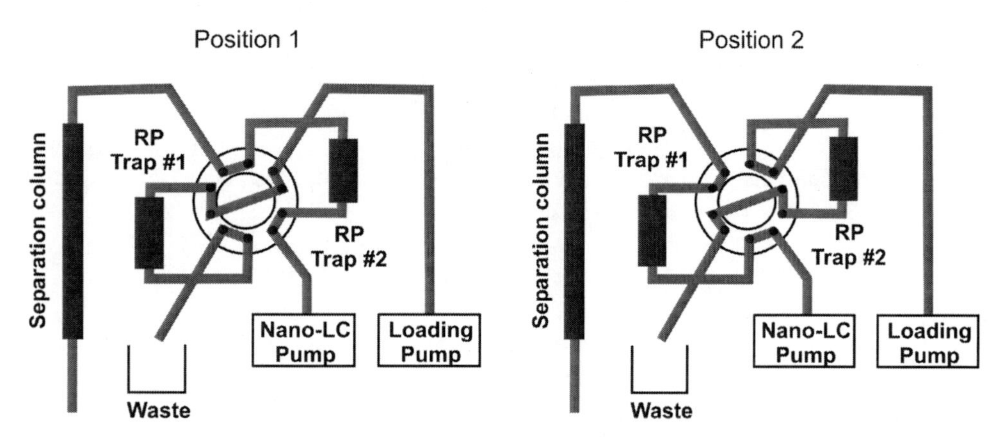


Fig. 3. Flow path of the sample trapping system used in our lab.

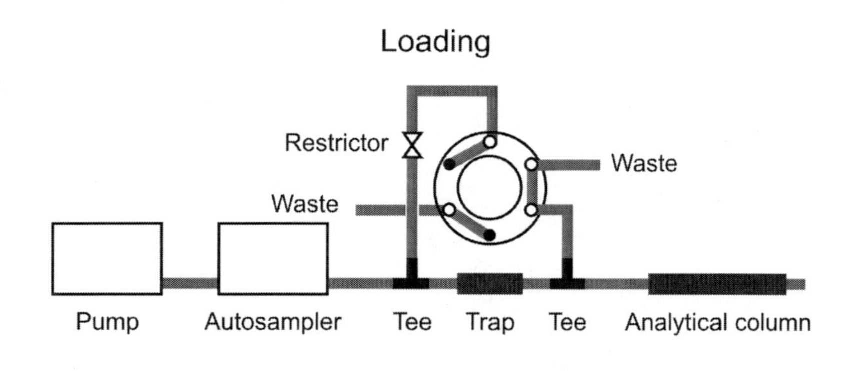


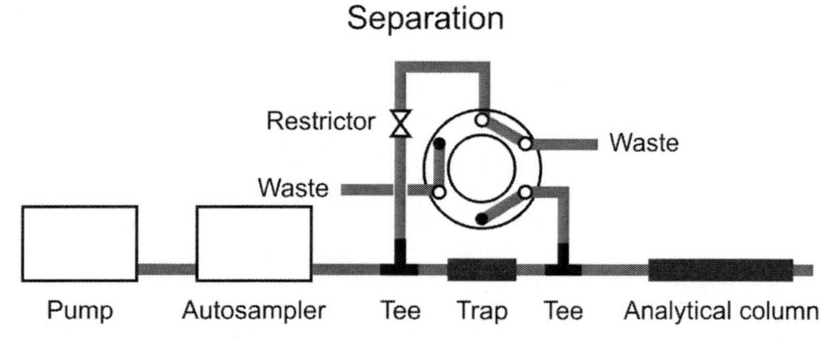


Fig. 4. Flow path of a "vented" trapping injection system as proposed by Meiring et al. *(10). Black circles* in the valve are representing plugs.

*2.1.4. Useful Sensors/UV Detector*

Commercially available nano-LC systems are equipped with several sensors being helpful for troubleshooting, e.g., pressure sensors (usually one for each pump), flow-meters, and conductivity cells. To minimize peak broadening caused by additional death volume introduced by the sensors, these are mostly positioned upstream of the analytical column. How these sensors can be useful to detect and fix malfunction of the system is described in **Subheading 3**. Several systems are provided with UV detectors and capillary flow cells to measure the absorbance of eluting peptides. Since typically, only femtomol amounts of complex samples are applied to nano-LC systems, UV detectors are usually not sensitive enough to monitor elution profiles but can nevertheless be very useful for troubleshooting purposes.

### *2.2. Capillary Columns*

Capillary reversed-phase columns packed with a variety of stationary phases are available from several manufacturers. Most popular are columns with IDs between 50 and 100 μm and a length of 50–150 mm packed with 3–5 μm fused silica particles derivatized with C8 (octyl) or C18 (octadecyl) groups (*see* **Note 4**). A common strategy to minimize dead volume after the chromatographic column is to integrate the column directly into ESI emitters *(12)* (*see* **Fig. 5**). Besides ready-to-use columns, accessories facilitating self-packing of columns are available, e.g., pressure bombs, empty capillaries, and ESI-emitters with frits. As an alternative to columns containing porous particles, the so-called monolithic columns are available, with enhanced porosity of the stationary phase and reduced backpressure. These columns typically show increased performance at higher flow rates. Since the separation

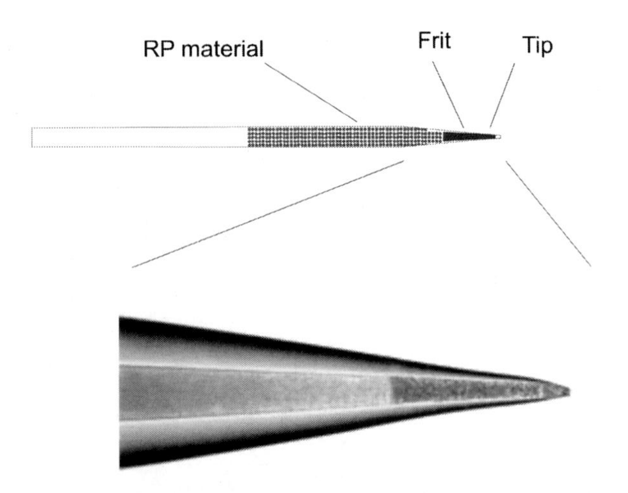


Fig. 5. Scheme and microscopic image of an ESI emitter packed with chromatography material (reproduced from www.newobjective.com with the kind permission of New Objective, Inc., Woburn, MA).

column is a very critical component for the overall performance of a nano-LC system, users are strongly encouraged to try several columns to optimize separation for their particular samples, experimental setups, and individual applications (*see* **Note 5**).

### *2.3. Solvents*

To avoid blockage and contamination of the nano-LC system, the use of high-purity solvents is recommended. Several manufacturers offer water, acetonitrile (MeCN), and other solvents in a special chromatography grade (*see* **Note 6**). As additive acids, 0.1% trifluouracetic acid (TFA) or formic acid (FA) are broadly used. TFA provides better chromatographic separation and sample recovery when compared with FA. However, TFA is especially unsuitable for online LC-ESI-MS setups, because it causes signal suppression, e.g., by the formation of ion pairs between peptide cations and trifluoracetate anions *(13)*.

## 3. Methods

### *3.1. Mounting the System and Making Connections*

To minimize dead volume and to avoid leakage or blockage of the system changes in the hardware configuration and assemblies of capillaries, the column and detectors have to be made with care using appropriate tools. Furthermore, it is a good practice to log any changes made in a given system. This may help to troubleshoot problems introduced by faulty mounted system components.

#### *3.1.1. Cutting Capillaries*

To cut capillaries, it is essential to use appropriate cutting tools to obtain leak-free connections with low dead volume. For polymer tubing, several types of cutters are available. The tube cutters fix the tubing at a 90° angle to the blade so that a clean flat cut can be achieved. After cutting, the capillary should be inspected using a loupe or a stereo microscope.

Cutting the fused silica tubing is more critical, because in case of badly cut capillaries, polyimide particles from the coating material or microparticles of cracked fused silica walls may be introduced into the nano-LC system. A microscopic image of an inappropriately cut and a perfectly cut capillary is shown in **Fig. 6**. Several tools like sapphire cutters, diamond scribes, and circumscribing cutters are available. With usual diamond and sapphire cutters, first the polyimide surface is scored by gently pressing down the cutter on the fused silica capillary. The cleavage is finally performed by pulling the capillary longitudinally. Too high pressure with the tool crushes the fused silica tubing, which may lead to the formation of microparticles, which can plug the nano-LC system. An easy-to-use but more expensive alternative are tools

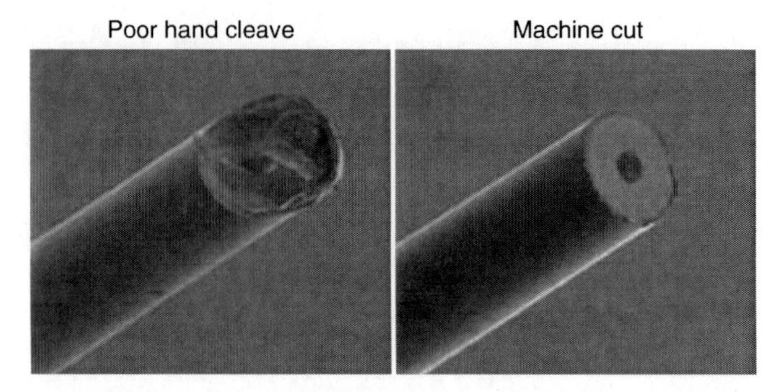


Fig. 6. Microscopic image of a badly and machine cut fused silica capillary (reproduced from www.newobjective.com with the kind permission of New Objective, Inc., Woburn, MA).

with cutting wheels circumscribing the capillary. Regardless of which tool is used to cut the fused silica tubing, it is absolutely recommended to check the quality of the cut using a loupe or a stereo microscope. Although self-cutting of capillaries is feasible using appropriate tools, high precision precut tubing with polished cut surface are available from a variety of vendors. Besides being convenient, they are especially useful for LC-systems where the requirements concerning dead volumes are particularly high.

*3.1.2. Connecting Capillaries to the System*

Usually fittings and tubing pockets used for pumps, valves, and detectors are designed for tubing with an OD of 1/16 or 1/32 in. To avoid dead volumes, fused silica capillaries are usually connected using the so-called sleeves. To work properly, the sleeves ID should be approximately 0.002 in. greater than the OD of the capillary tubing. Teflon and peek sleeves are available in a variety of IDs and ODs. Connections with PEEK sleeves resist higher pressures when compared with Teflon, whereas the latter is softer, and the fittings may be tightened by hand ("Fingertight" sleeves). An alternative is the so-called PEEKsil tubing (**Fig.** 7) consisting of fused silica tubing sheathed with PEEK polymer, available in several IDs and in ODs of 1/32 in., 1/16 in., and 360 μm. This material is as robust and convenient as PEEK material frequently used in conventional HPLC; however, it cannot be cut with standard equipment and has therefore to be purchased as precut material.

*3.1.3. Connecting Capillaries to Each Other*

To connect capillaries to each other, the usage of so-called true zero-dead-volume (ZDV) unions is recommended. A schematic view of a connection with a ZDV union is shown in **Fig. 8**. To precisely assemble ZDV unions, the obedience of the following procedure is recommended:

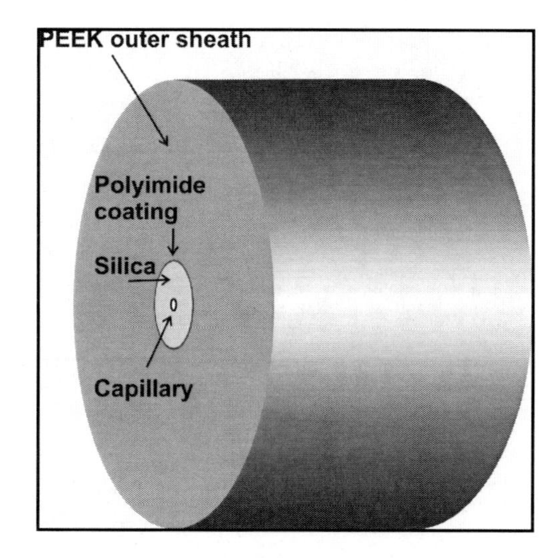


Fig. 7. Scheme of a PEEKsil capillary.

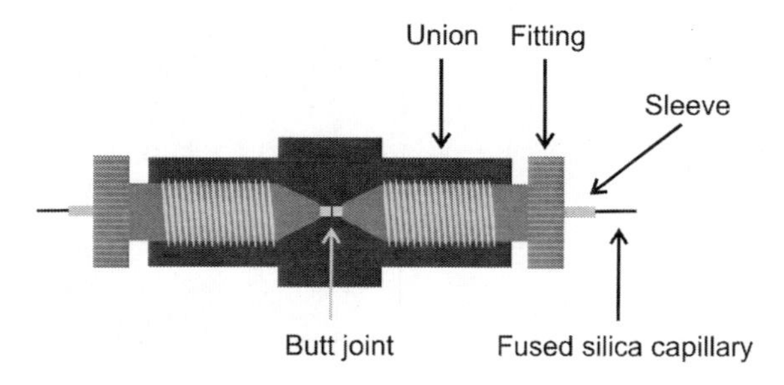


Fig. 8. Schematic view of a ZDV union.

1. Screw and tighten the gauge plug into one side of the union.
2. Insert the tubing and the appropriate sleeve into the fitting.
3. Gently push the tubing and sleeve from the opposite side of the union against the gauge.
4. Tighten the fitting.
5. Unscrew and remove the gauge plug.
6. Mount the second sleeved tubing in the same way as the first one.
7. Take care to keep the tubing and sleeve gently pushed against the already mounted tubing/sleeve while tightening the second fitting.

Although dead volumes of correctly mounted unions are relatively low, the number of connections should be kept to a minimum to avoid needless dead volumes (*see* **Note 7**).

### 3.2. Determination of Void Volumes

Generally, it has to be kept in mind that during separation, the total volume of the nano-flow system has to be passed by the solvent gradient. Although volumes are kept as small as possible, typical overall volumes of nano-LC systems are in the lower microliter range. For example, a tubing of 50-cm length with an ID of 50 μm by itself has a volume of about 1 μL. To estimate equilibration times, optimize LC methods, and detect undesired dead volumes, it is important to know the volumes between the components of the system and the overall void volume. Theoretical calculation of the void volume, which is easily possible (*see* **Note 8**), should be performed and compared with the experimentally determined void volume. The latter can be measured by injections of water containing 0.1% TFA into the nano-flow system equipped with a UV monitor: since UV absorption of 0.1% TFA in water is lower then 0.1% FA in water, the injected water/TFA mixture can be detected as a "negative" peak in the UV trace *(8, 14)*. Another method being used in our lab is the measurement of the mobile phase conductivity. The electrical resistance of the mobile phase rises with increasing MeCN concentration, thereby facilitating the determination of the time point at which the gradient front reaches the conductivity cell. The difference to the programmed start time of the gradient corresponds to the delay time, which is multiplied by the flow rate to determine the systems void volume. An example is shown in **Fig. 9**.

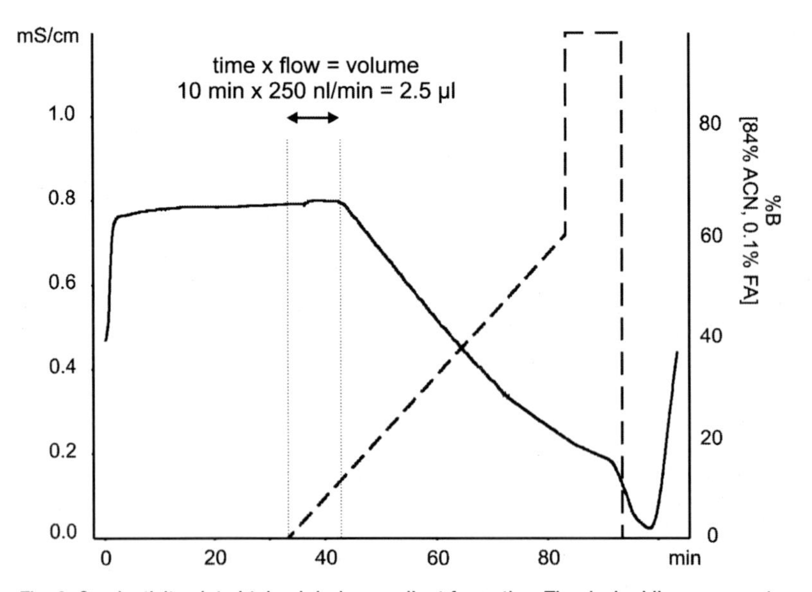


Fig. 9. Conductivity plot obtained during gradient formation. The dashed line represents the programmed gradient.

### 3.3. Determination of the Flow Rate

To ensure reliability and reproducibility of the results obtained with nano-LC systems, actual flow rates have to be determined regularly. An easy and precise way to measure flow rates is the usage of electronic flow sensors. Many state-of-the-art nano-LC systems are equipped with flow sensors, but they can also be purchased separately to upgrade the present systems. A manual way to measure flow rates is to connect the capillary outlet to a Hamilton microsyringe (e.g. 0.5–5 μL) using appropriate unions. The mobile phase then fills in the capillary or syringe and the volume can be read off after certain time intervals. Another possibility is to allow the system for a given time to form a droplet at the capillary outlet and to collect the droplet with a calibrated micropipette.

### 3.4. Testing the System

To test and to monitor the actual performance of nano-LC systems, the application of standardized test mixtures is recommended. In our lab, the test mixture consists of tryptic digests of proteins like bovine carboanhydrase II, bovine serum albumin, bovine casein, β-lactoglobulin, and cytochrome *c* (*see* **Note 9**). As an alternative, a variety of peptide standards and protein digests can be purchased from several suppliers.

### 3.5. Sample Preparation

In cases where sample amounts are rather limited, individual separation steps should be kept to a minimum. However, if samples are contaminated with buffer salts, detergents, or other chemicals, it is strongly recommended to purify and desalt the peptides before injection, e.g., by using C8 or C18 purification tips. After tip purification, it should be considered that organic solvent used for elution from the tip may inhibit binding of the peptides to the trap column or separation column. A common solution is a partial evaporation by a Speedvac to remove the organic solvent. Finally, the sample should be diluted with the starting solvent of the gradient to an appropriate volume. It is a good laboratory practice to centrifuge the sample (10,000*g*, 15 min) to sediment insoluble material and to gently pipette the supernatant into an appropriate sample vial.

### 3.6. Optimizing the Separation Method

Besides the used separation column, the applied gradient has a major influence on the quality of peptide separation. As the elution profile often varies from sample to sample, individual optimizations for the specific separation problem are recommended. To demonstrate the impact of the gradient shape, **Fig. 10** shows the MS base peak chromatograms of LC runs with the same peptide standard but with different gradients. The result reflects our experience that most peptides elute from a C18 reverse-phase column before the gradient contains 30% acetonitrile. Therefore, in cases of limited sample amounts preventing sample specific optimization, we typically use the following chromatographic method for LC-MS/MS analysis:

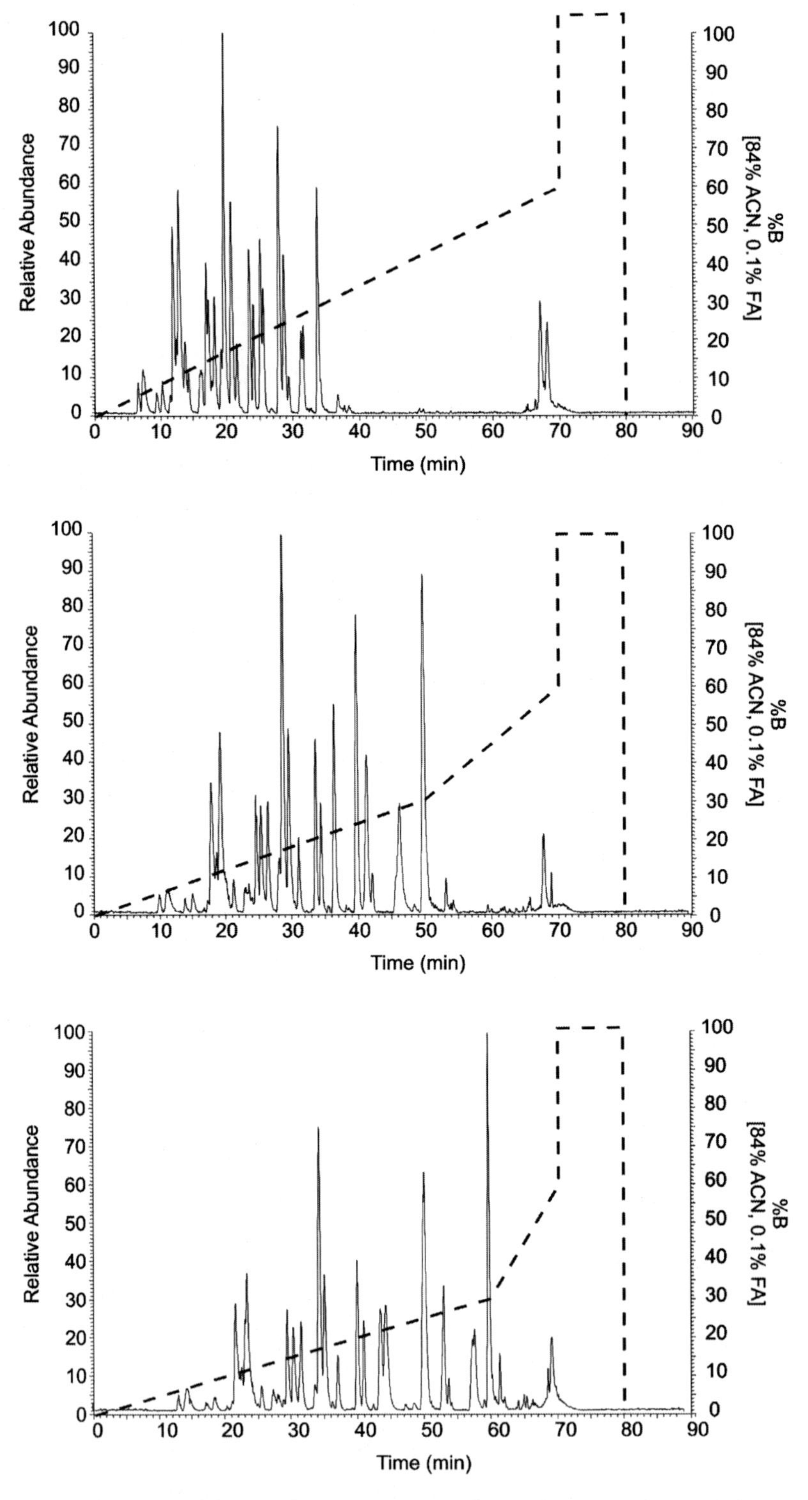


Fig. 10. Base peak chromatograms of nano-LC-MS runs of 100 fmol bovine carboanhydrase 2 using different separation gradients. The *dashed line* indicates the programmed gradient. LC was performed on an Ettan MDLC (GE Healthcare). Mass spectra were acquired on a LTQ linear ion trap (Thermo Fisher Scientific).

- Mobile phase A: water, 0.1% FA
- Mobile phase B: 84% MeCN, 16% water, 0.1% FA
- Trap column: PepMap 100, C18, 5 μm, 100 Å, 300-μm i.d. × 5 mm (Dionex Corporation, Sunnyvale, CA, USA)
- Separation column: PepMap 100, C18, 3 μm, 100 Å, 75-μm i.d. × 15 cm (Dionex Corporation, Sunnyvale, CA, USA)
- Sample loading flow: 5 μL/min
- Flow rate: 260 nL/min

1. Depending on the expected complexity, we use gradients of 0.2% MeCN/min up to a final concentration of 30% MeCN (e.g., total cell lysates) and 1% MeCN/min up to a final concentration of 30% MeCN (e.g., proteins from 2D gel spots).
2. To elute the more hydrophobic peptide fraction, a subsequent gradient of 1% MeCN/min up to a final concentration of 50% MeCN is applied.
3. A short (10 min) 84% MeCN isocratic step is performed.
4. Finally, a 15 min 1% MeCN isocratic step is performed for column equilibration.

This method should only be regarded as a first point of reference, since other gradients may lead to much better separations, depending on the individual sample and the chromatography column.

### 3.7. Troubleshooting

Troubleshooting of nano-LC systems is often annoying and time-consuming. However, most manufacturers of nano-LC systems provide useful troubleshooting guides for their systems, the compliance of which is strongly recommended. In this chapter, we try to discuss the most common problems and give hints and tips on how they can be identified.

#### 3.7.1. Pressure Curves

Pump pressure curves are one of the most important diagnostic tools indicating plugging or leaks if the pressure is too high or too low, respectively. Therefore, it is a good idea to compare the system pressure curve of individual runs with a reference profile monitored under standard operating conditions. However, in systems using flow -splitters, it has to be considered that plugs in the nano-flow path may not lead to a strong increase in the pressure because the main flow passes through the restriction capillary to the waste. Also, leaks in the nano-flow path behind components with high backpressure are often not reflected in pressure curves. Usually, the pressure undergoes smooth variations during gradient formation, since the viscosity of the liquid phase changes with the MeCN concentration. An example of a regular pressure curve and a pressure curve derived from a damaged pump is shown in

**Fig. 11**. Another possible cause for irregular pressure curves are air bubbles passing the pump or being trapped in it.

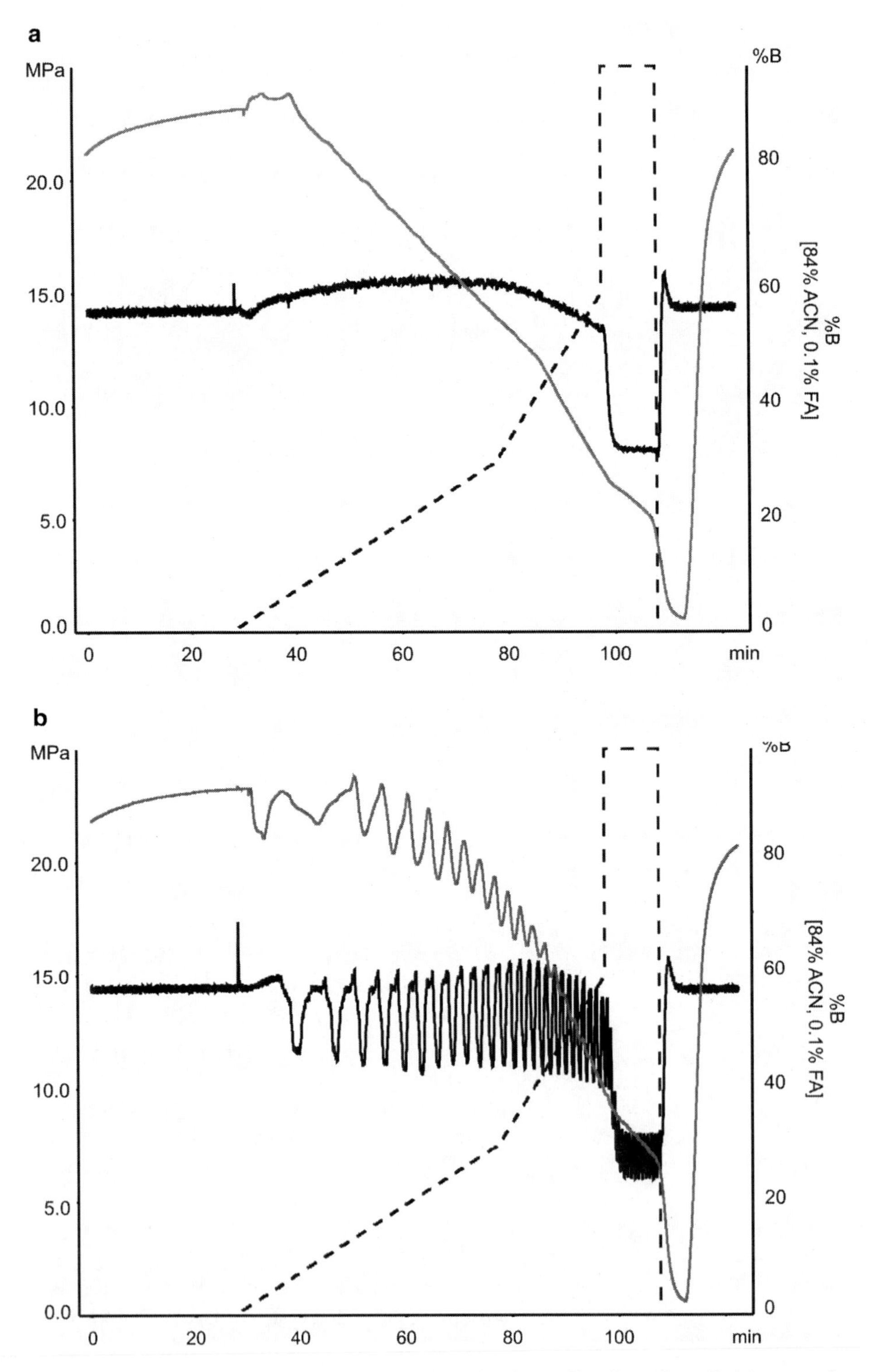


Fig. 11. Pressure (*black*) and conductivity (*gray*) plot during the gradient formation with **(a)** a correctly working pump and **(b)** with a broken pump.

#### 3.7.2. Clogging

A common problem of nano-LC is partial or total plugging of the nano-flow path. In systems without flow-splitter, the problem should lead to an increased pump pressure, whereas in systems with flow-splitters, the nano-flow rate should be reduced. However, in systems with flow-splitters, a partial leak upstream in the flow-path may also reduce the flow rate. The most efficient way to find a blockage is to check successively each component (*see* **Note 10**). In cases of blocked consumables such as capillaries, ESI emitters, or columns, the component should be exchanged. If sensors or valves are affected, specific procedures proposed by the manufacturer should be performed.

#### 3.7.3. Leaks

A widespread annoyance in nano-flow chromatography is the appearance of leaks. At very low flow rates, leaks may be hard to detect by visual inspection, because droplets may be very small and may evaporate before becoming visible. Major leaks in the nano-flow before components creating the major backpressure (e.g., analytical columns) may come along with a diminished pump or column pressure. In case of partial leaks between the column and detector, the measurement of the actual flow at appropriate sites of the nano-flow path could be helpful, since the flow after the leakage is usually diminished. Besides leaky connections, damaged rotor seals of switching valves may cause partial leaks. In case of damaged valves, the instructions given by the manufacturer should be followed to fix the problem.

#### 3.7.4. Dead Volumes

Large dead volumes mostly occur if connections are not made properly. Therefore, we recommend again to cut capillaries and to make connections with maximum care. The number of connections should be minimized, since each connection may contribute to the total dead volume. Dead volumes may cause either higher retention times or lead to peak broadening and, in some cases, corrupt the separation result totally. An extreme example is demonstrated in **Fig. 12**. When compared with a run with a regular setup (Panel A), Panel B shows a run with an artificial dead volume of about 2 μL introduced upstream of the analytical column. Retention times increased substantially, whereas the separation quality remained almost unaffected. In the run represented in Panel C, the same dead volume was introduced downstream to the separation column, where it acted as a remixing chamber for the already separated substances. In such cases, connections should be carefully checked (*see* **Note 11**). An effective but time-consuming way to detect dead volumes is to determine the delay times at several sites of the nano-flow path as described in **Subheading 3.2**.

#### 3.7.5. Carry-Over

In nano-LC-MS setups, carry-over can be very troublesome because of the extremely high sensitivity of modern mass spectrometers.

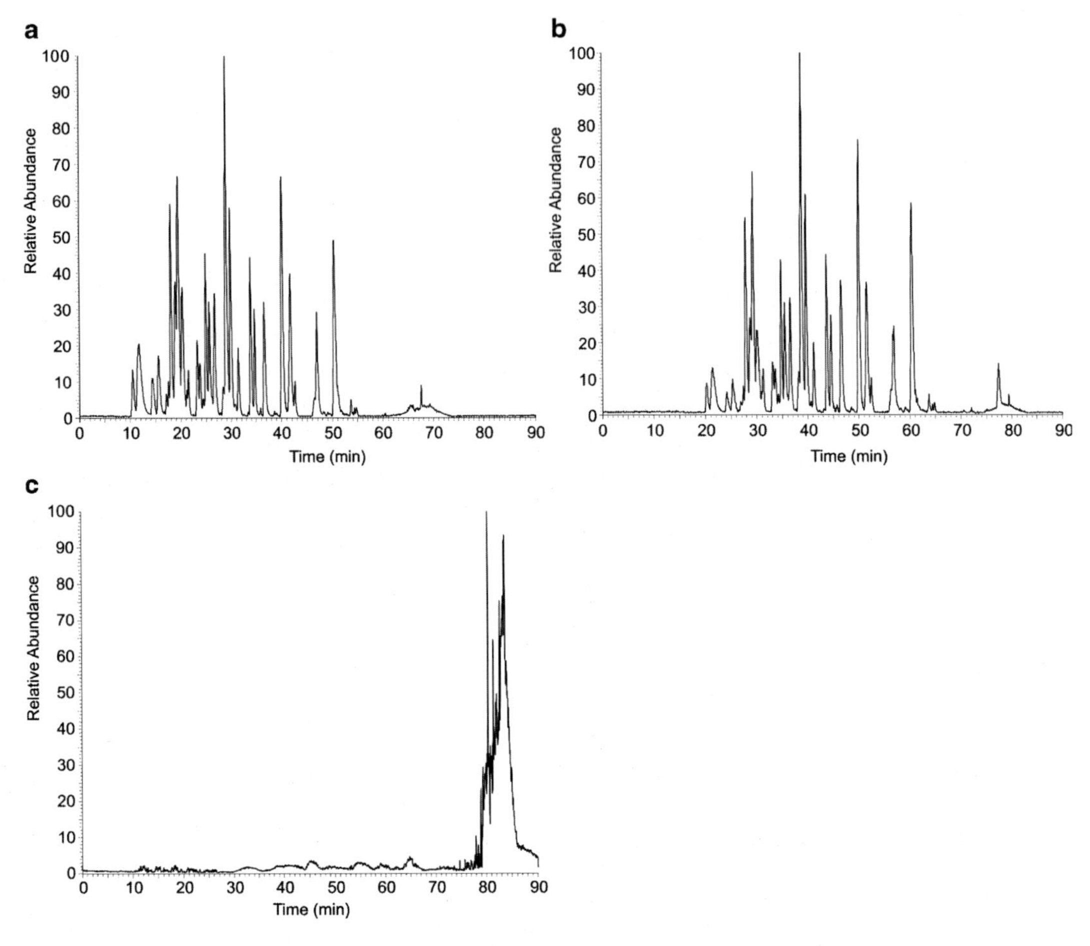


Fig. 12. Base peak intensity chromatograms (BIC) of nano-LC-MS runs of 100 fmol bovine carboanhydrase 2. **(a)** BIC without artificially introduced dead volume, **(b)** BIC with an artificially introduced 2 µL dead volume prior to the columns, leading to a shift of retention times, and **(c)** BIC with an artificially introduced 2 µL dead volume after the columns, acting as mixing chamber leading to corrupted results. LC was performed on an Ettan MDLC (GE Healthcare). Mass spectra were acquired on an LTQ linear ion trap (Thermo Fisher Scientific).

Carry-over of the sample between individual runs can be caused by improperly connected tubing, large dead volumes, damaged valves, or be an effect of auto-sampler malfunction. However, most commonly, carry-over is observed as a consequence of an LC system overloading. And example of the UV profile from a run of 10 pmol carboanhydrase 2 and of the following blank run is shown in **Fig. 13**. Overloading of the system usually does not lead to permanent effects, but it can be necessary to perform several blank runs to finally get rid of carry-over (*see* **Note 12**).

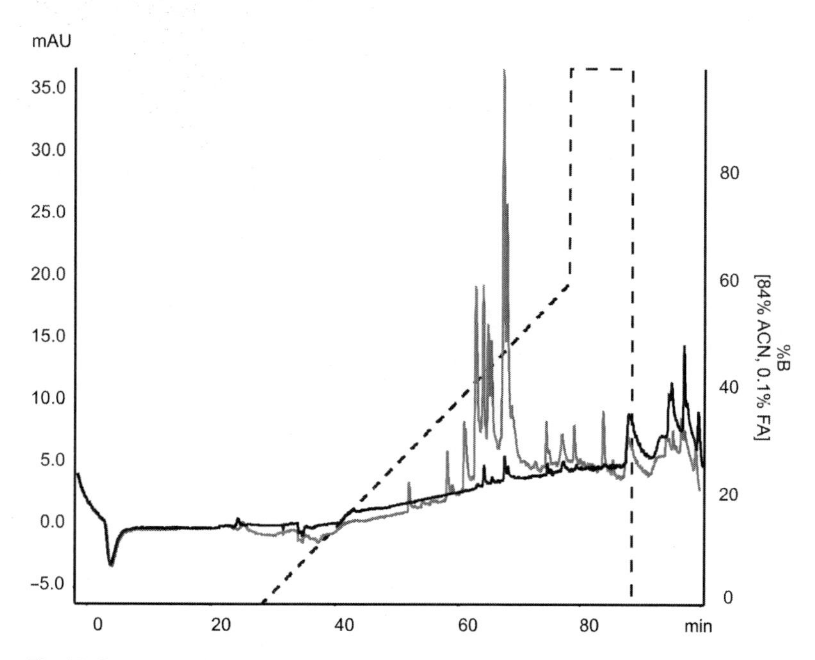


Fig. 13. Carry-over after chromatography of 10 pmol bovine carboanhydrase 2. *Gray line*: UV chromatogram of 10 pmol bovine carboanhydrase 2. *Black line*: UV chromatogram of the following blank run, showing significant carry-over. *Dashed line*: programmed gradient. LC was performed on an Ettan MDLC (GE Healthcare).

## 4. Notes

1. To our experience, nano-LC systems using conventional flow-splitters can be operated in a robust and reproducible way. However, it is a good idea to periodically control nano-flow-rates. Smaller variations of nano-flow-rates in the range of ±10% are routinely compensated by slightly increasing or decreasing the pumps' flow rate.
2. As an example, nano-LC systems in our lab are equipped with 75 μm ID columns and driven with 200-300 nL/min. They are predominantly tubed with 20 μm ID capillaries, except the tubing used to load samples on trap columns (**Subheading 2.1.3**). Since we use a high flow rate in the range of 5 μL/min to load the sample, tubing from the autosampler to the valve with the trap columns has an ID of 75 μm. The trap columns are connected to the valve with short capillaries of 30 μm ID.
3. To our experience, loss of hydrophilic peptides becomes especially relevant if loading is performed with organic content of >2% MeCN in the loading solvent.

4. When mounting the column, it is strongly recommended to follow the flow direction specified by the manufacturer because several columns only have frits at the column outlet. Mounting the column in the wrong direction may lead to the transfer of particles of the stationary phase to the nano-LC system.
5. Before the first experiments, several blank runs should be performed with fresh columns to remove potential contaminations. In addition, it is recommended to perform first runs with standard mixtures to test the performance of the column and to saturate binding sites, which may otherwise lead to partial sample loss.
6. In our lab, we do not store HPLC grade liquids in plastic containers because chemicals like plasticizers may diffuse into the solvents and contaminate them.
7. In our lab, ZDV unions are exclusively used to connect the capillary ESI-MS emitter with the fused-silica tubing coming from the analytical column. After connection of a new ESI emitter, the time until a droplet can be observed at the tip of the needle was measured. Typically, the time does not exceed 10 s.
8. To easily calculate the volume of a tubing, several "HPLC Tubing Volume Calculators" are available online, e.g., http://www.ionsource.com/Card/loop/loop.htm or http://www.upchurch.com/TechInfo/conversions.asp.
9. To ensure a constant quality of the test mixtures, these standards are prepared as a stock solution divided into several aliquots and kept frozen until utilization.
10. To our experience, blockage is more frequently found at components with narrow aperture such as ESI tips, analytical and trap columns, but also in capillaries with small IDs. Valves, flow-meters, conductivity cells, and UV cells with broader flow passages are rarely blocked.
11. To our experience, most dead volumes are introduced by inappropriately connected capillaries, damaged fittings, or ZDV unions. Therefore, the system should be checked with standard mixtures after each assembly of capillaries.
12. It is a good laboratory praxis to run blanks between each analytical run. In our lab, blanks are monitored by tandem mass spectrometry, and the obtained MS/MS data screened for carry-over from previous analytical runs. To get rid of carry-over, it is in most cases not sufficient to only equilibrate the system for a while. Instead, blank injections and valve switchings have to be performed to rinse the entire system. Therefore, we perform several short blank runs with short gradients in these cases.

## Acknowledgments

We thank Dr. S. Krebs for critical reading of the manuscript. Research work in our laboratory is granted by the "Deutsche Forschungsgemeinschaft," DFG research unit FOR 478 AR 362/1-4, DFG clinical research unit KFO 128 1-2, BMBF FUGATO project AZ 0313388, DFG research unit FOR 585 AR 362/3-2.

### References

1. Ishihama, Y. (2005) Proteomic LC-MS systems using nanoscale liquid chromatography with tandem mass spectrometry. *J. Chromatogr. A* **1067**, 73–83.
2. Frohlich, T. and Arnold, G. J. (2006) Proteome research based on modern liquid chromatography-tandem mass spectrometry: separation, identification and quantification. *J. Neural Transm.* **113**, 973–994.
3. Link, A. J., Eng, J., Schieltz, D. M., Carmack, E., Mize, G. J., Morris, D. R., Garvik, B. M. and Yates, J. R., III (1999) Direct analysis of protein complexes using mass spectrometry. *Nat. Biotechnol.* **17**, 676–682.
4. Shen, Y., Zhao, R., Berger, S. J., Anderson, G. A., Rodriguez, N. and Smith, R. D. (2002) High-efficiency nanoscale liquid chromatography coupled on-line with mass spectrometry using nanoelectrospray ionization for proteomics. *Anal. Chem.* **74**, 4235–4249.
5. Shen, Y. and Smith, R. D. (2005) Advanced nanoscale separations and mass spectrometry for sensitive high-throughput proteomics. *Expert Rev. Proteomics* **2**, 431–447.
6. Wilm, M., Shevchenko, A., Houthaeve, T., Breit, S., Schweigerer, L., Fotsis, T. and Mann, M. (1996) Femtomole sequencing of proteins from polyacrylamide gels by nano-electrospray mass spectrometry. *Nature* **379**, 466–469.
7. Wilm, M. and Mann, M. (1996) Analytical properties of the nanoelectrospray ion source. *Anal. Chem.* **68**, 1–8.
8. Mitulovic, G., Smoluch, M., Chervet, J. P., Steinmacher, I., Kungl, A. and Mechtler, K. (2003) An improved method for tracking and reducing the void volume in nano HPLC-MS with micro trapping columns. *Anal. Bioanal. Chem.* **376**, 946–951.
9. Pascual, G. J. (1996) Development of a precolumn capillary liquid chromatography switching system for coupling to mass spectrometry. *J. Microcolumn Sep.* **8**, 383–387.
10. Meiring, H. D., van der Heeft, E., ten Hove, G. J. and de Jong, A. P. J. M. (2002) Nanoscale LC-MS: technical design and applications to peptide and protein analysis. *J. Sep. Sci.* **25**, 557–568.
11. Licklider, L. J., Thoreen, C. C., Peng, J. and Gygi, S. P. (2002) Automation of nanoscale microcapillary liquid chromatography-tandem mass spectrometry with a vented column. *Anal. Chem.* **74**, 3076–3083.
12. Emmett, M. R. and Caprioli, R. M. (1994) Micro-electrospray mass spectrometry: ultra-high-sensitivity analysis of peptides and proteins. *J. Am. Soc. Mass Spectr.* **5**, 605–613.
13. Temesi, D. and Law, B. (1999) The effect of LC eluent composition on MS responses using electrospray ionization. *LC-GC* **17**, 626–632.
14. Noga, M., Sucharski, F., Suder, P. and Silberring, J. (2007) A practical guide to nano-LC troubleshooting. *J. Sep. Sci.* **30**, 2179–2189.

# Chapter 8

# Multidimensional Protein Identification Technology

Katharina Lohrig and Dirk Wolters

## Summary

Over the past years, large-scale analysis of proteomes gained increased interest to obtain a fast but nevertheless comprehensive overview about cellular protein content. While a complete proteome cannot be covered using current technologies because of its enormous diversity, subfractionation to reduce the complexity has become mandatory. While 2D-PAGE is well established as a high-resolution protein separation technique, it suffers from drawbacks, which can be overcome by using peptide separation methods based on multidimensional liquid chromatography. One of these technologies is multidimensional protein identification technology (MudPIT). It consists of two orthogonal separation systems – strong cation exchange (SCX) and reversed phase (RP) – coupled online in an automated fashion to mass spectrometric detection. This method offers the possibility to analyze high-complex peptide mixtures in a single experiment.

**Key words:** MudPIT, Liquid chromatography, Peptide separation, Mass spectrometry

## 1. Introduction

Upon completion of the human genome and many other genome projects, the elucidation of the cellular protein content has triggered renewed interest, since much information cannot be gained by genomic analysis alone. Changes in the highly dynamic proteome of cells as well as post-translational modifications, e.g. phosphorylations and glycosylations, require primarily high-throughput identification of proteins and peptides from complex cell lysates. For this purpose, mass spectrometry has been the method of choice for nearly a decade. However, owing to limitations of the mass spectrometric instrumentation, the process of identification has to be proceeded by rather sophisticated separation

Jörg Reinders and Albert Sickmann (eds.), *Proteomics,* Methods in Molecular Biology, Vol. 564
DOI: 10.1007/978-1-60761-157-8_8, © Humana Press, a part of Springer Science+Business Media, LLC 2009

mechanisms. Since mass spectrometers cannot resolve more than a certain number of ion signals within a specific timeframe, reduction of sample complexity is strongly recommended. For this task, several different strategies are applicable.

First of all, subcellular fractionation is the most commonly used initial step for protein prefractionation. To further reduce sample complexity on the protein level, separation methods based on PAGE (polyacrylamide gel electrophoresis), e.g., 2D-PAGE *(1)*, are very popular in proteomic approaches. However, those separation methods are biased against certain subclasses of proteins like hydrophobic membrane proteins, proteins with extreme size or p*I* limitations *(2–4)*. Therefore, alternative methods based on peptide separation instead of protein separation have been introduced because peptides in complex mixtures exhibit a more uniform behavior than individual subclasses of proteins. In many multidimensional liquid chromatography approaches, offline SCX chromatography or strong anion exchange (SAX) chromatography based prefractionation of complex peptide mixtures are followed by a reversed-phase (RP) chromatography directly coupled to mass spectrometry *(5, 6)*. These approaches offer several advantages. First of all, compatibility of samples and buffers between the first and the second dimension can be ensured because of sample cleanup steps during offline fractionation. This enables the combination of nearly all chromatographic separation modes as long as they exhibit different properties. Examples would be SCX/RP, SAX/RP, hydrophilic liquid interaction chromatography (HILIC)/RP, or high-pH-RP/low-pH-RP. Moreover, sample amounts can be readily adjusted, since the column dimensions of both separations are not readily connected to each other. Thereby, an SCX-separation on an analytic 4.6 mm inner diameter column can be offline coupled to a nano-LC-RP-separation interfaced with mass spectrometry. However, despite obvious advantages, offline multidimensional chromatography is also prone to several drawbacks. It is rather time-consuming and, although efforts have been made, it is not to be automated without manual steps. Furthermore, sample losses can occur during offline fractionation, thereby limiting the number of identified proteins/peptides. This in turn can lead to problems regarding the reproducibility of results.

To circumvent some of the problems connected with offline fractionation approaches, Washburn et al. introduced the multidimensional protein identification technology, termed MudPIT, in 2001 *(7)*. The MudPIT system consists of a triphasic column packed into an ESI-emitter tip directly coupled to a mass spectrometer. Most commonly and as originally published, a RP-precolumn is followed by a SCX-precolumn, and finally the main RP-separation column. The principle of a MudPIT analysis is depicted in **Fig. 1**. Briefly, peptides are trapped initially on

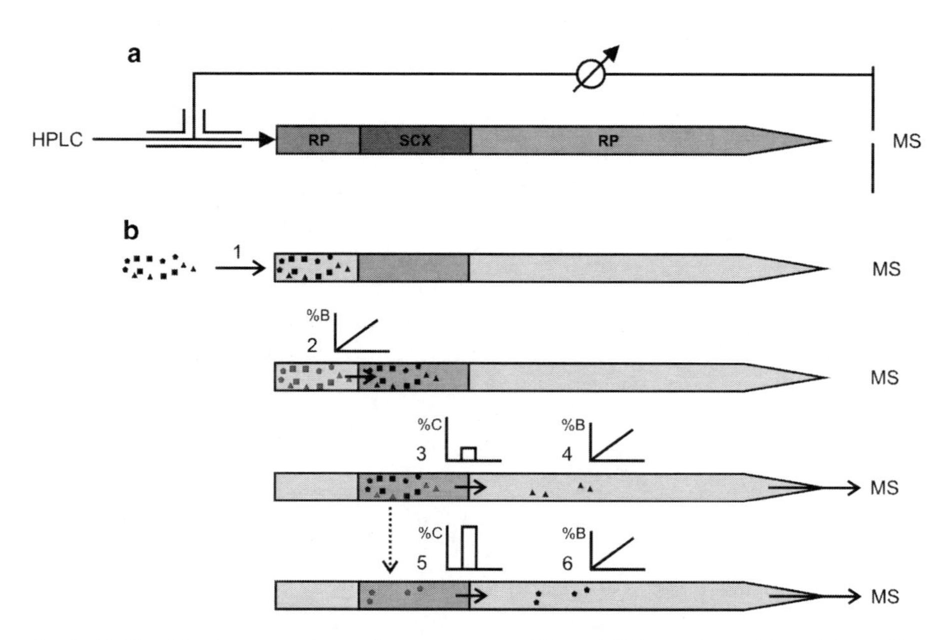


Fig. 1. (**a**) MudPIT column consists of three main parts: RP- and SCX-precolumn followed by an RP-separation phase in the emitter tip. High voltage is applied by a T-split in front of the column. (**b**) Schematic view of a MudPIT analysis. *(1)* Sample is trapped onto RP-precolumn during loading process. *(2)* Peptides are eluted onto the SCX-precolumn while salts and other contaminants are washed away. *(3)* Trapped peptides are eluted from the SCX- onto RP-separation column by applying a single salt step. *(4)* Based on their hydrophobicity, peptides are separated on the RP-phase and elute by an acetonitrile gradient directly into the MS for analysis. *(5, 6)* Increasing salt concentrations lead to stepwise release of peptide from SCX-phase onto the RP-column.

the RP-precolumn and are eluted by a single RP-gradient onto the SCX-phase. By applying increasing concentrations of salt, peptides are washed stepwise from the SCX-phase onto the RP-separation column where a normal RP-gradient separates the peptides directly and without any additional dead volumes by their hydrophobicity into the mass spectrometer. In contrast to offline fractionation mode, MudPIT can easily be automated with minimal manual interference. Analysis of high-complex sample mixtures can thereby be enabled within 24 h, which results in time-saving in comparison with offline MDLC. Furthermore, MudPIT avoids loss of low-abundant peptides by limiting contact with glass or plasticware, as caused in the offline fractionation mode. Identification of low- and high-abundant peptides in the same approach is thereby enabled. A combined approach of 2D-PAGE and MudPIT analysis for the detailed screening of metabolism pathways in rice, e.g., could only identify 22% of 2,528 proteins by 2D-PAGE, while using MudPIT alone 2,363 proteins were verified *(8)*. This high identification rate is a hint to a common advantage of multidimensional peptide separation. In principle, a single fragmentation spectrum is sufficient for unambiguous identification of a protein in most cases. Without multidimensional separation of a complex digest, low-abundant peptides can be masked or quenched during the ESI process. However, in MudPIT analysis, the peptides of a low-abundant protein are spread over several salt steps, thereby increasing the possibility for successful fragmentation and identification.

Since MudPIT is commonly used in a nanoscale setup with flow rates of about 250 nL/min, it also has some disadvantages to deal with. In contrast to offline multidimensional separations, the MudPIT column can only cope with a limited amount of sample, and consequently the RP- as well as the SCX-column can exhibit overloading effects. In addition, the columns have to be home-built in a reproducible fashion, since virtually no commercial supplier is available. This may somewhat limit the use of MudPIT in tightly controlled laboratory environments. Nevertheless, the reproducibility of two consecutive MudPIT runs – although strongly sample-dependent – has been estimated between 60% and 80% *(9, 10)*. Liu et al., for example, identified 1,751 proteins from a complex yeast cell lysate in a cumulate data set of nine consecutive MudPIT experiments; 1,331 (76%) of these proteins were identified in at least two or more analysis, whereas the reproducibility is highly dependent on the protein abundance (protein copy number per cell). In conclusion, two to three MudPITs can be analyzed during the same time frame as a single offline MDLC (typically 60 fractions collected during the first dimension, with 60 consecutive RP-separations in the second dimension).

Consequently, MudPIT analysis has been used in an increasing number of publications covering many of the main proteomic topics, such as protein identification, quantification and post-translational modifications. The application of MudPIT also to specialized samples such as membrane proteins led to an ever-increasing number of identified proteins. Fischer et al. identified 326 transmembrane spanning proteins from *C. glutamicum* by using MudPIT in combination with specialized digest conditions for membrane proteins *(11)*. In addition to qualitative data, MudPIT can as well be coupled to established labeling techniques such as iTRAQ *(12)* or metabolic labeling strategies as, e.g., $^{14}N/^{15}N$ labeling *(13)* or SILAC. Thereby, a quantitative element can be gained of MudPIT analysis as shown by Maurya et al. *(12)*, whereas novel bioinformatics tools are currently developed. *(14)*.

The identification of post-translational modification most often requires direct mapping of the modified residue by mass spectrometry. Since modified peptides such as phosphopeptides can be quenched in the presence of a surpassing number of unmodified species, enhanced prefractionation in combination with MudPIT may increase the yield of identified phosphorylation sites *(15)*.

In the following, we provide a simple to use protocol for a basic MudPIT setup. It briefly states buffer conditions as well as used gradients and sample preparation procedures. However, the success of any MudPIT analysis strongly depends on the nature of the applied sample. Therefore, sample pretreatment, employed separation phases/modes, buffer conditions, salt steps, gradient slopes as well as the mass spectrometric conditions may need to be adjusted to the analytic question to obtain optimal results.

## 2. Materials

### 2.1. Chemicals

In general, solvents should be of the highest purity available, but at least p.a. grade.

1. Water (18 MΩ).
2. Acetonitrile.
3. Methanol.
4. Formic acid (99.9%).
5. Ammonium acetate.

### 2.2. HPLC Buffers

1. Buffer A: 0.1% (v/v) formic acid and 2% (v/v) acetonitrile.
2. Buffer B: 0.1% (v/v) formic acid and 80% (v/v) acetonitrile.
3. Buffer C: 250 mM ammonium acetate and 0.1% (v/v) formic acid.
4. Buffer D: 1.5 M ammonium acetate and 0.1% (v/v) formic acid.

### 2.3. Other Materials

1. Capillaries are available from a range of suppliers such as Polymicro and Agilent. The use of non-deactivated polyimide-coated capillaries is recommended.
2. HPLC resins used in this example were ProntoSIL 200-3-C18 (Bischoff Analysentechnik, Leonberg, Germany) and PartiSphere SCX, 5 µm (Whatman GmbH, Dassel, Germany).

## 3. Methods

### 3.1. Sample Preparation

1. Samples for MudPIT analysis should be in an adequate loading buffer system concerning column arrangement without any perturbing salts (*see* **Note 1**). If samples are not in adequate loading buffer system, precleaning is recommended

(*see* **Note 2**). The choice of loading buffers can vary depending on the nature of the sample, e.g., for digests of cytosolic proteins 0.1-5% formic acid is recommended, while digests of primarily hydrophobic proteins can be supplemented with small amounts of acetonitrile (2–5%) in addition. Consequently, the loading buffer system should have acidic pH, if RP is the first phase used in a triphasic setup. Recommended buffer in this case is buffer A of the MudPIT system (*see* **Note 3**).

2. Loading amount of sample should be around 250 μg for the system described below. Anyway, changing column arrangements or dimensions may require more or less material to be applied. Furthermore, the capacity of the employed packing material has to be considered, since some SCX-materials, for example, show larger capacity than others.

### 3.2. Column Preparation and Packing Material

1. In general, column preparation and dimensions should match to the analytic question. Low sample amounts, for example, do not require excessive lengths of precolumns in this context.
2. The column described in this chapter consists of three parts (RP- and SCX-precolumns followed by RP-separation column) packed into a 100 μm ID × 360 μm OD capillary with a nano-emitter tip at one end (*compare* **Fig. 1** and **Note 4**). For both RP-phases, ProntoSil C18, 3 μm particle size, 200 A pore size was used. Whatman SCX, 5 μm particle size, 200 A pore size was chosen for the intermediary SCX-phase.
3. Prepare slurries of 50 mg/400 μL methanol for each packing material, sonicate and vortex thoroughly. Start with preparation of main separation column by placing the RP-slurry within a 1.5 mL reaction tube into a pressure-packing bomb (*see* **Notes 4** and **5**). Introduce the column with the non-emitter end into the bomb such that the end reaches into the slurry. Tightly fix the capillary and pressurize the bomb with argon (assure high purity of argon). Thereby, the material is filled into the column. Interrupt the packing procedure if desired length of 10 cm has been reached. Next, add the SCX-precolumn in the same way, stopping at 4 cm length (14 cm total column length) and add the RP-precolumn stopping at 3 cm (17 cm in total) (*see* **Note 5**).
4. Prior to initial use, the MudPIT column should be preconditioned and checked by applying a digest of a standard protein or a peptide mixture of known composition. Samples are applied as stated below and a simplified HPLC gradient is applied simulating a one-step MudPIT experiment. Following a 10-min wash with buffer A at a flow rate of 250 nL/min, 100% B are applied for 10 min before returning to 100% A. Subsequently, the column is rinsed with 100% D for 10 min,

followed by 10 min 100% A and a short, steep gradient to 100% B (e.g., 5% B/min) (*see* **Note 6**).

### 3.3. Sample Application

1. Prior to use, MudPIT columns should be equilibrated in buffer A for 15 min either by plugging the column to an HPLC system at 250 nL/min or again by placing the column into packing bomb with buffer A in the reaction tube.
2. Place up to 50 μL of sample into a reaction tube and load sample within the bomb-packing device to the column by pressurizing. Loading proceeds until column runs dry.

### 3.4. HPLC Setup

1. HPLC system consists of nano-LC, e.g., Ultimate™ (Dionex, Idstein, Germany) or equal with four different solvent channels (for solvents, *see* **Subheading 2** and **Note 7**). Flow rate for direct coupling to MS should be 250 nL/min with the MudPIT column fitted to a T-split to apply high voltage (**Fig. 1a**).
2. The first gradient serves two purposes. First, the column is washed with buffer A for 30 min to remove salts and to condition the SCX-phase. Second, the sample is transferred onto the SCX-precolumn by a linear gradient of 0.36% B/min until 55% B is reached followed by a 10-min washing step at 100% B (*see* **Note 8**).
3. Subsequently, peptides are stepwise eluted from the SCX-column onto the RP-separation column by applying increasing amounts of buffers C and D. In detail, for each run, the column is equilibrated for 5 min with buffer A followed by a salt step for 2 min and a 10-min re-equilibration step with buffer A. Peptides are separated by a linear RP-gradient with 0.75% B/min until 55% B. On a steep gradient from 55% to 100% buffer B in 15 min, the column is washed for 5 min at 100% buffer B. For an example of several gradients after individual salt steps, *see* **Fig. 2**. It displays selected base peak chromatograms of a tryptic digest of 400 μg protein.

   Recommended salt steps during a complete MudPIT analysis are: 10%, 20%, 30%, 40%, 50%, 75%, 90%, and 100% buffer C and 33%, 66%, and 100% buffer D (*see* **Note 9**).
4. The MudPIT run is completed by a last salt step to rinse the column. It consists of the same basic gradient as described in the previous step, with a prolonged salt step for 10 min (instead of 2 min as previously) at 100% buffer D.

### 3.5. Mass Spectrometric Detection

For mass spectrometric detection, a LTQ linear ion trap with a nano-ion spray source is used (LTQ XL, ThermoFisher Scientific, Dreieich, Germany). Instrument parameters may vary for different setups; however, in this case, spray voltage of 2 kV is used. All spectra are acquired with normal scan speed. Most commonly, a survey spectrum recorded from 400 to 2,000 is recorded, of

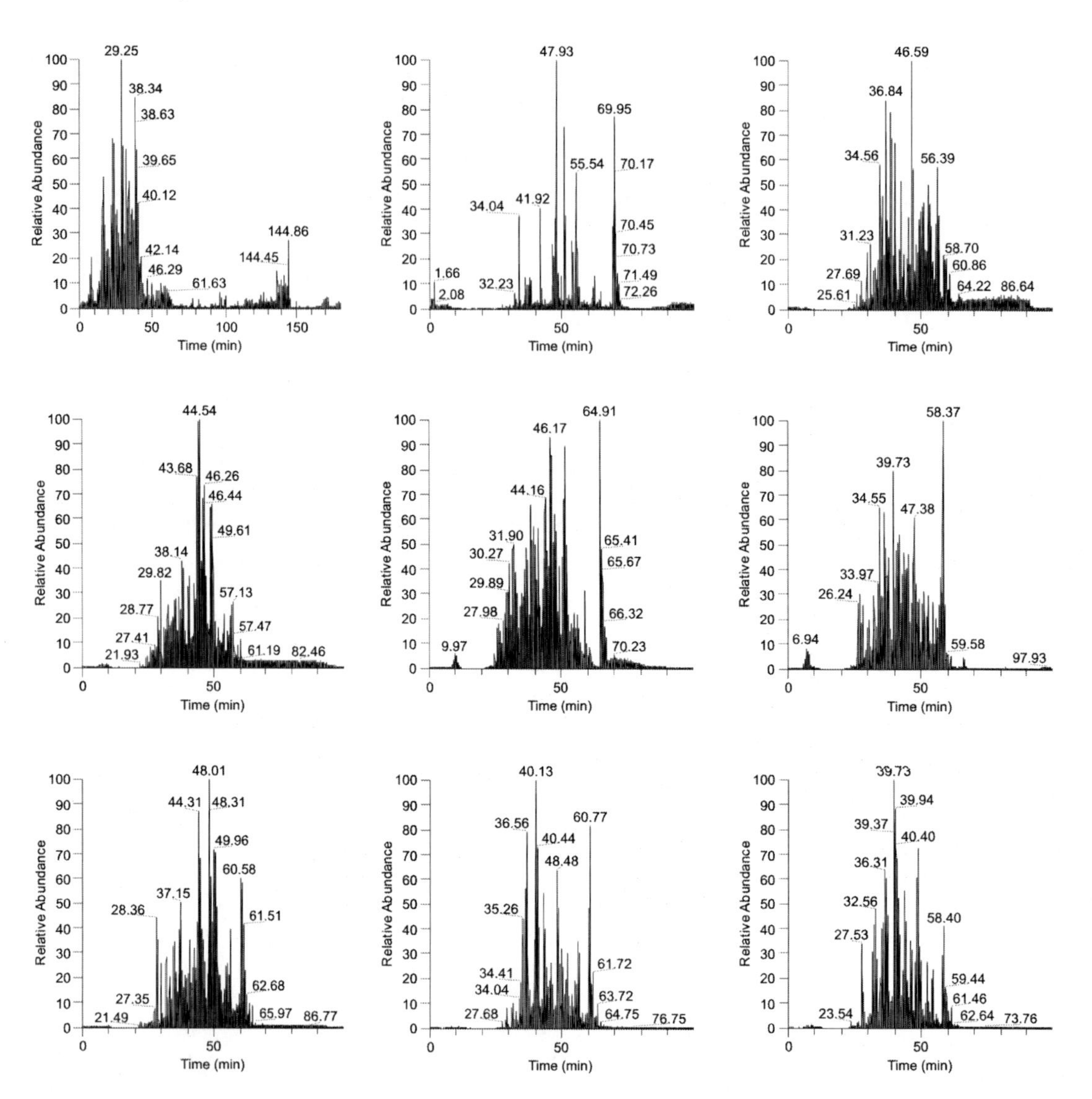


Fig. 2. Several selected base peak chromatograms for a MudPIT analysis of a tryptic digest with 400 μg protein are shown: First washing RP-gradient, 10%; 20%, 30%, 40%, 50%, 60%, 80%, and 100% buffer C.

which the three most intense ion signals are chosen for subsequent MS/MS fragmentation with a collision energy of $q = 35\%$. Dynamic exclusion time is set to 30 s after a single occurrence of the respective ion (*see* **Note 10**).

### 3.6. Data Interpretation

In general, peptides of a given protein can elute all over a MudPIT experiment. Therefore, it is mandatory to either search all individual runs and combine the results or to directly search all the combined raw data at once. However, because of the high number of acquired spectra during a MudPIT experiment, usually the former option is chosen, since many algorithms cannot deal with the increased number of spectra.

In case of LTQ data sets, raw-files can be directly submitted to database searches using BioWorks™ (e.g., version 3.3) with the implemented Sequest search algorithm. Using the "Load MultiConsensus Results" option, individual results files can be combined and thus provide an overview of the complete MudPIT experiment and its results. As an alternative, result files (*.dta- and *.out-files) can also be combined by DtaSelect *(16)* or ProtQuant *(14)*. The latter also offers a bioinformatic solution for label-free quantification of data sets resulting from MudPIT experiments. Other possibilities include the use of alternative search algorithms, such as Mascot™ (MatrixScience, London, UK) and/or combination with data-management systems, e.g., Integra (MatrixScience) or MS-LIMS (http://genesis.ugent.be/ms_lims/).

## 4. Notes

1. Samples should not contain high concentrations of either ion pairing reagents, such as trifluoroacetic acid, or detergents, such as sodium dodecylsulfate. While the first can be tolerated in small amounts (0.1%) to enhance binding of peptides to the first RP-phase, SDS can severely limit the separation efficiency of the RP-columns and should be avoided.
2. Precleaning of samples can be achieved by various methods. Precipitations of proteins or dialysis can be used for detergent depletion. Non-digested proteins may be removed by ultrafiltration using molecular weight cutoffs from 5 to 10 kDa. Moreover, the use of micro-C18 columns is recommended for the removal of salts.
3. Furthermore, the use of loading buffer A enables trapping of peptides on the SCX-precolumn in case of overloading of the RP-precolumn and thus provides the possibility for increasing the applicable sample amount.
4. Preparations of capillaries with emitter tips have been described elsewhere *(17)* (www.proteomcenter.org). Parameters for reproducible fabrication of tips can vary for each commercial pipette puller and are, therefore, not readily transferable to any other system. However, the inner diameter of the emitter tip opening should be smaller than the particle size of the packing material (usually around 3–5 μm). In addition, emitter tips containing prefabricated tips are commercially available (PicoFrits, New Objectives).

5. Pressure-packing devices can either be self-constructed or are commercially available, e.g., High Pressure MicroLoader/ Injection Cell (Next Advance Inc., Averill Park, NY, USA).
6. The quality of online MudPIT column as described cannot be assessed by UV chromatograms. Therefore, peak shapes as well as resolutions need to be checked within the base peak chromatogram of the coupled mass spectrometer. Note that high-salt concentrations within the MudPIT chromatography can lead to inferior spray conditions and in turn a high current of the spray itself.
7. In general, the used buffers and resins need to fulfill two basic requirements. First, they need to be compatible with both employed separation phases, neither providing damage to the resins, e.g., by extreme pH, nor interfering with binding of analyst to either of the two phases. Second, the employed separation systems need to be compatible with each other under online conditions. While several systems may provide the required orthogonality in offline multidimensional chromatography, e.g., RP- and HILIC-mode, it is not possible to re-buffer or change the solvents directly in an online system.

   Ensure compatibility of the chosen buffer system with mass spectrometric conditions. Some additives might interfere with a stable electrospray or can even permanently impair the sensitivity of the mass spectrometer, e.g., in case of phosphate buffers used for application in SCX with direct coupling to mass spectrometry.
8. The initial RP-gradient also separates and elutes peptides, which do not bind to the SCX-phase. Therefore, it should already be recorded by mass spectrometry.
9. Number and percentages of salt steps strongly depend on the sample composition as well as its complexity and have to be adjusted for each sample individually. Also, the choice of appropriate RP-gradients can influence the separation quality by using either shallower or steeper gradients.
10. Fast-acquisition mass spectrometry systems are mandatory for MudPIT analysis. Despite the fractionation by salt steps prior to RP-separations, the resulting survey spectra during RP-gradients may still contain high numbers of coeluting peptides. To prevent undersampling, the scan speed of the used mass spectrometer has to match this number as close as possible. Otherwise, only signals with high intensity will be chosen for fragmentation, while low-intensity signals may be lost.

## References


1. Gorg, A., Weiss, W., and Dunn, M. J. (2004) Current two-dimensional electrophoresis technology for proteomics. *Proteomics* **4**, 3665–85.
2. Rabilloud, T., Adessi, C., Giraudel, A., and Lunardi, J. (1997) Improvement of the solubilization of proteins in two-dimensional electrophoresis with immobilized pH gradients. *Electrophoresis* **18**, 307–16.
3. Righetti, P. G., Bossi, A., Gorg, A., Obermaier, C., and Boguth, G. (1996) Steady-state two-dimensional maps of very alkaline proteins in an immobilized pH 10-12 gradient, as exemplified by histone types. *J Biochem Biophys Methods* **31**, 81–91.
4. Harder, A., Wildgruber, R., Nawrocki, A., Fey, S. J., Larsen, P. M., and Gorg, A. (1999) Comparison of yeast cell protein solubilization procedures for two-dimensional electrophoresis. *Electrophoresis* **20**, 826–9.
5. Davis, M. T., Beierle, J., Bures, E. T., McGinley, M. D., Mort, J., Robinson, J. H., Spahr, C. S., Yu, W., Luethy, R., and Patterson, S. D. (2001) Automated LC–LC–MS–MS platform using binary ion-exchange and gradient reversed-phase chromatography for improved proteomic analyses. *J Chromatogr B Biomed Sci Appl* **752**, 281–91.
6. Wagner, Y., Sickmann, A., Meyer, H. E., and Daum, G. (2003) Multidimensional nano-HPLC for analysis of protein complexes. *J Am Soc Mass Spectrom* **14**, 1003–11.
7. Washburn, M. P., Wolters, D., and Yates, J. R., 3rd (2001) Large-scale analysis of the yeast proteome by multidimensional protein identification technology. *Nat Biotechnol* **19**, 242–7.
8. Hunter, T. C., Andon, N. L., Koller, A., Yates, J. R., and Haynes, P. A. (2002) The functional proteomics toolbox: methods and applications. *J Chromatogr B* **782**, 165–81.
9. Liu, H., Sadygov, R. G., and Yates, J. R., 3rd (2004) A model for random sampling and estimation of relative protein abundance in shotgun proteomics *Anal Chem* **76**, 4193–201.
10. Sadygov, R. G., Eng, J., Durr, E., Saraf, A., McDonald, H., MacCoss, M. J., Yates, J. R. (2002) Code developments to improve the efficiency of automated MS/MS spectra interpretation. *J Proteome Res* **1**, 211–5.
11. Bachelot, C., Cano, E., Grelac, F., Saleun, S., Druker, B. J., Levy-Toledano, S., Fischer, S., and Rendu, F. (1992) Functional implications of tyrosine protein phosphorylation in platelets Simultaneous studies with different agonists and inhibitors. *Biochem J* **284** (Pt 3), 923–8.
12. Maurya, P., Meleady, P., Dowling, P., and Clynes, M. (2007) Proteomic approaches for serum biomarker discovery in cancer. *Anticancer Res* **27**, 1247–55.
13. Washburn, M. P., Koller, A., Oshiro, G., Ulaszek, R. R., Plouffe, D., Deciu, C., Winzeler, E., and Yates, J. R., 3rd (2003) Protein pathway and complex clustering of correlated mRNA and protein expression analyses in *Saccharomyces cerevisiae. Proc Natl Acad Sci U S A* **100**, 3107–12.
14. Bridges, S. M., Magee, G. B., Wang, N., Williams, W. P., Burgess, S. C., and Nanduri, B. (2007) ProtQuant: a tool for the label-free quantification of MudPIT proteomics data. *BMC Bioinformatics* **8 Suppl** 7, S24.
15. MacCoss, M. J., McDonald, W. H., Saraf, A., Sadygov, R., Clark, J. M., Tasto, J. J., Gould, K. L., Wolters, D., Washburn, M., Weiss, A., Clark, J. I., and Yates, J. R., 3rd (2002) Shotgun identification of protein modifications from protein complexes and lens tissue. *Proc Natl Acad Sci U S A* **99**, 7900–5.
16. Tabb, D. L., McDonald, W. H., and Yates, J. R., 3rd (2002) DTASelect and contrast: tools for assembling and comparing protein identifications from shotgun proteomics. *J Proteome Res* **1**, 21– 6.
17. Delahunty, C. and Yates, J. R., 3rd (2005) in Encyclopedia of Genetics, Genomics, Proteomics and Bioinformatics (Dunn, M. J., Ed.), Wiley.

# Chapter 9

# Characterization of Platelet Proteins Using Peptide Centric Proteomics

Oliver Simon, Stefanie Wortelkamp, and Albert Sickmann

## Summary

In modern proteomics, undersampling of low abundant, cumbersome, and hydrophobic proteins states one of the major problems. To overcome this, especially in two 2D-PAGE (two-dimensional polyacrylamide gel electrophoresis) eminent drawbacks, the so-called peptide-centric techniques have been developed. These approaches do not separate proteins prior to digestion, but instead proteolytically generate peptide mixtures after it. However, by this procedure already complex protein mixtures become even more extensive peptide mixtures. Particularly, when dealing with large proteomes, the generated sample complexity is vast and therefore difficult to analyze. When separated and analyzed by LC/MS, too many peptides may enter the mass spectrometer at a certain time point, and only a small fraction of ions is selected for subsequent MS/MS analysis. Although protein hydrophobicity and size play minor roles (as long as protease cleavage sites are accessible), low copy number can severely limit identification rates. To reduce the amount of peptides entering the mass spectrometer simultaneously without the loss of overall proteomic information, different techniques have been developed. Among these, an approach is represented by COFRADIC (**Co**mbined **Fr**actional **Di**agonal **C**hromatography).

COFRADIC is a chromatography-based technique enabling the sorting of peptides due to retention time shifts after a specific modification step. In the original approach, a complex peptide mixture is separated by a primary RP-HPLC (reversed-phase high-performance liquid chromatography) run and fractions are retained. Subsequently, these fractions are modified to specifically change retention times of peptides and separated in one or more secondary RP-HPLC runs. In this chapter, COFRADIC approaches for methionine or cysteine containing as well as N-terminal peptides are described.

Besides the reduction of sample complexity, the major advantage of COFRADIC might be seen in its versatility. Nearly every feature unique for a subset of peptides, which can be specifically modified by a sorting reaction, is accessible for COFRADIC. Among these are protein phosphorylation, N-glycosylation, and in vivo protein processing sites. Finally, COFRADIC allows the analysis of large numbers of samples and is highly automatable.

**Key words:** Chromatography, Mass spectrometry, Proteomics, Platelets, Peptide centric, Proteome, Database

Jörg Reinders and Albert Sickmann (eds.), *Proteomics,* Methods in Molecular Biology, Vol. 564
DOI: 10.1007/978-1-60761-157-8_9, © Humana Press, a part of Springer Science+Business Media, LLC 2009

## 1. Introduction

Modern proteomics can be divided into a number of general areas. Among these, one may differentiate between classic gel-based 2D-PAGE (two-dimensional polyacrylamide gel electrophoresis) procedures *(1, 2)* and furthermore the so-called gel-free approaches.

Although being the most frequently used technique, 2D-PAGE has some inherent drawbacks. Owing to limited loading capacity of the gels, only proteins with more than 1,000 copies can be detected *(3, 4)*. Moreover, protein size itself is an important factor. Usually, proteins smaller than 10 kDa or larger than 150 kDa evade detection, e.g., by not entering the gel. A third class of proteins that is difficult to analyze are hydrophobic ones *(3, 4)*. These proteins are difficult to solubilize and very often precipitate near their isoelectric point. The aforementioned pitfalls and the fact that 2D-PAGE is not accessible for high-throughput studies led to the development of additional gel-free approaches.

Here we present a so called peptide-centric approach. The term peptide-centric indicates the initial step not to be a separation of proteins, but the digestion of a whole-protein mixture with a specific protease, such as trypsin. Especially in the analysis of whole proteomes, the generated amount of peptides is therefore enormous. Despite the use of modern mass spectrometers with short duty cycles and increased sensitivity, only a small fraction of peptides entering the mass spectrometer can be selected for fragmentation in an MS/MS analysis. Because of the data-dependent selection of precursor ions for fragmentation, high-abundant proteins are more likely to be fragmented, leading to an under-representation of low-abundant proteins in the subsequent analysis.

To overcome this, all peptide-centric approaches comprise a step for the reduction of sample complexity before MS analysis.

A major concern regarding this reduction of sample complexity is the potential loss of proteome coverage. Therefore, two different techniques are used. On the one hand, MudPIT (multidimensional protein identification technology) *(5)* has been used, which reduces the complexity by orthogonal liquid chromatography steps and on the other hand, different sorts of affinity purification approaches, such as ICAT (**I**sotope **C**oded **A**ffinity **T**ag) *(6)* and the herein described COFRADIC (**Co**mbined **Fr**actional **Di**agonal **C**hromatography) have been developed.

The COFRADIC principle (**Fig. 1**) (for a recent review *see* **ref.** *7)* was adapted from the diagonal electrophoresis *(8)* and adjusted to modern gel-free proteomics in 2003 by Vandekerckhove et al. *(9)*.

The backbone of COFRADIC is the separation of peptide mixtures by RP-HPLC (reversed-phase high-performance liquid

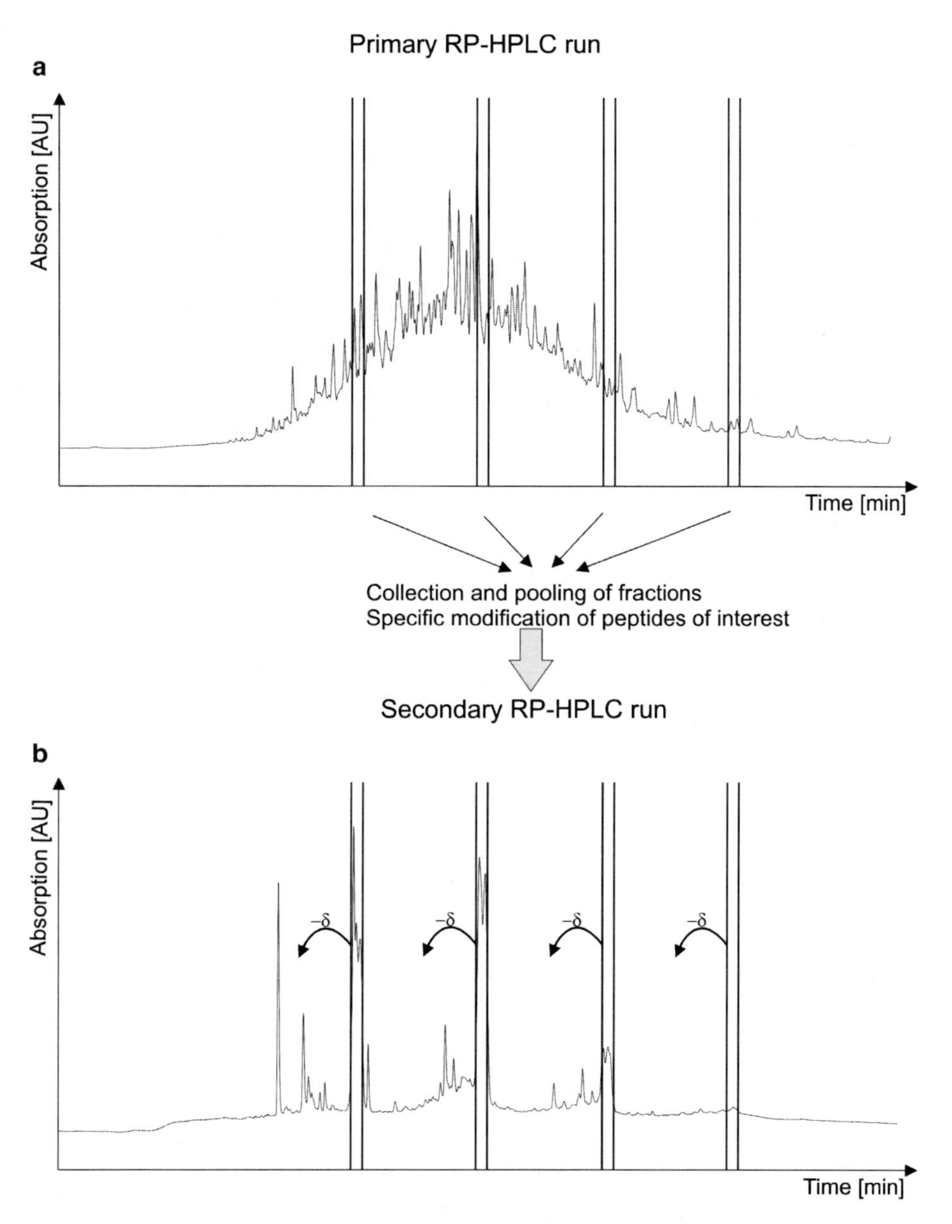


Fig. 1. The principle of COFRADIC. Peptides are separated by a primary RP-HPLC (**a**) run and fractions are collected in distinct time intervals. Afterwards, fractions are pooled, peptides chemically or enzymatically modified, and fractions are reinjected on the same column. The second run (**b**) is performed under identical chromatographic conditions. In this run, peptides that have been modified show a characteristic hydrophobic (+δ) or hydrophilic (–δ) shift and can be collected for further analysis. Whether shifted or not shifted, peptides are collected in the secondary run, depending on the chosen approach.

chromatography). During RP-HPLC separation, peptides elute at different retention times depending on their hydrophobic properties. By chemically modifying different amino acids, peptides can be rendered more hydrophilic (–δ) or hydrophobic (–δ), respectively.

After a first chemical modification step, which is not necessary in all approaches, the protein mixture is digested with trypsin and separated by a primary run on an RP-HPLC column. During the primary RP-HPLC run, fractions are collected in relation to the gradient and further derivatization steps are performed with each fraction. The intention of this modification step is to change column retention time of modified peptides in contrast to the unmodified rest of the peptide mixture. During the subsequent secondary runs, modified peptides show a hydrophobic ($+\delta$) or hydrophilic ($-\delta$) shift resulting in earlier or later elution times, respectively.

There are two major prerequisites for COFRADIC: First, modified amino acid side chains have to be evenly distributed over the proteome. Second, the induced modification must lead to a change in chromatographic retention behavior. Therefore, every specific modification of a peptide can be exploited by this technique, rendering it extremely versatile. Another advantage of COFRADIC is its high automation potential also enabling high-throughput proteomics.

In this chapter, three different COFRADIC approaches for methionine *(9)* and cysteine *(10)* containing as well as N-terminal peptides *(11)* are described (*see* **Table 1**, *see* **Note 1**). As raw material, platelets from platelet apharesis concentrates were used. The platelet proteome also contains a large number of problematic proteins, such as cumbersome and hydrophobic ones. These proteins for most part are not suitable for 2D-PAGE because of

**Table 1**
**Summary of the key characteristics of current used COFRADIC approaches**

| | Short description | Reactive species | References |
|---|---|---|---|
| Methionine COFRADIC | Primary RP-HPLC run<br>Oxidation of methionines<br>Hydrophilic shift due to oxidation in secondary run | $H_2O_2$ | *(9)* |
| Cysteine COFRADIC | Attachment of a TNB group to cysteines before primary run<br>Removal of TNB group between primary and secondary runs<br>Hydrophilic shift in secondary run | TNB | *(10)* |
| N-term COFRADIC | Blocking of free α,ε-amines and cysteines before primary RP-HPLC run<br>Modification with TNBS<br>TNB peptides shift out of original time interval in secondary run | TNBS | *(11)* |

their physicochemical properties and are therefore ideally suited for the use of alternate COFRADIC approaches.

## 2. Materials

### *2.1. Platelet Preparation*

#### *2.1.1. Preparation of Human Platelets from Platelet Apheresis Concentrates*

1. HEPES-washing buffer (pH 7.2). Ten millimolar HEPES, 157 mM NaCl, 4.17 mM KCl, and 3 mM EDTA.
2. Citrate buffer. 0.3 M citric acid.

### *2.2. COFRADIC*

#### *2.2.1. Met-COFRADIC*

##### Lysis and Digestion

1. Dithiothreitol (DTT).
2. Iodoacetamide (IAA).
3. Trifluoroacetic acid (TFA).
4. Hydrogen peroxide 30% in $H_2O$ (v/v).
5. Acetonitrile HPLC grade.
6. Guanidine hydrochloride.
7. Sodium phosphate, monobasic and dibasic, analytical-reagent grade.
8. Ammonium bicarbonate.
9. Trypsin (from hog pancreas).
10. High-purity deionized water.
11. Platelets isolated from platelet concentrates.
12. Lysis buffer: 2 M guanidine hydrochloride with 50 mM sodium phosphate (pH 8.0).
13. Digestion buffer: 50 mM ammonium bicarbonate (pH 7.4), freshly prepared.

##### Purification by Solid Phase Extraction

1. Methanol HPLC grade.
2. Five percent (v/v) methanol in $H_2O$.
3. C18 columns with filter: SPEC 3 mL C18AR 30 mg (VARIAN) or comparable.

##### Chromatographic Separation

*Equipment*

1. UltiMate 3000 Liquid Chromatograph (Dionex).
2. LPG 3000 Quaternary Micro Pump (Dionex).
3. TCC 3400 Column Compartment (Dionex).
4. UVD 3400 Variable Wavelength Detector (Dionex).
5. WPS-T 3000 well plate sampler equipped with 8-port valve, 0.15 mm bore for. fractionation option, with 125 μL loop (Dionex).
6. Chromeleon software version 6.80 SP2 with Extended Fraction Collection licence (Dionex).

*Solvents*

1. HPLC-buffer A: 0.1%TFA in $H_2O$.
2. HPLC-buffer B: 0.08% TFA and 84% acetonitrile.
3. Solution of 3% aqueous hydrogen peroxide.

*Conditions*

1. Column: Zorbax 300SB-C18, 5 μm, 2.1 × 150 mm (Agilent).
2. Column oven temperature: 30°C.
3. Loop: 125 μL.
4. Injection volume: 100 μL.
5. Sampler plate temperature: 30°C.
6. Flow rate: 80 μL/min.
7. Detection: Absorbance, 214 nm, 180 nL flowcell.

*2.2.2. Cys-COFRADIC*

1. NAP5 column (GE healthcare or comparable).
2. 10 mM Tris buffer (pH 8.7).
3. Lysis buffer:
   2 M guanidine hydrochloride and 50 mM Tris buffer (pH 8.7).
4. 100 mM Tris buffer (pH 8.7).
5. 10 mM Ellman's reagent in 100 mM Tris (pH 8.7).
6. 50 mM DTT.

*2.2.3. N-Term-COFRADIC*

1. Acetyl NHS acetate (*see* Regnier et al. *[12]*, *see* **Note 2**).
2. Hydroxylamine solution in 50% $H_2O$ (v/v).
3. 1 M glycine solution.
4. 50 mM sodium borate buffer (pH 9.5).
5. 15 mM trinitrobenzene sulfonic acid solution (TNBS) in 50 mM sodium borate buffer (pH 9.5).

## 3. Methods

### 3.1. Platelet Preparation

For COFRADIC, an amount of 0.5–1.5 mg of protein is required as starting material. To obtain a sufficient amount of proteins, platelet concentrates are used instead of fresh blood.

The use of plastic vessels at all times is essential to avoid aggregation of platelets on glass surfaces. If not indicated otherwise, all steps are performed at room temperature. For details *see* **ref.** *13*

1. Fill fresh platelet concentrate into 50 mL tubes, centrifuge at 200 × *g* for 20 min.
2. Transfer the supernatant into new tubes and centrifuge again at 200 × *g* for 20 min.
3. Discard the sediment.

4. Adjust the pH of the suspension to pH 6.4 with citrate buffer.
5. Centrifuge at 1200 × *g* for 20 min.
6. Keep the pellet, transfer the supernatant to a new tube, and spin down again.
7. Resuspend the pellets in HEPES wash buffer, pool the pellets, and spin down at 1200 × *g* for 10 s. Transfer the supernatant to a fresh tube (repeat at least two times).
8. Keep the platelet pellets at –80°C till further use.

## 3.2. COFRADIC

### 3.2.1. Met-COFRADIC

For Met-COFRADIC *(9)*, only reduction and alkylation of disulfide bonds are recommended as modification steps before digestion to overall increase solubility and accessibility of trypsin cleavage sites.

After the digestion step, peptides are purified by C18 solid-phase extraction columns, lyophilized, resuspended in 0.1% TFA, and separated in a primary RP-HPLC run. Fractions are collected, pooled and methionine residues are oxidized to their sulfoxide form with $H_2O_2$ (**Fig. 2**). In the secondary run, the methionine-containing peptides show a hydrophilic (–δ) shift toward earlier elution times and can thus be separated from unmodified peptides. Furthermore, methionine is a rare amino acid (1–2%). Shifting of the methionine-containing peptides therefore leads to a distinct reduction of sample complexity.

#### Lysis and Digestion

*see* **Note 3**

1. Lyse the platelet pellet in sufficient lysis buffer.
2. Reduce the disulfide bonds by adding 10 mM DTT to the sample and incubate for 30 min at 56°C.
3. Alkylate with 5 mM IAA and incubate for 30 min at room temperature, prepare fresh IAA every time and perform reaction in the dark.

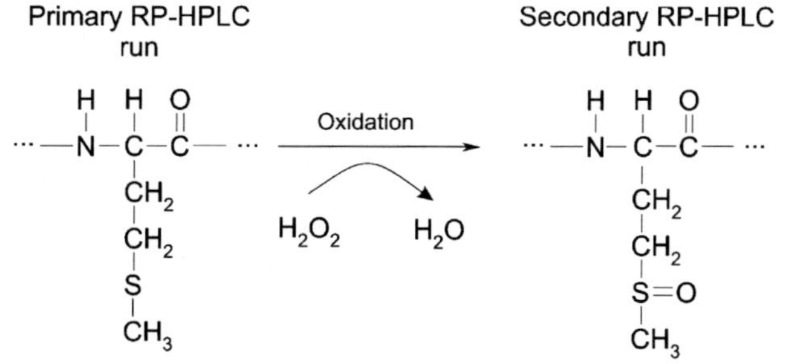

Primary RP-HPLC run | Oxidation ($H_2O_2$ → $H_2O$) | Secondary RP-HPLC run

Fig. 2. Modification principle in Met-COFRADIC. The peptide mixture obtained from tryptic digestion is separated by a primary RP-HPLC run. Fractions are collected, pooled and methionines are oxidized to their sulfoxide form with $H_2O_2$. In the secondary run, the methionine-containing peptides show a hydrophilic (–δ) shift towards earlier elution times and can thus be isolated.

4. Dilute sample 1:10 (v/v) with digestion buffer.
5. Digest sample with trypsin (protease to protein ratio 1:20 (w/w)) at 37°C overnight.

#### Purification and Concentration of Peptides by SPE C18AR Columns (VARIAN Protocol)

1. Apply 250 μL of methanol to the column, apply vacuum, and discard flow-through.
2. Apply 250 μL water or buffer, apply vacuum, and discard the flow-through.
3. Apply total sample volume on column (do not exceed the binding capacity!).
4. Wash with 5% (v/v) methanol in $H_2O$.
5. Elute with 250 μL methanol.
6. Dry fractions under vacuum.

#### Chromatographic Separation

1. Reconstitute dried sample in HPLC buffer A.
2. Separate digest by a primary RP-HPLC run (gradient *see* **Table 2**).
3. Collect 48 fractions of one minute each during the gradient, starting at 10% B.
4. Pool the 48 fraction according to the pool scheme in **Table 3** to a final number of 12 fractions.
5. Dry fractions under vacuum.
6. Reconstitute the fractions in 90 μL 0.1% TFA.

**Table 2**
**HPLC gradient method used for COFRADIC separations. This method is used for both primary and secondary run**

**HPLC method: Linear, 5-80% B in 80 min**

| Time | A (%) | B (%) | Comments |
|---|---|---|---|
| Initial | 95 | 5 | Equilibration |
| 0 | 95 | 5 | Sample injection |
| 20 | 95 | 5 | Start gradient |
| 100 | 20 | 80 | End gradient |
| 101 | 10 | 90 | Regeneration |
| 106 | 10 | 90 | Regeneration |
| 107 | 95 | 5 | Re-equilibration |
| 120 | 95 | 5 | Re-equilibration |

**Table 3**
**Pooling schema for Methionine COFRADIC fractions. Left part of the table sums the elution time of the pooled primary fractions. Right part of the table shows the elution time of individual oxidized fractions in the second run**

| Fractions | | | | | In primary run | | | | Start | | | | End | | | |
|---|---|---|---|---|---|---|---|---|---|---|---|---|---|---|---|---|
| L | 12 | 24 | 36 | 48 | 51 | 63 | 75 | 87 | 43 | 55 | 67 | 79 | 50 | 62 | 74 | 86 |
| K | 11 | 23 | 35 | 47 | 50 | 62 | 74 | 86 | 42 | 54 | 66 | 78 | 49 | 61 | 73 | 85 |
| J | 10 | 22 | 34 | 46 | 49 | 61 | 73 | 85 | 41 | 53 | 65 | 77 | 48 | 60 | 72 | 84 |
| I | 9 | 21 | 33 | 45 | 48 | 60 | 72 | 84 | 40 | 52 | 64 | 76 | 47 | 59 | 71 | 83 |
| H | 8 | 20 | 32 | 44 | 47 | 59 | 71 | 83 | 39 | 51 | 63 | 75 | 46 | 58 | 70 | 82 |
| G | 7 | 19 | 31 | 43 | 46 | 58 | 70 | 82 | 38 | 50 | 62 | 74 | 45 | 57 | 69 | 81 |
| F | 6 | 18 | 30 | 42 | 45 | 57 | 69 | 81 | 37 | 49 | 61 | 73 | 44 | 56 | 68 | 80 |
| E | 5 | 17 | 29 | 41 | 44 | 56 | 68 | 80 | 36 | 48 | 60 | 72 | 43 | 55 | 67 | 79 |
| D | 4 | 16 | 28 | 40 | 43 | 55 | 67 | 79 | 35 | 47 | 59 | 71 | 42 | 54 | 66 | 78 |
| C | 3 | 15 | 27 | 39 | 42 | 54 | 66 | 78 | 34 | 46 | 58 | 70 | 41 | 53 | 65 | 77 |
| B | 2 | 14 | 26 | 38 | 41 | 53 | 65 | 77 | 33 | 45 | 57 | 69 | 40 | 52 | 64 | 76 |
| A | 1 | 13 | 25 | 37 | 40 | 52 | 64 | 76 | 32 | 44 | 56 | 68 | 39 | 51 | 63 | 75 |

7. Add $H_2O_2$ to a final concentration of 0.5% (v/v) and incubate for 30 min at 30°C. To avoid over-oxidation of amino acids, perform this reaction just before the secondary RP-HPLC run.
8. Separate each fraction by a secondary RP-HPLC run, using identical chromatographic conditions as in the primary one.
9. Because of the hydrophilic shift, methionine-containing peptides elute roughly 4–8 min earlier than in the primary run *see* **Note 4**.
10. Dry the collected fractions and prepare for LC-MS/MS.

*3.2.2. Cys-COFRADIC*

For this approach, proteins are modified prior to digestion. First proteins are reduced with DTT and afterwards the formation of TNB–cysteine is performed with DTNB (Ellman's reagent). Moreover, methionines are oxidized with $H_2O_2$ to prevent random oxidation between the two RP-HPLC runs, which would as well lead to a hydrophilic shift (−δ). Following tryptic digestion, a primary RP-HPLC run is performed. To shift cysteine-containing peptides, the sample is again reduced with DTT, which leads to the loss of the TNB (trinitrobenzene) group, rendering the peptides more hydrophilic (−δ) (compare **Fig. 3**).

In the secondary run, cysteine-containing peptides undergo a hydrophilic shift (−δ) and can therefore be collected 3–10 min

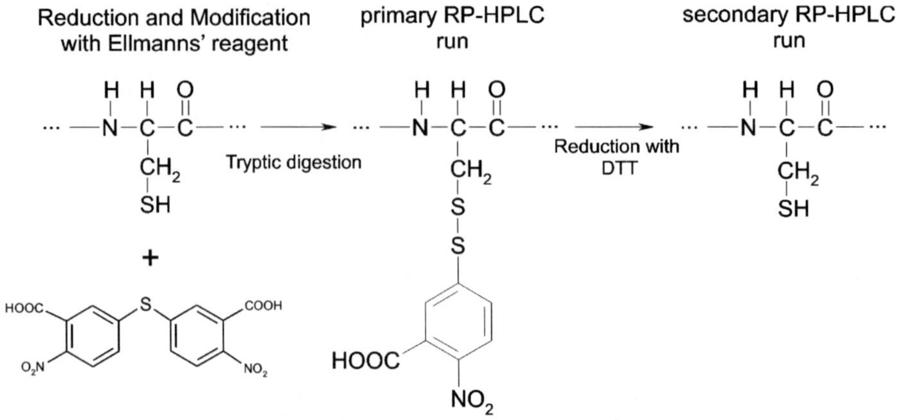

Fig. 3. Modification principle in Cys-COFRADIC. In the intact proteins, cysteines are reduced and afterwards modified with Ellman's reagent. Following a tryptic digestion, a primary RP-HPLC run is performed. Because of the modification, the cysteine-containing peptides show a hydrophobic (+δ) shift. Before the second run, cysteines are once again reduced with DTT, which leads to the loss of the attached hydrophobic group. Cysteine-containing peptides now exhibit a shift (–δ) towards earlier elution times.

earlier. Like Met-COFRADIC, Cys-COFRADIC leads to a reduction of sample complexity by a factor of 5 *(10)*, depending on the sample itself.

#### Lysis and Digestion

*see* **Note 3**

1. Lyse the platelet pellet in sufficient lysis buffer.
2. Reduce disulfide bonds by adding 10 mM DTT to the sample and incubate for 30 min at 56°C.
3. Desalt the sample by a NAP-5 column, use lysis buffer for elution.
4. Reduce sample to complete dryness by vacuum drying.
5. Reconstitute the sample in 10 mM Ellman's reagent, 100 mM Tris (pH 8.7) and incubate for 1 h at 37°C.
6. Desalt by a NAP-5 column, elute with 10 mM Tris (pH 8.7).
7. Digest sample with trypsin (protease to protein ratio 1:20 (w/w)) overnight at 37°C.

#### Purification and Concentration of Peptides by SPE C18AR Columns (VARIAN Protocol)

*See* above.

#### Chromatographic Separation

1. Reconstitute dried sample in HPLC buffer A and add $H_2O_2$ solution to a final concentration of 0.5% (v/v) in $H_2O$, and incubate for 30 min at 30°C. To avoid over-oxidation of amino acids, perform this reaction just before the primary RP-HPLC run.
2. Separate the digest by a primary RP-HPLC run (Gradient *see* **Table 2**).

3. Collect 48 fractions of 1 min each during the gradient, starting at 10% B.
4. Pool the 48 fraction according to the pool scheme in **Table 4** to final number of 16 fractions.
5. Dry fractions under vacuum.
6. Reconstitute fractions in 10 mM Tris, 50 mM DTT, and incubate for 30 min at 56°C.
7. Stop the reaction by adding 0.1% TFA (till pH 2).
8. Apply all fractions to a secondary RP-HLC run, using identical chromatographic conditions as in the primary one.
9. Recollect the fractions in a time interval of 3–10 min before the elution time in the primary run.
10. Dry the collected fractions and prepare for LC-MS/MS.

**Table 4**
**Pooling schema for Cysteine COFRADIC fractions. Left part of the table displays the fractions to be pooled, the middle sums the elution time of the pooled primary fractions and the right part of the table shows the elution time of individual shifted fractions n the second run**

| Pooled primary fractions | | | | Elution time in primary run | | | Elution time window in secondary run | | | | | | | | |
|---|---|---|---|---|---|---|---|---|---|---|---|---|---|---|---|
| P | 16 | 32 | 48 | 55 | 71 | 87 | 45 | to | 52 | 61 | to | 68 | 77 | to | 84 |
| O | 15 | 31 | 47 | 54 | 70 | 86 | 44 | to | 51 | 60 | to | 67 | 76 | to | 83 |
| N | 14 | 30 | 46 | 53 | 69 | 85 | 43 | to | 50 | 59 | to | 66 | 75 | to | 82 |
| M | 13 | 29 | 45 | 52 | 68 | 84 | 42 | to | 49 | 58 | to | 65 | 74 | to | 81 |
| L | 12 | 28 | 44 | 51 | 67 | 83 | 41 | to | 48 | 57 | to | 64 | 73 | to | 80 |
| K | 11 | 27 | 43 | 50 | 66 | 82 | 40 | to | 47 | 56 | to | 63 | 72 | to | 79 |
| J | 10 | 26 | 42 | 49 | 65 | 81 | 39 | to | 46 | 55 | to | 62 | 71 | to | 78 |
| I | 9 | 25 | 41 | 48 | 64 | 80 | 38 | to | 45 | 54 | to | 61 | 70 | to | 77 |
| H | 8 | 24 | 40 | 47 | 63 | 79 | 37 | to | 44 | 53 | to | 60 | 69 | to | 76 |
| G | 7 | 23 | 39 | 46 | 62 | 78 | 36 | to | 43 | 52 | to | 59 | 68 | to | 75 |
| F | 6 | 22 | 38 | 45 | 61 | 77 | 35 | to | 42 | 51 | to | 58 | 67 | to | 74 |
| E | 5 | 21 | 37 | 44 | 60 | 76 | 34 | to | 41 | 50 | to | 57 | 66 | to | 73 |
| D | 4 | 20 | 36 | 43 | 59 | 75 | 33 | to | 40 | 49 | to | 56 | 65 | to | 72 |
| C | 3 | 19 | 35 | 42 | 58 | 74 | 32 | to | 39 | 48 | to | 55 | 64 | to | 71 |
| B | 2 | 18 | 34 | 41 | 57 | 73 | 31 | to | 38 | 47 | to | 54 | 63 | to | 70 |
| A | 1 | 17 | 33 | 40 | 56 | 72 | 30 | to | 37 | 46 | to | 53 | 62 | to | 69 |

*3.2.3. N-Term-COFRADIC*

With regard to the reduction of sample complexity, sorting N-terminal peptides with COFRADIC *(11)* is the most powerful of the three described approaches (reduction of sample complexity by a factor of 10). Theoretically, each protein is represented only by a single peptide, the N-terminal one. Taking protein degradation into consideration, this is of course not the case. In "real life", every protein is represented by the N-terminal peptides of the protein and its degradation products. Prior to digestion, disulfide bonds are reduced and cysteines are alkylated. All free amino groups are acetylated. Following tryptic digest, the peptide mixture is separated by a primary RP-HPLC run. During digestion, two kinds of peptides are formed. N-terminal peptides show an acetylated N-terminus, whereas internal peptides display free amino groups. A modification of these groups with TNBS after the primary run renders all internal peptides more hydrophobic (+δ) and shifts them to later elution times during secondary RP-HPLC runs. Therefore, N-terminal peptides can be collected in the same time interval as in the first run (**Fig. 4**). Pooling of samples is not possible in this approach because of unpredictable shifts of modified internal peptides.

Lysis and Digestion *see* **Note 3**

1. Lyse the platelet pellet in sufficient lysis buffer.
2. Reduce disulfide bonds by adding 10 mM DTT to the sample and incubate for 30 min at 56°C.
3. Alkylate cysteines with 5 mM IAA incubating for 30 min in the dark, prepare fresh IAA every time.
4. Acetylate free amino groups with 25 mM acetyl NHS acetate for 1 h at 37°C.
5. Reverse acetylation of serine and threonine by adding hydroxylamine in fourfold excess of acetyl NHS acetate.
6. Stop the reaction by adding glycine in twofold excess to acetyl NHS acetate.
7. Dilute sample 1:10 with digestion buffer.
8. Digest sample with trypsin (protease to ratio 1:20 (w/w)) overnight at 37°C *see* **Note 5**.

Purification and Concentration of Peptides by SPE C18AR Columns (VARIAN Protocol)

*See* above.

Chromatographic Separation

1. Reconstitute dried samples in HPLC buffer A and add $H_2O_2$-solution to a final concentration of 0.5% (v/v) in $H_2O$, incubate at 30°C for 30 min. To avoid over-oxidation of amino acids, perform this reaction just before the primary RP-HPLC run.
2. Separate the digest by a primary RP-HPLC run (Gradient *see* **Table 2**).

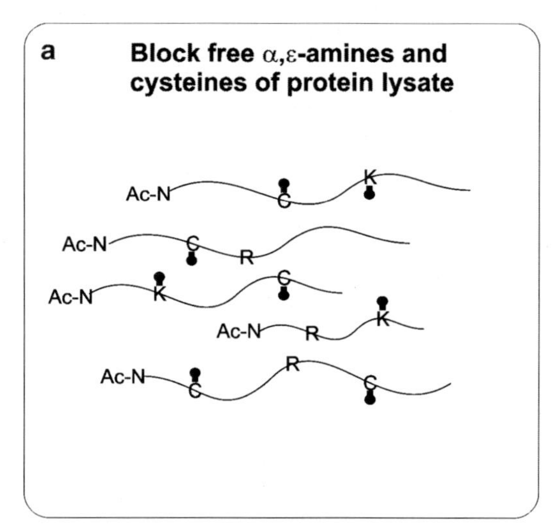


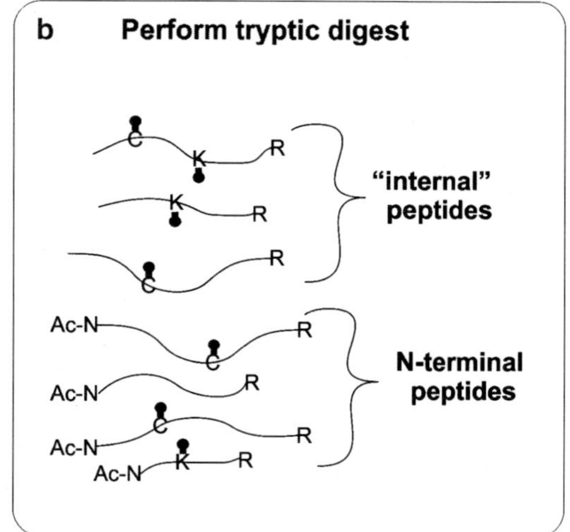


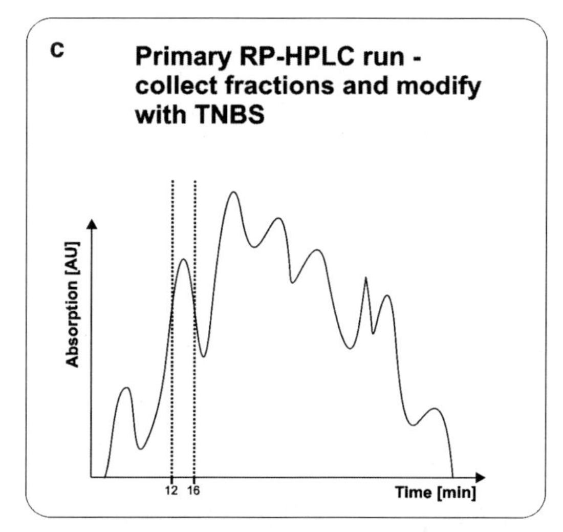


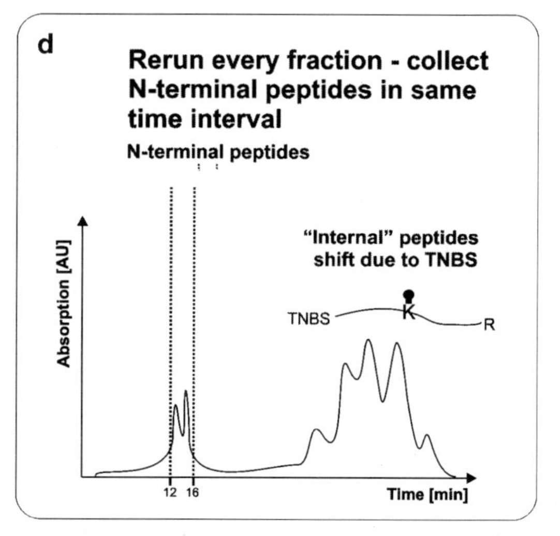


Fig. 4. Modification principle in N-term COFRADIC. In the intact proteins, cysteines are alkylated and N-termini as well as lysines are acetylated (**a**). Following tryptic digestion (**b**), the peptide mixture is separated by a primary RP-HPLC run (**c**). Fractions are collected and modified with TNBS. By this reaction, internal peptides are transformed to TNP peptides. These peptides are more hydrophobic than the acetylated N-terminal ones (which cannot react with TNBS) and show a shift towards later elution times in the second RP-HPLC run (**d**). N-terminal peptides can therefore be collected in the same time interval as in the first run.

3. Collect 16 fractions of 4 min each during the gradient, starting at 10% buffer B.
4. Dry fractions under vacuum.
5. Reconstitute fractions in 50 μL 50 mM sodium borate (pH 9.5).
6. Add 10 μL 15 mM TNBS and incubate for 30 min at 37°C, repeat this step three times.
7. Stop the reaction by adding 20 μL 10% TFA.
8. Apply all fractions to a secondary RP-HLC run, using identical chromatographic conditions as in the primary one *see* **Note 6**.

9. Recollect the fractions in a time interval from 4 min before to 4 min after the elution time in the primary run.
10. Dry the collected fractions and prepare for LC-MS/MS.

### 3.3. Data Analysis

Samples obtained from the COFRADIC experiments are measured by means of LC/MS (liquid chromatography tandem mass spectrometry). MASCOT generic files (*.mgf) are created from the raw MS files and used to search the human subset of the SwissProt Database by the MASCOT algorithm *see* **Note** 7.

Here, a QTOF is used for MS analysis (QStar Elite, Applied Biosystems).

#### 3.3.1. MASCOT Search Parameters

The following general parameters are used: Both peptide mass tolerance and peptide fragment tolerance are set to ±0.2 Da; the "instrument setting" of MASCOT is set to "ESI-QUAD TOF." A maximum number of one missed cleavage is allowed and peptide charge is set to doubly and triply charged precursors. Depending on the type of peptides, which was sorted for the MASCOT modification, parameters were set differently (see below).

##### N-term COFRADIC Modifications

For the N-term COFRADIC approach, fixed modifications are oxidation of methionine to its sulfoxide, trideutero-acetylation of lysine and carbamidomethylation of cysteine. Pyroglutamate formation (N-terminal Gln), pyrocarbamidomethyl cysteine formation (N-terminal alkylated cysteines), acetylation, and trideutero-acetylation of the α-N-terminus are the variable modifications. Endoprotease Arg-C/P is set as enzyme *see* **Note 8**.

##### Met COFRADIC and Cys COFRADIC Modifications

For the Met-COFRADIC and Cys-COFRADIC approach, fixed modifications are oxidation of methionine to its sulfoxide and carbamidomethylation of cysteine. Pyroglutamate formation (N-terminal Gln), pyrocarbamidomethyl cysteine formation (N-terminal alkylated cysteines), acetylation of the α-N-terminus are the variable modifications. Trypsin is set as enzyme.

#### 3.3.2. Useful Tools for Interpreting the Results

It is useful to parse or export the MASCOT results to receive a peptide list, which can be further sorted to create a subset peptide list.

A number of (bioinformatic) tools have been developed regarding the COFRADIC evaluation. One is the ms_lims software (http://genesis.ugent.be/ms_lims), which provides a MySQL relational database to store the mass spectra and link the peptides identified by MASCOT to the spectra. Then, SQL queries within a list of peptides from a COFRADIC experiment can be done and easily allow for extraction of identifications.

However, to set up this software, some knowledge about MySQL databases and working with SQL is required.

The MascotDatfile library (http://genesis.ugent.be/mascotdatfile) allows the fast analysis of peptide identifications from MASCOT dat files. Thereby, detailed analysis of raw data inside a large and complex data set becomes possible. The tool provides a set of java classes, which can be run from the command line for extracting information about sequence coverage, modified peptide hits, amino acid scoring, and the spectra of respective MASCOT results.

When working with a peptide-centric approach, peptide sequence databases instead of protein sequence databases are required. A peptide sequence database contains, e.g., all cysteine-containing peptide sequences from a given proteome. The dbtoolkit project (http://genesis.ugent.be/dbtoolkit/) is a user-friendly tool to parse sequence databases into derived, FASTA-formatted sequence databases.

## 4. Notes

1. It is recommended to apply all three COFRADIC approaches to a complex sample. When applying all three COFRADIC approaches on a proteome, the majority of proteins will only be detected by one approach. Reasonable coverage can therefore only be achieved using several approaches.
2. Deutero-acetyl NHS can be used instead of acetyl-NHS. The heavy-isotopic variant for acetylation distinguishes in vivo free (trideutero-acetylated) protein N-termini. For details of how to prepare *see* **ref.** ***12***.
3. Especially when working with hydrophobic membrane proteins, solubility represents a major problem. To avoid interference with the following digestion and LC, the addition of detergents such as SDS should be avoided. To solubilize proteins, use ultrasonication and glass beads.
4. Although not necessary, automation of the COFRADIC approach significantly reduces labor input. It is important to notice that automated sample collection and derivatization requires reproducible retention behavior of peptides.
5. Among the N-terminal peptides derived from tryptic digestion, also very short peptides can occur. These peptides are too small to be retained by the RP-HPLC column and therefore escape analysis. By the use of a second protease (e.g., V8), other peptides with different size distributions are generated providing higher-proteome coverage.

6. A recently published study suggests a set of modifications to increase the amount of α-amino-acetylated peptides among the COFRADIC sorted ones in N-term COFRADIC: Strong cation exchange (SCX) segregation of α-amino-blocked and α-amino-free peptides and an enzymatic step liberating pyroglutamyl peptides for TNBS modification, and thus COFRADIC sorting. This new approach is supposed to result in more protein identifications. For further details, *see* **ref. *14***.
7. When rerunning the same sample in the same approach several times, every time new proteins will be identified and already identified proteins will not be detected again. This is due to the still occurring undersampling effect. Not all peptides entering the MS at a given time point are selected for fragmentation and identification. To further reduce the sample complexity, different approaches are suggested, e.g., **ref. *15***.
8. In the N-terminal COFRADIC approach, trypsin only cleaves at the carboxylic side of arginine because of the acetylation of lysine residues. This has to be considered in database search parameters.

## References


1. O'Farrell, P. H. (1975) High resolution two-dimensional electrophoresis of proteins. *J Biol Chem* **250**, 4007–21.
2. Klose, J. (1975) Protein mapping by combined isoelectric focusing and electrophoresis of mouse tissues. A novel approach to testing for induced point mutations in mammals. *Humangenetik* **26**, 231–43.
3. Wilkins, M. R., Gasteiger, E., Sanchez, J. C., Bairoch, A., and Hochstrasser, D. F. (1998) Two-dimensional gel electrophoresis for proteome projects: the effects of protein hydrophobicity and copy number. *Electrophoresis* **19**, 1501–5.
4. Gygi, S. P., Corthals, G. L., Zhang, Y., Rochon, Y., and Aebersold, R. (2000) Evaluation of two-dimensional gel electrophoresis-based proteome analysis technology. *Proc Natl Acad Sci U S A* **97**, 9390–5.
5. Washburn, M. P., Wolters, D., and Yates, J. R., 3rd (2001) Large-scale analysis of the yeast proteome by multidimensional protein identification technology. *Nat Biotechnol* **19**, 242–7.
6. Gygi, S. P., Rist, B., Gerber, S. A., Turecek, F., Gelb, M. H., and Aebersold, R. (1999) Quantitative analysis of complex protein mixtures using isotope-coded affinity tags. *Nat Biotechnol* **17**, 994–9.
7. Gevaert, K., Van Damme, P., Ghesquiere, B., Impens, F., Martens, L., Helsens, K., and Vandekerckhove, J. (2007) A la carte proteomics with an emphasis on gel-free techniques. *Proteomics* **7**, 2698–718.
8. Brown, J. R., and Hartley, B. S. (1966) Location of disulphide bridges by diagonal paper electrophoresis. The disulphide bridges of bovine chymotrypsinogen A *Biochem J* **101**, 214–28.
9. Gevaert, K., Van Damme, J., Goethals, M., Thomas, G. R., Hoorelbeke, B., Demol, H., Martens, L., Puype, M., Staes, A., and

Vandekerckhove, J. (2002) Chromatographic isolation of methionine-containing peptides for gel-free proteome analysis: identification of more than 800 *Escherichia coli* proteins. *Mol Cell Proteomics* **1**, 896–903.

10. Gevaert, K., Ghesquiere, B., Staes, A., Martens, L., Van Damme, J., Thomas, G. R., and Vandekerckhove, J. (2004) Reversible labeling of cysteine-containing peptides allows their specific chromatographic isolation for non-gel proteome studies. *Proteomics* **4**, 897–908.
11. Gevaert, K., Goethals, M., Martens, L., Van Damme, J., Staes, A., Thomas, G. R., and Vandekerckhove, J. (2003) Exploring proteomes and analyzing protein processing by mass spectrometric identification of sorted N-terminal peptides. *Nat Biotechnol* **21**, 566–9.
12. Ji, J., Chakraborty, A., Geng, M., Zhang, X., Amini, A., Bina, M., and Regnier, F. (2000) Strategy for qualitative and quantitative analysis in proteomics based on signature peptides. *J Chromatogr B Biomed Sci Appl* **745**, 197–210.
13. Clark, E. A., Brugge, J. S. (1996) Platelets, Oxford University Press, Oxford.
14. Staes, A., Van Damme, P., Helsens, K., Demol, H., Vandekerckhove, J., and Gevaert, K. (2008) Improved recovery of proteome-informative, protein N-terminal peptides by combined fractional diagonal chromatography (COFRADIC). *Proteomics* **8**, 1362–70.
15. Gevaert, K., Pinxteren, J., Demol, H., Hugelier, K., Staes, A., Van Damme, J., Martens, L., and Vandekerckhove, J. (2006) Four stage liquid chromatographic selection of methionyl peptides for peptide-centric proteome analysis: the proteome of human multipotent adult progenitor cells. *J Proteome Res* **5**, 1415–28.

# Chapter 10

# Identification of the Molecular Composition of the 20S Proteasome of Mouse Intestine by High-Resolution Mass Spectrometric Proteome Analysis

**Reinhold Weber, Regina Preywisch, Nikolay Youhnovski, Marcus Groettrup, and Michael Przybylski**

## Summary

In the last years, intracellular protein degradation by the proteasome has become a focus area of scientific interest. Here, we describe a proteomics approach for the molecular mapping of the constituents of the proteolytically active core particle, the constitutive 20S proteasome from mouse intestine. In addition to the proteomics workflow widely used for protein isolation, gel electrophoretic separation, in-gel digestion, and UV-MALDI mass spectrometry, high-resolution Fourier transform ion cyclotron resonance mass spectrometry using infrared-MALDI ionisation (IR-MALDI FTICR-MS) has been employed as an efficient method for protein identification by peptide mass fingerprint. The 20S proteasome subunits $\alpha1$–$\alpha7$ and $\beta1$–$\beta7$ were completely and unambiguously identified. In addition to subunits $\beta1$ and $\beta2$, the corresponding inducible subunits being part of the immuno-proteasome were identified. The subunit $\beta5i$ was found to completely replace the corresponding constitutive subunit, suggesting a high proteolytic activity of the intestinal proteasome leading to increased production of antigenic peptides. The high mass accuracy in the low ppm range and resolution of FTICR-MS provide direct identifications of individual proteins as mixtures such as components resulting from incomplete electrophoretic separation. In addition, the comparison of UV- and IR-MALDI FTICR-MS may provide details of fragmentation and rearrangement reactions that may occur under UV-MALDI ionisation conditions.

**Key words:** 20S-Proteasome, High-resolution FTICR mass spectrometry, Proteome analysis, Proteasome constituents, UV- and IR-MALDI ionisation

## 1. Introduction

Protein degradation is essential for cellular viability and the recovery of amino acids for protein biosynthesis. Misfolded proteins and proteins with altered amino acid sequences have to be

Jörg Reinders and Albert Sickmann (eds.), *Proteomics,* Methods in Molecular Biology, Vol. 564
DOI: 10.1007/978-1-60761-157-8_10, © Humana Press, a part of Springer Science+Business Media, LLC 2009

degraded, and protein degradation exerts a key regulatory function in many cellular processes. Degradation can take place in two fundamentally different ways: (1) In the ATP-independent, lysosomal pathway, proteins are internalised into the lysosome and degraded in a relatively unspecific manner and (2) the ATP-dependant ubiquitin-proteasome pathway localised in the cytosol and the nucleus *(1–4)*. The importance of the latter process has been recognised by the 2004 Nobel Prizes in Chemistry awarded to Ciechanover, Hershko, and Rose. In proteasomal degradation, ubiquitin, a protein of 76 amino acids highly conserved in all eukaryotic cells, is activated by E1 (ubiquitin-activating enzyme) followed by conjugation to the protein substrate to be degraded by action of ubiquitin-conjugating enzyme (E2) and ubiquitin-protein ligase (E3). The labelled protein is recognised via the poly-ubiquitin tag and degraded by the 26S proteasome, a 2.5 MDa protease complex present in the cytoplasm and nucleus, generating peptide ligands for MHC class I molecules. The 26S proteasome consists of two distinct parts, a barrel-shaped proteolytic core complex (20S proteasome) that is capped at both ends by regulatory subunits (19S proteasome). The 19S proteasome recognises ubiquitin-conjugated target proteins and is involved in their unfolding and translocation to the proteolytic core, where proteins are degraded to small (4–15 amino acids) peptides.

The eukaryotic 20S proteasome is composed of two copies, each comprising seven different α-subunits and seven β-subunits *(1)*. The α- and β-subunits each form a heterooligomeric ring with the rings stacked on each other leading to the general assembly $(\alpha 1{-}\alpha 7)(\beta 1{-}\beta 7)(\beta 1{-}\beta 7)(\alpha 1{-}\alpha 7)$. The subunits can be separated by gel electrophoresis, and 2-D gel electrophoresis has been shown to be an efficient method for component separation *(5, 6)*. Since basic proteins tend to migrate out of the gel in immobilised pH-gradients during isoelectric focussing, non-equilibrium pH-gradient gel electrophoresis (NEPHGE) has been used to separate proteasome subunits in the first dimension. After SDS-PAGE and staining, the gel spots are "in-gel" digested and the peptide fragment mixtures subjected to mass spectrometric proteome analysis. Both electrospray ionisation (ESI) and MALDI-MS have been employed as efficient "soft" ionisation-MS methods *(7–9)*. Recently, high-resolution FTICR mass spectrometry (FTICR-MS) has been introduced as a powerful tool in proteome analysis. Because of its unrivalled mass resolving power ($>10^6$) and accuracy of mass determination (low- to sub-ppm range), FTICR-MS enables unambiguous protein identifications (1) with only a minimum number of peptide fragments required and (2) with a high efficiency to detect possible structure modifications *(5, 6)*. Furthermore, the high mass determination accuracy of FTICR-MS allows the application of low mass tolerance thresholds, thus substantially improving the selectivity of database search procedures and the quality of the search results. In addition, specific fragmentation experiments can be carried out in FTICR-MS to ascertain protein identifications, particularly using infrared multiphoton-dissociation (IRMPD) of isolated ions *(10–12)*.

In this study, the proteome analysis of the 20S proteasome from mouse intestine is described using both UV- and IR-MALDI FTICR-MS. IR laser ionisation (using an Er-YAG laser at 2.94 μm emission) has been recently introduced in MALDI-MS and employed, in addition to the standard nitrogen UV-(337 nm) laser *(13–19)*. Owing to lower irradiation energy, IR-MALDI may provide "softer" ionisation conditions when compared with UV-MALDI, which may be advantageous in the proteome analysis of large biomolecules and the identification of labile, post-translationally modified peptides. Since in MALDI-FTICR-MS, relatively long lifetimes of ions between ionisation and detection are required and molecular ions excited by high-energy laser pulses may undergo metastable fragmentation, IR-MALDI-MS may provide advantages because of the lower energy for ionisation.

## 2. Materials

### 2.1. Isolation of 20S Proteasomes from Mouse Intestine

20S Proteasome from BALB/c mouse intestine was isolated as previously described *(20)*.

1. Intestines were cut and grinded in 100 mM KCl + 0.1% Triton X-100 in a dounce homogeniser.
2. The tissues were further homogenised under cooling with ice using a Polytron homogeniser.
3. Following ultracentrifugation, proteins were batch purified by DEAE-Sephacel chromatography and precipitated with $(NH_4)_2SO_4$ (35–80%) at 3°C on ice.
4. The pellet was redissolved in 100 mM KCl, and the protein components were pre-separated by ultracentrifugation on a sucrose gradient (15–40% sucrose in 100 mM KCl).
5. The activity of supernatant fractions was determined using a fluorogenic Leu- or Tyr-substrate test and photometric detection *(20, 21)*.
6. Positive fractions were pooled, diluted in a ratio of 1:9 with 100 mM KCl, and subjected to FPLC.

7. The 20S Proteasomes were finally isolated by FPLC; following confirmation of purity by mini-gel electrophoresis, the pure proteasomes were precipitated with 50% TCA on ice. After washing with ethanol and acetone at −20°C, the pellet was dissolved in the sample buffer for NEPHGE.

### 2.2. 2-D Gel Electrophoresis

Aliquots of 60–80 μg of solubilised proteasome were separated by two-dimensional NEPHGE/SDS-PAGE as described in **ref.** *22.* After the run gels were stained with Coomassie Brilliant blue G 250 using either normal or colloidal staining *(23)*.

**Figure 1** illustrates the electrophoretic separation and summarises the identifications for the 20S proteasome (**Table 1**). The subunits $\beta 1_i$, $\beta 2_i$ (MECL-1), and $\beta 5_i$, which are inducible subunits and therefore part of immuno-proteasomes, were clearly identified. Upon infection, cytokines such as interferon γ (IFN-γ) and tumour necrosis factor-α (TNFα) are released, leading to changes in the architecture of the proteasome: The subunits β1 (δ), β2 (Z), and β5 (MB-1) bearing the catalytically active sites of the constitutive proteasome are replaced by the subunits $\beta 1_i$ (LMP2), $\beta 2_i$ (MECL-1), and $\beta 5_i$ (LMP7). These subunits provide catalytic activity, but different cleavage specificity. For example, the exchange of β1 to $\beta 1_i$ was shown to modify the preferred cleavage site from C-terminal peptide bonds of glutamic acid residues to C-terminal bonds of hydrophobic amino acid residues *(24)*. This change leads to an increased number of viral peptides to be presented to MHC class I, as a larger number of peptides with a suitable C-terminal anchor amino acid become available. In this example, the subunits β1 and $\beta 1_i$ are both present; however, constitutive subunits appear to be stronger expressed than the induced subunits, as indicated by the 2D gel pattern. The subunits β2 and β2i are also identified, while the subunit $\beta 5_i$ was found to completely replace subunit β5. These results

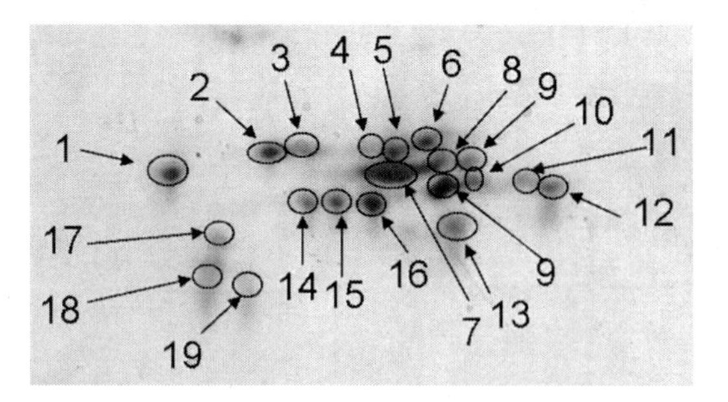


Fig. 1. 2-D gel electrophoresis separation of subunits of the 20S proteasome from mouse intestine. Identified subunits are indicated and numbered as listed in Table 1. The systematic nomenclature according to Groll et al. *(27)* is used.

**Table 1**
**Identified subunits from 2-D gel electrophoretic separation of the constitutive 20S proteasome from mouse intestine. The systematic nomenclature according to Groll et al. (27) is used**

| Spot No. | Proteasome subunit | m.w. (kDa) | SwissProt Accession No. |
|---|---|---|---|
| 1 | α5 | 26.4 | Q9Z2U1 |
| 2 | α7 | 28.3 | O70435 |
| 3 | α1+ | 27.3 | Q9QUM9+ |
| | α4 | 27.8 | Q9Z2U0 |
| 4 | α4 | 27.8 | Q9Z2U0 |
| 5 | α6 | 29.5 | Q9R1P4 |
| 6 | α1 | 27.3 | Q9QUM9 |
| 7 | α3+ | 29.5 | Q9R1P0+ |
| | α1 | 27.3 | Q9QUM9 |
| 8 | α2 | 25.8 | P49722 |
| 9 | β2i | 29.1 | O35955 |
| 10 | β6+ | 26.4 | O09061+ |
| | β2i | 29.1 | O35955 |
| 11 | β6 | 29.1 | O09061 |
| 12 | β2 | 29.9 | P70195 |
| 13 | β5i+ | 30.3 | P28063+ |
| | β4 | 22.9 | Q9R1P3 |
| 14 | β7 | 29.1 | P99026 |
| 15 | β7 | 29.1 | P99026 |
| 16 | β3 | 23.0 | Q9R1P1 |
| 17 | β1 | 25.4 | Q60692 |
| 18 | β1i | 23.4 | P28076 |
| 19 | β1i | 23.4 | P28076 |

suggest a high immunoproteasome-type of activity of the 20S proteasome from intestine of uninfected mice. The functional implications of this increased production of antigenic peptides remain a subject for future investigation.

## 3. Methods

### 3.1. In-Gel Digestion and Peptide Extraction

Protein spots were cut out with a scalpel and prepared for tryptic in-gel digestion either manually *(25)*, or by an automated procedure using a DigestPro 96 robot (Intavis Bioanalytical Instruments, Köln, Germany) *(26)*.

1. For manual in-gel digestion, gel spots were washed with water, then dehydrated by addition of 200 μL acetonitrile/water 3/2 (all values v/v), and dried after removal of supernatant in a SpeedVac (Eppendorf, Germany).
2. The gel pieces were rehydrated in 20 mM $NH_4HCO_3$, destained, and dried. This procedure was repeated until the gel pieces were completely destained (2–3 cycles, depending on the size of the gel piece and the intensity of staining).
3. The gel pieces were incubated in a solution of 12.5 ng/μL Trypsin (Promega, Mannheim, Germany) in 20 mM $NH_4HCO_3$ at 4°C for 45 min.
4. Following replacement of the supernatant by 20 mM $NH_4HCO_3$, in-gel digestion was carried out for 12 h at 37°C.
5. The tryptic peptides were eluted from the gel pieces by treatment each for 1–2 h with acetonitrile/20 mM $NH_4HCO_3$ (1/1) and acetonitrile/5% HCOOH (1/1).
6. The combined extracts were lyophilised to dryness and taken up in 10 μL 0.1% aqueous $CF_3COOH$.
7. Sample desalting was carried out using ZipTip® microcolumns (Millipore) according to supplier's recommendation.

### 3.2. UV-MALDI-FTICR Mass Spectrometry

Mass spectrometric measurements were carried out on an Apex II FT-ICR mass spectrometer (Bruker Daltonik, Bremen, Germany) equipped with an actively shielded 7 T superconducting magnet and a cylindrical Infinity analyser cell. The pulsed nitrogen laser of the Scout 100 MALDI source is operated at 337 nm. Ions were cooled immediately after desorption by pulsed Ar gas and accumulated in a hexapole ion guide situated 1 mm in front of the target. In UV-MALDI, 15–20 laser shots per scan were used. A solution of 100 mg/mL 2,5-dihydroxybenzoic acid in acetonitrile/0.1% aqueous TFA (2/1) was used as matrix. The IRMPD fragmentation experiments were carried out using a $CO_2$ laser (Synrad, Mukilteo, WA) at a wavelength of 10.6 μm. Ions of interest were isolated using SWIFT with a correlation sweep according to 100–4,000 *m/z* and an ejection safety belt of 500 Hz. After cooling with pulsed Ar gas for 80–120 ms, the parent ions were fragmented by IR-irradiation for 0.4–0.75 s.

All components of the 20S proteasome of mouse intestine separated by 2-D gel electrophoresis were unambiguously identified by UV-MALDI FTICR-MS following tryptic in-gel digestion. The high mass determination accuracy of the FTICR-MS enabled the setting of low mass tolerance thresholds in the subsequent database search, leading to high scores equivalent with identifications with high probability. Representative examples of spectra and identifications of proteins are shown in **Fig. 2**.

The high mass accuracy determination and high resolution of the FTICR-MS method can provide the direct identification of individual protein components in mixtures, as illustrated by the example of two proteasome components in **Fig. 3**. In some cases, the resolution of the gel electrophoresis was found insufficient to separate individual proteasome subunits, which is frequently indicated by a cloudy spot shape. In the example in **Fig. 3** (Spot No. 13 in **Fig. 1**), unequivocal identification of the two proteasome components β5i and β4 was obtained, thus indicating that high mass accuracy of FTICR-MS and minimum number of peptides required for database identification may overcome the problem of insufficient electrophoretic resolution of protein complexes.

The identification of proteins by peptide mass fingerprint can be ascertained by tandem mass spectrometric fragment analysis. As shown in **Fig. 4**, the proteasome subunit C5 (β6) was identified by database search with the monoisotopic masses of the tryptic peptide ions. For confirmation, the peptide with $m/z = 1545.75$ was isolated and fragmented by IRMPD as described above. In the IRMPD-spectra, the typical, abundant series of b- and y-ions could be observed. In addition, the fragment ions at $m/z$ 1527.7277 and 788.4041 can be assigned as condensation products of the parent ion and the y 7 ion, respectively. Similar to the IR-laser ionisation described later, the loss of water is favoured at elevated temperatures. All mass values could be assigned to the corresponding sequences, thus confirming the identifications obtained by peptide mass fingerprint.

### 3.3. IR-MALDI-FTICR Mass Spectrometry

The Scout 100 MALDI source of the FTICR-mass spectrometer was modified to accommodate an additional Q-switched 2.94 μm Er:YAG laser (originally produced by Bioptic, Berlin, Germany), as previously described in **ref.** *19*. The maximum pulse energy of the laser (7 mJ) was decreased by an attenuator at a repetition rate of 5 Hz until signals could be obtained. In IR-MALDI, only 5–10 laser shots were used because of strong ablation of the matrix. Although matrices such as succinic acid can be employed for IR-MALDI-MS, a solution of 100 mg/mL 2,5-dihydroxybenzoic acid in acetonitrile/0.1% aqueous TFA (2/1) was found to give best performance for IR-MALDI FTICR-MS.

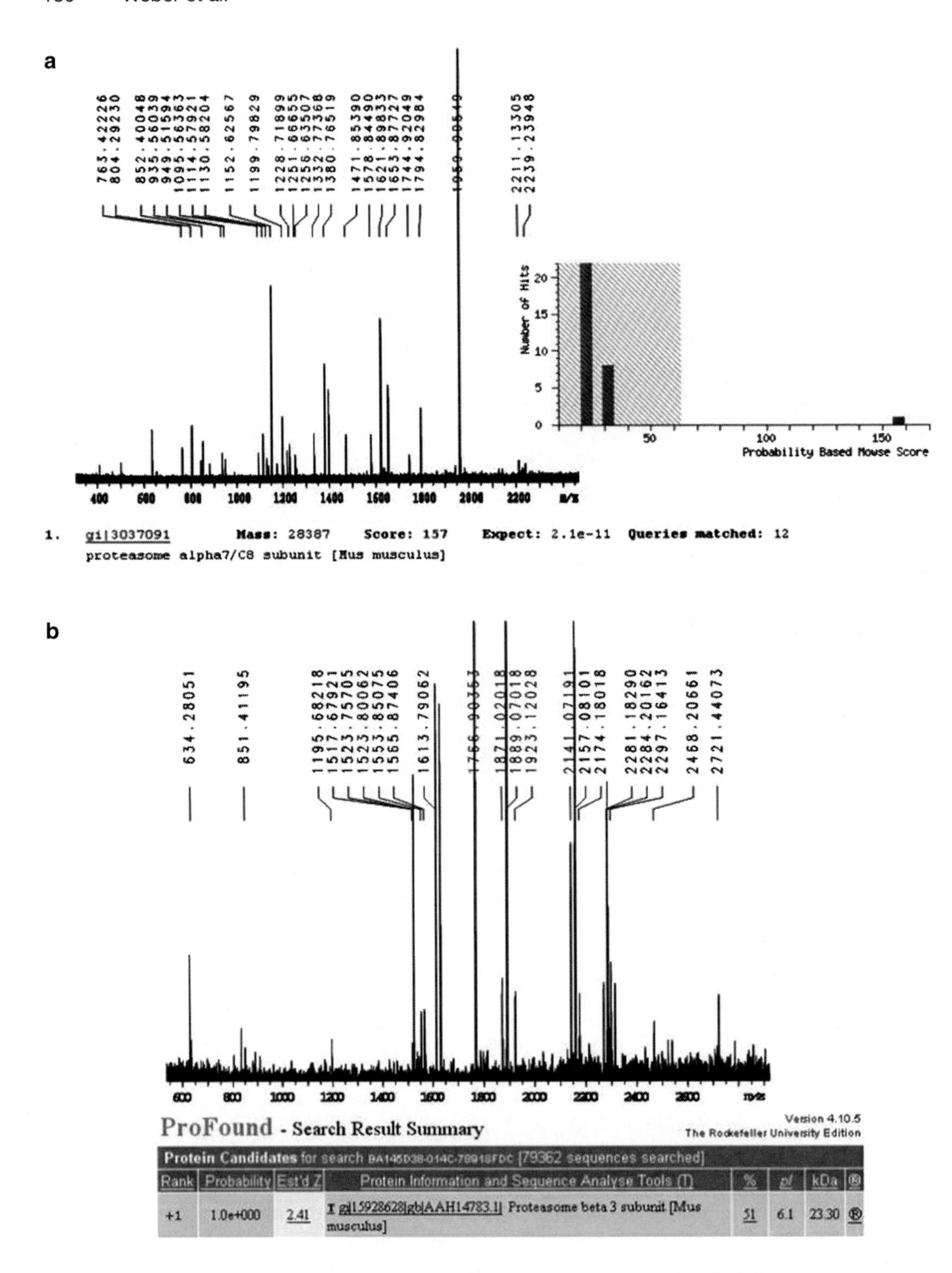


Fig. 2. MALDI FT-ICR-mass spectrum and identification of proteasome subunits α7 (a) and β3 (b).

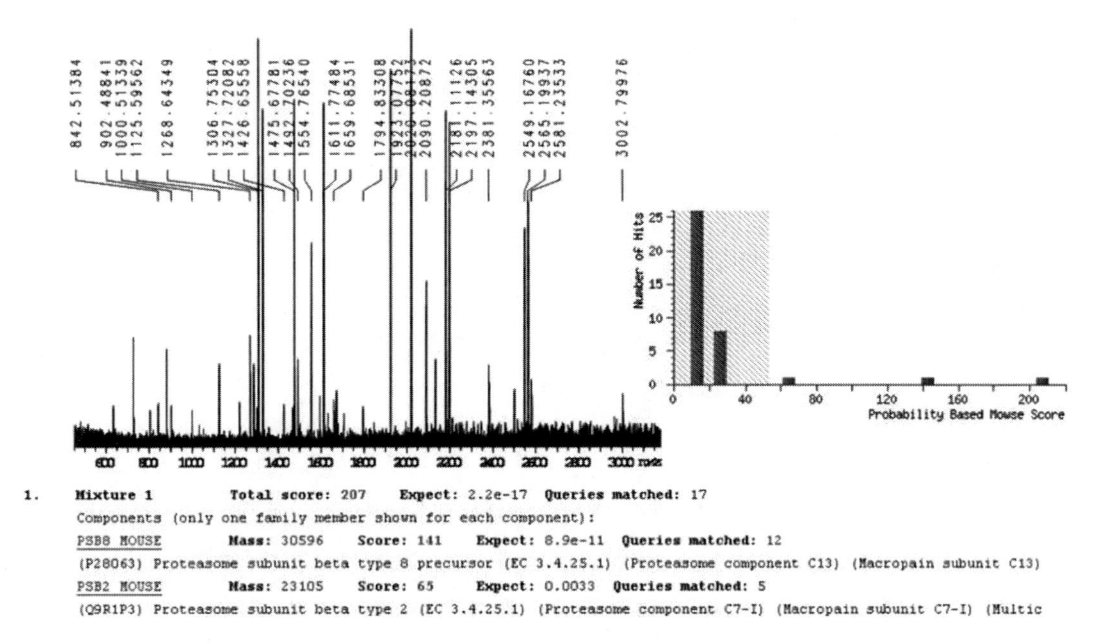


Fig. 3. MALDI FT-ICR mass spectrum and identification of a mixture of two different proteasome subunits.

The subunits α1, α6, β1, and β6 were unambiguously identified with high mass accuracies by peptide mass fingerprinting. Although UV- and IR-MALDI spectra generally showed a similar pattern, some characteristic differences were noted (*see* **Table 3**). A comparison of sequence coverages obtained in protein identifications by UV- and IR-MALDI FTICR-MS on tryptic in-gel digestion is given in **Table 2**; in both UV- an IR-MALDI, sequence coverages ranged from 15% to approximately 55%.

The comparative search for isobaric peptides in UV- and IR-MALDI FTICR-MS revealed more detailed information about fragmentation reactions, leading to modified peptides, which have been found to occur even under "soft ionisation" MALDI conditions. Generally, enhanced thermal fragmentation reactions were observed by IR-MALDI owing to elevated temperature in IR-irradiation. In contrast, photochemical fragmentation reactions are found enhanced by UV-MALDI, which are absent in IR-MALDI. The comparison of UV- and IR-MALDI FTICR spectra of tryptic peptides from subunit β1 reveals details of fragmentation and rearrangement reactions (**Fig. 5**, **Table 3**). For example, the higher abundance of the peptide (pyroE156 – R177) in comparison with the unmodified (Q156 – R177) in the IR-MALDI spectrum can be explained by the preferential release of ammonia at elevated IR-MALDI temperature conditions. In contrast, the peptide ion, *m/z* 2265, 28 was identified

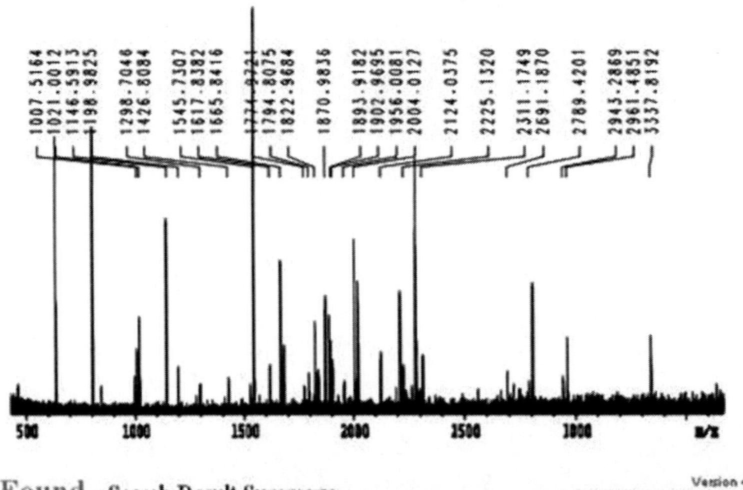


ProFound - Search Result Summary

Version 4.10.5
The Rockefeller University Edition

Protein Candidates for search 8920 16A7-0468-7586800C [66008 sequences searched]

| Rank | Probability | Est'd Z | Protein Information and Sequence Analyse Tools (T) | % | pI | kDa | |
|---|---|---|---|---|---|---|---|
| +1 | 1.0e+000 | 2.43 | T gi\|1165123\|emb\|CAA56702.1\| component C5 of proteasome [Mus musculus] | 36 | 8.7 | 24.66 | |

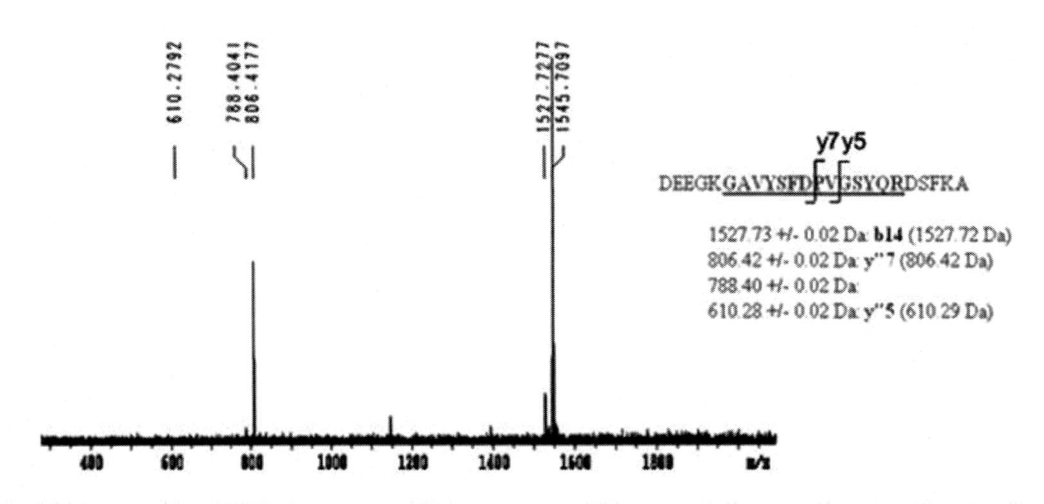


gi|1165123|emb|CAA56702.1| component C5 of proteasome [Mus musculus] mass = 24947.6 Da, pI = 8.7

Fig. 4. MALDI FT-ICR mass spectrum and identification of proteasome subunit C5 (β6) (**a**) and IRMPD of the parent ion *m/z* 1545.73 (**b**). The spectrum is dominated by sequence-specific b- and y-ions. Owing to the high accuracy of mass determination, direct peptide sequence assignment of the mass values is possible. Using this complementary technique, the results from peptide mass fingerprints could be confirmed.

as a photochemical deamidation product of the peptide fragent, (209–229) (LAAIQESGVERQVLLGDQIPK) in the UV-MALDI spectrum. This peptide and corresponding peptides were found unmodified in IR-MALDI-MS.

**Table 2**
**Comparison of percent sequence coverage values of proteasome subunits identified by UV-MALDI and IR-MALDI-MS**

| Proteasome subunit | Sequence coverage UV-MALDI (%) | Sequence coverage IR-MALDI (%) |
|---|---|---|
| α1 | 40 | 18 |
| α6 | 28 | 20 |
| β1 | 15 | 32 |
| β6 | 56 | 40 |

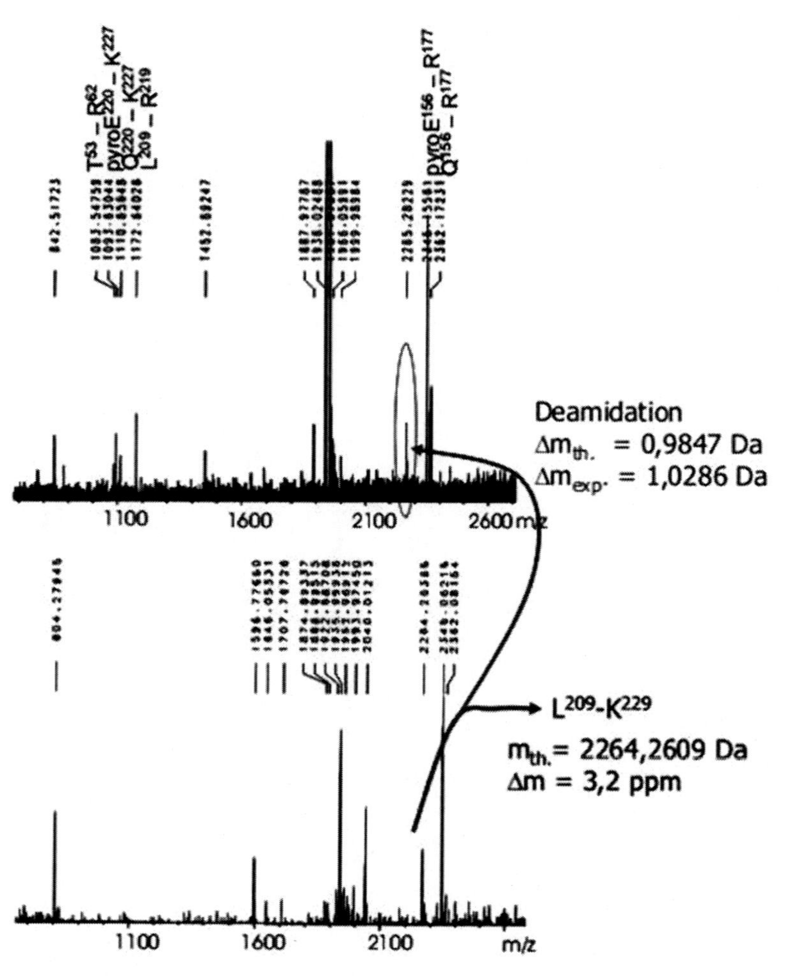


Fig. 5. Comparison of (**a**) UV- and (**b**) IR-MALDI FT-ICR mass spectra of subunit α1 reveals details of fragmentation and rearrangement reactions (*see* text and Table 3). The ion signal at $m/z$ = 2265,28 in the UV-MALDI-spectrum is a photochemical deamidation product of the peptide $L^{209}$AAIQESGVERQVLLGDQIP$K^{229}$.

**Table 3**
**Comparison of identified peptides from proteasome subunit β1 using UV- and IR-MALDI FT-ICR mass spectrometry. The peak with the mass 2264.2376 amu in the UV-MALDI spectrum is the result of a photochemical deamidation product of the peptide $L^{209}$AAIQESGVERQVLLGDQIPK$^{229}$, which can be observed intact in IR-MALDI. In the higher mass range, IR-MALDI provides better mass accuracies than UV-MALDI**

| Sequence | $m_{th}$ (amu) | $m_{exp}$ (amu) (UV) | Δm (ppm) | $m_{exp}$ (amu) (IR) | Δm (ppm) | Modification |
|---|---|---|---|---|---|---|
| $T^{53}-R^{62}$ | 1082.5356 | 1082.5397 | 3.8 | | | |
| pyro$E^{220}-K^{227}$ | 1092.6180 | 1092.6225 | 4.1 | | | Q → pyroE |
| $Q^{220}-K^{227}$ | 1109.6445 | 1109.6486 | 3.6 | | | |
| $L^{209}-R^{219}$ | 1171.6197 | 1171.6324 | 10.8 | | | |
| $L^{209}-K^{229}$ | 2263.2536 | | | 2263.2458 | 3.5 | |
| $L^{209}-K^{229}$ | 2264.2376 | 2264.2458 | 7.4 | | | Q deamidated |
| pyro$E^{156}-R^{177}$ | 2344.0648 | 2344.1177 | 22.5 | 2344.0543 | 4.4 | Q → pyroE |
| $Q^{156}-R^{177}$ | 2361.0913 | 2361.1444 | 22.5 | 2361.0736 | 7.4 | |

### *3.4. Database Search Procedures*

The monoisotopic masses of singly charged ions were directly used for database search using the following publicly available programmes: Mascot (http://www.matrixscience.com/cgi/search_form.pl?FORMVER = 2&SEARCH = PMF), MS-Fit (http://prospector.ucsf.edu/ucsfhtml4.0/msfit.htm), and ProFound (http://prowl.rockefeller.edu/profound_bin/WebProFound.exe). Both SwissProt and NCBInr were employed as databases. For MS/MS, PepFrag (http://prowl.rockefeller.edu/prowl/pepfragch.html) was used for database search.

## 4. Notes

In IR-MALDI-MS, the laser power and number of laser shots must be carefully adjusted. Best results were obtained in cases when matrix ablation occurs at a visible rate. A too high laser power or focus will lead to a drastic decrease of the number of ions produced. In UV-MALDI-MS, attention should be given to the possible formation of photochemical reactions such as deamidation, and laser energies should be adjusted accordingly.

## Acknowledgements

This work was supported by grants from the Deutsche Forschungsgemeinschaft (DFG), Bonn (1517/4-2 to M.G., M.P.; 175/10-2 to M.P).

## References

1. Voges, D., Zwickl, P., and Baumeister, W. (1999) The 26S proteasome: A molecular machine designed for controlled proteolysis. *Annu Rev Biochem* **68**: 1015–68.
2. Hershko, A. and Ciechanover, A. (1992) The ubiquitin system for protein degradation. *Annu Rev Biochem* **61**: 761–807.
3. Ciechanover, A., Orian, A., and Schwartz, A.L. (2000) The ubiquitin-mediated proteolytic pathway: Mode of action and clinical implications. *J Cell Biochem* 77(S34): 40–51.
4. Ciechanover, A. (2005) Proteolysis: From the lysosome to ubiquitin and the proteasome. *Nat Rev Mol Cell Biol* **6**(1): 79–87.
5. Damoc, E., Youhnovski, N., Crettaz, D., Tissot, J.D., and Przybylski, M. (2003) High resolution proteome analysis of cryoglobulins using Fourier transform-ion cyclotron resonance mass spectrometry. *Proteomics* **3**(8): 1425–33.
6. Bai, Y., Galetskiy, D., Damoc, E., Paschen, C., Liu, Z., Griese, M., Liu, S., and Przybylski, M. (2004) High resolution mass spectrometric alveolar proteomics: Identification of surfactant protein SP-A and SP-D modifications in proteinosis and cystic fibrosis patients. *Proteomics* **4**(8): 2300–9.
7. Smith, R.D., Loo, J.A., Edmonds, C.G., Barinaga, C.J., and Udseth, H.R. (1990) New developments in biochemical mass spectrometry: Electrospray ionization. *Anal Chem* **62**(9): 882–99.
8. Glocker, M.O., Bauer, S.H., Kast, J., Volz, J., and Przybylski, M. (1996) Characterization of specific noncovalent protein complexes by UV matrix-assisted laser desorption ionization mass spectrometry. *J Mass Spectrom* **31**(11): 1221–7.
9. Przybylski, M., and Glocker, M.O. (1996) Electrospray mass spectrometry of biomacromolecules complexes with non-covalent interactions – new analytical perspectives for supramolecular chemistry and molecular recognition processes. *Angew Chem Int Ed Engl* **35**(8): 806–26.
10. van der Rest, G., He, F., Emmett, M.R., Marshall, A.G., and Gaskell, S.J. (2001) Gas-phase cleavage of PTC-derivatized electrosprayed tryptic peptides in an FT-ICR trapped-ion cell: Mass-based protein identification without liquid chromatographic separation. *J Am Soc Mass Spectrom* **12**(3): 288–95.
11. Wigger, M., Eyler, J.R., Benner, S.A., Li, W., and Marshall, A.G. (2002) Fourier transform-ion cyclotron resonance mass spectrometric resolution, identification, and screening of non-covalent complexes of Hck Src homology 2 domain receptor and ligands from a 324-member peptide combinatorial library. *J Am Soc Mass Spectrom* **13**(10): 1162–9.
12. Laskin, J., and Futrell, J.H. (2005) Activation of large ions in FT-ICR mass spectrometry. *Mass Spectrom Rev* **24**(2): 135–67.
13. Laiko, V.V., Taranenko, N.I., Berkout, V.D., Yakshin, M.A., Prasad, C.R., Lee, H.S., and Doroshenko, V.M. (2002) Desorption/ionization of biomolecules from aqueous solutions at atmospheric pressure using an infrared laser at 3 microm. *J Am Soc Mass Spectrom* **13**(4): 354–61.
14. Feldhaus, D., Menzel, C., Berkenkamp, S., Hillenkamp, F., and Dreisewerd, K. (2000) Influence of the laser fluence in infrared matrix-assisted laser desorption/ionization with a 2.94 microm Er: YAG laser and a flat-top beam profile. *J Mass Spectrom* **35**(11): 1320–8.
15. Menzel, C., Dreisewerd, K., Berkenkamp, S., Hillenkamp, F. (2001) Mechanisms of energy deposition in infrared matrix-assisted laser desorption/ionisation mass spectrometry. *Int J Mass Spectrom* **207**: 73–96.
16. Berkenkamp, S., Menzel, C., Hillenkamp, F., and Dreisewerd, K. (2002) Measurements of mean initial velocities of analyte and matrix ions in infrared matrix-assisted laser desorption ionization mass spectrometry. *J Am Soc Mass Spectrom* **13**(3): 209–20.
17. Cramer, R., Richter, W.J., Stimson, E., and Burlingame, A.L. (1998) Analysis of phospho- and glycopolypeptides with infrared matrix-assisted laser desorption and ionization. *Anal Chem* **70**(23): 4939–44.
18. Budnik, B.A., Jensen, K.B., Jorgensen, T.J., Haase, A., and Zubarev, R.A. (2000) Benefits of 2.94 micron infrared matrix-assisted laser desorption/ionization for analysis of

labile molecules by Fourier transform mass spectrometry. *Rapid Commun Mass Spectrom* **14**(7): 578–84.

19. Petre, B.A., Youhnovski, N., Lukkari, J., Weber, R., and Przybylski, M. (2005) Structural characterisation of tyrosine-nitrated peptides by ultraviolet and infrared matrix-assisted laser desorption/ionisation Fourier transform ion cyclotron resonance mass spectrometry. *Eur J Mass Spectrom (Chichester, Eng)* **11**(5): 513–8.
20. Schmidtke, G., Emch, S., Groettrup, M., and Holzhutter, H.G. (2000) Evidence for the existence of a non-catalytic modifier site of peptide hydrolysis by the 20S proteasome. *J Biol Chem* **275**(29): 22056–63.
21. Schwarz, K., van den Broek, M., Kostka, S., Kraft, R., Soza, A., Schmidtke, G., Kloetzel, P.M., and Groettrup, M. (2000) Over expression of the proteasome subunits LMP2, LMP7, and MECL-1, but not PA28 alpha/beta, enhances the presentation of an immunodominant lymphocytic choriomeningitis virus T cell epitope. *J Immunol* **165**(2): 768–78.
22. Groettrup, M., Kraft, R., Kostka, S., Standera, S., Stohwasser, R., and Kloetzel, P.M. (1996) A third interferon-gamma-induced subunit exchange in the 20S proteasome. *Eur J Immunol* **26**(4): 863–9.
23. Neuhoff, V., Arold, N., Taube, D., and Ehrhardt, W. (1988) Improved staining of proteins in polyacrylamide gels including isoelectric focusing gels with clear background at nanogram sensitivity using Coomassie Brilliant Blue G-250 and R-250. *Electrophoresis* **9**: 255–62.
24. Groettrup, M., Ruppert, T., Kuehn, L., Seeger, M., Standera, S., Koszinowski, U., and Kloetzel, P.M. (1995) The interferon-gamma-inducible 11 S regulator (PA28) and the LMP2/LMP7 subunits govern the peptide production by the 20S proteasome in vitro. *J Biol Chem* **270**(40): 23808–15.
25. Mortz, E., Vorm, O., Mann, M., and Roepstorff, P. (1994) Identification of proteins in polyacrylamide gels by mass spectrometric peptide mapping combined with database search. *Biol Mass Spectrom* **23**(5): 249–61.
26. Houthaeve, T., Gausepohl, H., Ashman, K., Nillson, T., and Mann, M. (1997) Automated protein preparation techniques using a digest robot. *J Protein Chem* **16**(5): 343–8.
27. Groll, M., Ditzel, L., Lowe, J., Stock, D., Bochtler, M., Bartunik, H.D., and Huber, R. (1997) Structure of 20S proteasome from yeast at 2.4 A resolution. *Nature* **386**(6624): 463–71.

# Part IV

## Quantitative Proteomics

# Chapter 11

# Liquid Chromatography–Mass Spectrometry-Based Quantitative Proteomics


## Michael W. Linscheid, Robert Ahrends, Stefan Pieper, and Andreas Kühn



## Summary

During the last decades, molecular sciences revolutionized biomedical research and gave rise to the biotechnology industry. During the next decades, the application of the quantitative sciences – informatics, physics, chemistry, and engineering – to biomedical research brings about the next revolution that will improve human healthcare and certainly create new technologies, since there is no doubt that small changes can have great effects. It is not a question of "yes" or "no," but of "how much," to make best use of the medical options we will have.

In this context, the development of accurate analytical methods must be considered a cornerstone, since the understanding of biological processes will be impossible without information about the minute changes induced in cells by interactions of cell constituents with all sorts of endogenous and exogenous influences and disturbances.

The first quantitative techniques, which were developed, allowed monitoring relative changes only, but they clearly showed the significance of the information obtained. The recent advent of techniques claiming to quantify proteins and peptides not only relative to each other, but also in an absolute fashion, promised another quantum leap, since knowing the absolute amount will allow comparing even unrelated species and the definition of parameters will permit to model biological systems much more accurate than before. To bring these promises to life, several approaches are under development at this point in time and this review is focused on those developments.



**Key words:** Isotope dilution concept, Stable-isotope labeling, Label-free quantitation, LC–/MS


## 1. Introduction

The analysis of a complete proteome (which means the analysis of all protein species in a cell, a tissue, or any biological system at a given time) calls not only for a comprehensive qualitative analysis,


Jörg Reinders and Albert Sickmann (eds.), *Proteomics,* Methods in Molecular Biology, Vol. 564
DOI: 10.1007/978-1-60761-157-8_11, © Humana Press, a part of Springer Science+Business Media, LLC 2009

that is which is the protein, but for the quantitative analysis as well, which means the relative amounts of the proteins have to be measured at a given time and then, as a consequence, the changes in time. Only the quantity of proteins or the change in quantity is significant for status or, even more relevant, the dynamics of a biological system. Based on these data, the effect of a disturbance of the "normal" situation can be analyzed. Examples, which immediately highlight the importance, are first "toxicoproteomics" reflecting the interfering effect of toxic substances onto the proteome, or second, which proteins are expressed in the lifecycle of a bacterium and why, and third the analysis of cell cycles to allow the development of drugs with well-targeted effects, e.g. for children, women, and men. To achieve this, we need to be able to analyze the complete situation in the target cells accurately. Moreover, the quest for significant biological markers indicating cancer, tumors, and more diseases and disorders in early stages of development needs robust and reliable quantitative data as well.

Thus, it is not surprising that numerous attempts have been made to develop strategies for the quantitative analysis of proteins or, for that matter, of peptide mixtures that are assumed to reflect the proteomic status. During the last 5 years, several hundred papers were published claiming new techniques or new applications and many reviews have been published *(1–10)* up to now. The challenges are manifold: The dynamic range of proteins in a biological system extends from one copy to more than a million copies per cell, which asks for detection strategies capable of similar dynamic ranges *(11)*. Such a biological system can be ultimately a single cell with a volume of about 500 fL and a total protein content of about 50 pg (c. 2 fmol), which requires extreme sensitivity for the detection of such proteins. Then, proteins are vastly different in their physicochemical behavior, which results in a wide range of difficulties during handling such as very different solubility, stability, isoelectric points, optimal pH-ranges for digests, and the like. All this taken together shows clearly that the determination of a reliable and accurate amount of protein among proteins is close to impossible *(12, 13)*.

Till date, several approaches have been developed for comprehensively assessing proteomes and despite the mentioned complex situation (**Fig. 1**), the field has made great strides, brought about primarily by technological advances, particularly in separation techniques and mass spectrometry *(14)*. The separation techniques are mainly one- and two-dimensional polyacrylamide gel electrophoresis (1D or 2D-PAGE), but for the quantitative detection they suffer from a lack of sensitivity and reproducibility *(15)*. Concepts to replace especially 2D-PAGE by LC techniques were described, but are not applicable in many cases *(16)*. The regularly included MALDI mass spectrometry and modifications thereof are quite informative, but not reliable enough for quantitative analysis, even though there is potential especially when MALDI is combined with LC separation and MS/MS for identification *(17)*.

As already mentioned, the techniques which will be discussed in this review are generally based on peptide determinations, since this was, and still is, believed to reflect the proteins in the sample sufficiently (**Fig. 2**).

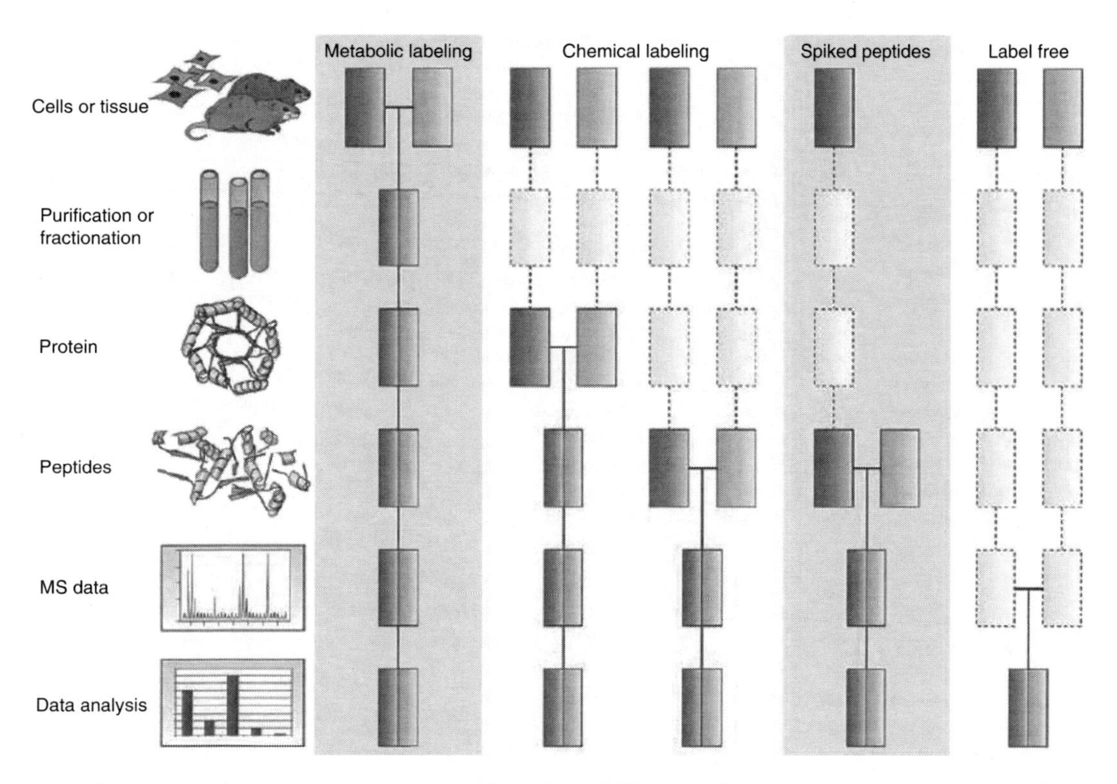


Fig. 1. Common quantitative mass spectrometry workflows. *Boxes in blue* and *yellow* represent two experimental conditions. *Horizontal lines* indicate when the samples are combined. *Dashed lines* indicate the points at which experimental variation and thus quantification errors can occur (from **ref.** *13*, with kind permission of Springer Science + Business Media).

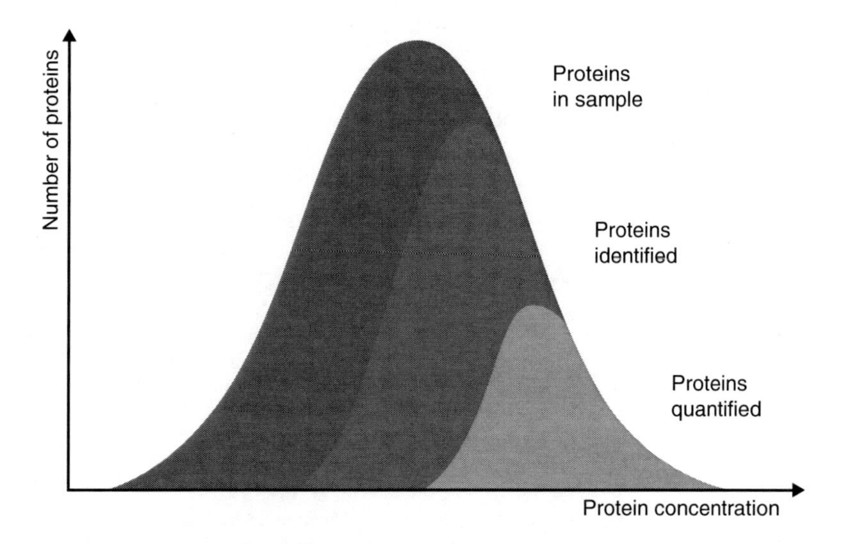


Fig. 2. Representation of the fraction of a proteome that can be identified and quantified by mass-spectrometry-based proteomics. Cellular proteins span a wide range of expression and current mass spectrometric technologies typically sample only a fraction of all the proteins present in a sample. Because of limited data quality, only a fraction of all the identified proteins can be reliably quantified (from **ref. 13**, with kind permission of Springer Science + Business Media).

## 2. ICAT: The Start

In this situation, LC/MS has emerged as the method of choice for characterizing complex protein mixtures quantitatively and the protocol setting the stage came to be known as ICAT (isotope-coded-affinity-tag) *(18)*. In the meantime, many new applications and technical modifications have been described *(19)*, but to illustrate the generic approach for relative quantification, ICAT is discussed here.

The quantification is based on the well-known isotope dilution concept. The rationale is that by means of a chemical tag attached to peptides (e.g. from a tryptic digest of protein mixtures), the differentiation between two different statuses of the same biological system becomes feasible, when each sample is characterized by a specific set of isotopes. The two samples are mixed and analyzed at once. Since the tagging chemistry needs targets, here often cysteine, the isotopic labeling, reduces the number of peptides, which can be analyzed along with the complexity of the sample (**Fig. 3**). When the label has not only the isotopes, but an affinity anchor such as biotin built in, it allows for an enrichment protocol with avidin affinity columns. The reagent has been optimized since its first introduction to give improved quantitative reactions, greater mass difference between the two tags, and better sensitivity in mass spectrometry *(20)*, but major problems are intrinsic. Incomplete

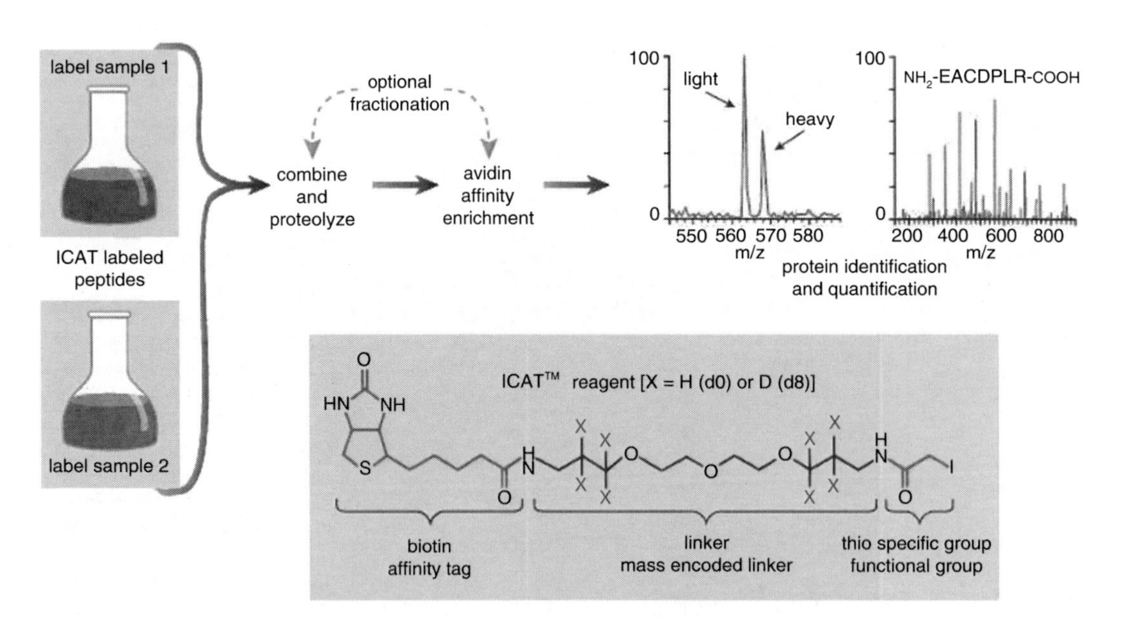


Fig. 3. Structure of the original ICAT reagent with reactive group, labeled linker, and affinity tag, after Patterson and Aebersold (from **ref. 21**, reprinted by permission from Macmillan Publishers Ltd., copyright 2004).

reaction yield induces loss of minor compounds and may bias the detection; therefore, careful design of the handling protocol is mandatory. However, even more importantly, the approach allows only differential quantitation and even this in rather large steps.

To improve this, several more strategies have been devised more recently and certainly we have not seen the end of that development.

## 3. General Remarks

To cope with the immense complexity of biological samples, a separation into fractions or species prior to MS analysis is mandatory. The techniques are especially useful when a high capacity can be combined with excellent separation capability; techniques such as gel electrophoresis, HPLC, and capillary electrophoresis have been interfaced to mass spectrometry in an online or offline fashion. For proteins, the 2D gel electrophoresis *(22, 23)* proved to be the most efficient high-resolution technique with dynamic ranges, best cases, up to $1 \times 10^4$. For a differential quantitative analysis, densitometry *(24)*, fluorescence *(24, 25)*, scintillation counting *(26, 27)*, and ESI or MALDI MS *(28)* and, most recently, ICP-MS *(29)* has been used. The combination of different LC columns (2D-LC) using, e.g., ion exchange, reversed phase, and affinity columns *(30–34)* show very efficient separation and specifically tailored characteristics. The dynamic ranges in concentration determined here can be as high as four orders of magnitude.

Despite the revolution brought about by the new mass spectrometric approaches, some specifics of mass spectrometry for quantitative determinations need to be addressed. Ionization efficiencies in electrospray and MALDI may vary and can be a function of the chemical nature of the analytes, which makes the application of internal standards indispensable. These suppression effects have been observed in many instances and are of major concern. The components in mixtures and the eluents may influence the ionization depending on their respective concentrations, which may eventually lead to complete loss of minor components. In MALDI-MS, crystallization effects or impurities may have the same detrimental results rendering quantitative measurements difficult *(35, 36)*. Thus, the linearity of calibration and responses in the high and the low concentration range may vary considerably. This is now, where recently inductively coupled plasma mass spectrometry (ICP-MS) came into play with the restriction that this type of instrumentation is not typical for biological laboratories.

Before quantitative, especially ***absolute*** quantitative, determinations will be discussed, a few aspects of affinity sample preparation are necessary, since this technique is central for proteomics. This strategy – tagging the analytes with a group, which has highly specific binding capacity with another compound, often an immobilized protein or antibody – was used with ICAT already and allows the purification and preconcentration of very low level analytes in complex mixtures, which are intractable otherwise. Some examples are antibody–antigen complexes such as the flag/c-myc-tag, enzyme–substrate complexes like ATP/ATPase, IMAC-materials such the already mentioned hexa-His- and Hat-Tag, lectine–polysaccharide complexes, or glutathion-S-transferase/glutathione or the well-known biotin–avidin (streptavidin) complex. All of these have their scope and limitations *(37)*; the binding constants are usually in the range of nanomoles *(38)*. Whetstone et al. *(39)* developed an antibody against the metal carrying DOTA complex, which could be used for affinity preconcentration; we rely on the well-known biotin/streptavidin complex ($K_d$ = $10^{-15}$ M), which shows great promise to reduce the complexity without changing the metal composition in the process; it has been used with mass spectrometry several times *(28, 40, 41)*.

After the advent of ICAT, more strategies were developed with the intention to overcome shortcomings of ICAT in terms of detection capabilities, increased number of comparable states, or using different tagging chemistry, if any. Methods without synthetic tagging were proposed as well. The most prominent new approaches are iTRAQ, SILAC, AQUA, and QconCAT (for a short review *see* **ref.** ***13***, more comprehensive reviews are **refs.** ***12, 42, 43)***. All these methods rely on ESI or MALDI mass spectrometry, which limit the precision, the detection limits, and the dynamic ranges. Therefore, we recently proposed MeCAT, a new tagging strategy using metal tags with lanthanides to overcome some of the discussed shortcomings by using ICP-MS in combination with liquid chromatography *(29)*.

Generally, most techniques can have one step in common: They use the shot-gun-proteomics strategy *(30)*. Proteins are chemically hydrolyzed or digested, yielding complex mixtures of peptides with a wide variety of chemical properties, which need to be separated into their components using liquid chromatography or other means. There is one weakness, though, since the use of such methods results generally in loss of information, especially with reference to protein species carrying post-translational modifications *(44)*.

## 4. Techniques

#### *4.1. iTRAQ*

The iTRAQ strategy uses isotopic labeling again, but relies solely on LC/MS/MS measurements. Up to four samples can be labeled simultaneously with chemical tags having an amino specific reactive group and a combination of "reporter" and "balance" groups. The rationale is that the mass of the combined groups is always the same (145 amu), but in MS/MS data of the labeled peptide mixture, the reporter group, after fragmentation between balance and reporter, appears at four different masses ($m/z$ 114–117) according to the respective isotopic labeling (**Fig. 4**).

In a recent study, a protocol for the quantitative analysis of four different phosphorylated proteins in different states has been proposed *(45)* (**Fig. 4**) using standard proteins. After gel separation, the proteins were digested, labeled, and mixed prior to LC/MS/MS analysis. The strategy has been used in many studies *(45–56)* and most recently, the number of channels has been increased to eight *(55)*, allowing the analysis of up to eight states simultaneously.

#### *4.2. SILAC*

The application of labeled peptides made biologically by growing cells or cell cultures spiked with labeled amino acids (for reviews, *see* **refs.** *8, 57)* seems promising, because spiking with synthetic peptides can be used for the absolute quantitation of some proteins *(58, 59)* (*see* **Fig. 5** as an illustration for the SILAC approach). The limit must be seen in the number of cell cultures, plants, or animals, which may be grown on labeled nutrition ingredients.

#### *4.3. QconCAT*

Another technique that came to be known as QCAT (or QconCAT) *(60)* makes direct use of biotechnology in so far as an artificial concatenated gene is constructed, which is included in the DNA of the species under investigation by means of a specifically designed vector. The gene is constructed from pieces that code for one of the tryptic peptides of each protein to be quantified. After expression, the amounts of peptides in the sample are strictly the same for each peptide owing to the design. Since a his-tag and one peptide carrying one cysteine are included, the artificial protein can be affinity-purified, digested, and quantified. When labeled metabolites are included, quantification may be extended to different system states as well.

#### *4.4. Label-Free Quantitation*

Currently, under the term label-free, two vastly different strategies are subsumed *(61)*. The first is based on ion intensities of peptides derived from proteins and the intensities can be measured using extracted mass chromatograms *(62 )*. The quantitation can

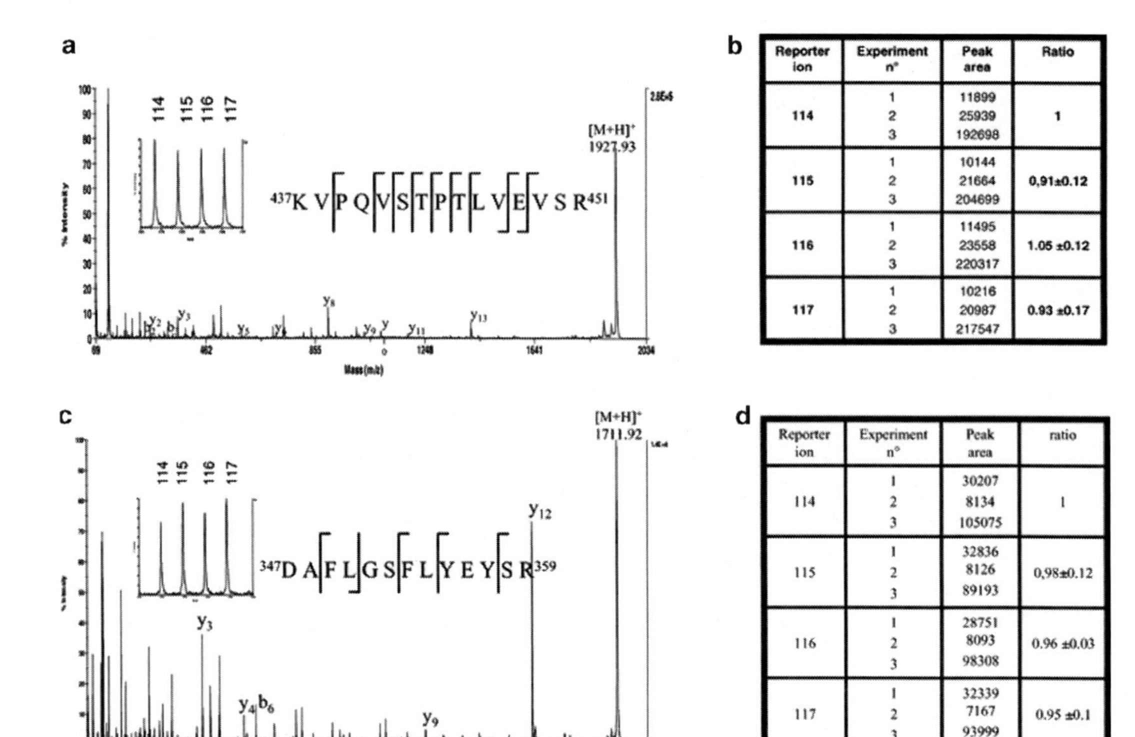


| Reporter ion | Experiment n° | Peak area | Ratio |
|---|---|---|---|
| 114 | 1<br>2<br>3 | 11899<br>25939<br>192698 | 1 |
| 115 | 1<br>2<br>3 | 10144<br>21664<br>204699 | 0,91±0.12 |
| 116 | 1<br>2<br>3 | 11495<br>23558<br>220317 | 1.05 ±0.12 |
| 117 | 1<br>2<br>3 | 10216<br>20987<br>217547 | 0.93 ±0.17 |

| Reporter ion | Experiment n° | Peak area | ratio |
|---|---|---|---|
| 114 | 1<br>2<br>3 | 30207<br>8134<br>105075 | 1 |
| 115 | 1<br>2<br>3 | 32836<br>8126<br>89193 | 0,98±0.12 |
| 116 | 1<br>2<br>3 | 28751<br>8093<br>98308 | 0.96 ±0.03 |
| 117 | 1<br>2<br>3 | 32339<br>7167<br>93999 | 0.95 ±0.1 |

Fig. 4. Tandem mass spectra of BSA tryptic peptides prepared by labeling with each of the four isobaric iTRAQ reagents and combining the reaction mixtures in a 1:1:1:1 ratio. Peptides were separated by capillary LC and analyzed on a 4700 MALDI-TOF/TOF analyzer. The precursor ion and b- and y-type ions are labeled. (**a**) Tandem mass spectrum of the precursor ion at *m/z* 1927.93 [437KVPQVSTPTLVEVSR451] derivatized with the iTRAQ reagents (at the N-terminus and the $\varepsilon NH_2$ of the K437 residue). The *inset* shows the low-mass region containing the reporter ions at *m/z* 114, 115, 116, and 117. The intensity scale was set to 60% of the base peak (*m/z* 114.10). (**b**) Peak areas of the four reporter ions at *m/z* 114, 115, 116, and 117, and the relative abundance ratios. These values are the average of three experiments consisting of three independent sample treatments and MS experiments. (**c**) Tandem mass spectrum of the precursor ion at *m/z* 1711.92 [347DAFLGSFLYEYSR359] derivatized with the iTRAQ reagents. The *inset* shows the low-mass region containing the reporter ions at *m/z* 114, 115, 116, and 117. The intensity scale was set to 70% of the base peak (*m/z* 114.10). (**d**) Peak areas of the four reporter ions at *m/z* 114, 115, 116, and 117, and relative abundance ratios. These values are the average of three independent sample treatments and MS experiments (from **ref.45**, reproduced with permission, copyright Wiley-VCH Verlag GmbH & Co. KGaA).

be extended to isotopic clusters to get more reliable data, but the rationale remains the same *(62–65)*. The second strategy relies on spectral counting under the assumption that more MS/MS spectra can be acquired when more peptide is available. For this reason, simply counting spectra in an LC/MS/MS analysis may be used to relatively quantify the amount of peptides in a mixture *(30, 66)*. This rather undemanding rationale seems to be valid to some extent, but is still disputed. Fast mass spectrometers may improve the data owing to larger scan numbers and less error in counting, but the common data-dependent scanning techniques, which can create unpredictable bias, may be counterproductive and should be handled with care *(61)*.

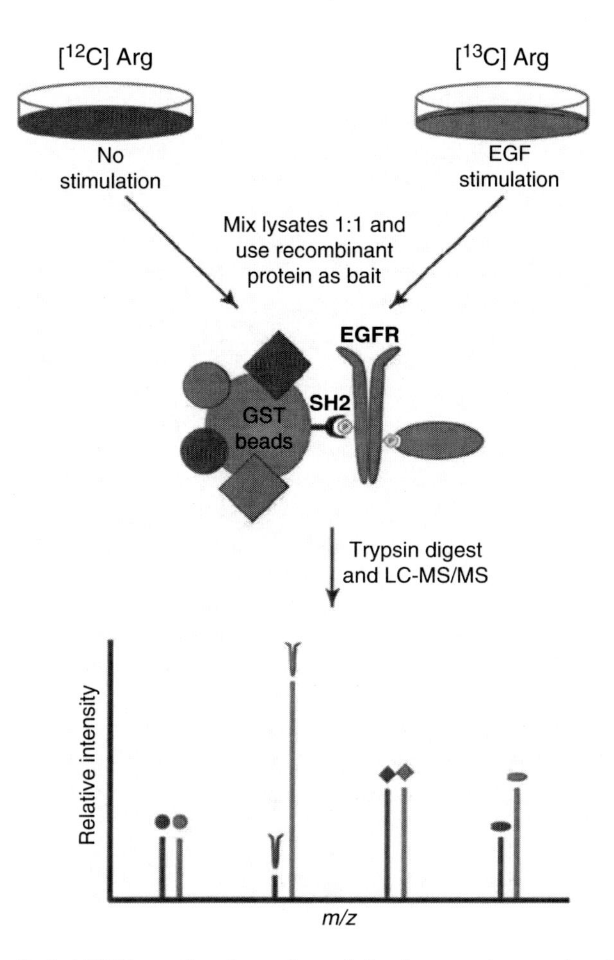


Fig. 5. Strategy to study activated EGFR complex. One cell population is grown in normal arginine-containing medium (*blue dish*), whereas another population is grown in [$^{13}$C]arginine ([$^{13}$C]Arg)-containing medium (*red dish*), encoding all proteins with this 6 Da heavier amino acid (*red color*). The encoded cells are stimulated with EGF, lysed, and mixed 1:1 with the lysate of unlabeled, untreated cells. The combined cell lysates are affinity-purified with GST fused to the SH2 domain of Grb2, which specifically interacts with the activated, phosphorylated form of the receptor and its associated binders. Proteins eluted from the beads are digested with trypsin and the resulting peptides analyzed by mass spectrometry. Arginine-containing peptides occur in doublets, separated by 6 Da. Proteins that interact with the bait in a stimulation-dependent manner are mostly in their encoded form and therefore have a more intense peak for the labeled arginine-containing peptide (*red peaks*). These proteins can be distinguished from a large excess of background binding proteins, which manifest themselves by their 1:1 ratio between stimulated and unstimulated states (**ref. 67**, reprinted by permission from Macmillan Publishers Ltd., copyright 1997).

#### 4.5. $^{18}$O Labeling

Labeling of peptides is also possible using $H_2{}^{18}O$ and trypsin, which introduces $^{18}$O in the hydrolysis step *(68)*. The advent of FTICR mass spectrometry for routine work will enhance the reliability of such approaches since common overlaps between analytes, internal standards, and matrix components can be avoided *(69)* (*see* **Fig. 6**).

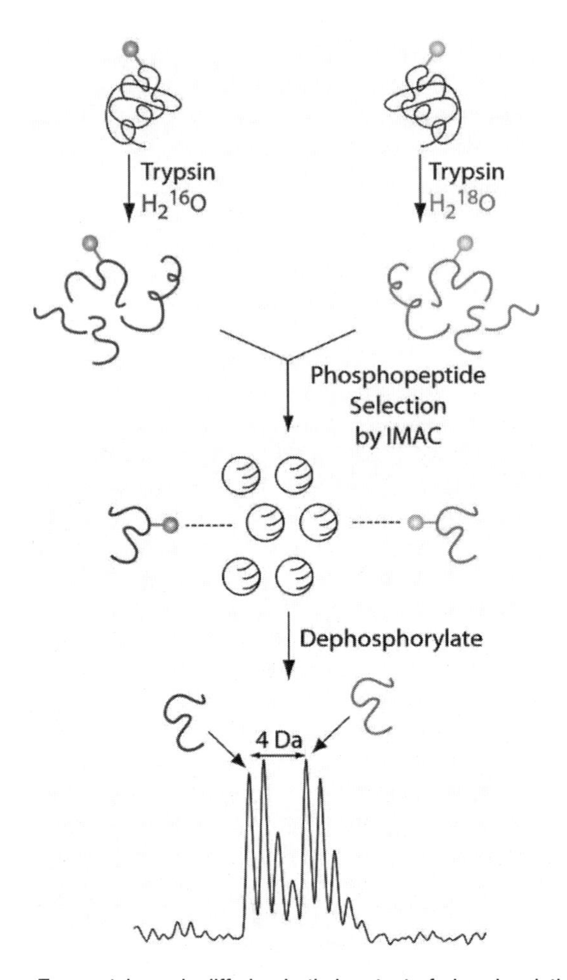


Fig. 6. Outline of the strategy: Two protein pools differing in their extent of phosphorylation are digested with trypsin either in $H_2{}^{16}O$ or in $H_2{}^{18}O$ to obtain differential mass labeling. Equal amounts of the two pools are mixed, and phosphopeptides are selected with IMAC beads charged with $Fe^{3+}$ . To enable mass determinations of the bound peptides with high accuracy, the peptides are eluted from the IMAC beads with alkaline phosphatase. The two peptide pools can be distinguished by a shift of 4 Da in the isotope cluster, and the difference in the extent of phosphorylation is reflected by the peak areas of the two monoisotopic peaks (**ref.** ***70***, copyright 2003, National Academy of Science, USA).

#### *4.6. Endogenous Element-Based Quantitation (Phosphorus, Sulfur, etc.)*

Some time ago, we could demonstrate that the use of endogenous elements as "quantitation tags" for the analysis of nucleotides is very elegant, since nucleotides as per definition have always well-defined phosphorus content *(71–73)* and a powerful detector, namely mass spectrometry based on inductively coupled plasma (ICP-MS) is available. Analytes are atomized and ionized inside of the 8,000–10,000 K plasma flame, which serves as the ion source. This temperature assures that for many heavy atoms, the ionization efficiency is close to 100% *(74)* and mutual suppression effects are unknown. The successful transfer of the approach to peptides and proteins has been published more recently *(75)*, in combination with sulfur detection, but very accurate and reliable measurements have been achieved as shown in **Fig. 7** *(76)*.

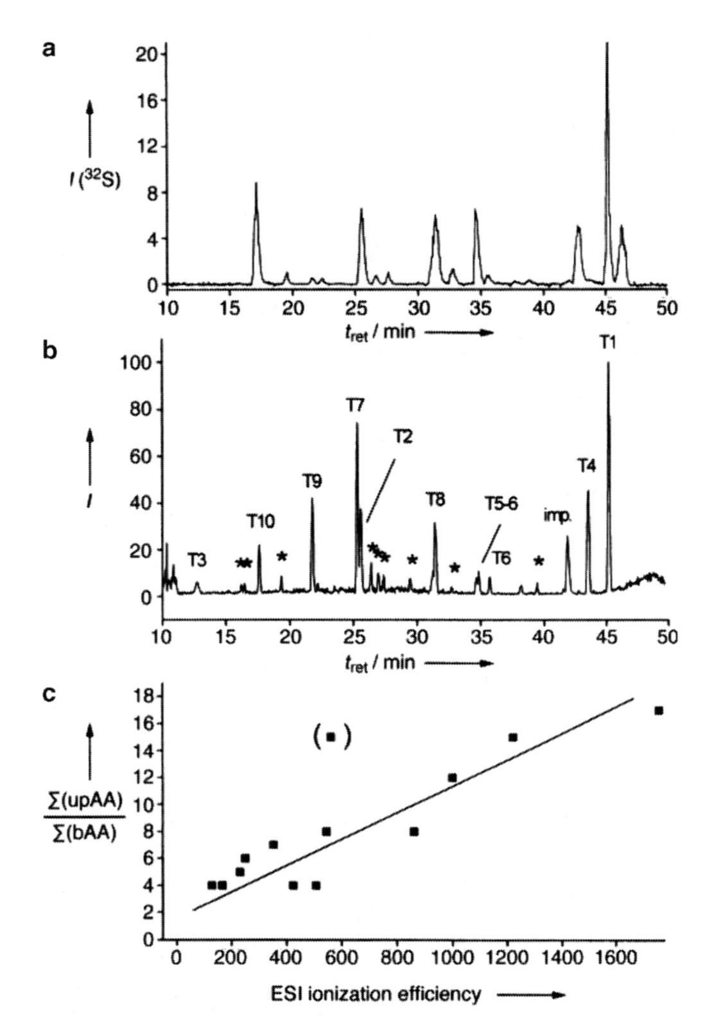


Fig. 7. Tryptic digest of a mixture containing 90% cheA-H and 10% cheA-C. (**a**) μLC/ICP– MS of this digest with $^{32}S$ detection. (**b**) μLC/ESI–MS of the same digest displayed as base peak chromatogram. Tryptic fragments of cheA-C are marked with an asterisk, those of cheA-H have T labels. (**c**) Correlation of the ESI ionization efficiency with the ratio calculated from the number of nonpolar residues divided by the number of basic residues in the corresponding peptide (R = 0.94, the data point in brackets was not included (from **ref. 76**, reproduced with permission, copyright Wiley-VCH Verlag GmbH & Co. KGaA).

The advantage is that the detection capability is structure–independent, allowing easy access to internal standards. Thus, absolute quantitation comes in reach. In addition, the detection limits for especially the heavy elements are very low (for Lu, LOD of 2 fg/g was measured), making LC/ICP–MS a very powerful LC–MS technique. Another benefit of ICP-MS is that modern instruments may have a nominal dynamic range of up to 12 orders of magnitude *(77)*.

#### *4.7. Protein and Peptide Tagging with Metal Complexes*

When the detection of metals is combined with chemical tagging, new strategies can be envisaged. For example, Baranov et al. developed an ELISA-like technique with ICP-MS as detector *(78)*.

They were not only able to quantify disease-relevant proteins, but also able to advance this method to a multiplex cell surface bioassay *(79)*. That assay allows the simultaneous analysis of six different proteins, using a specific monoclonal antibody. Each of these monoclonal antibodies bears a different metal tag, so every tagged protein could be distinguished from others. Using this technique, a cellular surface expression profiling of different proteins was possible. Developments are underway to combine this with a flow cytometer for cell analysis *(80)*.

Wheatstone et al. *(39)* used the "common" tagging chemistry to label peptides and to preconcentrate them in affinity columns with antibodies raised against a lanthanide complex. They obtained ESI spectra of the tagged peptides as well. DOTA (1,4,7,10-tetraazacyclododecane *N*′, *N*″, N‴, *N*⁗-tetraacetic acid) was used for complexation of triply charged lanthanide and rare earth element ions; the synthetic access to DOTA metal complexes has been already shown *(39)*. Besides the polydental macrocyclic DOTA, the open-form DTPA (diethylene triamine pentaacetic acid) has been employed *(81, 82)*. Both reagents are characterized by extreme stability; DOTA *(83)* reaches stability constants (log $\kappa$) up to 25.4 *(84, 85)*. Thus, this complex is suited for many analytical and bioanalytical applications without the threat of metal loss or metal exchange. Liu et al. demonstrated the labeling of peptides using yttrium (Y) and terbium (Tb) DTPA complexes *(82)*. They observed no elution time change of the peptides in liquid chromatography caused by different metals and identification of differential tagged peptides using tandem-MS as possible. A difference in ionization efficiency between Tb- and Y-labeled peptides was observed but could be corrected for; thus, the relative quantification of the tagged peptides was achieved.

In our laboratory, we used a different approach *(29)*. On the basis of DOTA-complexes, two metal-bearing reagents directed against cysteine residues were designed. In this complex, different lanthanide metal ions could be loaded for tagging and quantification. The two reagents differ in the spacer region, which is short for $MeCAT_{Bnz,}$ whereas $MeCAT_{Bio}$ possesses a biotin-bearing spacer to allow affinity purification ***(40, 41)***. The different masses of the complexes *(86)* allow detection of differently labeled species using ESI-MS and with LC/ICP-MS, absolute quantitation is feasible.

**Figure 8** shows the current workflow of the MeCAT technique. Cysteine thiol residues of proteins from different samples were differentially labeled with metal-coded M-MeCAT reagents. After combination of the samples, the protein mixture can be analyzed by any separation method such as SDS-PAGE, 2-D electrophoresis, or HPLC. Using electrophoresis for absolute quantification, the slices/spots were dissolved and analyzed for lanthanide content using FIA/ICP-MS. Since the proteins from

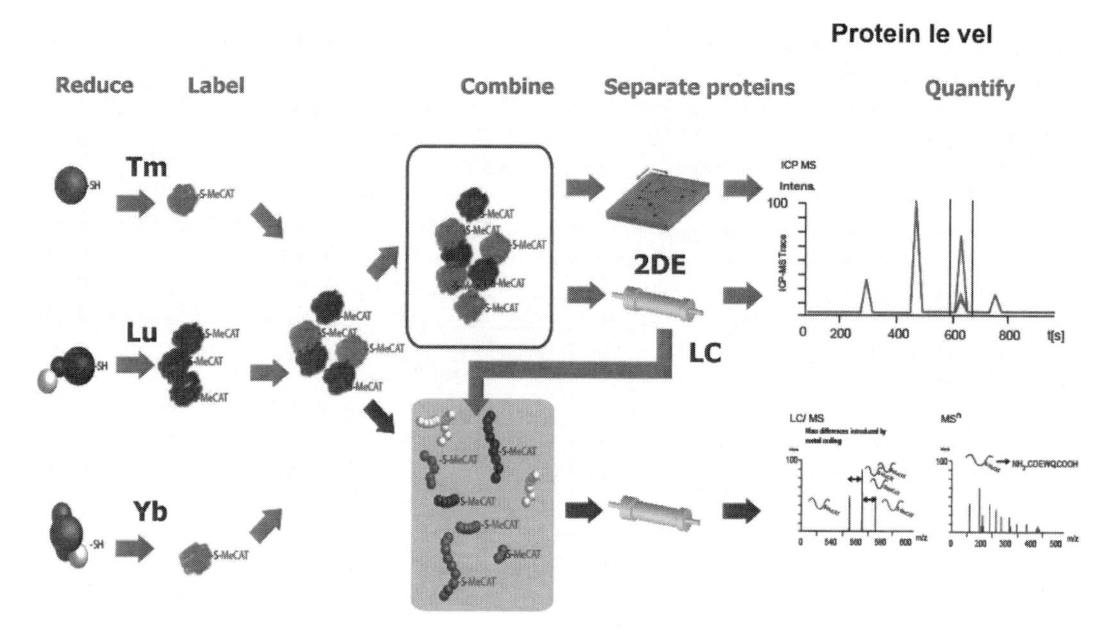


Fig. 8. Current workflow of the MeCAT labeling strategy.

the different samples are modified with a specific M-MeCAT label, the products differ in mass. This allows for the determination of the intact proteins using MALDI and ESI-MS$^n$ as well as for the metal detection with ICP-MS (**Fig. 9**). After proteolysis, any necessary identification may follow at the peptide level using LC/ESI-MS$^n$.

Recently, we demonstrated a complete labeling on peptide and protein level and we quantified peptides and proteins in a relative and absolute manner. The successful identification of tagged peptides using tandem-MS experiments was included in the analysis.

On protein level, 110 attomol BSA could be quantified, but presumably this is not the limit of detection. The quantification could be performed over a dynamic range of four orders of magnitude with a probability score of $R^2 = 0.9997$. Besides establishment using standard proteins and peptides, we quantified different eye lens proteins out of silver-stained dissolvable 2D gels. Absolute quantities between 30 and 1,000 ng were distinguished *(40)*.

Also, quantification of different proteins using LC/ICP-MS is demonstrated in **Fig. 8**.

One can state that in terms of accuracy, dynamic range, sensitivity, and the prospect of absolute quantitative measurements, metal tagging shows great potential.

There is no doubt that the future will bring about even more techniques for quantitation, but it seems that first a sense of quality assurance needs to be established, because the meaning of

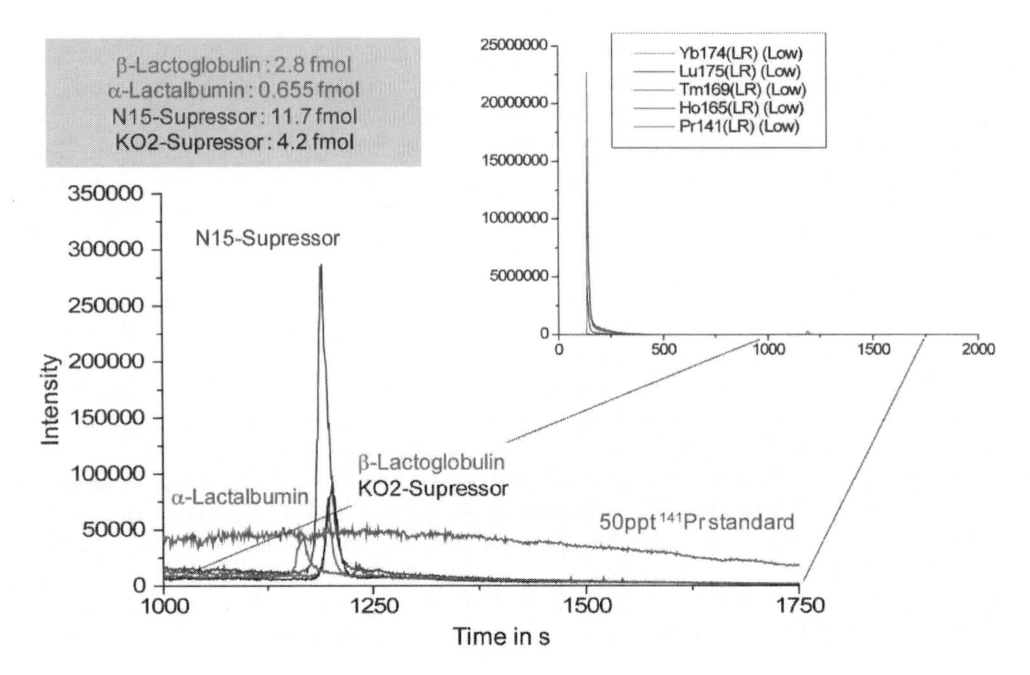


Fig. 9. LC/ICP-MS separation and quantification of proteins after tagging with MeCAT reagents carrying four different metals is shown. In addition the trace of the internal standard for absolute quantification and correction for possible changes in the sensitivity of the detection is shown.

"quantitative" measurement deteriorated to some extent owing to the great difficulties and needs (see, for example, a first step **ref.** 87). When figures of merit on reliable, robust, and accurate data can be assessed, standards and reference materials will become available and then, the fastidious goals especially in medicine and biology can be met.

## References

1. Sanz-Medel, A., et al., Elemental mass spectrometry for quantitative proteomics. *Anal Bioanal Chem*, 2008. **390**(1): 3–16.
2. Mueller, L.N., et al., An assessment of software solutions for the analysis of mass spectrometry based quantitative proteomics data. *J Proteome Res*, 2008. 7(1): 51–61.
3. Veenstra, T.D., Global and targeted quantitative proteomics for biomarker discovery. *J Chromatogr B Analyt Technol Biomed Life Sci*, 2007. **847**(1): 3–11.
4. Nakamura, T. and Y. Oda, Mass spectrometry-based quantitative proteomics. *Biotechnol Genet Eng Rev*, 2007. **24**: 147–63.
5. Miyagi, M. and K.C. Rao, Proteolytic 18O-labeling strategies for quantitative proteomics. *Mass Spectrom Rev*, 2007. **26**(1): 121–36.
6. Karp, N.A. and K.S. Lilley, Design and analysis issues in quantitative proteomics studies. *Proteomics*, 2007. **7** Suppl 1: 42–50.
7. Chen, X., et al., Amino acid-coded tagging approaches in quantitative proteomics. *Expert Rev Proteomics*, 2007. **4**(1): 25–37.
8. Mann, M., Functional and quantitative proteomics using SILAC. *Nat Rev Mol Cell Biol*, 2006. 7(12): 952–8.
9. Lilley, K.S. and P. Dupree, Methods of quantitative proteomics and their application to plant organelle characterization. *J Exp Bot*, 2006. **57**(7): 1493–9.
10. Ivakhno, S. and A. Kornelyuk, Quantitative proteomics and its applications for systems biology. *Biochemistry (Mosc)*, 2006. **71**(10): 1060–72.

11. Linscheid, M.W., Quantitative proteomics. *Anal Bioanal Chem*, 2005. **381**(1): 64–6.
12. Zhang, H., W. Yan, and R. Aebersold, Chemical probes and tandem mass spectrometry: a strategy for the quantitative analysis of proteomes and subproteomes. *Curr Opin Chem Biol*, 2004. **8**(1): 66–75.
13. Bantscheff, M., et al.Quantitative mass spectrometry in proteomics: a critical review. *Anal Bioanal Chem*, 2007. **389**(4): 1017–31.
14. Baldwin, M.A., Protein identification by mass spectrometry: issues to be considered. *Mol Cell Proteomics*, 2004. **3**(1): 1–9.
15. Stein, R.C. and M.J. Zvelebil, The application of 2D gel-based proteomics methods to the study of breast cancer. *J Mammary Gland Biol Neoplasia*, 2002. 7(4): 385–93.
16. Washburn, M.P., R.R. Ulaszek, and J.R. Yates, 3rd, Reproducibility of quantitative proteomic analyses of complex biological mixtures by multidimensional protein identification technology. *Anal Chem*, 2003. **75**(19): 5054–61.
17. Hofmann, S., et al., Rapid and sensitive identification of major histocompatibility complex class I-associated tumor peptides by Nano-LC MALDI MS/MS. *Mol Cell Proteomics*, 2005. **4**(12): 1888–97.
18. Gygi, S.P., et al., Quantitative analysis of complex protein mixtures using isotope-coded affinity tags. *Nat Biotechnol*, 1999. **17**(10): 994–9. java/Propub/biotech/nbt1099_994. fulltext java/Propub/biotech/nbt1099_994. abstract.
19. Hansen, K.C., et al., Mass spectrometric analysis of protein mixtures at low levels using cleavable 13C-isotope-coded affinity tag and multidimensional chromatography. *Mol Cell Proteomics*, 2003. **2**(5): 299–314.
20. Turecek, F., Mass spectrometry in coupling with affinity capture-release and isotope- coded affinity tags for quantitative protein analysis. *J Mass Spectrom*, 2002. **37**(1): 1–14.
21. Patterson, S.D. and R.H. Aebersold, Proteomics: the first decade and beyond. *Nat Genet*, 2003. **33** Suppl: 311–23.
22. Klose, J., Protein mapping by combined isoelectric focusing and electrophoresis of mouse tissues. A novel approach to testing for induced point mutations in mammals. *Humangenetik*, 1975. **26**: 231–43.
23. O'Farrell, P., High resolution two-dimensional electrophoresis of proteins. *J Biol Chem* 1975. **250**: 4007–4021.
24. Miller, I., J. Crawford, and E. Gianazza, Protein stains for proteomic applications: which, when, why? *Proteomics*, 2006. **6**(20): 5385–408.
25. Patton, W.F., A thousand points of light: the application of fluorescence detection technologies to two-dimensional gel electrophoresis and proteomics. *Electrophoresis*, 2000. **21**(6): 1123–44.
26. Wozny, W., et al., Differential radioactive quantification of protein abundance ratios between benign and malignant prostate tissues: cancer association of annexin A3. *Proteomics*, 2007. 7(2): 313–22.
27. Schrattenholz, A. and K. Groebe, What does it need to be a biomarker? Relationships between resolution, differential quantification and statistical validation of protein surrogate biomarkers. *Electrophoresis*, 2007. **28**(12): 1970–79.
28. Gygi, S.P., et al., Quantitative analysis of complex protein mixtures using isotope-coded affinity tags. *Nat Biotechnol*, 1999. **17**(10): 994–9.
29. Ahrends, R., et al., A metal-coded affinity tag approach to quantitative proteomics. *Mol Cell Proteomics*, 2007. **16**: 1907–16.
30. Washburn, M.P., D. Wolters, and J.R. Yates, 3rd, Large-scale analysis of the yeast proteome by multidimensional protein identification technology. *Nat Biotechnol*, 2001. **19**(3): 242–7.
31. Sheng, S., D. Chen, and J.E. Van Eyk, Multidimensional liquid chromatography separation of intact proteins by chromatographic focusing and reversed phase of the human serum proteome: optimization and protein database. *Mol Cell Proteomics*, 2006. **5**(1): 26–34.
32. Mallik, R. and D.S. Hage, Affinity monolith chromatography. *J Sep Sci*, 2006. **29**(12): 1686–704.
33. Azarkan, M., et al., Affinity chromatography: a useful tool in proteomics studies. *J Chromatogr B Analyt Technol Biomed Life Sci*, 2007. **849**(1-2): 81–90.
34. Johnson, R.D. and R.J. Lewis, Quantitation of atenolol, metoprolol, and propranolol in postmortem human fluid and tissue specimens via LC/APCI-MS. *Forensic Sci Int*, 2006. **156**(2-3): 106–17.
35. Tang, K., J.S. Page, and R.D. Smith, Charge competition and the linear dynamic range of detection in electrospray ionization mass spectrometry. *J Am Soc Mass Spectrom*, 2004. **15**(10): 1416–23.
36. Knochenmuss, R., et al., Secondary ion-molecule reactions in matrix-assisted laser desorption/ionization. *J Mass Spectrom*, 2000. **35**(11): 1237–45.
37. Bauer, A. and B. Kuster, Affinity purification-mass spectrometry. Powerful tools for the characterization of protein complexes. *Eur J Biochem*, 2003. **270**(4): 570–8.

38. Lichty, J.J., et al., Comparison of affinity tags for protein purification. *Protein Expr Purif*, 2005. **41**(1): 98–105.
39. Whetstone, P.A., et al., Element-coded affinity tags for peptides and proteins. *Bioconjug Chem*, 2004. **15**(1): 3–6.
40. Ahrends, R., et al., Identifying an interaction site between MutH and the C-terminal domain of MutL by crosslinking, affinity purification, chemical coding and mass spectrometry. *Nucleic Acids Res*, 2006. **34**(10): 3169–80.
41. Girault, S., et al., Coupling of MALDI-TOF mass analysis separation of biotinylated peptides streptavidin beads. *Anal Chem*, 1996. **68**(13): 2122–6.
42. Prange, A., Pröfrock, D., Chemical labels and natural element tags for the quantitative analysis of bio-molecules. *J. Anal. At. Spectrom.*, 2008. **23**(4): 432–59.
43. Julka, S. and F., Regnier, Quantification in proteomics through stable isotope coding: a review. *J Proteome Res*, 2004. **3**(3): 350–63.
44. Hoehenwarter, W., et al., The necessity of functional proteomics: protein species and molecular function elucidation exemplified by in vivo alpha A crystallin N-terminal truncation. *Amino Acids*, 2006. **31**(3): 317–23.
45. Sachon, E., et al., Phosphopeptide quantitation using amine-reactive isobaric tagging reagents and tandem mass spectrometry: application to proteins isolated by gel electrophoresis. *Rapid Commun Mass Spectrom*, 2006. **20**(7): 1127–34.
46. Yang, Y., et al., A comparison of nLC-ESI-MS/MS and nLC-MALDI-MS/MS for GeLC-based protein identification and iTRAQ-based shotgun quantitative proteomics. *J Biomol Tech*, 2007. **18**(4): 226–37.
47. Wiese, S., et al., Protein labeling by iTRAQ: a new tool for quantitative mass spectrometry in proteome research. *Proteomics*, 2007. **7**(3): 340–50.
48. Sui, J., et al., iTRAQ-coupled 2D LC-MS/MS analysis on protein profile in vascular smooth muscle cells incubated with S- and R-enantiomers of propranolol: possible role of metabolic enzymes involved in cellular anabolism and antioxidant activity. *J Proteome Res*, 2007. **6**(5): 1643–51.
49. Skalnikova, H., et al., Relative quantitation of proteins fractionated by the ProteomeLab PF 2D system using isobaric tags for relative and absolute quantitation (iTRAQ). *Anal Bioanal Chem*, 2007. **389**(5): 1639–45.
50. Li, Z., et al., Shotgun identification of the structural proteome of shrimp white spot syndrome virus and iTRAQ differentiation of envelope and nucleocapsid subproteomes. *Mol Cell Proteomics*, 2007. **6**(9): 1609–20.
51. Griffin, T.J., et al., iTRAQ reagent-based quantitative proteomic analysis on a linear ion trap mass spectrometer. *J Proteome Res*, 2007. **6**(11): 4200–9.
52. Dean, R.A. and C.M. Overall, Proteomics discovery of metalloproteinase substrates in the cellular context by iTRAQ labeling reveals a diverse MMP-2 substrate degradome. *Mol Cell Proteomics*, 2007. **6**(4): 611–23.
53. Bantscheff, M., et al., Quantitative chemical proteomics reveals mechanisms of action of clinical ABL kinase inhibitors. *Nat Biotechnol*, 2007. **25**(9): 1035–44.
54. Chong, P.K., et al., Isobaric tags for relative and absolute quantitation (iTRAQ) reproducibility: Implication of multiple injections. *J Proteome Res*, 2006. **5**(5): 1232–40.
55. Pierce, A., et al., Eight-channel iTRAQ enables comparison of the activity of 6 leukaemogenic tyrosine kinases. *Mol Cell Proteomics*, 2008. **7**(5): 853–63.
56. White, F.M., On the iTRAQ of kinase inhibitors. *Nat Biotechnol*, 2007. **25**(9): 994–6.
57. Ong, S.E. and M. Mann, Stable isotope labeling by amino acids in cell culture for quantitative proteomics. *Methods Mol Biol*, 2007. **359**: 37–52.
58. Ong, S.E., I. Kratchmarova, and M. Mann, Properties of 13C-substituted arginine in stable isotope labeling by amino acids in cell culture (SILAC). *J Proteome Res*, 2003. **2**(2): 173–81.
59. Foster, L.J., C.L. De Hoog, and M. Mann, Unbiased quantitative proteomics of lipid rafts reveals high specificity for signaling factors. *Proc Natl Acad Sci USA*, 2003. **100**(10): 5813–8.
60. Beynon, R.J., et al., Multiplexed absolute quantification in proteomics using artificial QCAT proteins of concatenated signature peptides. *Nat Methods*, 2005. **2**(8): 587–9.
61. Old, W.M., et al., Comparison of label-free methods for quantifying human proteins by shotgun proteomics. *Mol Cell Proteomics*, 2005. **4**(10): 1487–502.
62. Bondarenko, P.V., D. Chelius, and T.A. Shaler, Identification and relative quantitation of protein mixtures by enzymatic digestion followed by capillary reversed-phase liquid chromatography-tandem mass spectrometry. *Anal Chem*, 2002. **74**(18): 4741–9.
63. Ono, M., et al., Label-free quantitative proteomics using large peptide data sets generated by nanoflow liquid chromatography and mass spectrometry. *Mol Cell Proteomics*, 2006. **5**(7): 1338–47.

64. Wang, W., et al., Quantification of proteins and metabolites by mass spectrometry without isotopic labeling or spiked standards. *Anal Chem*, 2003. **75**(18): 4818–26.
65. Meng, F., et al., Quantitative analysis of complex peptide mixtures using FTMS and differential mass spectrometry. *J Am Soc Mass Spectrom*, 2007. **18**(2): 226–33.
66. Wolters, D.A., M.P. Washburn, and J.R. Yates, 3rd, An automated multidimensional protein identification technology for shotgun proteomics. *Anal Chem*, 2001. **73**(23): 5683–90.
67. Blagoev, B., et al., A proteomics strategy to elucidate functional protein-protein interactions applied to EGF signaling. *Nat Biotechnol*, 2003. **21**(3): 315–8.
68. Heller, M., et al., Trypsin catalyzed 16O-to-18O exchange for comparative proteomics: tandem mass spectrometry comparison using MALDI-TOF, ESI-QTOF, and ESI-ion trap mass spectrometers. *J Am Soc Mass Spectrom*, 2003. **14**(7): 704–18.
69. Page, J.S., C.D. Masselon, and R.D. Smith, FTICR mass spectrometry for qualitative and quantitative bioanalyses. *Curr Opin Biotechnol*, 2004. **15**(1): 3–11.
70. Bonenfant, D., et al., Quantitation of changes in protein phosphorylation: a simple method based on stable isotope labeling and mass spectrometry. *Proc Natl Acad Sci U S A*, 2003. **100**(3): 880–5.
71. Edler, M., N. Jakubowski, and M. Linscheid, Quantitative determination of melphalan DNA adducts using HPLC - inductively coupled mass spectrometry. *J Mass Spectrom*, 2006. **41**(4): 507–16.
72. Edler, M., N. Jakubowski, and M. Linscheid, Styrene oxide DNA adducts: quantitative determination using 31P monitoring. *Anal Bioanal Chem*, 2005. **381**(1): 205–11.
73. Siethoff, C., et al., Quantitative determination of DNA adducts using liquid chromatography/electrospray ionization mass spectrometry and liquid chromatography/high-resolution inductively coupled plasma mass spectrometry. *J Mass Spectrom*, 1999. **34**(4): 421–6.
74. Houk, R.S., Mass-spectrometry of inductively coupled plasmas. *Anal Chem*, 1986. **58**(1): A97–105.
75. Wind, M., et al., Analysis of protein phosphorylation by capillary liquid chromatography coupled to element mass spectrometry with P-31 detection and to electrospray mass spectrometry. *Anal Chem*, 2001. **73**(1): 29–35.
76. Wind, M., et al., Sulfur as the key element for quantitative protein analysis by capillary liquid chromatography coupled to element mass spectrometry. *Angew Chem Int Ed Engl*, 2003. **42**(29): 3425–7.
77. Thermo, Finnigan, and Bremen, Finnigan ELEMENT XR: Extended Dynamic Range High Resolution ICP-MS. Technical Note, 2005(TN30064_E 01/05C): 4.
78. Baranov, V.I., et al., A sensitive and quantitative element-tagged immunoassay with ICPMS detection. *Anal Chem*, 2002. **74**(7): 1629–36.
79. Tanner, S., et al., Multiplex bio-assay with inductively coupled plasma mass spectrometry: Towards a massively multivariate single-cell technology. *Spectrochim Acta Part B* 2007. **62**: 188–95.
80. Baranov, V.I., S.D. Tanner, and D.R. Bandura, Method and apparatus for flow cytometry linked with elemental analysis *(WO/2005/093784)*. 2005: US (CA).
81. Lee, S., et al., Method to site-specifically identify and quantitate carbonyl end products of protein oxidation using oxidation-dependent element coded affinity tags (O-ECAT) and nanoliquid chromatography Fourier transform mass spectrometry. *J Proteome Res*, 2006. **5**(3): 539–47.
82. Liu, H.L., et al., Method for quantitative proteomics research by using metal element chelated tags coupled with mass spectrometry. *Anal Chem*, 2006. **78**(18): 6614–21.
83. Byegard, J., G. Skarnemark, and M. Skahlberg, The stability of some metal EDTA, DTPA and DOTA complexes: Application as tracers in groundwater studies. *J Radioanal Nuclear Chem*, 1999. **241**(2): 281–290.
84. Moreau, J., et al., Complexing mechanism of the lanthanide cations Eu3 + , Gd3 + , and Tb3 + with 1,4,7,10-tetrakis(carboxymethyl)-1,4,7,10-tetraazacyclododecane (dota)-characterization of three successive complexing phases: study of the thermodynamic and structural properties of the complexes by potentiometry, luminescence spectroscopy, and EXAFS. *Chemistry*, 2004. **10**(20): 5218–32.
85. Bunzli, J.C., Benefiting from the unique properties of lanthanide ions. *Acc Chem Res*, 2006. 39(1): **53**–61.
86. Bohlke, J.K., et al., Isotopic compositions of the elements. *J Phys Chem Ref Data*, 2005. **34**: 57–67.
87. Carr, S.A., et al., The need for guidelines in publication of peptide and protein identification data: working group on publication guidelines for peptide and protein identification data. *Mol Cell Proteomics*, 2004. **3**(6): 531–3.

# Chapter 12

# iTRAQ-Labeling of In-Gel Digested Proteins for Relative Quantification


## Carla Schmidt and Henning Urlaub



## Summary

In addition to standard MS-based protein identification, quantification of proteins by mass spectrometry (MS) is rapidly gaining acceptance in proteomic studies. MS-based quantification involves either the incorporation of stable isotopes or can be performed label-free. Recently, more attention has been devoted to label-free quantification; however, this approach has not been fully established among the proteomic community yet. More common is still the introduction of stable isotopes, which can be done by metabolic (e.g., SILAC) or by chemical (e.g., ICAT, iTRAQ, etc.) labeling. Here, we present an overall quantification strategy for chemical labeling of in-gel digested proteins using iTRAQ reagents. This includes (1) protein separation by gel electrophoresis, (2) excision of protein bands, (3) in-gel digestion and extraction of peptides, (4) labeling of peptides, (5) pooling the samples to be compared, (6) LC-MS/MS of labeled peptides, and (7) database search. The presented workflow is well suited for protein samples of moderate complexity (i.e., protein samples of 300–400 components), and it is exemplified by using different amounts of 25S [U4/U6.U5] tri-snRNPs.



**Key words:** PAGE, iTRAQ, Mass spectrometry (MS), Liquid chromatography (LC)


## 1. Introduction

A typical proteomics workflow comprises protein separation using polyacrylamide-based gel electrophoresis or liquid chromatography (LC), digestion of proteins, and finally identification by mass spectrometry (MS) and database search. This workflow is widely accepted in the diverse field of proteomic research. In addition, much attention is currently being devoted to the relative quantitative comparison of different samples by chemical or metabolic labeling of samples and subsequent analysis of the labeled species in the mass spectrometer. By combining the protein identification


Jörg Reinders and Albert Sickmann (eds.), *Proteomics,* Methods in Molecular Biology, Vol. 564
DOI: 10.1007/978-1-60761-157-8_12, © Humana Press, a part of Springer Science+Business Media, LLC 2009

with quantitative analysis, reliable information about the abundance of a certain protein component in, e.g., different functional stages of a cell or in different cellular complexes are obtained.

Nowadays, many and diverse techniques are available for separation of proteins, and for their relative quantification by MS *(1)*. However, the choice of application depends on the biological/biochemical problem to be solved, the cost of the analysis, the time it is expected to require, and, last but not the least, on the individual experimental setup for proteomics in the laboratory concerned.

In this chapter, we describe in detail the proteomic workflow generally followed in our laboratory. It comprises sample separation by 1D gel electrophoresis, digestion of the proteins within the gel, chemical labeling of the extracted peptides by isotope-labeled reagents, and the ESI and MALDI techniques for MS analysis of the labeled peptides with a view to protein identification and subsequent quantification.

The protocols listed in this chapter should prove useful for the analysis of protein samples of moderate complexity, e.g., isolated protein complexes, or purified cellular compartments with a protein composition that does not exceed 300–400 components.

The approach is illustrated by the in-gel-digestion of different amounts of 25S [U4/U6.U5] tri-snRNPs *(2, 3)*, labeling of the extracted peptides with iTRAQ reagents *(4)*, and analysis of the differentially labeled peptides for quantification of the corresponding proteins by LC-coupled tandem MS (LC-MS/MS).

### *1.1. Protein Separation – General Remarks*

The final step in a proteomics workflow is the identification of proteins by mass-spectrometric sequencing of the peptides derived from hydrolysis of the proteins in the sample to be analyzed. The more proteins are present in a given sample, the more peptides are generated by hydrolysis. However, even state-of-the-art Mass Spectrometers can only analyze a limited number of peptides at a given time (see below). Therefore, apart from the strategic design of proteome experiments (analysis of entire cells, cellular compartments, isolated protein complexes), sample preparation and protein/peptide separation is one of the major prerequisites for a successful proteome analysis using MS. During this initial step, the complexity of the sample is reduced, which in turn leads to the acquisition of more and better-evaluable data in the MS analysis.

There are two well-established general strategies for reducing the complexity of a protein sample in preparation for MS: (1) Hydrolysis of the protein sample in situ, followed by separation of peptides by one- or multidimensional LC and subsequent elution of peptides via reversed-phase chromatography (RP-LC) with organic solvents into the tandem mass spectrometer (RP-LC-MS/MS) and (2) separation of proteins by gel electrophoresis (e.g., 1D *[5]*, 2D *[6, 7]*, Blue Native *[8, 9]*, 16BAC gel electrophoresis *[10, 11]*), followed by hydrolysis of proteins within the gel, and subsequent elution of peptides from the gel through a RP-LC column with organic solvents into the tandem mass spectrometer (LC-MS/MS).

The first approach has the advantage that samples are hydrolyzed into peptides *in situ*, which bypasses the (more time-consuming) protein-separation step. Owing to the great complexity of the peptide mixture thus generated, a further peptide-separation step is recommended. For this purpose, it is widely accepted that two-dimensional liquid chromatography – using strong cation-exchange (SCX) chromatography in the first dimension and RP-LC in the second dimension – is the best option. The two LC steps can be coupled directly to each other (online), or they can be separated from each other (offline). The former configuration has been successfully introduced in multidimensional protein identification technology (MuD-PIT), where the generated peptides are loaded onto a biphasic microcapillary column packed with SCX and RP material. The eluted peptides are detected and identified on a tandem mass spectrometer *(12)*. Although highly sensitive and extremely fast, such a workflow requires much experience in handling the samples, the multidimensional LC system and the mass spectrometer. Very recently, an interesting alternative approach has been described, which uses isoelectric focusing (IEF) on IPG strips to separate complex peptide mixtures. The IPG strips are cut into pieces, and peptides are eluted from them and subsequently analyzed by RP-LC-coupled MS *(13)*.

Separation of the protein mixtures by gel electrophoresis before digestion and LC-MS/MS has two advantages: (1) proteins are separated according to their size (using one-dimensional gel electrophoresis) and charge (two-dimensional gel electrophoresis) and (2) they can be visualized by staining. Both of these features make the subsequent MS analysis more reliable by allowing a rough check of molecular weights inferred from the proteins' migration behavior in gel electrophoresis.

The combination of IEF and sodium dodecylsulfate (SDS) polyacrylamide gel electrophoresis (PAGE) is currently the most common technique in proteomic analysis, with a resolution of up to 10,000 protein spots per gel *(14)*. IEF separates proteins according to their isoelectric point. Most often, IEF is performed with immobilized pH gradients *(15)*. Alternatively, carrier ampholytes are used to establish the pH gradient during the IEF process *(6, 16)*. In the second dimension, the focused proteins are separated by SDS-PAGE (see below) according to their apparent molecular weight. A drawback of 2D PAGE is the fact that it is less suitable for the separation of hydrophobic proteins, such as integral membrane proteins. Hydrophobic and membrane proteins exhibit incomplete solubility in the buffers used for 2D PAGE. Moreover, such proteins can precipitate, resulting in less efficient transfer of material from the first to the second dimension.

#### 1.1.1. 1D-page

When working in experimental systems that investigate protein samples with moderate complexity (e.g., purified cell compartments, purified protein complexes, or interaction between proteins– the so-called subcellular proteomics), a simpler alternative for 2D gel-based analysis of proteins is protein separation by 1D PAGE. The most common technique for 1D PAGE is the Laemmli system *(5)* that contains SDS as detergent and Tris-glycine as buffer. Although widely used, it has some drawbacks. The highly alkaline pH of the Laemmli system may cause band distortion, loss of resolution, or artifact bands. Laemmli-gels exhibit a short shelf life of 4–6 weeks that is limited by hydrolysis of polyacrylamide at the high gel pH. In addition, chemical modification of proteins, such as deamidation and alkylation, can occur. Since the redox state of the Laemmli-gels is not constant, reoxidation of reduced disulfides of cysteine-containing proteins is possible, and because of the need to heat the samples at 100°C in Laemmli sample buffer, cleavage of Asp–Pro bonds has been observed *(17)*.

Alternatively, pre-cast 1D polyacrylamide gels (e.g., the NuPAGE® system by Invitrogen, Carlsbad, USA) can be used for protein separation prior to MS analysis. NuPAGE® pre-cast gels have several advantages over Laemmli SDS gels *(18)*: (1) the neutral operating pH of pre-cast gels provides a longer shelf life of the gels (8–12 months) and improves protein stability during electrophoresis, resulting in sharper band resolution. (2) Manufacturers guarantee complete reduction of disulfides, and no cleavage of Asp–Pro bonds is observed. (3) Importantly, pre-cast gels are small in size and yield highly reproducible results. The commercially available polyacrylamide-gradient gels offer a separation range (i.e., from 1 to 120 kDa, or from 36 to 400 kDa) that is difficult to achieve with in-house-prepared polyacrylamide gels. (4) Separation of proteins, staining, and destaining can be achieved within a couple of hours. (5) The small dimensions and high resolution of the gels make them ideal for cutting the entire lanes into pieces of exactly the same size for comprehensive analysis and/or relative quantification of samples in two (or more) different gel lanes.

Pre-cast gels are compatible with silver-staining protocols *(19)*, with all the standard Coomassie staining procedures and with alternative staining methods such as copper or zinc and fluorescence staining (all reviewed, e.g., by Westermeier & Marouga *[20]*).

Staining is generally recommended for visualizing protein bands of interest, which can then be excised and hydrolyzed within the gel for MS analysis (see below). Alternatively, entire lanes can be cut into pieces and the proteins hydrolyzed. However, when this approach is used, staining is not necessary, or it should be restricted to Coomassie staining to guarantee a maximum yield of extracted peptides after in-gel digestion.

### *1.2. Protein Quantification Methods – General Remarks*

The power of proteomics has been greatly enhanced by the development of relative and absolute quantitative proteomic methods. This has been demonstrated by using traditional quantitative approaches such as 2D PAGE combined with differential staining techniques, such as DIGE *(21, 22)*. More recently, quantitative methods – including chemical, enzymatic, and metabolic labeling of specific protein populations with different isotopes and subsequent quantification by MS – were introduced, and they show increased sensitivity when compared with gel-based methods. They include cleavable isotope-coded affinity tags (cICAT) *(23)*, isobaric tags for relative and absolute quantification (iTRAQ) *(4)*, enzymatic labeling with protein hydrolysis in the presence of heavy (i.e., $^{18}O$-containing) water *(24, 25)* ($^{18}O/^{16}O$ proteolytic labeling), and stable isotope labeling with amino acids in cell culture (SILAC) *(26)*. Basically, in all chemical labeling approaches, proteins derived from various samples are hydrolyzed, and the corresponding peptide pool is chemically labeled with reagents that differ in mass because of incorporation of different heavy stable isotopes, such as $^{13}C$, $^{15}N$, or $^{2}D$. The samples thus labeled are mixed and then analyzed by MS. Importantly, the peak's signal intensity in the mass spectrum of the same, but differently labeled, peptides derived from the various samples can be directly compared, and this reflects the relative quantities of the particular peptide and therefore of the particular protein in the different samples. Chemical labeling is used most frequently at the peptide level, i.e. labeling takes place after the proteins have been digested with specific endoproteinases. cICAT was originally reported for labeling at the protein level and, in principle, all its reagents should also work at the protein level. The advantage of the chemical labeling strategy is the fact that it can be used for MS-based quantification approaches in tissues (e.g., brain). SILAC, although a highly efficient labeling approach, as $^{13}C$- and $^{15}N$-labeled amino acids (of which [$^{13}C$]-lysine and [$^{13}C$]-arginine are especially relevant on account of the specificity of trypsin for these residues) are incorporated into proteins to nearly 100% in cell cultures, is restricted to the investigation of cells growing in culture (e.g., HeLa cells).

More recently, label-free quantification by MS has also shown great promise; this uses techniques, such as spectral counting *(27)* and ion-current measurement *(28)*.

#### *1.2.1. Isobaric Tags for Relative and Absolute Quantification*

The iTRAQ method *(4, 29)* involves protein reduction and alkylation, enzymatic digestion, labeling up to four different samples (or nowadays eight samples) with heavy and light isotope-labeled reagents, sample pooling, HPLC separation, and finally detection and quantification by tandem MS. The iTRAQ reagent is an amine-specific reagent that labels peptide amino termini and lysine side chains with multiplex (4-plex or 8-plex) mass tags: i.e., tags with identical masses. These tags are isobaric (i.e., they have the same mass of 144.1 Da), but their fragmentation patterns differ, allowing species bearing them to be distinguished after fragmentation. Upon fragmentation, every isobaric tag produces a specific marker ion (114.1, 115.1, 116.1, and 117.1 Da, respectively) and a corresponding neutral fragment, which is not detectable (28, 29, 30, and 31 Da, respectively). Thus, in this method (in contrast to other chemical labeling methods described above) differentially labeled peptides show no difference in the mass spectrometer when intact: i.e., in MS mode of the instrument. Only after fragmentation of the peptides, and thus of the tag attached to the peptides, a low-molecular-mass reporter ion is generated, the mass of which is specific to the iTRAQ tag used to label the sample. Therefore, quantification of the differently labeled samples is achieved solely upon fragmentation of the correspondingly labeled peptides in the mass spectrometer. The MS measurement of the intensity of these reporter-ion peaks allows the relative quantification of the peptides in each sample.

iTRAQ has several advantages: (1) Multiplexing (i.e., mixing of more than two differentially labeled species) per se can save time during the quantification of samples taken at several time points or from different stages. (2) It offers the unique possibility to create an internal standard that contains a mixture of all samples, so that more than four (or eight) samples can be investigated and quantified in relation to the internal standard. (3) Since the mixed peptides carry an isobaric tag, the overall signal intensity of the peptide/precursor ions derived from different samples is enhanced in the MS mode (i.e., before fragmentation). (4) Importantly, the signal intensity of the peptide fragment ions in the MS/MS is enhanced as well, since the tag is completely cleaved from the labeled $NH_2$ group upon fragmentation.

For reliable quantification by chemical labeling, the following practical aspects need to be considered carefully: (1) (Exact) knowledge of the sample amount subjected to quantitative analysis. This is the prerequisite for a reliable analysis. However, differences in sample amounts can be monitored by the ratio of, e.g., marker proteins that are known to be present in equal amounts in the different samples; appropriate correction can be applied later by evaluation software. (2) Digestion efficiency. When sample labeling with iTRAQ is performed after digestion, reliable quantification is only possible when proteins in both samples are digested either completely or exactly to the same degree. (3) Labeling efficiency. This should be as high as possible (i.e., <90%). With iTRAQ, the labeling efficiency can be monitored by searching the fragment spectra of the labeled peptides against a database (using a search engine, see below) with the iTRAQ as variable (optional) extra mass. In this manner, the number of non-labeled peptides is also displayed. (4) The number of labeled peptides per identified protein by MS and subsequent database search should be high enough to ensure significant statistics on the labeling/quantification. (5) As the iTRAQ method does not reduce the sample complexity (in contrast to, e.g., cICAT, where only peptides that carry cysteine residues are labeled and quantified), and this places greater demands upon the separation step before tandem MS.

### 1.3. LC-Coupled Tandem MS

MS uses two main methods to ionize "softly" biomolecules, such as proteins and peptides: matrix-assisted laser desorption/ionization *(30–32)* and electrospray ionization *(33)*. Both techniques, when implemented with a state-of-the-art mass spectrometer, allow (1) accurate determination of the mass-to-charge ratio ($m/z$) of a peptide within a mixture (the so-called MS experiment) and (2) isolation of peptides (precursor) within the mass spectrometer and subsequent fragmentation of the isolated precursor in the mass spectrometer (the so-called MS/MS experiment). Mass analysis of the fragments (giving the sequence of the peptide) combined with the accurate monoisotopic precursor (peptide) mass allows the identification of corresponding protein sequence in the database. For reliable identification of proteins by tandem MS, several peptides from one protein should be selected, fragmented, and searched against a database (using search engines, e.g., Mascot *[34]*).

Coupling of nanoLC to an ESI mass spectrometer is a fast and powerful way to obtain peptide sequence data. The chromatography system greatly reduces the complexity of the sample, and the direct coupling to the source ensures that the peptides eluted are immediately ionized and analyzed in the instrument.

State-of-the-art ESI mass spectrometers use data-dependent acquisition (DDA) to analyze the peptide mixture that is eluted from the LC system. The DDA duty cycle consists of two main steps: First, an MS scan over a certain mass range (typically 350–1,500 $m/z$) is performed to detect the charge state of the precursor (peptide). In a routine online-coupled ESI MS approach only doubly, triply, and quadruply charged peptide precursors/peptides are selected for further MS/MS analysis (fragmentation). Importantly, after the initial MS scan, several peptide precursors can be selected for subsequent MS/MS analysis, which allows the most intense precursor signals in the MS spectrum to be sequenced. Note that while the instrument is occupied with MS/MS experiments, usually no MS scans can be performed (except for Orbitrap-FT mass analyzer); therefore, during this time, peptides that are eluted from the column into the instrument cannot be detected and sequenced. Consequently, instruments with a short duty cycle can sequence more peptides in a given time than instruments with a long duty cycle.

Despite the separation power of modern LC techniques, in complex samples, a multitude of peptides are still eluted every second from the column, and even modern mass spectrometers cannot register all of these in the time available. Therefore, an important factor is the acquisition speed (duration of the duty cycle) of the mass spectrometer. In general, ion-trap instruments have a shorter duty cycle than with qQ-ToF instruments, and the former are therefore frequently used for the analysis of very complex mixtures. On the other hand, the quality of the MS/MS spectra from qQ-ToF instruments is considerably better, so that the number of MS/MS spectra that can be used to assign a peptide sequence correctly is higher *(35)*. Importantly, in relative quantification with the iTRAQ reagent, a reliable result depends on the spectral quality in the low *m/z* range: i.e., between 110 and 120 *m/z*. However, 3D and/or linear ion-trap instruments have no – or only a limited – capability to monitor ions within this mass range and are thus not recommended for iTRAQ quantification approaches. In preference, qQ-ToF instruments, the Orbitrap mass-analyzer *(36)* or MALDI instruments should be chosen.

### 1.4. Combining 1D-PAGE with iTRAQ Labeling for Relative Quantification of Protein (RNA) Complexes

Because of the above-mentioned restrictions and drawbacks of a quantitative proteome study using iTRAQ, we designed an approach to reduce the complexity of the sample that combines 1D PAGE of protein samples with iTRAQ labeling of the extracted peptides derived from in-gel digested proteins. We use the following workflow to monitor quantitative differences in protein complexes (with or without bound RNA) derived from different functional stages of a cell:

1. Separation of purified protein (-RNA) complexes by 1D PAGE on pre-cast gels (8 × 8 cm × 1 mm); *see* **Subheading 3.1**
2. Cutting the entire sample lanes into pieces of exactly the same size; *see* **Subheading 3.2.1** and **Fig. 2**
3. Digestion of proteins with the endoproteinase trypsin; *see* **Subheading 3.2.2**
4. iTRAQ labeling of extracted peptides; *see* **Subheading 3.3**
5. Pooling of the differentially labeled samples and subsequent LC-MS/MS analysis (ESI MS); *see* **Subheading 3.3** paragraph 8, **Fig. 3**

The overall workflow is illustrated in **Fig. 1**.

On the one hand, the workflow – although time-consuming, owing to labeling of numerous separate gel pieces – showed several advantages over *in situ* digestion of samples and subsequent labeling: (1) The overall analysis yield is greatly enhanced. We obtain more labeled peptides, higher sequence coverage, and thus

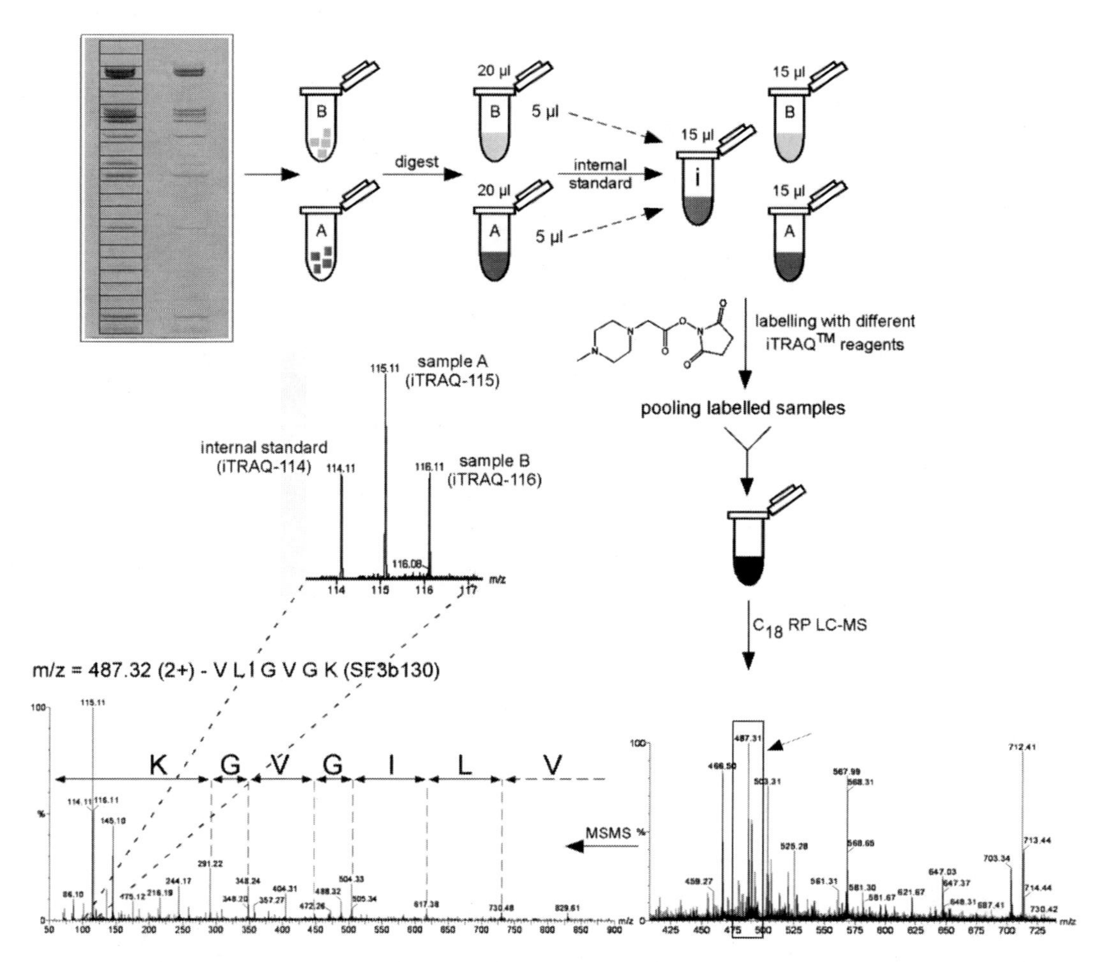


Fig. 1. Overall workflow of in-gel digestion and subsequent labeling with iTRAQ™ reagents. Entire gel lanes of different samples to be analyzed and quantified are cut into gel slices of equal size resulting in 23 samples per lane. Each gel slice further manually cut into small pieces and proteins within gel pieces are digested with trypsin as described in **Subheading 3.2.2**. Extracted peptides are re-dissolved in 20 µL 100 mM TEAB and an internal standard is prepared by pooling 5 µL of each sample. Samples and internal standard are labeled with iTRAQ™ reagents 114, 115, 116, respectively (*see* **Subheading 3.3**). After pooling, the samples are analyzed by LC-MS/MS and quantitation is done by calculating the peak areas of individual reporter ions (114, 115, 116, and 117) in MS/MS.

a higher confidence in quantification when compared with digestion of the samples *in situ* under semi-denaturing conditions followed by labeling of the peptides. We assume that the protein/peptide complexity is less within the single gel regions and, therefore, labeling of peptides is more efficient. On the other hand, the digestion efficiency might be enhanced as well. (2) Digestion, extraction, and labeling can be performed using the same buffer (*see* **Subheadings 3.2.2** and **3.3**), the volume can be kept much smaller and labeled species can be injected directly into the LC system coupled to the mass spectrometer without any further chromatography step. (3) If desired, the gel regions of interest can be investigated and compared selectively. (4) Importantly,

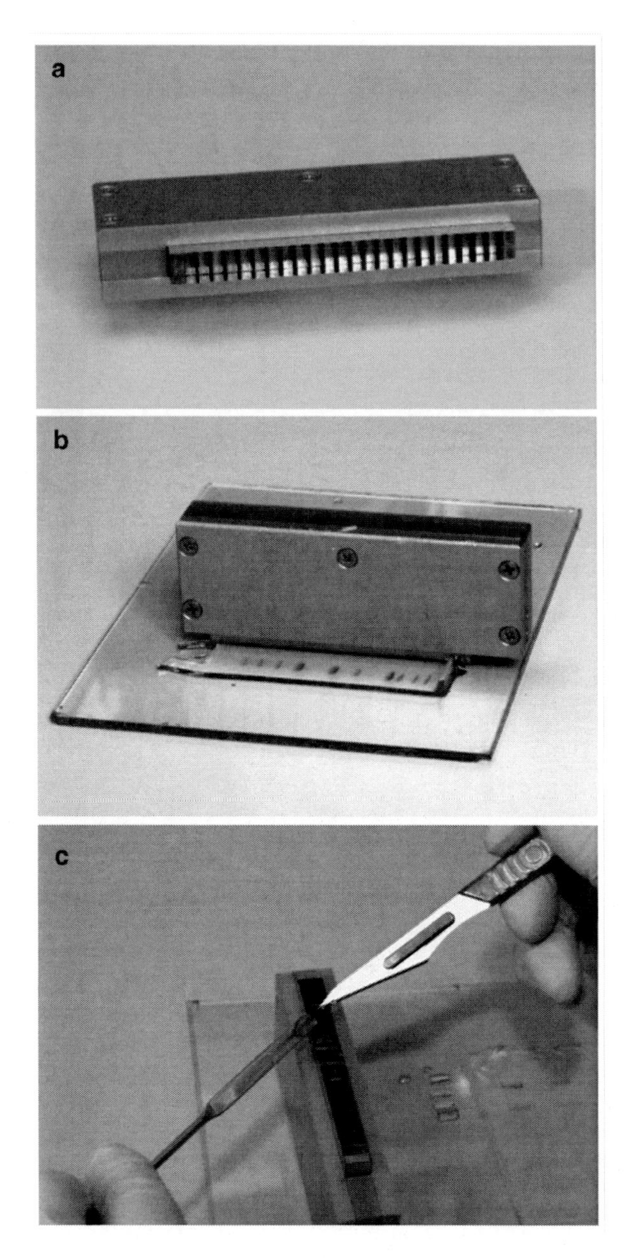

Fig. 2. Gel lane cutting device: (**a**) Site view of in-house manufactured gel-lane cutter for cutting lanes of pre-cast gels (e.g., NuPAGE® 10 cm in length) into 23 slices of equal size (3 × 7 mm). (**b**) Gel cutter attached to the gel. (**c**) Additional manual cutting of slices into gel pieces of approximately 1 × 1 mm by using spatula and scalpel.

the relative amounts of sample can be compared directly with the relative amounts seen in the staining of the samples after gel electrophoresis. Although the latter is not highly accurate, the investigator can at least estimate whether the samples drastically differ in amount by comparing the staining intensity of marker proteins (see also above).

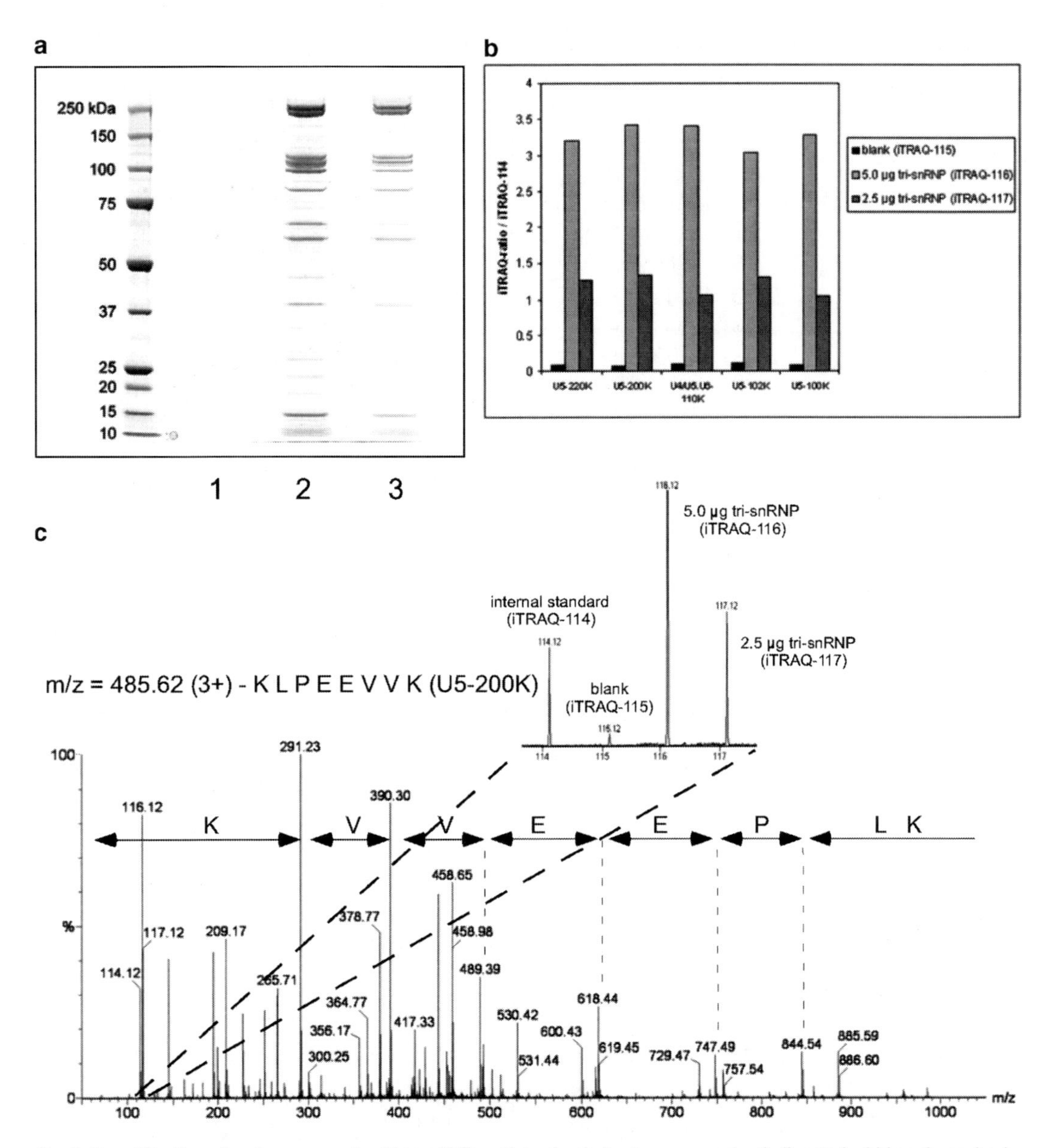


Fig. 3. Quantification of various amounts of tri-snRNP particles loaded onto a pre-cast gel, digested within gels, and subsequently labeled with iTRAQ. (**a**) Colloidal Coomassie stained NuPAGE® 4–12% Bis–Tris Gel; lane 1: blank, lane 2: 5.0 µg tri-snRNPs, lane 3: 2.5 µg tri-snRNPs. (**b**) Example of iTRAQ ratios for various identified proteins. Ratios are calculated by dividing the peak area of reporter ions of the blank sample (iTRAQ-115), 5.0 µg (iTRAQ-116), and 2.5 µg tri-snRNP (iTRAQ-117), respectively, by the peak area of the internal standard (iTRAQ-114). (**c**) Example of MS/MS spectrum of a labeled peptide (KLPEEVVK, U5-200K) derived after in-gel digestion and labeling procedure. The reported region (*m*/*z* = 114–*m*/*z* = 117) is magnified and shows the individual reporter ions used for quantification.

In contrast, when digestion is performed *in situ*, most protein (-RNA) complexes must be completely denatured before digestion, using a higher concentration of detergents/denaturants than is compatible with endoproteinase digestion and labeling. Consequently, the sample volume has to be adjusted before digestion/labeling, and this can interfere with the conditions of the labeling reaction. Furthermore, any additional intermediate offline chromatography step – i.e., between digestion, labeling, and LC-MS/MS – results in loss of sample material.

We illustrate this protocol by examining the increasing amounts of 25S [U4/U6.U5] tri-snRNP particles loaded onto 1D-PAGE. 25S [U4/U6.U5] tri-snRNP is a key component of the spliceosomal machinery that is responsible for the excision of the intron and the ligation of the exons of eukaryotic pre-mRNAs to yield mature mRNA *(3)*. Tri-snRNPs consist of 30 proteins with molecular weights ranging from approximately 8 to 280 kDa and three of the so-called uridine-rich small nuclear ribonucleoprotein particles (U snRNPs).

For in-gel digestion and subsequent iTRAQ labeling, we cut the entire lanes in gel slices of exactly the same size. For this purpose, we use an in-house manufactured gel lane cutter. **Fig. 2a–c** shows the handling of the gel-cutting device for in-gel-digestion.

**Fig. 3a** shows an example of the MS quantification using iTRAQ after in-gel digestion of various amounts of 25S [U4/U6.U5] tri-snRNP *(2)* (0, 5.0, and 2.5 μg) loaded onto a pre-cast gel and stained with colloidal Coomassie blue. **Fig. 3b** summarizes the results as a bar diagram for selected single proteins. The values found by iTRAQ quantification of the different samples match the amount of sample loaded onto the gel (5 and 2.5 μg, respectively).

## 2. Materials

### 2.1. Ethanol Precipitation

1. 100% (v/v) ethanol, stored at –20°C.
2. 80% (v/v) ethanol, stored at –20°C.
3. 3 M natrium acetate (NaOAc), pH 5.3. Stored at room temperature.

### 2.2. NuPAGE® Electrophoresis System Components

1. NuPAGE® Novex Bis-Tris pre-cast gels (4–12%, 1.0 mm, 8 × 8 cm). Stored at 4°C–25°C.
2. NuPAGE® LDS sample buffer (4×). Stored at 4°C.
3. NuPAGE® reducing agent (10×). Stored at 4°C.
4. NuPAGE® antioxidant. Stored at 4°C.
5. NuPAGE® MES SDS or MOPS SDS running buffer (20×). Stored at room temperature.
6. Gel chamber (Xcell *SureLock*™ Mini-Cell).
7. *See* also **Note 2**

**2.3. In-Gel Digestion and Extraction of Peptides**

1. Ultrapure water (LiChrosolv®).
2. Acetonitrile (ACN, LiChrosolv®).
3. 50 mM triethylammonium bicarbonate (TEAB, Sigma-Aldrich, Steinheim, Germany), pH 8.0. Prepared freshly prior to use.
4. 10 mM (m/v) dithiothreitol in 50 mM TEAB. Prepared freshly prior to use.
5. 55 mM (m/v) iodoacetamide in 50 mM TEAB. Prepared freshly prior to use.
6. 5% (v/v) formic acid (FA, p.a.). Prepared freshly prior to use.
7. Trypsin (Roche, sequencing grade, 0.1 μg/μL).
8. Buffer 1: 50 μL $H_2O$, 50 μL 50 mM TEAB, 15 μL trypsin. Prepared freshly prior to use.
9. Buffer 2: 50 μL $H_2O$, 50 μL 50 mM TEAB. Prepared freshly prior to use.
10. *See* also **Note 1** and **3**.

**2.4. iTRAQ-Labeling**

1. 100 mM TEAB, pH ≥ 8.0. Prepared freshly prior to use.
2. iTRAQ-reagents (iTRAQ™ Reagent Multi-Plex Kit, Applied Biosystems, Foster City, USA). Stored at –20°C.
3. Ethanol (iTRAQ™ Reagent Multi-Plex Kit, Applied Biosystems, Foster City, USA).
4. 50 mM (m/v) glycine. Prepared freshly prior to use.
5. *See* also **Note 1** and **3**.

**2.5. LC-Coupled ESI MS/MS**

1. Ten percent (v/v) ACN, 0.1% (v/v) formic acid (FA, p.A.).
2. *See* also **Note 4**.

## 3. Methods

**3.1. Sample Separation by 1D Gel Electrophoresis Using Pre-cast Gels**

All steps necessary for running the gels including the recommended sample buffer are according to the manufacturer's protocols (e.g., Invitrogen, Carlsbad, USA). Protocols are delivered on purchase. Note that all pre-cast gel systems require gel chambers from the same manufacturer.

Most pre-cast gels are available in different thicknesses (e.g., 1.0, 1.5 mm). It is advisable to use 1-mm gels for in-gel digestion (*see* Chapter 3.1), as the amount of gel material should be kept as small as possible. The sample amount depends on the gel's thickness. On 1-mm gels approximately 40 μL of sample (including 4 × sample buffer, e.g., NUPAGE® (Invitrogen, Carlsbad, USA)) can be loaded. Therefore, most samples should be concentrated before loading, either by precipitation (e.g., TCA, EtOH, and $CHCl_3$/MeOH) or by ultrafiltration (e.g., Sartorius stedim biotech, Aubagne Cedex, France, or Millipore Corporation, Billerica, USA).

#### 3.1.1. Sample Preparation for Gel Electrophoresis

25S [U4/U6.U5] tri-snRNP particles (harboring 30 proteins *[2]*) were obtained after glycerol-gradient centrifugation of immunoisolated total snRNPs as described elsewhere *(37–39)*. Protein concentration is determined according to Bradford *(40)*.

1. Precipitate glycerol-gradient fraction containing 2.5 and 5.0 µg, respectively, tri-snRNP particles with three volumes ice-cold EtOH and 1/10 volume 3 M NaOAc, pH 5.3.
2. Spin down for 30 min, 13,000 rpm, 4°C.
3. Wash with ice-cold 80% (v/v) EtOH.
4. Spin down for 30 min, 13,000 rpm, 4°C.
5. Dry pellets in vacuum centrifuge.
6. Dissolve samples in 30 µL sample buffer (mix of 4 × NuPAGE® LDS sample buffer, 10 × NuPAGE® reducing agent, and double-distilled water).
7. Heat samples to 70°C for 10 min.
8. Load sample onto Bis-Tris pre-cast 4–12% gradient gels (NuPAGE®, Invitrogen).

#### 3.1.2. Gel Electrophoresis

Electrophoresis is carried out constantly at 200 V (Bis-Tris Gels) using the appropriate gel chamber (Xcell *SureLock™* Mini-Cell) according to the manufacturer's protocols *(18)*.

#### 3.1.3. Staining

Pre-cast gels were stained with silver *(19)* or colloidal Coomassie *(41)* and stored in water (LiChrosolv®).

### 3.2. In-Gel Digestion

#### 3.2.1. Excision of Bands/Lanes

After resolution of proteins by gel electrophoresis, the protein spots/lanes of interest are excised from the gel, and the proteins within them are digested with endoproteinases.

Proteins can be excised from 1D PAGE by two different methods, i.e., excision of single bands or cutting entire lanes into equal-sized pieces.

In our laboratory, the routine method is analysis of entire lanes cut from pre-cast gels. For this purpose, we use a gel-cutter made in-house (**Fig. 2**) that cuts an entire lane from a NuPAGE® pre-cast gel into exactly 23 pieces. Cutting the whole lane ensures that the entire sample will be analyzed. Furthermore, owing to the identical sizes of the gel pieces, higher reproducibility and more confidence in relative quantification between samples in different lanes is achieved. **Fig. 2** shows the gel-cutter together with the stained protein samples on NuPAGE®-gels.

The gel pieces have a size of 3 × 7 mm. Slices are further cut manually into pieces of 1 × 1 mm using a scalpel and a spatula, and the pieces are transferred into 0.5-ml Eppendorf tubes for in-gel digestion.

#### 3.2.2. In-Gel Digestion

Trypsin is normally the preferred endoproteinase for the in-gel digestion of proteins, for several reasons: (1) it cleaves proteins specifically at the C-terminal side of arginine and lysine residues. The peptides (precursors) generated thus always contain a charged amino acid that favors their ionization under acidic conditions. (2) Tryptic peptides have mass-to-charge ratios that are well suited for MS analysis ($m/z$ 350–1,400). (3) Trypsin is active in many buffers containing denaturing agents (urea, guanidinium chloride, and SDS).

The protocol is modified from Shevchenko et al. *(42)* for direct iTRAQ labeling. Note that during iTRAQ labeling, the presence of primary amines should be avoided. TEAB is the preferred buffer.

##### Digestion

1. Wash the gel pieces with 150 µL $H_2O$ and incubate for 5 min in a thermomixer at 26°C and 1,050 rpm.
2. Spin the gel pieces down and remove all liquid with thin gel-loader tips.
3. Add 150 µL ACN to shrink the gel pieces and incubate for 15 min in a thermomixer at 26°C and 1,050 rpm.
4. Spin the gel pieces down and remove all liquid with thin gel-loader tips.
5. Dry the gel pieces in a vacuum centrifuge for approximately 10 min.
6. Add 100 µL 10 mM dithiothreitol (in 50 mM TEAB) to reduce the disulfides and incubate for 50 min in a thermomixer at 56°C.
7. Spin the gel pieces down and remove all liquid with thin gel-loader tips.
8. Add 150 µL ACN to dehydrate and incubate for 15 min in a thermomixer in the dark at 26°C and 1,050 rpm.
9. Spin the gel pieces down and remove all liquid with thin gel-loader tips.
10. Add 100 µL 55 mM iodoacetamide (in 50 mM TEAB) to alkylate the reduced cysteines and incubate for 20 min in a thermomixer in the dark at 26°C and 1,050 rpm.
11. Wash the gel pieces with 150 µL 50 mM TEAB and incubate for 15 min in a thermomixer at 26°C and 1,050 rpm.
12. Add 150 µL ACN and incubate for 15 min in a thermomixer at 26°C and 1,050 rpm.

13. Spin gel pieces down and remove all liquid with thin gel-loader tips.
14. Add 150 µL ACN to dehydrate and incubate for 15 min in a thermomixer at 26°C and 1,050 rpm.
15. Spin gel pieces down and remove all liquid with thin gel-loader tips.
16. Dry the gel pieces in a vacuum centrifuge for approximately 10 min.
17. Rehydrate the dried gel pieces with buffer 1 (*see* Subheading 2.3) on ice; if necessary add more buffer 1 after approximately 30 min.
18. Cover the rehydrated gel pieces with buffer 2 (*see* Subheading 2.3) on ice; if necessary add more buffer 2 after approximately 30 min.
19. Incubate overnight at 37°C.

Extraction of the Tryptic Peptides

1. Add 25 µL $H_2O$ to the gel pieces and incubate for 15 min in a thermomixer at 37°C and 1,050 rpm.
2. Add 50 µL ACN and incubate for 15 min in a thermomixer at 37°C and 1,050 rpm.
3. Transfer the supernatant with thin gel-loader tips to new tubes.
4. Add 50 µL 5% (v/v) FA to the gel pieces and incubate for 15 min in a thermomixer at 37°C and 1,050 rpm.
5. Add 50 µL ACN and incubate for 15 min in a thermomixer at 37°C and 1,050 rpm.
6. Remove the supernatant and pool with the first supernatant.
7. Add 50 µL ACN and incubate for 15 min in a thermomixer at 37°C and 1,050 rpm.
8. Remove the supernatant and pool supernatants.
9. Dry the supernatant in a vacuum centrifuge for approximately 1.5 h.
10. The dried samples could be prepared for MS analysis or stored at –20°C for future labeling.

### 3.3. iTRAQ-Labeling

Labeling of peptides with iTRAQ™ reagents

1. Dissolve the extracted peptides in 20 µL 100 mM TEAB (sonication).
2. Prepare an internal standard by pooling 5 µL of each sample to a final volume of 15 µL, if there are fewer than three samples fill up with a corresponding volume (e.g., 5 µL) of 100 mM TEAB.

3. Reconstitute each iTRAQ™ reagent in 70 µL ethanol (vortex).
4. Add to each sample 5 µL of the different iTRAQ™ reagents resulting in a final concentration of 25% (v/v) ethanol; use iTRAQ™-114 for labeling the internal standard.
5. Incubate for 1 h in a thermomixer at 25°C and 550 rpm.
6. Quench excess of reagent by adding 5 µL 50 mM glycine resulting in a final concentration of 10 mM glycine.
7. Incubate for 30 min in a thermomixer at 25°C and 550 rpm.
8. Pool the labeled samples and dry down in a vacuum centrifuge.
9. The dried samples could be prepared for MS analysis directly or stored at –20°C.
10. For MS analysis, dissolve samples in appropriate volume (e.g., 20 µL) of 10% (v/v) ACN, 0.1% (v/v) FA.

### *3.4. Standard LC-Coupled ESI MS/MS of iTRAQ-Labeled Peptides and Data Evaluation*

Since every laboratory has its own very individual setup for LC-coupled MS/MS and uses different software, we do not provide any specific protocol but rather a view comments on more general aspects.

1. As mentioned above, Q-ToF, the Orbitrap-FT mass-analyzer, and MALDI-MS/MS instruments are most suitable for iTRAQ quantification.
2. Labeled samples can be analyzed either by LC-coupled ESI-MS/MS using qQ-ToF or with an Orbitrap-FT mass-analyzer or off-line LC MALDI-MS/MS instruments.
3. LC systems should be equipped with a trapping column (RP C18) on which the sample is extensively desalted (several minutes with elevated flow rates, e.g., 10 µL/min) to remove the excess of iTRAQ reagent.
4. Separation of labeled peptides is achieved by standard "peptide mapping" RP C18 columns, most frequently used with inner diameter of 75 µm working at a flow rate between 180 and 300 nL/min.
5. For MS/MS, the routine settings in the mass spectrometer MS can be used. However, sometimes it is recommended to enhance the detection of fragment ions in the lower *m/z* range in the mass spectrometer (e.g., in the qQ-ToF Waters Ultima by setting the appropriate radio-frequency values).
6. Before submission of the data for database search, the raw data-processing parameter (to generate, e.g., centroid data) should be adjusted, in particular when the lower *m/z* mass range of a spectrum is being monitored.
7. To analyze the iTRAQ data, several software tools are available, both commercially (Mascot 2.2 (Matrix Science), Protein pilot (ABI)) and freely (iTracker *[43]*, Multi-Q *[44]*,

etc.). However, not all available software (in particular, the freely available software on the internet) can handle the processed instrument-specific data formats without any difficulties.

Our laboratory uses Mascot 2.2 to quantify the data. At searches with iTRAQ™-labeled peptides, modifications at the N-term, at lysine side chains and at tyrosine side chains (side reaction of the iTRAQ reagent) are taken into account. The labeling efficiency is checked by searching the fragment spectra with iTRAQ as a variable (i.e., optional) modification. The labeling efficiency must be ≥90%. All parameters for data-processing and -searching should be tested with known standards (see elsewhere in this chapter).

### 3.5. Conclusions

iTRAQ labeling from in-gel digested samples is a reliable approach to compare the relative protein quantities between different samples, in particular when working with samples of moderate protein complexity. Since the labeling step does not require any further desalting and/or chromatography steps, it can be implemented on a routine basis within the well-established in-gel digestion procedure of proteins. It offers the opportunity to compare the relative protein amount in a multitude of purified protein (–RNA) complexes that were isolated under different conditions or from different functional stages of a cell.

## 4. Notes

1. For all buffers and solutions p.a. grade water and solvents were used.
2. All NuPAGE® electrophoresis system components were used according to manufacturer's protocols.
3. All buffers for in-gel digestion, extraction of tryptic peptides, and iTRAQ-labeling were prepared fresh prior to use.
4. For LC-coupled MS/MS the routine solvents and buffers as well as the routine settings in the MS can be used.

## Acknowledgments

We thank Monika Raabe und Uwe Plessmann for technical assistance and Reinhard Lührmann for providing purified tri-snRNP particles.

## References


1. Ong, S. E. and Mann, M. (2005) Mass spectrometry-based proteomics turns quantitative. *Nat Chem Biol* 1, 252–62.
2. Behrens, S. E. and Luhrmann, R. (1991) Immunoaffinity purification of a [U4/U6.U5] tri-snRNP from human cells. *Genes Dev* 5, 1439–52.
3. Will, C. L. and Luhrmann, R. (2006) Spliceosome Structure and Function. In: Gesteland, R. F., Cech, J. F., and Atkins, J. F., eds. *The RNA World*, 3rd Ed., pp. 369-400, Cold Spring Harbor Laboratory Press, Cold Spring Harbor.
4. Ross, P. L., Huang, Y. N., Marchese, J. N., Williamson, B., Parker, K., Hattan, S., Khainovski, N., Pillai, S., Dey, S., Daniels, S., Purkayastha, S., Juhasz, P., Martin, S., Bartlet-Jones, M., He, F., Jacobson, A., and Pappin, D. J. (2004) Multiplexed protein quantitation in *Saccharomyces cerevisiae* using amine-reactive isobaric tagging reagents. *Mol Cell Proteomics* 2004/09/24 Ed., pp. 1154-1169.
5. Laemmli, U. K. (1970) Cleavage of structural proteins during the assembly of the head of bacteriophage T4. *Nature* 227, 680–85.
6. Klose, J. (1975) Protein mapping by combined isoelectric focusing and electrophoresis of mouse tissues. A novel approach to testing for induced point mutations in mammals. *Humangenetik* 26, 231–43.
7. O'Farrell, P. H. (1975) High resolution two-dimensional electrophoresis of proteins. *J Biol Chem* 250, 4007–21.
8. Schagger, H. and von Jagow, G. (1991) Blue native electrophoresis for isolation of membrane protein complexes in enzymatically active form. *Anal Biochem* 199, 223–31.
9. Schagger, H., Aquila, H., and von Jagow, G. (1988) Coomassie blue-sodium dodecyl sulfate-polyacrylamide gel electrophoresis for direct visualization of polypeptides during electrophoresis. *Anal Biochem* 173, 201–5.
10. Macfarlane, D. E. (1989) Two dimensional benzyldimethyl-*n*-hexadecylammonium chloride – sodium dodecyl sulfate preparative polyacrylamide gel electrophoresis: a high capacity high resolution technique for the purification of proteins from complex mixtures. *Anal Biochem* 176, 457–63.
11. Hartinger, J., Stenius, K., Hogemann, D., and Jahn, R. (1996) 16-BAC/SDS-PAGE: a two-dimensional gel electrophoresis system suitable for the separation of integral membrane proteins. *Anal Biochem* 240, 126–33.
12. Washburn, M. P., Wolters, D., and Yates, J. R., 3rd (2001) Large-scale analysis of the yeast proteome by multidimensional protein identification technology. *Nat Biotechnol* 19, 242–47.
13. Krijgsveld, J., Gauci, S., Dormeyer, W., and Heck, A. J. (2006) In-gel isoelectric focusing of peptides as a tool for improved protein identification. *J Proteome Res* 5, 1721–30.
14. Klose, J. (1999) Large-gel 2-D electrophoresis. *Methods Mol Biol* 112, 147–72.
15. Gorg, A., Obermaier, C., Boguth, G., Harder, A., Scheibe, B., Wildgruber, R., and Weiss, W. (2000) The current state of two-dimensional electrophoresis with immobilized pH gradients. *Electrophoresis* 21, 1037–53.
16. Klose, J. and Kobalz, U. (1995) Two-dimensional electrophoresis of proteins: an updated protocol and implications for a functional analysis of the genome. *Electrophoresis* 16, 1034–59.
17. Kubo, K. (1995) Effect of incubation of solutions of proteins containing dodecyl sulfate on the cleavage of peptide bonds by boiling. *Anal Biochem* 225, 351–353.
18. Invitrogen (2003) General information and protocols for using the NuPAGE electrophoresis system. *NuPAGE™ Technical Guide.*
19. Blum, H., Beier, H., and Gross, H. J. (1987) Improved silver staining of plant proteins, RNA and DNA in polyacrylamide gels. *Electrophoresis* 8, 93–99.
20. Westermeier, R. and Marouga, R. (2005) Protein detection methods in proteomics research. *Biosci Rep* 25, 19–32.
21. Hu, Y., Malone, J. P., Fagan, A. M., Townsend, R. R., and Holtzman, D. M. (2005) Comparative proteomic analysis of intra- and interindividual variation in human cerebrospinal fluid. *Mol Cell Proteomics* 4, 2000–09.
22. Wilson, K. E., Marouga, R., Prime, J. E., Pashby, D. P., Orange, P. R., Crosier, S., Keith, A. B., Lathe, R., Mullins, J., Estibeiro, P., Bergling, H., Hawkins, E., and Morris, C. M. (2005) Comparative proteomic analysis using samples obtained with laser microdissection and saturation dye labelling. *Proteomics* 5, 3851–58.
23. Gygi, S. P., Rist, B., Gerber, S. A., Turecek, F., Gelb, M. H., and Aebersold, R. (1999) Quantitative analysis of complex protein mixtures using isotope-coded affinity tags. *Nat Biotechnol* 17, 994–99.
24. Yao, X., Freas, A., Ramirez, J., Demirev, P. A., and Fenselau, C. (2001) Proteolytic 18O labeling for comparative proteomics: model studies

with two serotypes of adenovirus. *Anal Chem* 73, 2836–42.

25. Staes, A., Demol, H., Van Damme, J., Martens, L., Vandekerckhove, J., and Gevaert, K. (2004) Global differential non-gel proteomics by quantitative and stable labeling of tryptic peptides with oxygen-18. *J Proteome Res* 3, 786–91.
26. Ong, S. E., Blagoev, B., Kratchmarova, I., Kristensen, D. B., Steen, H., Pandey, A., and Mann, M. (2002) Stable isotope labeling by amino acids in cell culture, SILAC, as a simple and accurate approach to expression proteomics. *Mol Cell Proteomics* 1, 376–86.
27. Old, W. M., Meyer-Arendt, K., Aveline-Wolf, L., Pierce, K. G., Mendoza, A., Sevinsky, J. R., Resing, K. A., and Ahn, N. G. (2005) Comparison of label-free methods for quantifying human proteins by shotgun proteomics. *Mol Cell Proteomics* 4, 1487–502.
28. Wiener, M. C., Sachs, J. R., Deyanova, E. G., and Yates, N. A. (2004) Differential mass spectrometry: a label-free LC-MS method for finding significant differences in complex peptide and protein mixtures. *Anal Chem* 76, 6085–96.
29. Wu, W. W., Wang, G., Baek, S. J., and Shen, R. F. (2006) Comparative study of three proteomic quantitative methods, DIGE, cICAT, and iTRAQ, using 2D gel- or LC-MALDI TOF/TOF. *J Proteome Res* 5, 651–58.
30. Karas, M., Bachmann, D., Bahr, U., and Hillenkamp, F. (1987) Matrix-assisted ultraviolet laser desorption of non-volatile compounds. *Int J Mass Spectrom Ion Proc* 78, 53–58.
31. Karas, M. and Hillenkamp, F. (1988) Laser desorption ionization of proteins with molecular masses exceeding 10,000 daltons. *Anal Chem* 60, 2299–301.
32. Tanaka, K., Waki, H., Ido, Y., Akita, S., Yoshida, Y., and Yoshida, T. (1988) Laser ionization Tome-of-flight mass spectrometry. *Rapid Comm Mass Spectrom* 2, 151–153.
33. Fenn, J. B., Mann, M., Meng, C. K., Wong, S. F., and Whitehouse, C. M. (1989) Electrospray ionization for mass spectrometry of large biomolecules. *Science* 246, 64–71.
34. Perkins, D. N., Pappin, D. J., Creasy, D. M., and Cottrell, J. S. (1999) Probability-based protein identification by searching sequence databases using mass spectrometry data. *Electrophoresis* 20, 3551–67.
35. Elias, J. E., Haas, W., Faherty, B. K., and Gygi, S. P. (2005) Comparative evaluation of mass spectrometry platforms used in large-scale proteomics investigations. *Nat Methods* 2, 667–75.
36. Hu, Q., Noll, R. J., Li, H., Makarov, A., Hardman, M., and Graham Cooks, R. (2005) The Orbitrap: a new mass spectrometer. *J Mass Spectrom* 40, 430–43.
37. Bach, M., Bringmann, P., and Luhrmann, R. (1990) Purification of small nuclear ribonucleoprotein particles with antibodies against modified nucleosides of small nuclear RNAs. *Methods Enzymol* 181, 232–57.
38. Bringmann, P., Rinke, J., Appel, B., Reuter, R., and Luhrmann, R. (1983) Purification of snRNPs U1, U2, U4, U5 and U6 with 2,2,7-trimethylguanosine-specific antibody and definition of their constituent proteins reacting with anti-Sm and anti-(U1)RNP antisera. *EMBO J* 2, 1129–35.
39. Kastner, B. (1998) Purification and Electron Microscopy of Spliceosomal snRNPs. In: Schenkel, J., ed. *RNA Particles, Splicing and Autoimmune Diseases*, pp. 95–140, Springer Lab Manual, Springer, Berlin, Heidelberg.
40. Bradford, M. M. (1976) A rapid and sensitive method for the quantitation of microgram quantities of protein utilizing the principle of protein-dye binding. *Anal Biochem* 72, 248–54.
41. Neuhoff, V., Arold, N., Taube, D., and Ehrhardt, W. (1988) Improved staining of proteins in polyacrylamide gels including isoelectric focusing gels with clear background at nanogram sensitivity using Coomassie Brilliant Blue G-250 and R-250. *Electrophoresis* 9, 255–62.
42. Shevchenko, A., Wilm, M., Vorm, O., and Mann, M. (1996) Mass spectrometric sequencing of proteins silver-stained polyacrylamide gels. *Anal Chem* 68, 850–58.
43 Shadforth, I. P., Dunkley, T. P., Lilley, K. S., and Bessant, C. (2005) i-Tracker: for quantitative proteomics using iTRAQ. *BMC Genomics* 6, 145.
44. Lin, W. T., Hung, W. N., Yian, Y. H., Wu, K. P., Han, C. L., Chen, Y. R., Chen, Y. J., Sung, T. Y., and Hsu, W. L. (2006) Multi-Q: a fully automated tool for multiplexed protein quantitation. *J Proteome Res* 5, 2328–38.

# Chapter 13

# Electrospray Mass Spectrometry for Quantitative Plasma Proteome Analysis

**Hong Wang and Sam Hanash**


## Summary

Electrospray ionization mass spectrometry (ESI-MS) is an efficient soft ionization procedure for macro biomolecules. However, it is a rather delicate process to produce charged molecules for mass-to-charge ratio ($m/z$) based measurement. In this chapter, the mechanism of ESI is briefly presented, and the experimental pipeline for quantitative profiling of plasma proteins (prefractionation immunodepletion, protein isotope tagging, 2D-HPLC separation of intact proteins, and LC-MS) is presented as applied by our group in studies of cancer biomarker discovery.


**Key words:** Electrospray ionization, Plasma protein sample fractionation, *Nano*-HPLC ESI-MS, Quantitative plasma proteome

## 1. Introduction

Since the development of soft ionization techniques, referred to as electrospray ionization mass spectrometry (ESI-MS), in late-1980s *(1)*, biological mass spectrometry has become of wide-spread use in proteomics. The measurement of the molecular weight of intact molecules, the determination of amino acid sequences, and the elucidation of the modified structures of peptides and proteins have become fairly robust procedures. The electrospray ionization process can be simply described as follows: The analyte solution is pumped through a needle to which a high voltage is applied. A Taylor cone with an excess of charge on its surface forms as a result of the electric field gradient between the ESI needle and the counter-electrode. When the solution that comprises the Taylor cone reaches the Rayleigh limit, the point

Jörg Reinders and Albert Sickmann (eds.), *Proteomics,* Methods in Molecular Biology, Vol. 564
DOI: 10.1007/978-1-60761-157-8_13, 

at which Coulombic repulsion of the surface charge is equal to the surface tension of the solution, the droplets that contain an excess of charge detach from its tip *(2)*. These droplets evaporate as they move toward the entrance to the mass spectrometer to produce free, charged analyte molecules that can be analyzed for their mass-to-charge ratio. There are two major mechanisms (Coulomb fission mechanism and ion evaporation mechanism) that depict the formation of the charged analyte molecules. The Coulomb fission mechanism assumes that the increased charge density due to solvent evaporation causes large droplets to divide into smaller and smaller droplets, which eventually consist only of single ions *(3)*. The ion evaporation mechanism, which assumes the increased charge density that results from solvent evaporation, causes Coulombic repulsion to overcome the liquid's surface tension, resulting in a release of ions from droplet surfaces *(4)*. The ESI process takes place at atmospheric pressure and generates vapor phase ions that can be analyzed for mass-to-charge ratio within the mass spectrometer. Since ESI is a gentle process without significant fragmentation of analyte ions in the gas phase, it presents two advantages for proteomics. One is MS (based on the $m/z$ of precursor) based quantification for profiling differential expression of proteins in biological samples, and the other is sequence prediction by using an enzyme that cleaves proteins at particular residues (i.e., trypsin) for peptide identification. The introduction of *nano*-flow (<500 nL/min) ESI has played an important role in MS-based proteomics studies, as this made analysis of minute amounts of sample feasible without compromising on signal intensity *(5)*. Moreover, *nano*-flow ESI generates smaller droplets, which makes the desolvation process more efficient.

Comprehensive profiling of the biofluid proteome has substantial relevance to the identification of circulating protein biomarkers that may have predictive value and utility for disease diagnosis and management. However, the extreme complexity and vast dynamic range of protein abundance in biological fluids, notably plasma, present a substantial challenge for comprehensive protein analysis. The dynamic range of plasma protein abundance spans at least 12 orders of magnitude. We have implemented an orthogonal multidimensional intact-protein analysis system (IPAS), coupled with protein tagging and immunodepletion of abundant proteins, to quantitatively profile the human plasma proteome *(6)*. The strength of the intact-protein-based multidimensional separation is its ability to resolve related proteins, such as differentially modified forms, and to reduce the complexity of samples generated for mass spectrometry analysis *(7)*. Our experimental pipeline is illustrated in **Fig. 1**. Briefly, following the immunodepletion (to remove the high abundance proteins), plasma proteins in each of the samples are concentrated and labeled with different tags, i.e., the isobaric 4-plex iTRAQ labeling for four different types of samples, or the isotopic 2-plex

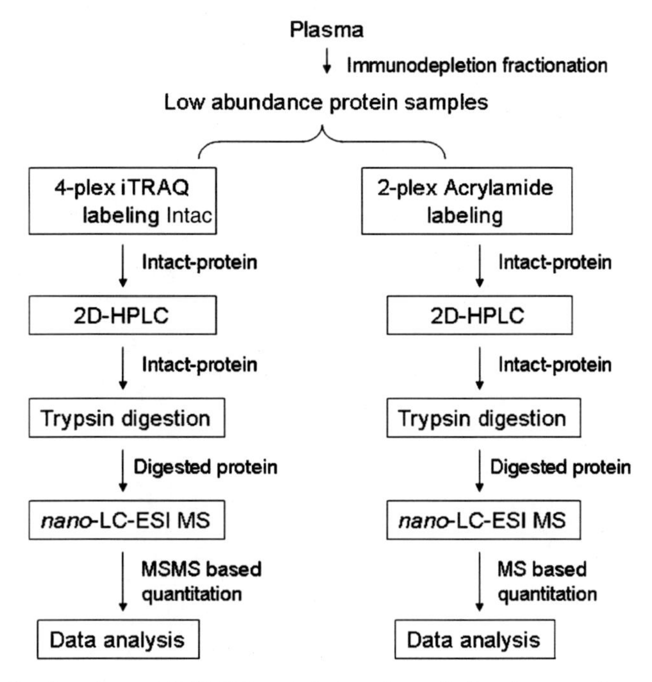


Fig. 1. Pipeline of *nano*-LC ESI-MS analysis-based quantitative plasma proteome.

acrylamide labeling for two different types of samples *(8)*. A mixture consisting of each of the intact-protein labeled samples is subjected to an orthogonal two-dimensional HPLC separation based on charge and hydrophobicity. Protein fractions from 2D-HPLC are dried by freeze-drying, followed by in-solution trypsin digestion. Then, the digests are subjected to LC-MS/MS analysis. Quantification is achieved by comparing the signal intensities of identified peptides in different samples (MS for the acrylamide labeling approach and MS/MS for the iTRAQ labeling approach). We have applied our current intact-protein-based multidimensional sample preparation approach with *nano*-LC ESI mass spectrometry to different types of cancers for which serum and plasma were analyzed disease study with the ultimate goal of discovering. In general, we have identified ~1,500 proteins with high confidence and obtained quantitative data for some 40% of identified proteins in a given experiment.

## 2. Materials

### 2.1. Sample

Samples (disease or control) are from individual subjects or pools. The protocol described here is designed for a total of 600 μL each of disease and control samples.

### 2.2. iTRAQ Labeling

1. Centricon YM-3 devices (millipore).
2. Exchange Buffer: 6.6 M Urea + 0.5 M TEAB (from the kit) +0.2% OG, pH 8.3.
3. Labeling buffer: 0.5 M TEAB (from the kit) +0.2% OG, pH 8.3.
4. Reduction solution: TCEP (50 mM) in the kit and ready to use.
5. Alkylation solution: iodoacetamide (200 mM), needs to be freshly prepared with water (74 mg/2 mL).
6. iTRAQ solvent: ETOH in the kit.
7. iTRAQ labeling reagent (reporter ion): 114, 115, 116, and 117 in the kit.

### 2.3. Acrylamide Isotope Labeling

1. Centricon YM-3 devices (millipore).
2. Labeling buffer: 8 M urea, 100 mM Tris pH 8.5, 0.5% OG (octyl-β-*d*-glucopyranoside).
3. Dithiothreitol (DTT): (ultrapure, USB).
4. Light acrylamide: acrylamide (>99.5% purity, Fluka).
5. Heavy acrylamide: 1,2,3–$^{13}C_3$-acrylamide (>98% purity, Cambridge Isotope Laboratories, Andover, MA).

### 2.4. Protein Fractionation

#### 2.4.1. Automated 1D-HPLC System

1D-HPLC system (Shimadzu Corporation, Columbia, MD) For immunodepletion, and the system consists of:

1. Pump System: two LC-10Ai pumps.
2. Degasser: DGU-20A3 Prominence Degasser.
3. Column oven: CTO-20 Prominence Column Oven.
4. Detector: SPD-20 UV/VIS Prominence Detector.
5. 5 Collector: FRC-10A Fraction Collector.
6. Controller: SCL-10A VP System Controller.
7. Workstation: EZStart 7.4 SP1.

#### 2.4.2. Automated Online 2D-HPLC System

Automatic online 2D-HPLC system (Shimadzu Corporation, Columbia, MD).

For intact protein 2D-HPLC fractionation (*see* **Fig. 2**), and the system consists of:

First dimension: anion-exchange chromatography.

Second dimension: reversed-phase (RP) chromatography.

1. Auto Sampler: SIL-10AP.
2. Pump system: one LC-10Ai pump for the First Dimension HPLC two LC-20AD pumps for the 2D-HPLC.
3. Degasser: DGU-20A5 Prominence Degasser.
4. Column oven: CTO-20AC Prominence Column Oven.
5. Detector: SPD-20 UV/VIS Prominence Detector.

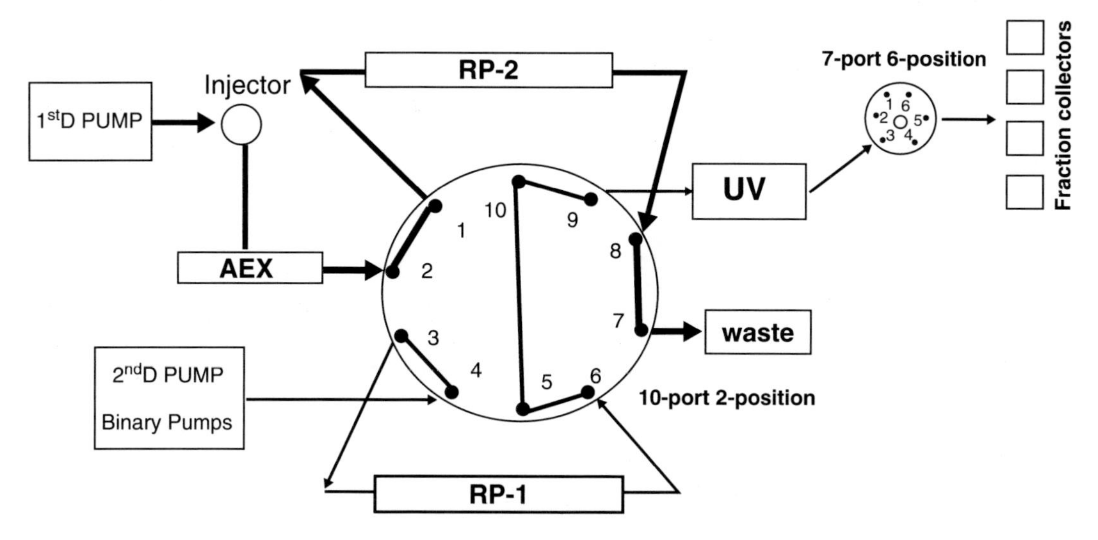


Fig. 2. Workflow diagram of automatic online 2D-HPLC system. (**a**) The First Dimension Pump loads the sample through the 10-mL sample loop onto the anion-exchange column with step elution **step 1**. (**b**) Through the 10-port 2-position valve, the flow-through proteins (unable to be absorbed within anion-exchange column) will be trapped within one of the two reversed-phase (RP) columns, here is the RP-2 column. (**c**) The Second Dimension Pumps starts the gradient elution within the other RP column; here is the RP-1 that is directed to the Fraction Collector through the 7-port 6-position valve. (**d**) At the end of the RP-1 gradient elution, the 10-port 2-position valve will switch to another position. (**e**) The First Dimension Pump will start the step elution **step 2**. (**f**) Through the 10-port 2-position valve, proteins eluted from the anion-exchange column will be trapped within the RP-1 column. (**g**) The Second Dimension Pumps start the gradient elution within the RP-2 that is directed to the Fraction Collector through the 7-port 6-position valve and starts to collect protein fractions. (**h**) The 10-port 2-position valve automatically switches the position at the end of RP gradient elution, and the First Dimension Pump will start the next step-elution. The 2D-HPLC system will repeat the online two dimension separation until it finishes the last step-elution of anion-exchange chromatography.

6. 6 Collector: four FRC-10A Fraction Collectors.
7. Controller: SCL-10A VP System Controller.
8. Workstation: Class-VP 7.4 (Shimadzu Client/Server).

*2.4.3. Immunodepletion Chromatography*

1. Hu-6 HC columns (10 ×100 mm; Agilent Technologies, Wilmington, DE).
2. Ms-3 HC columns (10 × 100 mm; Agilent Technologies, Wilmington, DE).
3. Buffer A (5185–6987), Equil/Load/Wash (Agilent Technologies, Wilmington, DE).
4. Buffer B (5185–5988), Elution (Agilent Technologies, Wilmington, DE).
5. 0.22 μm RC syringe filter (Fisher Scientific).

*2.4.4. Anion-Exchange Chromatography*

1. Poros HQ/10 column, 10 mm ID × 100 mmL (Applied Biosystems, Forster City, CA).
2. Mobile phase A: 20 mM Tris, 6% isopropanol, 4 M Urea, pH 8.5.

3. Mobile phase B: 20 mM Tris, 6% isopropanol, 4 M Urea, 1 M NaCl, pH 8.5.
4. Trizma, base (T1503) (Sigma, St. Louis, MO).
5. Hydrochloric acid (A508–500), Trace Metal Grade (Fisher Scientific).
6. Sodium chloride (S271-3), Fisher Scientific.
7. Urea (U15–3), Fisher Scientific.
8. Isopropanol (A451–1), HPLC grade (Fisher Scientific).

#### 2.4.5. Reversed-Phase Chromatography

1. Poros R2/10 column, 4.6 mm ID × 100 mmL (Applied Biosystems, Forster City, CA).
2. Mobile phase A: 95% $H_2O$, 5% Acetonitrile + 0.1% TFA.
3. Mobile phase A: 90% Acetonitrile, 10% $H_2O$, + 0.1% TFA.
4. Water (W5–1)), HPLC grade (Fisher Scientific).
5. Acetonitrile (A998–1), HPLC grade (Fisher Scientific).
6. Trifluoroacetic acid (TFA), 3–3077 (Supelco, Bellefonte, PA).

### 2.5. Protein Fractionation Process

FreeZone Plus 12 Liter Cascade Console Freeze Dry Systems (Labconco Corporation, Kansas City, MO).

### 2.6. Nano-LC ESI MS Analysis of Protein Fractionated Samples

#### 2.6.1. Tryptic Digestion

1. Sequence-grade modified trypsin (Porcine) V511A/20 µg (Promega, Madison, WI).
2. Digestion buffer: 0.25 M urea containing 50 mM ammonium bicarbonate and 4% of acetonitrile (v/v).
3. Ammonium bicarbonate (A643–500), Fisher Scientific.
4. Acetonitrile (A955–1), Optima LC/MS grade (Fisher Scientific).
5. Urea (U15–3), Fisher Scientific.
6. Formic acid (28905), 99+% (Pierce).

#### 2.6.2. Nano-LC ESI Mass Spectrometry Analysis

1. *Nano*-ESI Q-TOF Premier mass spectrometer (Waters Corporation).
2. *Nano*-ESI LTQ-FT/or Orbitrap *nano*-ESI mass spectrometer (Thermo Scientific).
3. *Nano*ACQUITY UPLC system (Waters Corporation).
4. Eksigent *nano*-LC-1D plus (Dublin, CA).
5. Capillary column: 25 cm Picofrit 75 µm ID (New Objectives) in house-packed with MagicC18 (100 A pore size/5 µm particle size).
6. Trap column: Symmetry $C_{18}$ 180 µm ×20 mm, particle size 5 µm (Waters Corporation).
7. Mobile phase A: 0.1% formic acid in HPLC grade water (v/v).

8. Mobile phase B: 0.1% formic acid in HPLC grade acetonitrile (v/v).
9. Water (W5–1), HPLC grade (Fisher Scientific).
10. Acetonitrile (A955–1), Optima LC/MS grade (Fisher Scientific).
11. Formic acid (28905), 99 +% (Pierce).

#### *2.6.3. Data Analysis*

##### Acrylamide Labeling Data Analysis

1. All the data management was performed using the Computational Proteomics Analysis System *(9)*.
2. Search Algorithm: X!Tandem *(10)*.
3. Scoring: PeptideProphet *(11)* and ProteinProphet *(12)*.
4. Quantitation: Q3 *(8)*.

##### iTRAQ Labeling Data Analysis

All the data management was performed using the ProteinLynx Global Server (version 2.25), Waters Corporation.

## 3. Methods

### *3.1. Immunodepletion Chromatography*

1. Filter plasma samples with a 0.22 μm RC syringe filter.
2. Equilibrate the Hu-6 HC column with buffer A at 3 mL/min for 10 min.
3. Load 300 μL of the plasma sample in 2 mL sample loop. Start the step-elution at 0.5 mL/min (*see* the step-elution in **Table 1**).
4. Collect the flow-through fractions for 19.5 min and immediately store at −80°C until use.
5. Elute the bound material and regenerate the column with buffer B at 3 mL/min for 5 min (collect if wanted).
6. The typical chromatogram is shown in **Fig. 3**.

### *3.2. Protein Tagging*

#### *3.2.1. Protein Labeling with Acrylamide*

1. Concentrate the control and disease immunodepleted plasma samples separately (only the flow-through low abundance protein fraction, *see* **Fig. 2**) in the Amicon YM-3 device until volume <100 μL. Do not mix or combine the control and disease samples.
2. Dilute each of the concentrated samples up to 1 mL with 8 M urea, 0.5 M TEAB (0.2% OG).
3. Measure protein concentration using Bradford method for disease and control samples, respectively (*see* **Note 1**).
4. Reduce protein disulfide bonds in each sample for 2 h at room temperature by adding 0.66 mg dithiothreitol (DTT) per milligram of protein.

**Table 1**
**Immunodepletion (UV = 280 nm)**

| Time (min) | Buffer B (%) | Flow rate (mL/min) | Fraction collection |
|---|---|---|---|
| 0.00 | 0.0 | 0.5 | |
| 6.30 | 0.0 | 0.5 | Start to collect the flow-through fraction (low abundance proteins) |
| 25.00 | 0.0 | 0.5 | |
| 25.10 | 100 | 0.5 | |
| 26.00 | 100 | 0.5 | Stop to collect the flow-through fraction (low abundance proteins) |
| 26.90 | 100 | 0.5 | |
| 27.00 | 100 | 3.0 | |
| 29.00 | 100 | 3.0 | Start to collect the binding fraction (high abundance proteins) |
| 34.00 | 100 | 3.0 | Stop to collect the binding fraction (high abundance proteins) |
| 35.00 | 100 | 3.0 | |
| 35.10 | 0.0 | 3.0 | |
| 45.00 | 0.0 | 3.0 | |
| 45.10 | 0.0 | 0.5 | |
| 45.20 | 0.0 | stop | |

5. Alkylate the samples with acrylamide isotopes immediately after the reduction by adding 7.1 mg/mg of total protein of light acrylamide to one of the samples (control or disease) and 7.4 mg/mg of total protein of heavy acrylamide to the other sample. The reaction is carried out for 1 h at room temperature. Keep the reaction in dark.
6. Mix or combine the control and disease samples, dilute it to 9 mL with Mobile-phase A of anion-exchange chromatography, and submit it immediately to 2D-HPLC protein separation.

*3.2.2. Protein Labeling with iTRAQ (see Note 2)*

1. Concentrate the control and disease immunodepleted plasma samples separately (only the flow-through low abundance protein fraction, *see* **Fig.2**) in the Amicon YM-3 device until

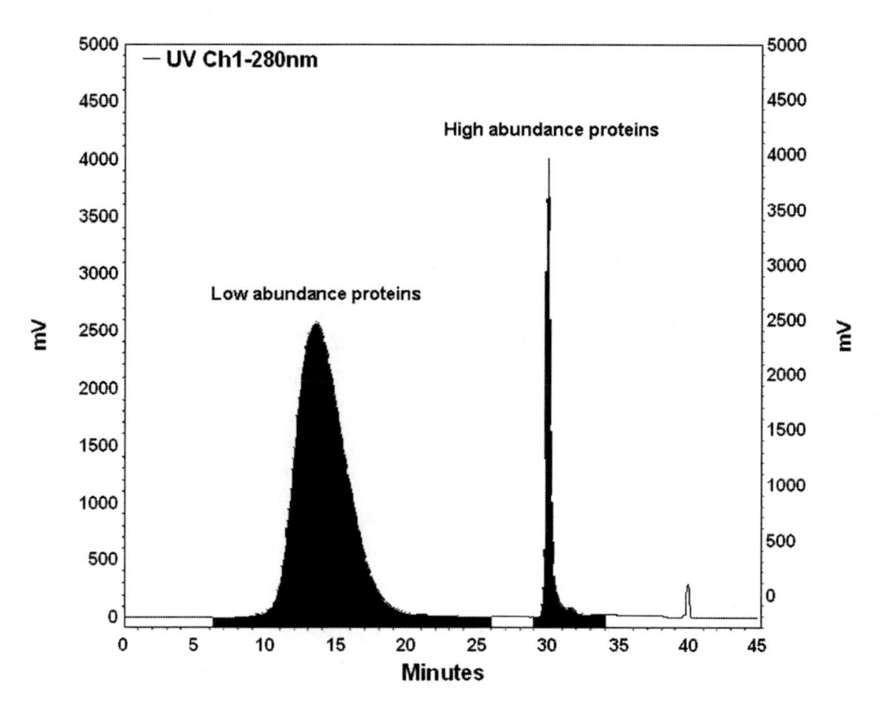


Fig. 3. Immunodepletion chromatogram.

volume <100 μL. Do not mix or combine the control and disease samples.

2. Dilute each of the concentrated samples up to 4 mL with the iTRAQ Exchange Buffer: 6.6M Urea +0.5 M TEAB +0.2% OG (w/v). Concentrate the sample in the same Amicon YM-3 device until volume <100 μL to replace the immunodepletion Buffer A.
3. Repeat **step 2** four times.
4. Measure protein concentration using Bradford method for disease and control samples, respectively.
5. Reduce protein disulfide bonds in each sample for 2 h at room temperature by adding 50 mM TCEP (20 μL/mg of protein).
6. Alkylate the samples immediately after the reduction by adding 200 mM iodoacetamide (10 μL/mg of protein). Keep the alkylation reaction in dark at room temperature for 1 h.
7. Add the iTRAQ solvent ETOH (from the iTRAQ kit) to each of the four iTRAQ reagent (corresponding to the reporter ion 114, 115, 116, and 117). Then, transfer them to each of the four under labeling samples. Each iTRAQ regent vial can be used for labeling 200 μg of protein.
8. Start the labeling reaction in dark at room temperature for 2 h.

9. Remove the organic solvent (ETOH) by Speed Vacuum Device and do not let the sample to dry totally.
10. Pool the four samples and dilute to 9 mL with anion-exchange Mobile Phase A.
11. Submit immediately to 2D-HPLC protein separation.

### 3.3. Protein Fractionation

#### 3.3.1. The First Dimension Anion-Exchange Chromatography

Step-elution is listed in **Table 2**

1. Equilibrate the anion-exchange column with Mobile-Phase A for 60 min at 0.8 mL/min.
2. Load the samples onto the 10 mL of sample-loop.
3. Start the run (*see* **Fig.2** for the online 2D-HPLC system diagram).

#### 3.3.2. The Second Dimensional Reversed-Phase Chromatography

Gradient-elution: Flow rate: 2.4/min, UV = 280 nm

1. Equilibrate the two RP columns with 95% of Mobile-Phase A for 15 min before start the run.
2. Desalt the column with 95% Mobile-Phase A for 10 min before starting the gradient elution (see the gradient-elution in **Table 3**).
3. There is an 8-step-elution from anion-exchange column, and the RP gradient-elution will apply to each of the 8 step-elution of anion exchange chromatography. See **Fig. 2** for the online 2D-HPLC system workflow diagram and a typical online 2D-HPLC chromatogram is shown in **Fig. 4**.
4. Collect the protein fractions from the RP column (3 fractions/min, 800 μL/fraction) and keep the protein fraction in –80°C.

**Table 2**
**Anion-exchange chromatography step-elution, flow rate 0.8 mL/min**

| Step-elution | Mobile phase B (%) |
|---|---|
| 1 | 0.0 |
| 2 | 5.0 |
| 3 | 10.0 |
| 4 | 15.0 |
| 5 | 20.0 |
| 6 | 30.0 |
| 7 | 50.0 |
| 8 | 100.0 |

**Table 3**
**Reversed-phase chromatography gradient elution**

| Time (min) | Buffer B (%) | Fraction collection |
|---|---|---|
| 0.00 | 5.0 | |
| 10.0 | 5.0 | Start to collect the fraction, 3 fractions/min, 800 μL/fraction |
| 12.00 | 20 | |
| 30.00 | 65 | |
| 32.00 | 65 | |
| 33.00 | 90 | |
| 35.00 | 90 | |
| 37.00 | 5.0 | |
| 40.00 | 5.0 | Stop to collect the fraction |
| 46.00 | Stop | |

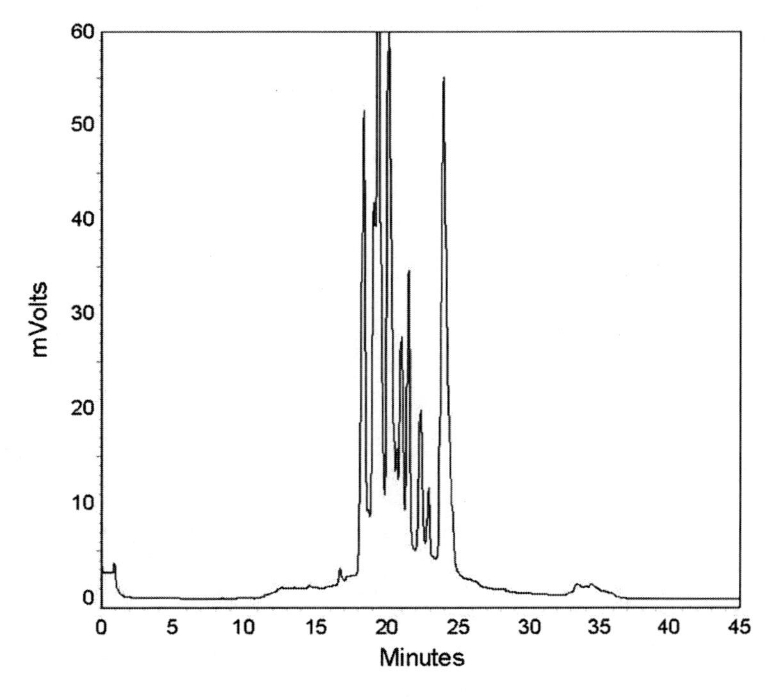


Fig. 4. 2D-HPLC Chromatogram (second dimension RP chromatography). Protein fractions collected between 10 and 40 min. A total of 84 fractions were collected.

5. Lyophilize the protein fractions using the Freeze Dry Systems (it usually takes 48 h to get the protein fraction dry completely).

### 3.4. Nano-LC ESI Mass Spectrometry Analysis

#### 3.4.1. In-Solution Protein

1. Individual lyophilized protein fractions were suspended with 200 ng of trypsin in the digestion buffer. The digestion was performed at 37°C overnight.
2. The digestion was quenched by adding 5 μL of 1.0% formic acid solution.
3. Based on the RP chromatography pattern, individual fractions were combined to get 12-pooled fractions for *nano*-LC ESI mass spectrometry analysis. **Table 4** lists the pooling sequence.

#### 3.4.2. Nano Flow-Rate LC ESI Mass Spectrometry Analysis

Waters *nano*-Acquity UPLC is coupled with *nano*-ESI QTOF Premier, and Eksigent *nano*-LC-1D plus is coupled with *nano*-ESI LTQFT/or LTQ Orbitrap. Both *nano*-HPLC systems directly deliver *nano* flow rate without using splitter. The gradient elution program is listed in **Table 5**.

1. Load 10 μL of each pooled digested protein fraction into 20 μL of sample-loop.
2. After injection, sample is trapped onto the trap column and desalted with 3% of Mobile-phase B for 10 min at 4 μL/min
3. Start the *nano*-HPLC gradient elution and acquire spectra in a data-dependent mode in $m/z$ range of 400 to 1,800. Select

**Table 4**
**Digested protein fraction pooling sequence for *nano*-LC ESI MS analysis**

| Pool # | Fraction # |
|---|---|
| 1 | From fraction #04 to #22 |
| 2 | From fraction #23 to #24 |
| 3 | From fraction #25 to #26 |
| 4 | From fraction #27 to #28 |
| 5 | From fraction #29 to #30 |
| 6 | From fraction #31 to #32 |
| 7 | From fraction #33 to #34 |
| 8 | From fraction #35 to #36 |
| 9 | From fraction #37 to #38 |
| 10 | From fraction #39 to #40 |
| 11 | From fraction #41 to #42 |
| 12 | From fraction #43 to #72 |

**Table 5**
***Nano* flow-rate reversed-phase HPLC gradient elution 300 nL/min**

| Time (min) | Mobile phase B (%) |
|---|---|
| 0.00 | 3.00 |
| 2.00 | 7.00 |
| 92.0 | 35.0 |
| 93.0 | 50.0 |
| 102.0 | 50.0 |
| 103.0 | 95.0 |
| 108.0 | 95.0 |
| 109.0 | 3.00 |
| 130.0 | 3.00 |

the five most abundant +2 or +3 ions of each MS spectrum for MS-MS analysis.

4. *nano*-ESI LTQFT mass spectrometer parameters are capillary voltage of 2.0 kV, capillary temperature of 200°C, resolution of 1,00,000, FT target value of 1,000,000, and ion trap target value of 10,000.
5. *nano*-ESI QTOF Premier mass spectrometer parameters are capillary voltage of 2.0 kV, *nano*-ESI source temperature of 150°C, resolution of 10,000, and MCP value of 2,300.

#### 3.4.3. Data Analysis

The acquired QTOF data (for 4-plex iTRAQ labeling) is automatically processed with ProteinLynx Global Server 2.25

1. For databank search: Carbamidomethylation (C) was set as the fix modification, Oxidation (M) and iTRAQ (K and N terms) were set as the variable modification.
2. Peptide tolerance was set as 100 ppm, and fragment tolerance was set as 50 ppm. Missed cleavages were set as 3.
3. **Figure 5** shows a representative *nano* flow-rate ESI MS (MS/MS) analysis, including the TIC chromatogram, MS spectrum, MS/MS sequence spectrum, and iTRAQ reporter ions for quantitation.

The acquired LTQFT data (for acrylamide labeling) is automatically processed by the Computational Proteomics Analysis System—CPAS *(9, 10)*.

1. For databank search, D0-acrylamide alkylation was considered a fixed modification, and D3 isotope-labeled peptides were

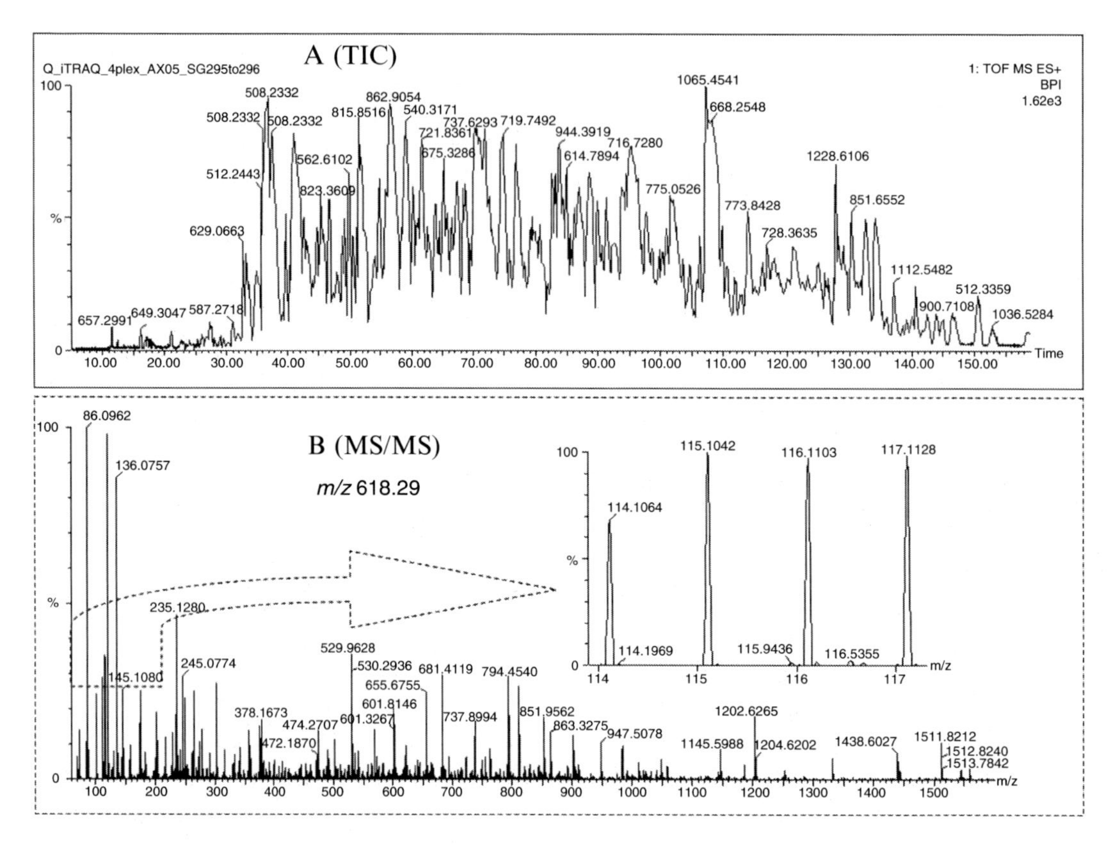


Fig. 5. 4-plex iTRAQ analysis by *nano*-LC ESI QTOF Premier mass spectrometer. Four different plasma samples (after immunodepletion) were labeled with 4-plex iTRAQ reagent, respectively (with four reporter ion: 114.1, 115.1, 116.1, and 117.1). (**a**) The reconstructed total ion current (TIC) chromatogram corresponds to ~2.5 µg of on-column loading (digested protein obtained from one pool of the 2D-HPLC fractionation, *see* **Subheading 3.2.2**). (**b**) The MS/MS sequence spectrum of precursor (*m/z* 618.29) and the quantifiable reporter ions from the low *m/z* range.

detected using a delta mass of 3.01884 Da *(8)*. Acrylamide ratios are obtained by a script called "Q3," which was developed in-house to obtain the relative quantification for each pair of peptides identified by MS/MS that contains cysteine residues.

2. Use the minimum criteria for peptide matching of Peptide Prophet score greater than 0.2 (16). Peptides that met these criteria are further grouped to protein sequences using the Protein Prophet algorithm at an error rate of 5% or less.

3. **Figure 6** shows a representative *nano* flow-rate ESI MS (MS/MS) analysis, including the TIC chromatogram, MS spectrum, MS/MS sequence spectrum, and acrylamide-labeled paired precursor ions for quantitation.

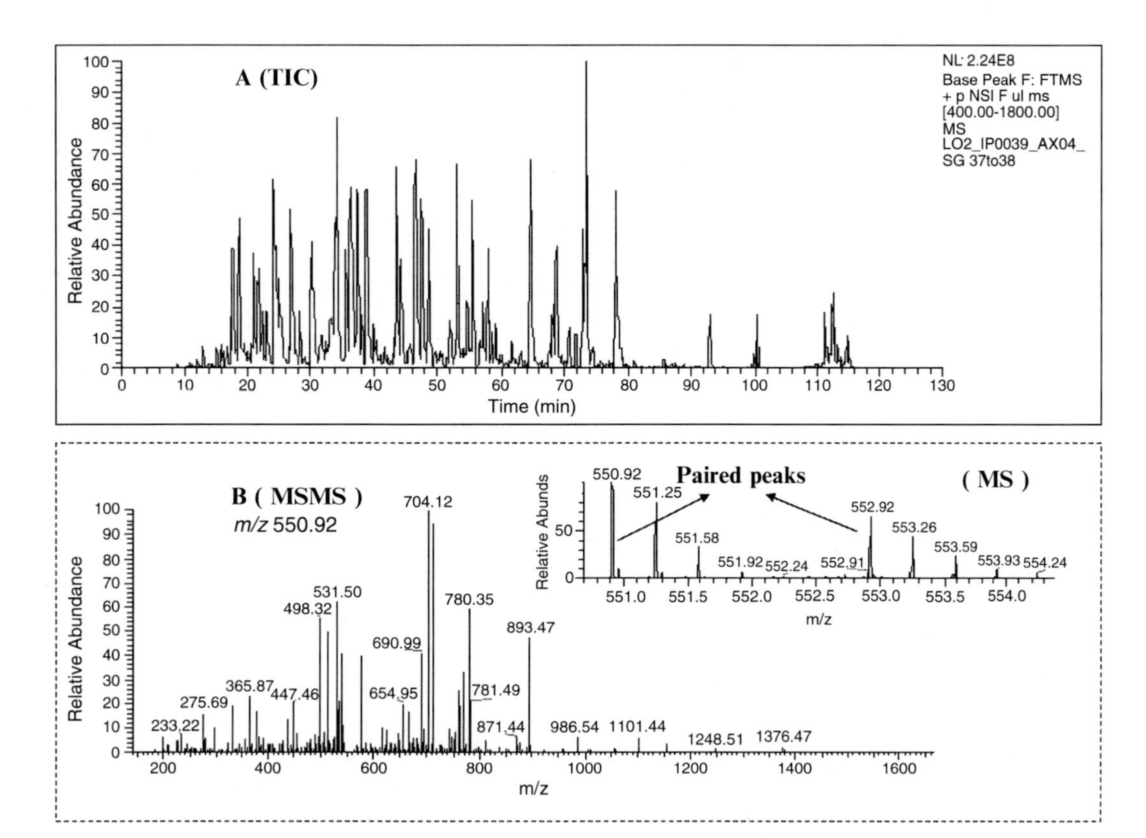


Fig. 6. 2-plex Acrylamide labeling analysis by *nano*-LC ESI LTQ-Orbitrap mass spectrometer. Two different plasma samples (after immunodepletion) were labeled with 2-plex Acrylamide (light and heavy) reagent, respectively. (**a**) The reconstructed total ion current (TIC) chromatogram corresponds to ~2.5 μg of on-column loading (digested protein obtained from one pool of the 2D-HPLC fractionation, *see* **Subheading 3.2.1**). (**b**) The MS/MS sequence spectrum of precursor (*m/z* 550.92) and the quantifiable paired precursor ions from the MS scan.

## 4. Notes

1. As a result of the immunodepletion step, the total protein concentration is reduced by 90%, from ~70 to 7 mg/mL.
2. Before the iTRAQ protein level labeling, the protein sample solution must be replaced with the exchange buffer with a pH 8.3 to avoid the labeling failure resulted from the free amine component in original sample solution and improper pH. Maintain 54% of organic solvent (ETOH) during the labeling reaction to keep 99% of labeling efficiency. After labeling reaction, remove ETOH before subject to 2D-HPLC separation because the higher percentage of organic solvent will result in sample loss during the loading step.

The feasibility of acrylamide labeling for protein identification and quantification has been demonstrated previously using MALDI-TOF for proteins separated by two-dimensional electrophoresis *(13)*. No protein precipitation was observed, indicating that acrylamide-based alkylation is compatible with intact-protein-based approaches. A very high yield of cysteine alkylation is achieved *(8)*. The interruption of reaction after 1 h is very important to avoid side reactions with free amines in the protein.

## References


1. Fenn, J.B., Mann, M., Meng, C.K.,Wong, S.F., Whitehouse, C.M. (1989). Electrospray ionization for mass spectrometry of large biomolecules. *Science* **246**, 64–71.
2. Taflin, D.C., Ward, T.L., Davis, E.J. (1989). Electrified droplet fission and the Rayleigh limit. *Langmuir* **5**, 376–384.
3. Dole, M., Mack, L.L., Hines, R.L., Mobley, R.C., Ferguson, L.D., Alice, M.B. (1968). Molecular beams of macroions. *J Chem Phys* **49**, 2240–2249.
4. Iribarne, J.V., Thomson, B.A. (1976). On the evaporation of charged ions from small droplets. *J Chem Phys* **64**, 2287–2294.
5. Wilm, M., Mann, M. (1996). Analytical properties of the nanoelectrospray ion source. *Anal Chem* **68**, 1–8.
6. Wang, H., Clouthier, S.G., Galchev, V., Misek, D.E., Duffner, U., Min, C.-K., Zhao, R., Tra, J., Omenn, G.S., Ferrara, J.L., Hanash, S.M. (2005). Intact-protein based high-resolution three-dimensional quantitative analysis system for proteome profiling of biological fluids. *Mol Cell Proteomics* **4**, 618–25.
7. Misek, D.E., Kuick, R., Wang, H., Galchev, V., Bin Deng, B., Zhao, R., Tra, J., Pisano, M.R., Amunugama, R., Allen, D., Walker, A.K., Strahler, J.R., Andrews, P., Omenn, G.S., Hanash, S.M. (2005). A wide range of protein isoforms in serum and plasma uncovered by a quantitative intact protein analysis system. *Proteomics* **5**, 3343–3352.
8. Faca, V., Coram, M., Phanstiel, D., Glukhova, V., Zhang, Q., Fitzgibbon, M., McIntosh, M., Hanash, S. (2006). Quantitative analysis of acrylamide labeled serum proteins by LC-MS/MS. *J Proteome Res* **5**, 2009–2018.
9. Rauch, A., Bellew, M., Eng, J., Fitzgibbon, M., Holzman, T., Hussey, P., Igra, M., Maclean, B., Lin, C.W., Detter, A., Fang, R., Faca, V., Gafken, P., Zhang, H., Whitaker, J., States, D., Hanash, S., Paulovich, A., McIntosh, M.W. (2006). Computational Proteomics Analysis System (CPAS): an extensible, open-source analytic system for evaluating and publishing proteomic data and high throughput biological experiments. *J Proteome Res* **5**, 112–121.
10. Maclean, B., Eng, J.K., Beavis, R.C., McIntosh, M. (2006). General framework for developing and evaluating database scoring algorithms using the TANDEM search engine. *Bioinformatics.*
11. Keller, A., Nesvizhskii, A.I., Kolker, E., Aebersold, R. (2002). Empirical statistical model to estimate the accuracy of peptide identifications made by MS/MS and database search. *Anal Chem* **74**, 5383–5392.
12. Nesvizhskii, A.I., Keller, A., Kolker, E., Aebersold, R. (2003). A statistical model for identifying proteins by tandem mass spectrometry. *Anal Chem* **75**, 4646–4658.
13. Sechi, S., Chait, B.T. (1998). Modification of cysteine residues by alkylation. A tool in peptide mapping and protein identification. *Anal Chem* **70**, 5150–5158.

# Part V

## Interpretation of Mass Spectrometry Data

# Chapter 14

# Algorithms and Databases

Lennart Martens and Rolf Apweiler

## Summary

The capacity of proteomics methods and mass spectrometry instrumentation to generate data has grown substantially over the past years. This data volume growth has in turn led to an increased reliance on software to identify peptide or protein sequences from the recorded mass spectra. Diverse algorithms can be applied for the processing of these data, each performing a specific task such as spectrum quality filtering, spectral clustering and merging, assigning a sequence to a spectrum, and assessing the validity of these assignments.

The key algorithms to mass spectral processing pipelines are the ones that assign a sequence to a spectrum. The most commonly used variants of these are crucially dependent on the information contained in the sequences database, which they use as a basis for identification. Since these sequence databases are constructed in different ways and can therefore vary substantially in the amount and type of data they contain, they are also discussed here.

**Key words:** Sequence database, Search algorithm, Mass spectrum, Clustering, Merging, Quality assignment, Tandem-MS, Identification, Protein, Peptide

## 1. Introduction

Proteomics experiments largely rely on mass spectrometers to analyze proteolytic digests of the proteins or protein mixtures obtained from samples. Two types of mass spectra are typically obtained from this analysis: spectra containing the mass-over-charge ($m/z$) and intensity peaks of intact peptides (so-called *peptide mass fingerprints*), and spectra that contain $m/z$ and intensity signals of fragmented precursors (so-called *peptide fragmentation fingerprints* (PFF)). Successful application of the peptide mass fingerprinting (PMF) approach typically requires purified proteins, such as those obtained after two-dimensional gel electrophoresis, whereas PFF can be obtained directly from very

Jörg Reinders and Albert Sickmann (eds.), *Proteomics,* Methods in Molecular Biology, Vol. 564
DOI: 10.1007/978-1-60761-157-8_14, © Humana Press, a part of Springer Science+Business Media, LLC 2009

complex peptide mixtures derived from protein mixtures. Fragmentation spectra are obtained by first selecting a specific peptide ion based on its $m/z$, subsequently breaking the selected peptide into parts, and finally analyzing the $m/z$ and intensity of the resulting fragment ions. A fragmentation spectrum is thus composed of the $m/z$ and intensity values of the fragments of a particular peptide (the *parent* or *precursor*) whose $m/z$ is also known. Ideally, the fragments form a ladder in which each successive fragment contains one additional residue, thus allowing the sequence to be read from the mass differences between the peaks. However, actual experimental PMF or fragmentation spectra are usually far from ideal and typically lack several expected peptide or fragment peaks while containing a considerable amount of unexpected peaks and noise (*see* **Fig. 1**). Peptides can also carry modifications, which change the mass of the intact peptide and of those fragment ions that include the affected residue(s). Additionally, modern proteomics technologies and instruments are able to generate thousands of spectra in a single run, necessitating the use of sophisticated algorithms for automatically assigning a protein or peptide sequence as the most likely source of the PMF or PFF spectrum. Identification algorithms are discussed in **Subheading 2**.

Although identification algorithms form the backbone of mass spectrometry data processing in proteomics, several other processing steps can be applied to the recorded spectra. Nonsensical, overly noisy, or contaminant-derived spectra can for instance be excluded from further analysis by first applying a quality filter to the spectra. Redundancy in the accumulated spectral data

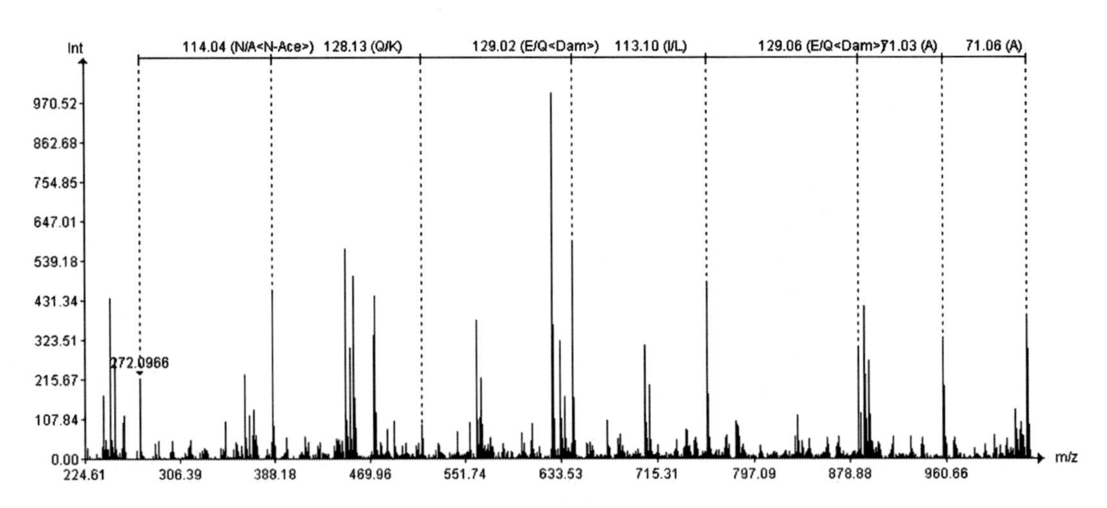


Fig. 1. Experimental fragmentation mass spectrum obtained from an ESI-Q-TOF instrument. A ladder series of ions has been manually highlighted, showing the ability to sequence (parts of) peptides directly from the spectrum. Note the presence of many unassigned noise peaks in the spectrum, as well as the ambiguities that can occur in the observed peak mass differences (e.g., a mass difference of 128.13 Da can, within the mass tolerance of 0.3 Da applied here, correspond to either a lysine (K) or a glutamine (Q) residue).

can be employed to cluster similar spectra, and reduce them to a single, representative spectrum instead. The latter can be chosen as the highest-quality spectrum in the cluster, or can be an artificially created consensus spectrum that attempts to optimize the information spread over the different cluster members. These approaches are discussed in **Subheading 3**.

Identification and subsequent biological interpretation of proteomics data ultimately depends on the availability of sequence databases and the annotations that they contain. The most commonly used sequence databases and their data contents are presented in **Subheading 4**.

## 2. Identification Algorithms

Identification algorithms can be roughly divided into two distinct categories: database search algorithms and de novo sequencing algorithms. Algorithms from the first category are most commonly used in the field, and these attempt to match an experimental spectrum with theoretical spectra generated from the sequences in a protein database. This overall approach is applied to both PMF and PFF spectra. The second category of algorithms, which can only be applied to fragmentation spectra, relies solely on the information found in these spectra to extract a partial peptide sequence. Finally, an indirect way of identifying an experimental spectrum is by matching it to a previously recorded and identified experimental spectrum, thus bypassing the need to create theoretical spectra from sequence databases.

The overall concepts of the different identification algorithms are briefly explained here, and more detailed discussions can be found in the following excellent review articles on the subject *(1–3)*.

### *2.1. Database Search Algorithms*

Common to all these algorithms is that they rely on the information in (protein) sequence databases to assign a sequence to a spectrum. To this end, the sequences in the database are digested *in silico* according to the known cleavage pattern for the protease that was used for the *in vitro* digest. This yields a collection of proteolytic peptides per database entry, which can directly be used to create a theoretical, ideal PMF spectrum. An experimental PMF can subsequently be compared with this theoretical spectrum. Two of the most popular peptide-mass fingerprinting algorithms are Mascot *(4, 5)* and ProFound *(6)*.

Algorithms that identify peptide fragmentation spectra perform an additional step on the *in silico* generated peptides before constructing theoretical spectra. Based on the peptide sequence, a list of all the potential fragment ions is calculated, which is then used to create a theoretical fragmentation spectrum. This theoretical spectrum is then matched with an experimental fragmentation spectrum in the next step. Popular fragmentation spectrum search engines include Mascot *(4, 5)*, SEQUEST *(7)*, and X!Tandem *(8)*.

The match between a theoretical and an experimental spectrum is given a score that typically takes into account the number of matching peaks, the number of peaks predicted in the theoretical spectrum but missing in the experimental spectrum, and the number of measured peaks in the experimental spectrum that have not been predicted in the theoretical spectrum. The actual matching of individual peaks is usually tolerant of a user-defined mass error to allow for imperfect mass accuracy and precision in the instrument. Since a protein can also carry post-translational modifications (or artefactual modifications, occurring *in vitro* rather than *in vivo*) that lead to mass shifts in the resulting spectra, search algorithms usually permit the user to specify a (limited) list of modifications that may occur on specified residues in the sequence.

To increase specificity or sensitivity (or both), it has become common practice to perform an additional postprocessing step of the reported scores for certain algorithms. If specificity is an issue, the reported assignments are further analyzed using stringent criteria to restrict the number of identifications to only the most reliable matches, whereas sensitivity is increased by rescuing assignments that had a low score. The most popular postprocessing algorithm, PeptideProphet *(9)*, tries to simultaneously optimize both sensitivity and specificity by essentially recalibrating the search engine score according to certain assumptions about the nature of the identifications.

Finally, it is important to realize that all database search algorithms are limited by the information contained in the databases they rely on. This means that they can only assign a sequence to a spectrum if that sequence was present in the database, and that they will fail to identify a spectrum derived from a sequence that is not in that database.

### 2.2. De Novo Sequencing Algorithms

*De novo* algorithms *(10–16)* extract sequence information directly from the mass differences between the fragment peaks in fragmentation spectra. This approach is particularly useful in cases where the sequence of the peptide precursor that led to the fragmentation spectrum is either not present in any sequence database or when that sequence carries unexpected modifications that alter its mass signature. Ideally, a spectrum subjected to *de novo* analysis should consist of at least one recognizable ladder of fragment ions, in which the mass differences between any two consecutive peaks corresponds to the mass of one modified residue.

In such an ideal case, the precursor peptide sequence can easily be read directly from the spectrum.

In reality, however, fragmentation spectra seldom carry complete fragment ion ladders, which cause gaps in the sequence. They usually also contain unexpected peaks or noise peaks that are scattered among the fragment ions, potentially leading to the extraction of incorrect or ambiguous sequences (*see* **Fig. 1**). As such, *de novo* analysis is usually restricted to spectra of sufficient quality and clarity, and even here the results often require manual filtering. They are therefore ill-suited to process data obtained from typical high-throughput proteomics experiments, where several thousands of fragmentation spectra are accumulated every hour. Finding out which *de novo* algorithm might work well in a particular context has become easier, thanks to the publication of performance reviews of several *de novo* algorithms in a variety of settings *(17, 18)*.

### 2.3. Tag-Based Algorithms: Meeting De Novo Sequencing and Database Search Algorithms Halfway

When only a small but usually quite reliable part of the peptide sequence is extracted from a PFF spectrum, it is called a *sequence tag*, or simply a *tag*. The process of extracting such a tag is very similar to a limited *de novo* analysis and also relies exclusively on the information contained in the spectrum. However, contrary to an actual *de novo* approach, the sequence tag (which in its final form consists of the extracted residues as well as the flanking masses corresponding to the unknown residues) is then searched against a sequence database. The approach allows three filters to be applied to the sequences obtained from the database. First, the calculated intact peptide mass (potentially including modifications) must match the observed precursor mass within a specified threshold. Second, the obtained stretch of sequence must be part of the peptide sequence retrieved from the database (potentially allowing for ambiguities). Third, the masses of the flanking residues in the database peptide must match the ones obtained from the experimental spectrum (again, potentially allowing for modifications). Note that these constraints can be relaxed in certain applications, for instance to allow identification of peptides containing single residue substitutions.

Interestingly, tag-based algorithms used to dominate the field initially. PeptideSearch, originally developed by Matthias Mann and coworkers *(19)*, was one of the first identification algorithms that employed a tag-based approach. The same approach was also shown to work for intact proteins when using high-accuracy mass spectrometers *(20)*. However, the advent of SEQUEST *(7)*, a database search algorithm discussed earlier, signaled a decreased interest in tag-based algorithms. A recent example of the sequence tag approach is given by the GutenTag algorithm developed by David Tabb and colleagues *(21)*, illustrating that this methodology remains an interesting option for the identification of PFF spectra.

### 2.4. Spectrum Matching Algorithms

The increased throughput of modern proteomics methods and instruments has led to a massive increase in the amount of available experimental data. Since these data are also increasingly captured in proteomics data repositories such as PRIDE *(22, 23)*, PeptideAtlas *(24)*, and GPMDB *(25)*, algorithms have been developed that make use of this accumulated information to speed up the identification of mass spectra. These algorithms bypass the sequence database by first matching new spectra with a spectrum library composed of older, already identified experimental spectra. Only those spectra that do not provide a reliable match against the spectral library are subsequently analyzed by more traditional database searching algorithms.

The creation of a spectral library can be achieved in many ways, from simply adding the highest scoring spectrum for each identified peptide sequence to a nonredundant set to the complex procedure of manually curating a set of carefully constructed consensus spectra, created by summing together several spectra that all led to the same identified sequence. Available spectrum matching algorithms include SpectraST *(26)* and X!Hunter (http://h201.thegpm.org/tandem/thegpm_hunter.html).

## 3. Pre-processing Algorithms

A number of processing steps are usually applied to mass spectra before submitting them to search algorithms, and these are therefore called preprocessing steps. The three most common preprocessing steps are deisotoping, centroiding, and charge deconvolution. The spectra output by such algorithms are usually called *peak lists*, as opposed to the *trace spectra* or *raw data spectra* they used as input.

Interestingly, certain instruments never record the actual raw data from the instrument to file, but automatically perform some form of preprocessing on the obtained signal before writing the result to disk. The most pressing points with regard to potential caveats associated with the usage of raw data instead of peak lists are highlighted in **ref.** *27*.

These common preprocessing steps are typically carried out by the vendor-supplied software on the instrument, and although some user-specified settings might be involved, these processing steps are typically considered a black box by most users. Third-parties (vendors or academics) sometimes provide specialized tools that can produce peak lists that are better suited to subsequent identification *(28, 29)*.

Slightly more involved preprocessing steps include spectrum filtering, and spectrum clustering and merging. These will be discussed in detail next.

### 3.1. Spectrum Quality Filtering

High-throughput proteomics experiments effectively exploit the ability of modern-day instruments to produce many thousands of fragmentation spectra from a single one-hour run. Despite efforts aimed at reducing the number of noninformative spectra (commonly grouped under the moniker *data-dependent acquisition*), a large amount of undesired spectra persist. Typical examples include fragmentation spectra of known contaminants such as the ubiquitous and easily ionized polymer polyethylene glycol (PEG; $HO–(CH_2–CH_2–O)n–H$, thus consisting of monomers with a nominal residue mass of 44 Da), or simply spectra derived from poorly fragmented precursor peptides.

On average, only about 10% to 20% of the spectra recorded during high-throughput proteomics experiments are successfully identified, and it may therefore seem that the mass spectrometer spends most of its analysis time (80% to 90%) fragmenting "junk." However, while a substantial amount of these unidentified spectra are indeed clearly derived from known contaminants, laborious manual inspection typically reveals that many unidentified spectra are in fact derived from peptide precursors and often even contain consecutive fragment ion peaks or *immonium ions* ($m/z$ peaks in the spectrum that correspond to certain individual amino acid residues). There can be several reasons why such spectra elude identification, including the presence of unexpected modifications, sequences that are not included in the search database (e.g., because of single residue substitutions), or unexpected proteolytic events (e.g., *in vivo* N-terminal truncation due to protein maturation *(30)*, or *in vitro* residual activity of resident proteases after sampling or cell lysis *(31)*). It is important to note that unidentified yet clearly peptide-derived spectra are often highly interesting precisely because they cannot be identified; indeed, understanding the origin and identity of such spectra gives us the opportunity to expand our knowledge about the biological system or the sample processing protocol. As already noted earlier, the most common way to attempt identification of such informative yet unidentified spectra would be to apply *de novo* sequencing algorithms, although there are alternative approaches available as well *(32, 33)*. However, common to all these methods is that they tend to be computationally intensive and more error-prone than regular database searching.

Rather than spending many computing cycles or man-hours on investigating the entire collection of unidentified spectra (which can amount to hundreds of thousands of spectra), many researchers have devised ways to assign a "quality score" to fragmentation spectra, which is taken to be indicative of the chances at assigning a peptide sequence to a spectrum using more exhaustive attempts at identification. Cut-offs can be applied to these scores to select only the most informative spectra for subsequent identification attempts. Different strategies to assign such quality scores have been described, based on a variety of algorithms and approaches *(34–41)*. Essentially, these algorithms all try to emulate or employ expert knowledge in deciding which spectra are of high quality. This expert knowledge can be fixed *a priori*, or can be learned from existing training sets. In the latter case, the training set is typically divided into identified and nonidentified spectra, with the identified spectra considered as high quality and the unidentified spectra as low quality. The different approaches are mainly distinguishable by the rules they use to emulate expert knowledge, the actual algorithms they employ to learn about weighting of these individual rules, and finally the way in which they then apply these rules to new data.

### 3.2. Spectrum Clustering and Merging

While the recording of contaminant spectra as outlined earlier clearly constitutes "lost time" on the instrument, a more subtle by-product of high-throughput proteomics is the acquisition of redundant fragmentation spectra. Although wasteful in theory, the fact that a single peptide precursor sometimes yields multiple fragmentation spectra can actually be put to good use. One example is the use of the spectrum or identification count of a peptide for so-called *label-free quantitation(42, 43)*, where the underlying idea is that more abundant proteins yield more abundant peptides, which in turn are more likely to be selected for fragmentation multiple times.

Another way to make use of redundant fragmentation spectra is to group similar spectra together in clusters and to replace each cluster either by a representative member *(44)* or by an artificial spectrum that aims to optimize the information present across the entire cluster *(45, 46)*. Since the initial step in a spectrum clustering approach consists of calculating similarity scores for spectrum pairs, it is very similar to the matching performed by spectrum library based identification algorithms (see above). Yet, while the identification-oriented spectrum library searches will essentially select the highest scoring match and retain it if it is deemed significant, the spectrum clustering approach further processes these similarity scores to define groups of related spectra.

The simplest way to construct such groups, called *single linkage* clustering, is based on the assumption that the concept of transitivity applies: if spectrum A is sufficiently similar to spectrum B, and spectrum B is sufficiently similar to spectrum C, then spectrum A and C must also be similar. This principle is then iteratively applied to expand the group. The clusters obtained from this approach thus only require sufficient similarity to a single other member of that cluster for any spectrum in that cluster. Single linkage is fast and does not require a lot of memory, which makes it easier to apply to large amounts of spectra, especially when compared with the more specific but extremely memory-intensive *complete linkage* clustering algorithm, which requires sufficient similarity between a cluster member and all other spectra in that cluster. A marked disadvantage of single-linkage clustering is that the concept of transitivity rarely applies to actual experimental spectra, making this approach prone to *chaining problems*, in which two unrelated spectra are grouped together in the same cluster by a third, "intermediate" spectrum. A typical example would be the case where a peptide spectrum is grouped with a PEG contaminant spectrum through a third spectrum that resulted from the cofragmentation of the peptide precursor and a PEG ion. To resolve the chaining issue, Flikka et al. *(46)* propose a more sophisticated approach they call *two-pass single linkage*. The first pass creates clusters based on single linkage clustering using a high-similarity threshold, and then reduces each cluster to a single, representative spectrum. These representative spectra are then clustered in the second single linkage step, this time using a more lenient similarity threshold. Another suggestion by Tabb and coworkers uses a slight deviation of complete linkage clustering by applying a *paraclique criterion* in which cluster members are allowed to be dissimilar from at most one other cluster member *(45)*.

As already outlined earlier, a cluster of spectra can be represented by a single member of that cluster, which could for instance be the spectrum that gets assigned the highest quality score or by a newly constructed artificial spectrum that is created by combining the information found in all the cluster members. In either case, the representative spectrum is usually the only one that is submitted for identification, and failure to identify this spectrum results in a failure to identify any spectrum in the cluster. However, this concept of transitive identification can also boost the identification efficiency. When a cluster consists of several relatively poor PFF spectra of a peptide and a few very good PFF spectra of that peptide, the poor spectra would not have been identified if submitted by themselves, yet can now be identified through inference since the representative spectrum for that cluster (one of the good spectra) has been identified. It is therefore important to select a good representative spectrum *(44)*, or to construct a good artificial one, and the latter is explained in detail in **ref.** *46*.

## 4. Databases

There are two types of databases that play a vital role in proteomics today, and they occur at opposite sides of a typical proteomics data processing pipeline. The first type of database is the sequence database, which plays an important role in the initial data processing as the search base for the most commonly used identification algorithms.

The second type of database, the proteomics data repository, comes into play at the end of the processing pipeline, when the processed and annotated data are deposited and become publicly available after publication. These repositories also perform a crucial function by allowing data that was generated by one lab to be reused effectively by other labs. The spectral libraries discussed earlier present one example of such data reuse.

Ultimately, a feedback loop can be closed by using the information gathered in the proteomics data repositories to increase the data content and level of annotation in the sequence databases, thus transparently carrying the combined output of many individual proteomics labs back into each individual lab in the form of better sequence data.

### 4.1. Sequence Databases

Sequence databases have played a significant role in the development of the field of proteomics. It was the increased availability of comprehensive archives of nucleotide and protein sequences that enabled the development of automated identification algorithms such as the tag-based PeptideSearch *(19)*, and later the purely database-based SEQUEST *(7)*. The importance of such algorithms on the ability to cope with the large amounts of data that are being obtained by present-day proteomics experiments has already been illustrated earlier. As such, sequence databases play a vital and often underestimated role in the successful completion of any proteomics experiment.

Although a large variety of sequence databases is available to the scientific community today, the vast majority of proteomics experiments rely on one of the following list of protein sequence databases: UniProt knowledgebase (UniProtKB), the NCBI nonredundant (NCBI nr) protein database, and the International protein index (IPI) database. These three databases all have extensive and readily accessible documentation available online, which details the myriad (and continuously expanding) ways in which their information can be accessed and employed, and these topics will therefore not be discussed here. Instead, the databases will be briefly described in terms of their construction and their information content, and the curious reader is invited to explore the resources of interest further via the provided web links.

#### 4.1.1. The UniProt Knowledgebase

The UniProtKB (http://www.uniprot.org) actually consists of a combination of efforts at creating annotated sequence databases. It consists of a combination of the Swiss-Prot, PIR, and TrEMBL databases, which contribute both expert manual annotation and automatic annotation to the sequences it contains. The end result is a database that not only provides protein sequences for use in, for instance, database search engines, but also contains a rich source of metadata and cross-references for each of these sequences. An important benefit to users is that an identification made in the UniProtKB automatically provides biological or biomedical context.

An additional benefit of the UniProtKB database is its vast sequence archive, aptly called the UniProt Archive (UniParc). It is a sequence-based archive that captures all of the unique a sequences that pass through UniProtKB. This is important since a sequence resulting from automated annotation might be removed after refining of the prediction algorithms, or because sequences may be updated in length or composition as new data becomes available. UniParc, which assigns an accession number to each unique sequence, thus allows the originals of removed or altered sequences to remain retrievable through time.

The UniProt Reference Clusters (UniRef) are provided as precalculated, non sequence redundant views of the sequence space at certain resolutions (100%, 90%, and 50% sequence identity). These resources are noticeably smaller than the complete list of sequences and can, therefore, speed up searches. Annotation loss within a cluster is minimized by ranking the level of annotation for each cluster member and then choosing the best annotated member as representative for the cluster annotation.

#### 4.1.2. NCBI Nonredundant Protein Database

The NCBI nr database (http://www.ncbi.nlm.nih.gov/sites/entrez?db = protein) is essentially a *metadatabase*, created by merging many different sequence databases (UniProtKB/Swiss-Prot, PIR, PRF, PDB, GenBank, and RefSeq) into a single collection of sequences. The affix "nonredundant" is added because identical sequences are merged into a single entry, thus creating a set of unique sequences. Differences in length (e.g., from one of the many protein fragments included in the database) or in only a few amino acids (e.g., in the hyper variable *Fab* region at the tips of immunoglobins) will constitute different sequences. This can be considered useful, especially if the objective is to identify start sites of fragments or (minor) sequence variations. On the other hand, the repetition of certain parts of sequences results in a large number of identical (tryptic) peptides. The information conveyed by these peptides is then redundant, and essentially superfluous.

An important by-product of the assembly of sequences from different source databases is that the level of available annotation for an NCBI nr derived sequence can vary dramatically. Ranging from thoroughly documented UniProtKB/Swiss-Prot entries to pure machine predictions from genomic sequences, the heterogeneity in sequence reliability, annotation, and peptide sequence redundancy is the price paid for obtaining a more comprehensive resource. A conservative largest common denominator of NCBI nr information thus consists of a sequence, its accession number and version, and one or more cross-references to source databases.

#### 4.1.3. The International Protein Index Database

The IPI database (http://www.ebi.ac.uk/IPI) is somewhat similar to the previously discussed NCBI nr database in that it is a metadatabase, composed of a combination of other databases (including UniProtKB, Ensembl, RefSeq, H-invDB, and VEGA). The main difference with the NCBI nr lies in the method used to reduce sequence redundancy in the final database. IPI uses a fairly complex algorithm to minimize sequence overlap while maintaining comprehensive proteome coverage for a given organism *(47)*. This algorithm can be incompletely summarized by the following three rules: (1) sequences that are subsets of other sequences (e.g., protein fragments) are grouped together with the longer sequence, (2) sequences that share 95% sequence identity across the length of the shortest of the compared sequences are grouped together, (3) entries that satisfy any of the above-mentioned requirements, and should therefore be grouped together, can be kept as individual entries if sufficient evidence for their separate existence is available (an example of such evidence is given by the UniProtKB/Swiss-Prot splice variant annotations). As highlighted earlier, the actual process is more involved, however. The details of the algorithm are documented on the IPI website and can be consulted there.

Each IPI cluster is represented by a single sequence (typically the longest sequence within that cluster), yet cross-references are available to all source sequences across all databases. Note that because of the nature of IPI, these cross-references need not refer to identical sequences. The annotation of this representative sequence is usually minimal and typically consists of the sequence description line from the most highly annotated source sequence. As with the NCBI nr database, annotation and sequence reliability in IPI can vary dramatically, and the conservative common denominator of information is the sequence, its unique accession number and version, and one or more cross-references to source sequences.

### 4.2. Proteomics Data Repositories

Proteomics data repositories have only come about fairly recently and come in a variety of forms. The most prominent of these repositories are the Open Proteomics Database (OPD, http://

bioinformatics.icmb.utexas.edu/OPD) *(48)*, the Global Proteome Machine Database (GPMDB, http://gpmdb.thegpm.org) *(25)*, the Proteomics Identifications Database (PRIDE, http://www.ebi.ac.uk/pride) *(22, 23)*, and PeptideAtlas (http://www.peptideatlas.org) *(24)*. These repositories currently have different objectives, with GPMDB and PeptideAtlas focused primarily on research topics and pipeline provisioning, while PRIDE and OPD provide data repositories for (annotated) deposition of data associated with publications. A review of the technical and functional properties of these repositories has also been published recently *(49)*.

The different repositories have already agreed in principle to exchange the data they capture, which should ultimately translate in propagating data across all databases regardless of the initial point of submission *(50)*. This collaboration would be closely modeled on existing efforts for nucleotide sequence data bases (http://www.insdc.org) and molecular interactions databases (http://imex.sourceforge.net).

## Acknowledgments

Lennart Martens thanks Prof. Dr. Joël Vandekerckhove and Prof. Dr. Kris Gevaert for sharing their extensive knowledge on proteomics, and Henning Hermjakob for support.

### References

1. Sadygov, R. G., Cociorva, D. and Yates, J. R. (2004) Large-scale database searching using tandem mass spectra: looking up the answer in the back of the book. *Nat Methods* **1**, 195–202.
2. Nesvizhskii, A. I., Vitek, O. and Aebersold, R. (2007) Analysis and validation of proteomic data generated by tandem mass spectrometry. *Nat Methods* **4**, 787–797.
3. Matthiesen, R. (2007) Methods, algorithms and tools in computational proteomics: a practical point of view. *Proteomics* 7, 2815–2832.
4. Perkins, D. N., Pappin, D. J., Creasy, D. M. and Cottrell, J. S. (1999) Probability-based protein identification by searching sequence databases using mass spectrometry data. *Electrophoresis* **20**, 3551–3567.
5. Cottrell, J. S. (1994) Protein identification by peptide mass fingerprinting. *Pept Res* 7, 115–124.
6. Zhang, W. and Chait, B. T. (2000) ProFound: an expert system for protein identification using mass spectrometric peptide mapping information. *Anal Chem* **72**, 2482–2489.
7. Eng, J. K., McCormack, A. L. and Yates, J. R. (1994) An approach to correlate tandem mass-spectral data of peptides with amino-acid-sequences in a protein database. *J Am Soc Mass Spectrom* **5**, 976–989.
8. Craig, R. and Beavis, R. C. (2004) TANDEM: matching proteins with tandem mass spectra. *Bioinformatics* **20**, 1466–1467.
9. Keller, A., Nesvizhskii, A. I., Kolker, E. and Aebersold, R. (2002) Empirical statistical

model to estimate the accuracy of peptide identifications made by MS/MS and database search. *Anal Chem* **74**, 5383–5392.

10. Zhang, Z. (2004) De novo peptide sequencing based on a divide-and-conquer algorithm and peptide tandem spectrum simulation. *Anal Chem* **76**, 6374–6383.
11. Taylor, J. and Johnson, R. (2001) Implementation and uses of automated de novo peptide sequencing by tandem mass spectrometry. *Anal Chem* **73**, 2594–2604.
12. Ma, B., Zhang, K., Hendrie, C., Liang, C., Li, M., Doherty-Kirby, A. et al. (2003) PEAKS: powerful software for peptide de novo sequencing by tandem mass spectrometry. *Rapid Commun Mass Spectrom* **17**, 2337–2342.
13. Grossmann, J., Roos, F., Cieliebak, M., Liptak, Z., Mathis, L., Muller, M. et al. (2005) AUDENS: a tool for automated peptide de novo sequencing. *J Proteome Res* **4**, 1768–1774.
14. Frank, A. and Pevzner, P. (2005) PepNovo: de novo peptide sequencing via probabilistic network modeling. *Anal Chem* **77**, 964–973.
15. Fernandez-de-Cossio, J., Gonzalez, J., Satomi, Y., Shima, T., Okumura, N., Besada, V. et al. (2000) Automated interpretation of low-energy collision-induced dissociation spectra by SeqMS, a software aid for de novo sequencing by tandem mass spectrometry. *Electrophoresis* **21**, 1694–1699.
16. Dancik, V., Addona, T., Clauser, K., Vath, J. and Pevzner, P. (1999) De novo peptide sequencing via tandem mass spectrometry. *J Comput Biol* **6**, 327–342.
17. Pitzer, E., Masselot, A. and Colinge, J. (2007) Assessing peptide de novo sequencing algorithms performance on large and diverse data sets. *Proteomics* **7**, 3051–3054.
18. Pevtsov, S., Fedulova, I., Mirzaei, H., Buck, C. and Zhang, X. (2006) Performance evaluation of existing de novo sequencing algorithms. *J Proteome Res* **5**, 3018–3028.
19. Mann, M. and Wilm, M. (1994) Error-tolerant identification of peptides in sequence databases by peptide sequence tags. *Anal Chem* **66**, 4390–4399.
20. Mørtz, E., O'Connor, P. B., Roepstorff, P., Kelleher, N. L., Wood, T. D. et al. (1996) Sequence tag identification of intact proteins by matching tanden mass spectral data against sequence data bases. *Proc Natl Acad Sci U S A* **93**, 8264–8267.
21. Tabb, D. L., Saraf, A. and Yates, J. R. (2003) GutenTag: high-throughput sequence tagging via an empirically derived fragmentation model. *Anal Chem* **75**, 6415–6421.
22. Martens, L., Hermjakob, H., Jones, P., Adamski, M., Taylor, C., States, D. et al. (2005) PRIDE: the proteomics identifications database. *Proteomics* **5**, 3537–3545.
23. Jones, P., Cote, R. G., Martens, L., Quinn, A. F., Taylor, C. F., Derache, W. et al. (2006) PRIDE: a public repository of protein and peptide identifications for the proteomics community. *Nucleic Acids Res* **34**, D659–D663.
24. Desiere, F., Deutsch, E. W., Nesvizhskii, A. I., Mallick, P., King, N. L., Eng, J. K. et al. (2005) Integration with the human genome of peptide sequences obtained by high-throughput mass spectrometry. *Genome Biol* **6**, R9.
25. Craig, R., Cortens, J. P. and Beavis, R. C. (2004) Open source system for analyzing, validating, and storing protein identification data. *J Proteome Res* **3**, 1234–1242.
26. Lam, H., Deutsch, E. W., Eddes, J. S., Eng, J. K., King, N., Stein, S. E. et al. (2007) Development and validation of a spectral library searching method for peptide identification from MS/MS. *Proteomics* **7**, 655–667.
27. Martens, L., Nesvizhskii, A. I., Hermjakob, H., Adamski, M., Omenn, G. S., Vandekerckhove, J. et al. (2005) Do we want our data raw? Including binary mass spectrometry data in public proteomics data repositories. *Proteomics* **5**, 3501–3505.
28. Gentzel, M., Köcher, T., Ponnusamy, S. and Wilm, M. (2003) Preprocessing of tandem mass spectrometric data to support automatic protein identification. *Proteomics* **3**, 1597–1610.
29. Zhang, X., Asara, J. M., Adamec, J., Ouzzani, M. and Elmagarmid, A. K. (2005) Data preprocessing in liquid chromatography-mass spectrometry-based proteomics. *Bioinformatics* **21**, 4054–4059.
30. Gevaert, K., Goethals, M., Martens, L., Van Damme, J., Staes, A., Thomas, G. R. et al. (2003) Exploring proteomes and analyzing protein processing by mass spectrometric identification of sorted N-terminal peptides. *Nat Biotechnol* **21**, 566–569.
31. Yi, J., Kim, C. and Gelfand, C. A. (2007) Inhibition of intrinsic proteolytic activities moderates preanalytical variability and instability of human plasma. *J Proteome Res* **6**, 1768–1781.
32. Creasy, D. M. and Cottrell, J. S. (2002) Error tolerant searching of uninterpreted tandem mass spectrometry data. *Proteomics* **2**, 1426–1434.

33. Falkner, J. and Andrews, P. (2005) Fast tandem mass spectra-based protein identification regardless of the number of spectra or potential modifications examined. *Bioinformatics* **21**, 2177–2184.
34. Salmi, J., Moulder, R., Filén, J., Nevalainen, O. S., Nyman, T. A., Lahesmaa, R. et al. (2006) Quality classification of tandem mass spectrometry data. *Bioinformatics* **22**, 400–406.
35. Bern, M., Goldberg, D., McDonald, W. H. and Yates, J.R.3rd (2004) Automatic quality assessment of peptide tandem mass spectra. *Bioinformatics* **20** Suppl 1, i49–i54.
36. Hoopmann, M. R., Finney, G. L. and MacCoss, M. J. (2007) High-speed data reduction, feature detection, and MS/MS spectrum quality assessment of shotgun proteomics data sets using high-resolution mass spectrometry. *Anal Chem* **79**, 5620–5632.
37. Wong, J. W. H., Sullivan, M. J., Cartwright, H. M. and Cagney, G. (2007) msmsEval: tandem mass spectral quality assignment for high-throughput proteomics. *BMC Bioinformatics* **8**, 51.
38. Nesvizhskii, A. I., Roos, F. F., Grossmann, J., Vogelzang, M., Eddes, J. S., Gruissem, W. et al. (2006) Dynamic spectrum quality assessment and iterative computational analysis of shotgun proteomic data: toward more efficient identification of post-translational modifications, sequence polymorphisms, and novel peptides. *Mol Cell Proteomics* **5**, 652–670.
39. Flikka, K., Martens, L., Vandekerckhove, J., Gevaert, K. and Eidhammer, I. (2006) Improving the reliability and throughput of mass spectrometry-based proteomics by spectrum quality filtering. *Proteomics* **6**, 2086–2094.
40. Xu, M., Geer, L. Y., Bryant, S. H., Roth, J. S., Kowalak, J. A., Maynard, D. M. et al. (2005) Assessing data quality of peptide mass spectra obtained by quadrupole ion trap mass spectrometry. *J Proteome Res* **4**, 300–305.
41. Purvine, S., Kolker, N. and Kolker, E. (2004) Spectral quality assessment for high-throughput tandem mass spectrometry proteomics. *OMICS* **8**, 255–265.
42. Liu, H., Sadygov, R. G. and Yates, J.R.3rd. (2004) A model for random sampling and estimation of relative protein abundance in shotgun proteomics. *Anal Chem* **76**, 4193–4201.
43. Ishihama, Y., Oda, Y., Tabata, T., Sato, T., Nagasu, T., Rappsilber, J. et al. (2005) Exponentially modified protein abundance index (emPAI) for estimation of absolute protein amount in proteomics by the number of sequenced peptides per protein. *Mol Cell Proteomics* **4**, 1265–1272.
44. Tabb, D. L., MacCoss, M. J., Wu, C. C., Anderson, S. D. and Yates, J. R. (2003) Similarity among tandem mass spectra from proteomic experiments: detection, significance, and utility. *Anal Chem* **75**, 2470–2477.
45. Tabb, D. L., Thompson, M. R., Khalsa-Moyers, G., VerBerkmoes, N. C. and McDonald, W. H. (2005) MS2Grouper: group assessment and synthetic replacement of duplicate proteomic tandem mass spectra. *J Am Soc Mass Spectrom* **16**, 1250–1261.
46. Flikka, K., Meukens, J., Helsens, K., Vandekerckhove, J., Eidhammer, I., Gevaert, K. et al. (2007) Implementation and application of a versatile clustering tool for tandem mass spectrometry data. *Proteomics* **7**, 3245–3258.
47. Kersey, P. J., Duarte, J., Williams, A., Karavidopoulou, Y., Birney, E. and Apweiler, R. (2004) The International Protein Index: an integrated database for proteomics experiments. *Proteomics* **4**, 1985–1988.
48. Prince, J. T., Carlson, M. W., Wang, R., Lu, P. and Marcotte, E. M. (2004) The need for a public proteomics repository. *Nat Biotechnol* **22**, 471–472.
49. Mead, J. A., Shadforth, I. P. and Bessant, C. (2007) Public proteomic MS repositories and pipelines: available tools and biological applications. *Proteomics* **7**, 2769–2786.
50. Hermjakob, H. and Apweiler, R. (2006) The Proteomics Identifications Database (PRIDE) and the ProteomExchange Consortium: making proteomics data accessible. *Expert Rev Proteomics* **3**, 1–3.

# Chapter 15

# Shotgun Protein Identification and Quantification by Mass Spectrometry

**Bingwen Lu, Tao Xu, Sung Kyu Park, and John R. Yates III**

## Summary

Shotgun proteomics is based on identification and quantification of peptides from digested proteins using tandem mass spectrometry. In this chapter, we discuss computational methods to analyze tandem mass spectra of peptides, including database searching, de novo peptide sequencing, hybrid approaches, library searching, and unrestricted modification search. A special focus is given to database searching programs since they are most widely used. The process of inferring proteins from identified peptides is then discussed. We also provide description of key steps in the quantitative analysis of mass spectrometry proteomics data.

**Key words:** Shotgun proteomics, LC-MS, Database searching, Search algorithm, Protein quantification

## 1. Introduction

A proteome is the collection of all proteins of an organism. Proteomics is the study of the diverse properties and functions of proteins as a function of time and conditions. Proteomes are dynamic responding to environmental and cellular challenges. Many different technologies have been developed, which are used for proteomic studies. These technologies include two-dimensional polyacrylamide gel electrophoresis (2D-PAGE) *(1)*, protein microarrays *(2)*, mass spectrometric (MS) characterization and identification of intact proteins (also known as "top-down" proteomics) *(3, 4)*, MS identification of digested proteins ("bottom-up" proteomics), and methods based on the use of genetics.

Jörg Reinders and Albert Sickmann (eds.), *Proteomics,* Methods in Molecular Biology, Vol. 564
DOI: 10.1007/978-1-60761-157-8_15, © Humana Press, a part of Springer Science+Business Media, LLC 2009

Among these tools, bottom-up MS currently plays a key role in qualitative and quantitative analysis in proteomics.

An overview of shotgun proteomic analysis is schematically illustrated in **Fig. 1** (or *see* **ref. 5**). A protein mixture is digested using proteolytic enzymes such as trypsin. The resulting peptides are then separated by one or multi-dimensional liquid chromatography (LC). The separated peptides are then ionized and introduced into the instrument for MS analysis. Traditional MS/MS data acquisition consists of two stages. The first stage is the full scanning of all peptide ions introduced into the instrument at a given time to obtain an MS spectrum. During the second stage, peptide ions, either selected by intensity from the full scan analysis (data-dependent acquisition), or selected by a predefined mass range (data-independent acquisition) *(6)*, are subjected to fragmentation by collision-induced dissociation (CID) before the MS/MS spectrum is acquired. The MS/MS spectrum acquired at the second stage is a record of all *m/z* values with associated intensities of the resulting fragment ions under a mass range such as 200–2,000 (**Fig. 2**). After the MS and MS/MS data are acquired, efforts are directed to the identification and quantification of peptides by computational analysis.

Computational analysis usually starts with interpretation of the acquired tandem mass spectra. The sequences of the peptides that gave rise to the tandem mass spectra are inferred by using various software tools, such as database searching tools (SEQUEST *[7]*, Mascot *[8]*), de novo peptide sequencing tools (PepNovo *[9]*, PEAKS *[10]*), sequence tag database search tools (Guten Tag *[11]*, Inspect *[12]*), spectrum library search tools (LIBQUEST

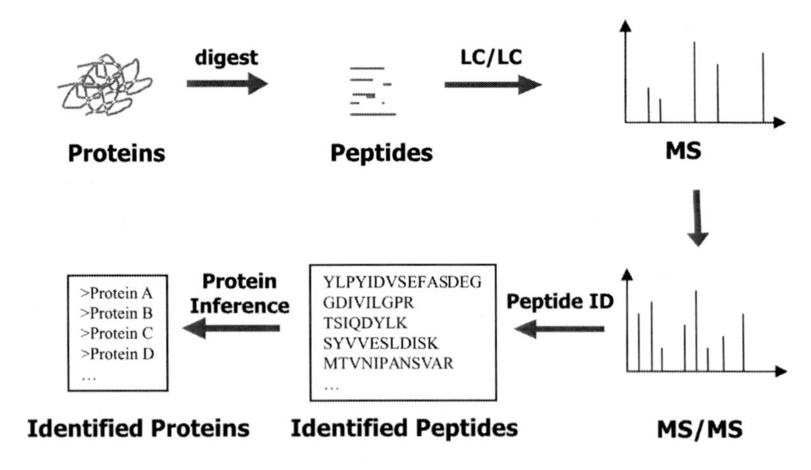


Fig. 1. Experimental procedure for MudPIT and data analysis. A protein mixture is digested with an endoprotease such as trypsin to generate digested peptides. The digested peptides are separated by multiple dimensional LC before being sprayed into a mass spectrometer. MS and MS/MS spectra are acquired by the mass spectrometer. Computer software tools are employed to infer peptide sequences and finally a list of identified proteins is produced.

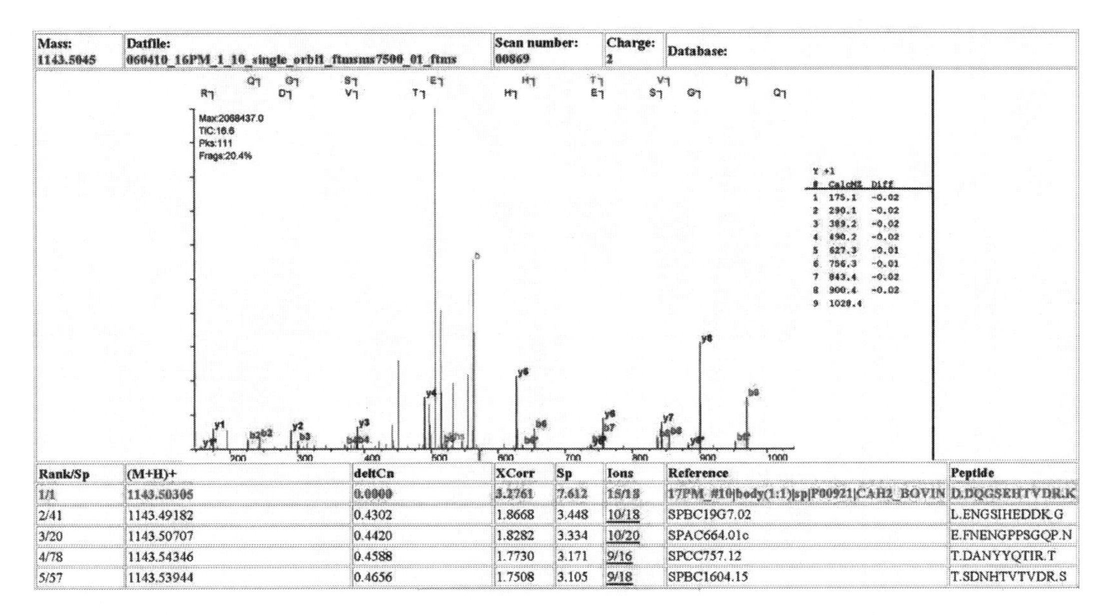


| Rank/Sp | (M+H)+ | deltCn | XCorr | Sp | Ions | Reference | Peptide |
|---|---|---|---|---|---|---|---|
| 1/1 | 1143.50306 | 0.0000 | 3.2761 | 7.612 | 15/18 | 17PM_#10\|body(1:1)\|sp\|P00921\|CAH2_BOVIN | D.DQGSEHTVDR.K |
| 2/41 | 1143.49182 | 0.4302 | 1.8668 | 3.448 | 10/18 | SPBC19G7.02 | L.ENGSIHEDDK.G |
| 3/20 | 1143.50707 | 0.4420 | 1.8282 | 3.334 | 10/20 | SPAC664.01c | E.FNENGPPSGQP.N |
| 4/78 | 1143.54346 | 0.4588 | 1.7730 | 3.171 | 9/16 | SPCC757.12 | T.DANYYQTIR.T |
| 5/57 | 1143.53944 | 0.4656 | 1.7508 | 3.105 | 9/18 | SPBC1604.15 | T.SDNHTVTVDR.S |

Fig. 2. An example of tandem mass spectrum.

*[13]*, BiblioSpec *[14]*), or unrestricted modification search (MSAlign *[15]*, TwinPeaks *[16]*). Among these tools, the most commonly used approach for peptide identification nowadays is still database searching. After peptides are identified, protein inference is carried out to obtain protein identifications. Peptide and protein quantification follows by the examination of the mass spectral information.

There are currently many approaches to identify peptides from MS/MS data. One needs to bear in mind that each of these peptide identification approaches has its own advantages and limitations. We discuss these approaches in the following sections, beginning with the discussion of database searching methods.

## 2. Peptide Identification Using MS/MS Database Searching

### 2.1. Basic Principles

The goal of a tandem mass spectral database search is to identify the best sequence match to the spectrum. All MS/MS database search tools operate in a similar way. They return the best matching peptide found in the database for each input spectrum, unless there is no candidate peptide in the database that satisfies the search parameters specified by the user. SEQUEST *(7)* and Mascot *(8)* are the two most widely used database search engines.

Generally, there are three major steps for a database search program to find the peptide in the database that is a best match to a given tandem mass spectrum: (1) candidate peptide selection; (2) preliminary scoring; and (3) final scoring and peptide ranking.

The first step is to select candidate peptides that are within some mass tolerance of the peptide represented in the tandem mass spectrum, enzyme specificity, and other constraints that are specified in the search parameter file. Some programs begin by placing an enzyme specificity constraint, while others begin with the peptide mass constraint, or short stretches of sequence interpreted from the fragmentation pattern. To speed up the candidate peptide selection process, some programs pre-index the sequence database based on peptide mass and the enzyme specificity. Preprocessing a database makes a search go much faster. For example, tryptic peptides can be indexed and mass lists can be pre-made. The downside is that a separate indexed database needs to be created for each enzyme or each time a database is updated.

The second step is a preliminary scoring step that is used to create a short list of candidate peptides. This preliminary scoring step is important for the speed of the identification process, since the final scoring algorithms are usually slow, making it impractical to score every candidate sequence. Preliminary scoring is typically performed based on the number of shared peaks between the experimental spectrum and a theoretical spectrum.

The third step is final scoring and ranking sequence matches. A final score is computed for each of the peptides in the candidate list selected by the preliminary scoring routine. The final scoring uses one of the following two methods to measure closeness of fit between spectra and peptide sequences. The first method uses a shared peak model to generate a quantitative measure of the fit while the second method uses fragment ion frequency to generate the probability that the sequence and spectrum are the best fit. The final scoring is usually more sophisticated and sensitive than the preliminary scoring method. All the candidate peptides are ranked by the final score and the sequence with the top final score is considered as the best match to the experimental spectrum.

A list of common MS/MS database search engines is given in **Table 1**. Two additional reviews *(17, 18)* are also helpful if the reader would like to know more about MS/MS database search algorithms.

### 2.2. Preliminary Scoring

The most commonly used preliminary scoring method is the number of shared peaks between the experimental spectrum and the theoretical spectrum. This overly simplified approach is used by most probability-based database search programs such as PepProbe *(19)*, OMSSA *(20)*, and X!Tandem *(21)*.

**Table 1**
**MS/MS database search programs**

| Program | Category | Website and/or reference |
|---|---|---|
| SEQUEST | C | http://www.thermo.com/, *(7)* |
| Mascot | C free online | http://www.matrixscience.com/, *(8)* |
| MasLynx | C | http://www.waters.com/, *(72)* |
| MS-Tag/MS-Seq | P | http://prospector.ucsf.edu/, *(72)* |
| Sonar | | *(23)* |
| OMSSA | P | http://pubchem.ncbi.nlm.nih.gov/omssa/, *(20)* |
| ProLuCID | N | http://fields.scripps.edu/prolucid/ *(22)* |
| PepProbe | P | http://bart.scripps.edu/public/search/pep_probe/search.jsp, *(19)* |
| X!Tandem | P | http://www.thegpm.org/TANDEM/, *(21)* |
| SpectrumMill | C | http://www.chem.agilent.com/ |

C, Commercial; P, Publicly available; N, Need to contact authors for availability.

SEQUEST employs a different preliminary scoring method to derive preliminary score ($S_p$). $S_p$ sums the peak intensity of fragment ions matching the predicted sequence ions and accounts for the continuity of an ion series and the length of a peptide. The original $S_p$ score is

$$S_p = \left(\sum_K I_K\right) m(1+\beta)(1+\rho)/L$$

where the first term in the product is the sum of ion abundances of all matched peaks, $m$ is the number of matches, $\beta$ is a "reward" for each consecutive match of an ion series (for example, 0.075), $\rho$ is a "reward" for the presence of an ammonium ion (for example 0.15) and $L$ is the number of all theoretical fragment ions for an amino acid sequence.

ProLuCID *(22)*, a database search algorithm recently developed in the Yates laboratory, uses an approximated binomial probability score for preliminary scoring. This binomial probability score is computed using the following formulae:

$$P(x \geq m) = \sum_{k=m}^{n} P(x=k) \tag{1}$$

where $P(x=k)$

$$= \frac{n!}{k!(n-k)!} p^k (1-p)^{(n-k)}$$

where $n$ is the number of theoretical peaks of the candidate peptide tested, which is determined by the peptide length together with the minimum and maximum $m/z$ in the tandem mass spectrum; $m$ is the number of theoretical peaks that match to a peak in the experimental spectrum and is guaranteed to be less than or equal to $n$; $p$ is the probability that any fragment ion matches a peak in the spectrum, which is determined by the mass tolerance for a fragment ion match and the density and distribution of peaks in the experimental spectrum. The binomial probability score $P(x >= m)$ is the probability of getting $m$ or more matches when $n$ theoretical peaks are tested. By design, the binomial probability score computed by ProLuCID is database-independent and is solely dependent on the characteristics of the spectrum and the peptide sequence.

### 2.3. Final Scoring

There are two major types of final scores. One is a similarity score that measures the closeness of fit between the theoretical spectrum and the experimental spectrum. A cross-correlation score (used in SEQUEST) or dot-product score (used in Sonar *[23]*) have been used for this purpose. Another type of final score is a probability or statistical score that measures how likely a given hit is a random hit.

The cross-correlation score is used by SEQUEST to measure the similarity between the theoretical spectrum and experimental spectrum:

$$\mathrm{Corr}(E,T) = \sum X_i Y_i + \tau$$

The correlation is processed and averaged to remove the periodic noise in the interval (–75 to 75).

The hypergeometric probability score and Poisson probability score are used by PepProbe *(19)* for the final scoring. The hypergeometric probability score can be expressed as

$$P_{K,N}(K_1, N_1) = \frac{C_K^{K_1} X C_{N-K}^{N_1-K_1}}{C_N^{N_1}}$$

The Poisson probability score is

$$P(x,u) = \frac{u^x}{x!} e^{-u}$$

where

$$u = \frac{C_K^{K^1} X C_{N-K}^{N_1-K_1}}{C_N^{N_1}}$$

Two recently developed open source database search programs, X!Tandem *(21)* and OMSSA *(20)*, also compute an E-value as its final score. An E-value for a hit is a score that is the expected number of random hits from a database to a given spectrum such that the random hits have an equal or better score than the hit.

$$E = n \sum_{s \geq s_m} P(s)$$

Alternatively, ProLuCID computes a Z score for each final candidate peptide. For each spectrum, there should only be one correct answer and all the other candidate peptides can be considered as random matches. We have found the distribution of XCorrs for the top 500 peptide hits for each spectrum to be very close to a normal distribution with the true hit being an obvious outlier and statistically significantly different from the other final candidates. There are many ways to detect outliers from normal distributions and the Z score of the Grubbs' test *(24)* is used in ProLuCID. The Z score is calculated as the difference between the outlier and the mean divided by the standard deviation SD (**Eq.2**). A large Z score means that the XCorr of the top hit is significantly different from the other hits so that the peptide is more likely to be a true hit.

$$Z = \frac{X - \mu}{SD} \quad \text{where } SD = \sqrt{\frac{\sum_{i=1}^{n} (x_i - \mu)^2}{n - 1}} \qquad (2)$$

X is the XCorr of the top hit, μ is the mean XCorr of all the final candidate peptides, and n is the number of final candidate peptides.

### 2.4. Database Search Parameters

The following are some parameters that are important for database search programs. The users of database search programs are usually allowed to configure these parameters.

1. Protein database – A protein database has to be specified by the user for the database search program to use. The search program will consider each peptide in the protein database that fulfills the user-specified search conditions, such as peptide mass tolerance, enzyme specificity, and consider modifications. The selection of protein databases is important and will be discussed in more details in the next subsection.

2. Types of sequence ions – Precursor ions can be fragmented using different dissociation mechanisms, such as CID, ECD (Electron Capture Dissociation) *(25)*, or ETD (Electron Transfer Dissociation) *(26)*. Different dissociation methods will produce different types of fragment ions. For example, the major fragment ion series in CID spectra are B- and Y-ions, while fragment ions in ETD spectra are mostly C- and Z-ions. The database search programs need to know the spectrum type to generate the theoretical spectrum and compare the theoretical spectrum with the experimental spectrum.
3. Peptide and fragment mass type – The user also needs to specify the method of calculating the peptide (precursor) and the fragment ion mass. This can be done using either monoisotopic or average masses.
4. Peptide and fragment ion mass tolerances – to select candidate peptides, the search program compares calculated peptide molecular weight with the measured precursor molecular weight. Generally, the narrower you can set the parent mass tolerance, the faster the search will go because fewer peptides will need to be correlated in the next step. Similarly, a fragment ion mass tolerance is used when the search program compares the peaks in the theoretical spectrum with the peaks in the experiment spectrum. One needs to be certain that the accuracy of the mass spectrometer is not exceeded when setting the parent and fragment ion mass tolerances in the search (e.g. tolerance set too narrow).
5. Enzyme specificity - Most biological samples are digested with a protease before MS and typically trypsin is used. Accordingly, a database search program can be configured to search for peptides that fulfill the enzyme specificity used to create the peptides, and any peptide in the database that does not fulfill both the molecular weight and enzyme specificity constraints will not be considered.
6. Static and differential modifications.
   a) A static or fixed modification is a mass shift that occurs on every specified residue in all proteins, e.g., C + 57 for carboamidomethylation on cysteine.
   b) Differential or variable modification is a mass shift of a post-translational modification (PTM) that can be present or not on certain amino acid residues. The user usually needs to specify the mass shift, residues where the mass shift may occur, together with the maximum number of differential modifications to consider per candidate peptide. The number of differential modifications to be considered

will usually dramatically affect the database search speed. The more the modification considered, the longer the search will take as the process scales $2^n$, where $n$ is the number of potential modification sites in the peptide.

### 2.5. Selection of Protein Sequence Database

Identification of proteins using MS relies on the existence of reliable protein sequence databases. Ideally, an organism-specific database should be used if the protein database is relatively complete. This is probably true for most of the model organisms. The most commonly used public databases are International Protein Index (IPI, http://www.ebi.ac.uk/IPI/) database maintained by European Bioinformatics Institute (EBI), the National Center for Biotechnology Informations's (NCBI) Entrez protein database (http://www.ncbi.nlm.nih.gov/sites/entrez?db=Protein), the NCBI Reference Sequence (RefSeq, http://www.ncbi.nlm.nih.gov/RefSeq/) database, and Universal Protein Resource (UniProt, http://www.pir.uniprot.org/) protein database, which combines Swiss-Prot, TrEMBL, and PIR. Currently, EBI_IPI database provides protein FASTA databases of seven model organisms, including human, mouse, rat, zebrafish, Arabidopsis, chicken, and cow. The flybase (http://flybase.bio.indiana.edu/), wormbase (http://www.wormbase.org/), and Saccharomyces Genome Database (SGD, http://www.yeastgenome.org/) are other commonly used public databases.

If the protein sequence of the target organism is incomplete but the genome is sequenced, a six-frame translation program or a gene-finding program can be used to generate the predicted proteins. For most organisms, however, the vast majority of sequence information for any given organism remains as cDNA sequence in the form of either single-pass sequenced expressed sequence tags (ESTs) or (more rare) complete full-length cDNA sequences. The use of EST databases should be cautioned since single-pass EST sequence data are usually of poorer quality.

If there is no genomic or transcript data available for the organism, the last resort would be using the protein non-redundant (NR) database from NCBI or the UniProt protein database. Alternatively, one can use de novo peptide sequencing and MS-BLAST to identify the proteins, as discussed below.

### 2.6. Evaluation of the Search Results

Hundreds of thousands of tandem mass spectra can be routinely generated in a shotgun proteomics experiment. A database search program will find one best matching peptide for each of these spectra even though many of the spectra may be from sequencing attempts on chemical noise, non-peptides, or are of poor quality. It is known that after the database searching is completed but before any filtering is performed, the majority of the identifications are false-positives that result from random hits to the database or matches of MS/MS of background noise to sequences and only a small proportion of peptide identification results are true hits. Post database search filtering programs, such as DTASelect *(27)* and PeptideProphet *(28)*, are essential for the optimal separation of true peptide/protein matches from random matches. For a peptide to be successfully identified by a database search algorithm, it has to pass the following three tests: (1) it must be ranked high enough in the preliminary scoring to be selected for final scoring, (2) it must be assigned to the top rank during the final scoring, and (3) its score or scores have to be high enough to pass the post-search filtering criteria. The major challenge to improving the overall performance of a database search algorithm is to increase the sensitivity of searches while maintaining adequate discrimination between correct answers and false-positives.

The DTASelect algorithm is a powerful tool for filtering database search results. Traditionally, filtering of SEQUEST results by DTASelect used threshold cutoffs for XCorr and DeltaCN where DeltaCN is the difference between the top hit XCorr and the second best hit XCorr divided by the XCorr of the top hit. In the latest version of DTASelect (DTASelect2 *[29]*), these two measurements are combined using a quadratic discriminant function to compute a confidence score to achieve a user-specified false discovery rate.

### *2.7. Sources of Failure*

Here, we discuss some sources of failure to correctly assign peptide sequences to MS/MS spectra that truly represent peptides and not simply chemical noise or other background contaminants. The readers are also referred to the review by Nesvizhskii *(18)* for a similar discussion.

1. Poor MS/MS quality – Low abundant peptide precursors usually result in poor quality MS/MS spectra, and some peptides may not fragment well because of their chemical properties.
2. Search constraints – In most of the shotgun proteomics experiments, it is true that the majority of peptides are fully cleaved by trypsin digestion. However, half- and non-tryptic peptides can be detected in biological samples, in part because of nonspecific cleavage by the enzyme, the presence of in vivo protease activities, or in-source fragmentation of peptides. Thus, fully tryptic searches will miss half- or non-tryptic peptides, especially the peptides close to the N-terminus/amino terminus of the protein, since the protein sequences in the database usually may or may not contain the initial Met, and often protein sequences in the database may also contain additional sequence like that of a signaling peptide.
3. Unanticipated modifications – If the spectrum is generated from a peptide with some modifications that are not specified in the search parameter, then the mass of the peptide will not match the precursor mass, thus it will not be considered by the

database search algorithm and of course will not be identified. These modifications include chemical modifications (such as sodium and potassium adducts, in vitro carbamylation, loss of water) and biological PTMs (phosphorylation, in vivo carbamylation, methylation, acetylation, etc.).

4. Spectrum charge state determination errors – For high mass accuracy data such as those produced by LTQ-Orbitrap instruments, the charge state of a precursor ion usually can be correctly assigned. However, for low-resolution data, the charge state of multiply charged precursor ions may not be unambiguously assigned. A common practice is searching the spectrum twice, once assuming +2 charge and then +3 charge. However, this approach will miss the peptides with precursor charge states higher than +3.
5. Peptide sequence not in the database. Most protein databases are created by translations of genes. It is possible that some of the proteins may not be present in the database or the stop and start sites were incorrectly interpreted or alternate splicing incorrectly predicted. If a protein sequence is not in the database, due to a single nucleotide polymorphism (SNP), a mutation, sequencing, or prediction errors, then there is no way that the database search program can identify its peptide sequences.
6. Deficiencies of the scoring scheme or post database search filtering. Even when the correct peptide sequence is in the data, the database search program may fail to identify it.

## 3. Other Methods for Peptide Identification from MS/MS

If database searching fails to identify a sequence for a tandem mass spectrum, there are currently several other approaches to identify the peptides that generated the MS/MS, such as de novo sequencing, combining sequence tags and database searching, library searching, and unrestricted blind search.

### 3.1. De Novo Peptide Sequencing

De novo peptide sequencing is used to infer peptide sequences from tandem mass spectra data without using a sequence database to aid in the interpretation. Usually, searching against a sequence database is the first choice for peptide identification. However, de novo peptide sequencing comes into play in various situations. First, the protein of interest might not be present in the sequence database. For example, the sequence database might be incomplete, which is still the situation for many animals and plants, or the protein of interest might be a novel protein whose sequence information is not available from the sequence database. Second, there could be prediction errors in gene-finding programs. Thus, it might not be possible to find the true protein from the predicted protein database. Third, genes might undergo alternative splicing, which would result in novel proteins. The occurrence of SNPs in coding regions may also lead to different protein variants. In a fourth case, de novo sequencing can be helpful for studying amino acid mutations and protein modifications. Finally, when a database search generates ambiguous results, de novo sequencing can be used as a potential validation tool.

De novo peptide sequencing has its limitations. A basic limitation is the necessity of reasonably complete backbone cleavage between each pair of adjacent amino acids. In other words, one needs a complete ladder of ion series (b, y, c, or z) to infer the complete peptide sequence accurately and CID does not usually produce a complete ion series. Furthermore, internal peptide fragmentation is not uncommon. The more recent fragmentation technology ECD *(25)* and ETD *(26)* tends to generate more complete ion series for tandem mass spectra and could potentially be more suitable for de novo peptide sequencing. Researchers also tried to combine complementary fragmentation techniques, CID and ECD, for de novo sequencing *(30)*. Mass spectrometer mass accuracy also plays an important role in de novo peptide sequencing *(9, 31)*. Since high mass accuracy mass spectrometers are usually more expensive, this also places another limitation on de novo peptide sequencing.

Because de novo peptide sequencing from tandem mass spectra is a difficult problem and confident identifications are difficult to achieve, follow-up verification of peptide sequences from de novo analysis takes additional effort. For this reason, de novo peptide sequencing is rarely applied to complex mixtures. Only recently has de novo peptide sequencing been shown to conceptually work on a complex mixture by employing both high mass accuracy CID and high mass accuracy ECD techniques *(30)*.

Various de novo sequencing programs have been developed over the years (**Table 2**). Some of these programs are publicly available, such as PepNovo and Lutefisk. Some are only available as commercial software, such as PEAKS. Most of these programs are geared towards tandem mass spectra generated by CID fragmentation technology. Some of the earlier algorithms have previously been reviewed *(32)*.

### 3.2. Hybrid Approach

The hybrid approach combines de novo peptide sequencing and database searching for peptide identification. This approach starts by inferring short peptide sequences from MS/MS by de novo sequencing, followed by a database search constrained by the sequence tag information. The short peptide sequences are usually referred to as sequence tags. There are two methods that

**Table 2**
**De novo peptide sequencing programs**

| Program | Category | Website and/or reference |
|---|---|---|
| Lutefisk | P | http://www.hairyfatguy.com/lutefisk/, *(38)* |
| MSNovo | P | http://msms.cmb.usc.edu/supplementary/msnovo/, *(73)* |
| NovoHMM | P | http://people.inf.ethz.ch/befische/proteomics, *(74)* |
| PepNovo | P | http://peptide.ucsd.edu/pepnovo.html, *(75)* |
| PEAKS | C | http://www.bioinfor.com/products/peaks/, *(10)* |
| MassSeq | C | Micromass/Waters |
| DeNovoX | C | http://www.thermo.com |
| EigenMS | N | *(76)* |
| PILOT | N | *(77)* |

C, Commercial; P, Publicly available; N, Need to contact authors for availability.

these sequence tags can be used for database searching. De novo interpretation of short stretches of sequence is often more accurate and reliable than trying to sequence full length peptides from a tandem mass spectrum.

For the first method, the sequence tags are used to eliminate candidate peptides for MS/MS database searching *(11, 12, 33)*. The database searching process can be potentially sped up since the sequence tags obtained from de novo inference can eliminate unwanted candidate peptides, thus saving time scoring the experimental spectrum against these candidate peptides. Another advantage of the sequence tag approach is that it allows error-tolerant database searches as explored by Mann and Wilm *(33)*. Tanner and colleagues showed that the filtering using sequence tags obtained from de novo sequencing can be very efficient so that it allows fast identification of post-translationally modified peptides *(12)*. Hybrid approaches using sequence tags and MS/MS database searching were implemented in the software tools GutenTag *(11)*, Inspect *(12)*, SPIDER *(34)*, and OpenSea *(35)*.

For the second method, the sequence tags obtained are not used for MS/MS database searching. Instead, the sequence tags are used for sequence-based database searching such as BLAST *(36)* or FASTA *(37)*. In practice, even if the sequence of the studied protein is not in a protein database, chances are that the homologues of this protein are in the database. In such a situation, a sequence-based database searching program that handles homology mutations will be able to use the sequence tags to identify the protein homologues. Two common homology search programs, FASTA and BLAST, are usually employed for sequence-based database searching using sequence tags inferred from tandem mass spectra. For example, FASTA has been employed in CIDentify *(38)* and FASTS *(39)*, while BLAST has been implemented in MS-BLAST *(40)*. Similar to de novo sequencing, the follow-up verification of identified peptides from sequence-based homology searches will usually take additional effort.

A list of software tools for MS/MS data analysis using the hybrid approach are given in **Table 3**.

### 3.3. Library Search

Library searching matches experimental spectra to a library of peptide spectra collected from previous peptide identification studies. The library contains peptide MS/MS that have been matched to peptides and filtered using stringent filtering criteria. Spectral library searching is a natural way to take advantage of previous peptide identifications. Actually, library searching is a commonly practiced method for identifying mass spectra of small organic molecules, by comparisons of the newly collected experi-

**Table 3**
**Programs employing hybrid approaches**

| Program | Category | Website and/or reference |
|---|---|---|
| GutenTag | P | http://fields.scripps.edu/GutenTag/, *(11)* |
| Inspect | P | http://peptide.ucsd.edu/inspect.html, *(12)* |
| OpenSea | N | *(35)* |
| SPIDER | P | http://bif.csd.uwo.ca/spider/, *(34)* |
| ByOnic | P | http://bio.parc.xerox.com/home, *(78)* |
| CIDentify | P | http://www.hairyfatguy.com/lutefisk/, *(38)* |
| MS-BLAST | P | http://genetics.bwh.harvard.edu/msblast/, *(40)* |

P, Publicly available; N, Need to contact authors for availability.

mental spectra with spectra of known molecules *(41)*. About 10 years ago, Yates et al. *(13)* proposed and explored the possibility of carrying out peptide spectra library searching using tandem mass spectra of peptides. Recently, with more and more MS/MS data collected and identified, this approach has drawn attention from other research groups *(14, 42–44)*.

For traditional database search, experimental spectra are matched with theoretical mass spectra generated from a sequence database. The theoretical mass spectra are generated anew each time an experimental MS/MS spectrum is searched, thus wasting a large amount of computational time. Furthermore, even though efforts have been put to model peptide fragmentation patterns in a mass spectrometer *(45)*, theoretical mass spectra are usually generated with the peptide fragmentation patterns not being fully modeled. Library searching provides a solution to address the above issues. By searching against the peptide spectra library, which is usually smaller, the computational time is saved. Furthermore, there is no need to model the fragmentation patterns of respective identified peptides since they are intrinsic to the experimental spectra. Another advantage of library searching is the inclusion of spectra corresponding to modified peptides in the library. Identification of modified peptides by database search is known to be very time-consuming. Furthermore, modified peptides usually go through a more sophisticated fragmentation model, for example, precursor neutral losses and product ion neutral losses are frequently observed from CID phosphorylated peptides *(46)*. However, once the post-translationally modified peptides are identified, the spectra can be included in the library. Searching against the spectra of modified peptides and spectra of unmodified peptides should take the same amount of time, and the fragmentation pattern of the modified peptides is again modeled in the experimental spectra.

Library searching has its limitations. Since the search is against a library of identified peptide spectra, this approach is not suitable for discovery of previously unobserved peptides. Furthermore, constructing a spectral library for library searching is not a trivial process. Extreme precautions should be taken to not include incorrectly identified peptide spectra. Additional efforts should also be spent to re-annotate the peptide spectra in the library once errors are found.

Nevertheless, once the peptide spectra libraries are established, library searching could be very useful for some studies such as targeted proteomics, where the primary goal is to repeatedly investigate a predefined set of proteins and there is no need to discover novel peptides. A list of library search tools is given in **Table 4**.

**Table 4**
**Library searching programs**

| Program | Category | Website and/or reference |
|---|---|---|
| BiblioSpec | P | http://proteome.gs.washington.edu/bibliospec/documentation/, *(14)* |
| LIBQUEST | N | *(13)* |
| SpectraST | P | http://www.peptideatlas.org/spectrast/, *(43)* |
| X! Hunter | P | http://h201.thegpm.org/tandem/thegpm_hunter.html, *(42)* |

P, Publicly available; N, Need to contact authors for availability.

### 3.4. Unrestricted Modification Search

Unrestricted modification search is the identification of post-translationally modified peptides without knowing the modification types. Protein PTMs greatly increase the complexity of the proteome. PTMs often regulate activity and function of proteins. The study of protein PTMs is very important and is one of the current focuses of proteomic studies. There are two common ways of identifying modified peptides from MS/MS data: restricted modification search and unrestricted modification search.

The first approach to identify PTMs is by restricted modification search based on a selected set of modifications by multiple considerations of potentially modified database peptides *(47)*. This approach has been widely applied in the identification of peptides with known modifications and are implemented in common database search engines such as SEQUEST, Mascot, X!Tandem, and OMSSA. One limitation of restricted modification search is that only known modifications (mass shifts) can be considered.

The second approach, unrestricted modification search, has the capability of identifying peptides of unknown modifications. The first unrestricted modification search was demonstrated by Pevzner and colleagues, using a blind search based on a Smith-Waterman-like spectral alignment algorithm called MSAlgin *(15)*. Havilio and Wool also showed that a variant of the SEQUEST algorithm, TwinPeaks, has the capability to carry out unrestricted modification search *(16)*. An issue arising in unrestricted modification searches is justifying the mass shifts observed. The follow-up to such findings is not usually clear.

It should be noted that identification of PTMs is computationally expensive, for either restricted or unrestricted modification searches. The unrestricted modification search will usually take even more computational time.

## 4. Protein Summary and Protein Inference

In most proteomic studies, the interest of researchers lies in the identification of proteins, although the interpretation starts with peptide inference from tandem mass spectra. Thus, it is necessary to group peptides together to produce the identified proteins list, a process usually referred to as protein inference.

There are several issues associated with protein inference. First of all, a small false-positive rate at the spectrum level usually will translate into a higher false-positive rate at the peptide level, which can become even higher again at the protein level. The reason for this phenomenon is that the incorrectly assigned peptide-spectral matches can be described as random matches. Because of the random nature of this process, almost every new incorrectly assigned spectrum adds one new incorrect peptide identification. However, the correctly assigned peptide-spectrum matches are not random matches. Thus, correctly identified peptides tend to have multiple copies of associated MS/MS, while peptides incorrectly identified tend to have one or a few associated MS/MS. The propagation of error rate at the peptide level to the protein level can be explained similarly. This issue can be address to some extent by requiring two or more identified peptides for each identified protein. If a user chooses to use reverse databases for the estimation of false-positive rates, one can see that the protein identification false-positive rate drops dramatically when one increases the number of identified peptides for each identified protein (**Table 5**).

The second issue in protein inference is the shared peptides problem. Owing to the presence of homologous proteins, alternative splicing forms, or redundant protein entries in the database, an identified peptide can correspond to multiple entries in the protein database. One approach is to use the principle of Occam's razor to address this issue. The principle of Occam's razor states that the explanation of any phenomenon should make as few assumptions as possible, eliminating those that make no difference in the observable predictions of the explanatory hypothesis or theory. Both DTASelect *(27, 29)* and ProteinProphet *(48)* implement this principle. Under this principle, if two different proteins have identical sequence coverage, the proteins should be grouped together. For example, if both protein A and protein B have three peptides identified 1, 2, and 3, these two proteins should be treated the same. Under this principle, subset proteins should also be removed. For example, if protein A has three peptides identified 1, 2, and 3, while protein B only has two peptides identified 1 and 2, then B is treated as a subset protein of A. However, to allow users to be able to identify a processed protein as well as its precursor, DTASelect provides an option to show subset proteins on the inferred proteins report. **Figure 3** shows an example of DTASelect output where several proteins are grouped together since the identified peptides are common to all the proteins. A list of protein summary tools is given in **Table 6**.

**Table 5**
**False-positive rates drop when the required number of identified peptides increases to infer a protein**

| Required peptides # | | 1 | 2 | 3 | 4 |
|---|---|---|---|---|---|
| Spectrum | FP | 5.00% | 1.01% | 0.12% | 0.00% |
| | Forward | 14,736 | 13,872 | 13,298 | 12,743 |
| | Reverse | 737 | 140 | 16 | 0 |
| Peptide | FP | 11.43% | 2.32% | 0.32% | 0.00% |
| | Forward | 5,861 | 5,140 | 4,727 | 4,424 |
| | Reverse | 670 | 119 | 15 | 0 |
| Protein | FP | 40.21% | 7.21% | 0.86% | 0.00% |
| | Forward | 1,512 | 791 | 580 | 475 |
| | Reverse | 608 | 57 | 5 | 0 |

Locus Key:

| Validation Status | Locus | Sequence Count | Spectrum Count | Sequence Coverage | Length | MolWt | pI | Descriptive Name |
|---|---|---|---|---|---|---|---|---|

Similarity Key:

| Locus | # of identical peptides | # of differing peptides |
|---|---|---|

| | | | | | | | | |
|---|---|---|---|---|---|---|---|---|
| U | YDR098C-B | 12 | 19 | 7.6% | 1755 | 198594 | 8.4 | YDR098C-B SGDID:S000007391, Chr IV from 651121-649817,649815-645853, reverse complement, transposable_element_gene, "TyB Gag-Pol protein; proteolytically processed to make the Gag, RT, PR, and IN proteins that are required for retrotransposition" |
| U | YER160C | 12 | 19 | 7.6% | 1755 | 198568 | 8.5 | YER160C SGDID:S000000962, Chr V from 498119-496818,496816-492851, reverse complement, transposable_element_gene, "TyB Gag-Pol protein; proteolytically processed to make the Gag, RT, PR, and IN proteins that are required for retrotransposition" |
| U | YER138C | 12 | 19 | 7.6% | 1755 | 198561 | 8.1 | YER138C SGDID:S000000940, Chr V from 449020-447719,447717-443752, reverse complement, transposable_element_gene, "TyB Gag-Pol protein; proteolytically processed to make the Gag, RT, PR, and IN proteins that are required for retrotransposition" |
| U | YDR365W-B | 12 | 19 | 7.6% | 1755 | 198658 | 8.0 | YDR365W-B SGDID:S000007401, Chr IV from 1206987-1208291,1208293-1212255, transposable_element_gene, "TyB Gag-Pol protein; proteolytically processed to make the Gag, RT, PR, and IN proteins that are required for retrotransposition" |
| U | YDR316W-B | 12 | 19 | 7.6% | 1755 | 198526 | 8.2 | YDR316W-B SGDID:S000007399, Chr IV from 1096059-1097363,1097365-1101327, transposable_element_gene, "TyB Gag-Pol protein; proteolytically processed to make the Gag, RT, PR, and IN proteins that are required for retrotransposition" |
| U | YDR261C-D | 12 | 19 | 8.4% | 1604 | 181660 | 7.5 | YDR261C-D SGDID:S000007395, Chr IV from 992343-991039,991037-987528, reverse complement, transposable_element_gene, "TyB Gag-Pol protein; proteolytically processed to make the Gag, RT, PR, and IN proteins that are required for retrotransposition" |

| Filename | XCorr | DeltCN | Conf% | ObsM+H+ | CalcM+H+ | PPM | SpR | SpScore | Ion% | # | Sequence |
|---|---|---|---|---|---|---|---|---|---|---|---|
| 0331602 SCX 1 05 itms.04042.04042.2 | 3.4392 | 0.4852 | 100.0% | 1639.7972 | 1639.7927 | 2.8 | 2 | 368.9 | 53.6% | 1 | K.EVHTNQDPLDVSASK.I |
| 0331602 SCX 1 13 itms.06207.06207.2 | 2.4487 | 0.1061 | 98.6% | 2125.186 | 2124.1929 | -4.8 | 5 | 309.5 | 36.8% | 1 | K.FLQNSNLGGIIPTVNGKI |
| 0331602 SCX 1 08 itms.06936.06936.2 | 3.6085 | 0.2443 | 100.0% | 1412.7021 | 1412.6984 | 2.7 | 1 | 1303.9 | 81.8% | 5 | K.DILSVDYTDIMK.I |
| 0331602 SCX 1 08 itms.04182.04182.2 | 3.3061 | 0.3657 | 100.0% | 990.5224 | 990.5229 | -0.5 | 1 | 1058.0 | 92.9% | 3 | K.VACQLIMR.G |
| 0331602 SCX 1 10 itms.01627.01627.2 | 1.7524 | 0.1134 | 98.0% | 1039.543 | 1039.5424 | 0.6 | 12 | 406.2 | 62.5% | 1 | R.SYTNTTKPK.V |
| 0331602 SCX 1 04 itms.03302.03302.2 | 5.2663 | 0.3421 | 100.0% | 1866.8362 | 1866.8317 | 2.4 | 1 | 1188.5 | 67.6% | 1 | H.NVSTSNNSPSTDNDSISI |
| 0331602 SCX 1 04 itms.03264.03264.2 | 2.9523 | 0.2433 | 100.0% | 1752.7915 | 1752.7888 | 1.5 | 1 | 332.5 | 46.9% | 1 | N.VSTSNNSPSTDNDSISK. |
| 0331602 SCX 1 05 itms.03624.03624.2 | 2.2295 | 0.1547 | 99.9% | 802.41296 | 802.41327 | -0.4 | 2 | 924.9 | 91.7% | 1 | C.PDCLIGK.S |
| 0331602 SCX 1 04 itms.03463.03463.1 | 1.9144 | 0.2281 | 100.0% | 841.39496 | 841.39435 | 0.7 | 6 | 342.6 | 66.7% | 1 | F.TDETTKF.R |
| 0331602 SCX 1 06 itms.04601.04601.2 | 2.9741 | 0.3023 | 100.0% | 1451.7769 | 1451.7745 | 1.6 | 1 | 693.0 | 66.7% | 1 | K.TVDTTNYVILQGK.E |
| 0331602 SCX 1 10 itms.06885.06885.2 | 4.6191 | 0.5087 | 100.0% | 2088.963 | 2088.9514 | 5.5 | 1 | 1007.8 | 59.4% | 2 | R.LDQFNYDALTFDEDLNR. |
| 0331602 SCX 1 09 itms.03978.03978.2 | 2.323 | 0.2628 | 100.0% | 894.5405 | 894.54126 | -0.8 | 12 | 498.7 | 78.6% | 1 | K.LNVPLNPK.G |

Fig. 3. DTASelect example output where several proteins are grouped together since the identified peptides are common to all the proteins.

**Table 6**
**Protein Summary Tools**

| Program | Category | Website and/or reference |
|---|---|---|
| DTASelect | P | http://fields.scripps.edu/DTASelect/, *(11)* |
| ProteinProphet | P | http://tools.proteomecenter.org/software.php, *(48)* |
| Scaffold | C | http://www.proteomesoftware.com/ |

C, Commercial; P, Publicly available.

## 5. Protein Quantification

### 5.1. Basic Principles

Protein quantification is used to measure the amount of a protein in a sample. MS-based protein quantification is achieved either through stable isotope labeling or label-free strategies. Stable isotope labeling strategies typically compare naturally abundant stable isotope peptides to physicochemically identical peptides with atoms enriched with a heavy stable isotope *(49)*. Several quantitative labeling technologies have been developed by in vivo labeling via metabolic incorporation or in vitro labeling via chemical reactions. The most popular in vivo labeling technologies are $^{15}N$ labeling and stable isotope labeling with amino acids in cell culture (SILAC) *(50)*. $^{15}N$ labeling requires that cells be grown in either $^{14}N$ (light) or $^{15}N$ (heavy) media to incorporate each isotope into the proteins in the two different samples, respectively. Mixtures of two cell cultures are analyzed by MS and the mass shift of heavy isotope labeled peptides are observed from light peptides. Both elution profiles from light and heavy peptides are compared to determine relative abundance. SILAC is gaining popularity utilizing a single isotope or a combination of $^{2}H$, $^{13}C$, and $^{15}N$ isotopes.

A limitation of metabolic labeling is that tissues or body fluids that cannot be grown in culture usually are not applicable to this technology. In vitro labeling via chemical reaction is an alternative approach to incorporate stable isotope tags onto specific sites such as N-terminal, C-terminal, cysteine, lysine, and tyrosine. Chemical reaction labeling strategies include ICAT *(51)*, iTRAQ *(52)*, and $^{18}O$ labeling *(53)*.

**Figure 4** illustrates a general scheme for stable isotope labeling combined with MS. Labeling with stable isotopes is expensive and sometimes achieving highly enriched samples takes a large

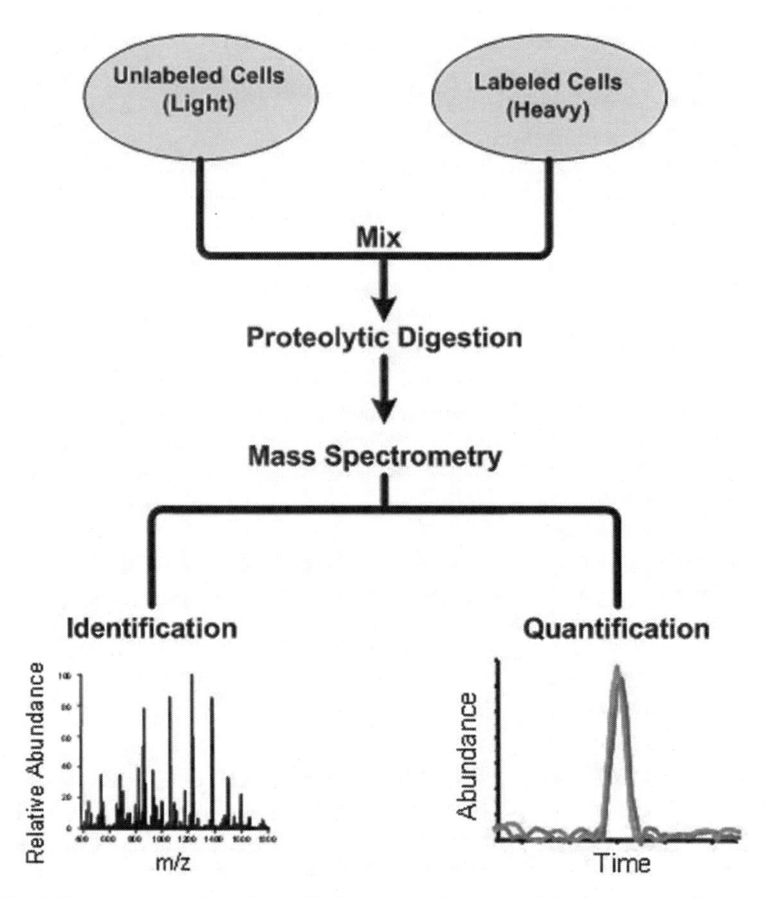


Fig. 4. General schematics of quantitative proteomics using stable isotope labeling.

amount of time and effort. As an alternative approach, label-free strategies have been presented by a number of groups, and with its simplicity and cost-effectiveness, it is gaining momentum. There are several different label–free quantitative strategies including spectral counting *(54, 55)*, LC-MS peak area based strategies *(56–59)*, LC-MS strategies *(60, 61)*, and differential MS *(62)*.

In the following, we are going to discuss the computational analysis of quantitative proteomics data using stable isotope labeling. The analysis usually contains the following steps: (1) extraction of peptide ion chromatograms, (2) smoothing of chromatograms and noise reduction, (3) ion current ratio calculation to infer peptide abundance ratio, (4) expression of peptide ratios as relative protein abundance. We will then discuss two special quantification techniques using tandem mass spectra (instead of using full scan MS) and spectral counting.

### *5.2. Extraction of Peptide Ion Chromatograms*

Ion chromatogram extraction process is a computationally intensive step in quantitative analysis. Before the quantitative analysis, protein database search results are generally required, because the identified peptide information from database search results including scan number, retention time, and sequence are often used for quantitative applications to locate specific peptide elution time in chromatograms to rebuild a peak profile. In addition, the amino acid elemental composition data is also needed to calculate both light and heavy isotope masses. **Figure 5** shows an example of reconstructed chromatogram generated by Census. However, label-free strategies often extract chromatograms from spectral files of multiple samples without pre-identification database search and perform chromatogram alignments to accommodate peak shifts using several different types of algorithms including dynamic time warping, correlation optimized warping, parametric time warping, and peak alignment by a genetic algorithm.

### 5.3. Smoothing and Noise Reduction

Finding correct start and end points of peptide elution peaks in chromatograms is crucial for accurate and precise quantification results. Smoothing techniques and baseline removal algorithms based on characteristics of individual peptide elution profiles are commonly utilized before identifying peptide peaks. In cases where the intensity of a peptide is low while its corresponding light or heavy isotopomer is high, algorithms using a single isotopomer may not correctly find the peak range to accommodate both light and heavy peaks. An improvement in accuracy for peak finding can be achieved with normalization of both light and heavy elution profiles *(63, 64)*.

### 5.4. Ion Current Ratio Calculation

After defining start and end points of peptide elution peaks, the next step is to calculate relative peptide abundances for the light and heavy peptides. There are several different algorithms to calculate relative peptide abundances – peak area *(65–67)*, linear regression *(64, 68)*, and principal component analysis *(63)*. The peak area approach simply calculates the area of both light and heavy peaks and compares their ratios. The linear regression approach converts peak profiles of the two isotopologues into

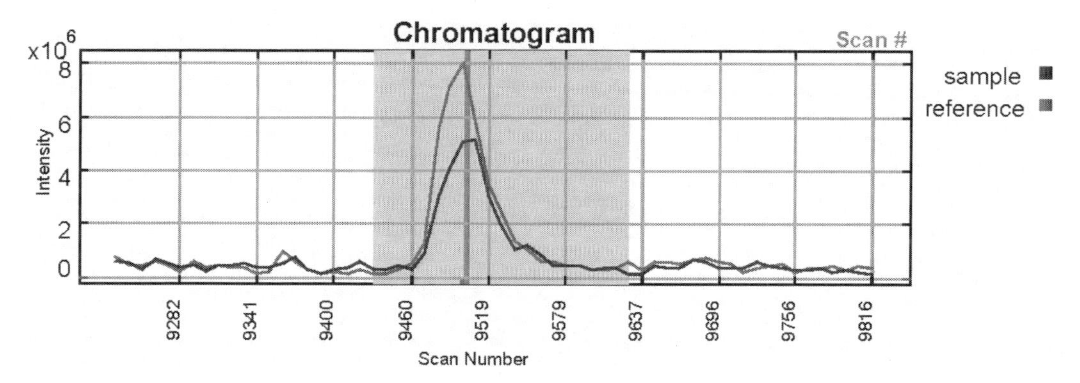


Fig. 5. Reconstructed chromatograms of both labeled and unlabeled peptides.

a scatterplot based on their ion intensities. The slope of linear regression of the data points indicates the relative abundance of light and heavy peaks and the correlation coefficient shows the quality of the elution profiles. The advantage of linear regression is that it is tolerant to poor signal-to-noise (S/N) ratio data. Principal component analysis generates a similar scatterplot of ion intensities from both light and heavy isotopologues, and calculates two principal components and their eigenvalues. The slope of the first principal component indicates the peptide abundance ratio.

### *5.5. Expression of Peptide Ratios as Relative Protein Abundances*

After the peptide ratios are obtained, there are two different approaches for calculating protein ratios. One approach is to simply calculate the mean of all peptide measurements. The second method is a weighted average where peptides with given weight based on scores such as the quality or the standard deviation are used to derive a protein abundance ratio. By using known mixtures of $^{15}N$ labeled yeast samples, Park et al. showed that weighted average using quality scores of peptides provides more accurate protein abundance measurements. **Figure 6** shows an example protein quantification output using the software Census *(64)*.

A list of protein quantification software tools is provided in **Table 7**.

```
H    CenSus v. 0.9 census -out.txt file
H    Created by
H    John Venable   jvenable@scripps.edu
H    "Robin, Sung Kyu Park rpark@scripps.edu "
H    Michael J. MacCoss
H    "The Scripps Research Institute, La Jolla, CA"
H    created date    Wed Mar 08 14:06:17 PST 2006
H    PLINE    LOCUS    AVERAGE_RATIO    STANDARD_DEVIATION    PEPTIDE_NUM    DESCRIPTION
H    SLINE    UNIQUE    SEQUENCE    RATIO    REGRESSION_FACTOR    DETERMINANT_FACTOR    FILE_NAME
H    Total peptides    528
H    Quantified peptides    384
H    Quantification efficiency    72.73%
P    YGR192C    0.792    0.05    14    "TDH3 SGDID:S0003424,   Chr VII from 883815 - 882817, reverse complement, Verified ORF "
                                                        FALSE    FALSE
S    U    R.IALSRPNVEVVALNDPFITNDYAAYMFK.Y    0.822    0.998    0.995    120205_yeast_1to1_LTQ_FTMSComp    -06_itms.07951.07951.2
                                                        TRUE    TRUE
S    U    R.IALSRPNVEVVALNDPFITNDYAAYMFK.Y    0.878    0.999    0.999    120205_yeast_1to1_LTQ_FTMSComp    -06_itms.07948.07948.3
                                                        TRUE    TRUE
S    U    R.PNVEVVALNDPFITNDYAAYMFK.Y    0.738    0.895    0.8    120205_yeast_1to1_LTQ_FTMSComp    -02_itms.11067.11067.2
                                                        TRUE    TRUE
S    U    K.KIATYQERDPANLPWGSSNVDIAIDSTGVFK.E    0.77    0.852    0.725    120205_yeast_1to1_LTQ_FTMSComp    -06_itms.17830.17830.3
                                                        TRUE    TRUE
S    U    K.IATYQERDPANLPWGSSNVDIAIDSTGVFK.E    0.737    0.851    0.724    120205_yeast_1to1_LTQ_FTMSComp    -03_itms.17972.17972.3
                                                        TRUE    TRUE
S         K.VVITAPSSTAPMFVMGVNEEK.Y    0.876    0.986    0.972    120205_yeast_1to1_LTQ_FTMSComp    -02_itms.09330.09330.2
                                                        TRUE    TRUE
S         K.IVSNASCTTNCLAPLAK.V    0.804    0.968    0.938    120205_yeast_1to1_LTQ_FTMSComp    -02_itms.07738.07738.2
                                                TRUE    TRUE
S    U    K.VINDAFGIEEGLMTTVHSLTATQK.T    0.81    0.979    0.959    120205_yeast_1to1_LTQ_FTMSComp    -03_itms.08284.08284.3
                                                        TRUE    TRUE
S    U    K.VINDAFGIEEGLMTTVHSLTATQK.T    0.818    0.977    0.955    120205_yeast_1to1_LTQ_FTMSComp    -03_itms.08234.08234.2
                                                        TRUE    TRUE
S         R.TASGNIIPSSTGAAK.A    0.785    1    0.999    120205_yeast_1to1_LTQ_FTMSComp    -02_itms.21642.21642.2
                                                TRUE    TRUE
S         R.TASGNIIPSSTGAAK.A    0.758    0.977    0.955    120205_yeast_1to1_LTQ_FTMSComp    -02_itms.07405.07405.1
                                                TRUE    TRUE
S         R.VPTVDVSVVDLTVK.L    0.783    0.994    0.988    120205_yeast_1to1_LTQ_FTMSComp    -04_itms.08023.08023.1
                                                TRUE    TRUE
S    U    K.GVLGYTEDAVVSSDFLGDSHSSIFDASAGIQLSPK.F    0.704    0.865    0.748    120205_yeast_1to1_LTQ_FTMSComp    -06_itms.11258.11258.3
                                                        TRUE    TRUE
S         R.VVDLVEHVAK.A    0.806    0.975    0.951    120205_yeast_1to1_LTQ_FTMSComp    -02_itms.07633.07633.2
                                                TRUE    TRUE
P    YBR072W    1.249    0.458    5    "HSP26 SGDID:S0000276,   Chr II from 382027 - 382671, Verified ORF "
                                                        FALSE    FALSE
S    U    R.PNNYAGALYDPRDETLDDWFDNDLSLFPSGFGFPR.S    1.524    0.901    0.812    120205_yeast_1to1_LTQ_FTMSComp    -02_itms.12068.12068.3
                                                        TRUE    TRUE
S    U    R.DETLDDWFDNDLSLFPSGFGFPR.S    1.449    0.988    0.976    120205_yeast_1to1_LTQ_FTMSComp    -02_itms.12622.12622.3
                                                        TRUE    TRUE
S    U    R.DETLDDWFDNDLSLFPSGFGFPR.S    1.467    0.99    0.981    120205_yeast_1to1_LTQ_FTMSComp    -02_itms.12819.12819.2
                                                        TRUE    TRUE
S    U    R.SVAVPVDILDHDNNYELK.V    1.369    0.851    0.725    120205_yeast_1to1_LTQ_FTMSComp    -02_itms.21313.21313.2
                                                TRUE    TRUE
S    U    K.NQILVSGEIPSTLNEESKDK.V    0.436    0.943    0.889    120205_yeast_1to1_LTQ_FTMSComp    -02_itms.08459.08459.2
```

Fig. 6. Census example output.

**Table 7**
**List of protein quantification tools**

| Program | Category | Website and/or reference |
|---|---|---|
| Census | P | http://fields.scripps.edu/download.php, *(64)* |
| ProRata | P | http://www.msprorata.org/, *(63)* |
| MSQuant | P | http://msquant.sourceforge.net/, |
| XPRESS | P | http://tools.proteomecenter.org/XPRESS.php, *(61)* |
| ASAPRatio | P | http://tools.proteomecenter.org/ASAPRatio.php, *(67)* |

P, publicly available.

### 5.6. Quantification from Tandem Mass Spectra

While full scan mass spectra have been used predominantly for quantitative analyses, the use of tandem mass spectra has some benefits over single-stage MS. The use of tandem mass spectra can improve sensitivity, specificity, enhance dynamic range, and increase the S/N ratio. There are several different types of strategies that take advantage of tandem MS for protein quantification, including selective reaction monitoring (SRM), and data-independent acquisition *(6)*. SRM analyses are achieved by monitoring user-defined precursors and collecting a fragment ion in the tandem mass spectra. Another approach is a data-independent strategy that uses successive isolation windows (typically in the range of 10~20 m/z) throughout the mass range, and the ions in each isolation window are fragmented to create a tandem mass spectrum.

### 5.7. Spectral Counting

The total number of spectra matching to each protein obtained from data-dependent data acquisition is called "spectral count." Washburn et al. noted in 2001 that more spectra matched with abundant proteins than with less abundant proteins *(5)*. Pang et al. *(69)* and Gao et al. *(70)* suggested that spectral count is related to protein abundance and therefore could be used as a measure of abundance. A number of groups now use spectral count as a semi-quantitative approach, and even in complex mixtures, it correlates well with protein abundances *(44)*. Because small proteins tend to have fewer spectral counts than large proteins, spectral count normalization is often used to minimize quantitative errors. Zybailov et al. *(71)* proposed a normalization approach using normalized spectral abundance factors (NSAFs):

$$(\mathrm{NSAF})_K = \frac{(S_pC)_K \,/\, (\mathrm{Length})_K}{\sum_{i=1}^{N} (S_pC)_i \,/\, (\mathrm{Length})_i}$$

where $(SpC)i$ is the number of spectral counts for a protein $i$, $(\mathrm{Length})i$ is the length of protein $i$, $N$ is the total number of proteins in the experiment. Briefly, NSAF for a protein $k$ is a normalized spectral count taking into consideration the spectral counts and protein length of all proteins identified in an experiment. The relative abundance proteins within an experiment can be compared by using the NSAF of each protein.

Spectral counting offers a way of protein quantification without carrying out stable isotope labeling. However, spectral counting is usually less sensitive and less accurate, compared with protein quantification using stable isotope labeling.

## 6. Concluding Remarks

With significant progress in protein chemistry, protein separation methods, and MS technologies, it is not uncommon for a mass spectrometer to acquire mass spectral data that contains information about thousands of proteins in a single day. Data analysis is a critical step in MS-based shotgun proteomics. Accurate and automated approaches to analyze protein information contained in large data sets makes shotgun proteomics a powerful analytical approach.

Data analysis of shotgun proteomics data usually begins with peptide identifications from tandem MS data. A variety of approaches, such as database searching, de novo peptide sequencing, hybrid approach combining de novo peptide sequencing and database searching, spectral library searching, unrestricted PTM searching, can be employed to identify peptides from tandem mass spectra. Each approach has its own strength and weaknesses. Generally speaking, database searching is still the most widely used approach for identifying peptides from tandem mass spectra. Sometimes, a combination of several peptide identification approaches can help mine the data more thoroughly. Proteins are then inferred from the peptide information derived from the MS/MS data.

With steady advances in MS technologies, quantitative proteomics also has progressed dramatically. Current approaches for quantitative proteomics rely primarily on stable isotope labeling, including in vivo metabolic labeling and in vitro chemical reaction labeling. In some situations, label-free strategies are also employed in quantitative proteomics study. Comprehensive qualitative and quantitative proteomics analysis will ultimately lead to a more thorough understanding of complex biology systems.

## References


1. Herbert, B.R., Sanchez, J.-C., Bini, L. (1997) Two-dimensional electrophoresis: the state of art and future directions. In *Proteome Research: New Frontiers in Functional Genomics*. pp. 13–33. Springer Berlin, Germany.
2. Zhu, H. and Snyder, M. (2003) Protein chip technology. *Curr Opin Chem Biol* 7, 55–63.
3. Little, D.P., Speir, J.P., Senko, M.W., O'Connor, P.B., McLafferty, F.W. (1994) Infrared multiphoton dissociation of large multiply charged ions for biomolecule sequencing. *Anal Chem* **66**, 2809–2815.
4. Senko, M.W., Speir, J.P., McLafferty, F.W. (1994) Collisional activation of large multiply charged ions using Fourier transform mass spectrometry. *Anal Chem* **66**, 2801–2808.
5. Washburn, M.P., Wolters, D., Yates, J.R. 3rd (2001) Large-scale analysis of the yeast proteome by multidimensional protein identification technology. *Nat Biotechnol* **19**, 242–247.
6. Venable, J.D., Dong, M.Q., Wohlschlegel, J., Dillin, A., Yates, J.R. (2004) Automated approach for quantitative analysis of complex peptide mixtures from tandem mass spectra. *Nat Method* **1**, 39–45.
7. Eng, J.K., McCormack, A.L., Yates, J.R. 3rd (1994) An approach to correlate tandem mass spectral data of peptides with amino acid sequences in a protein database. *J Am Soc Mass Spectrom* **5**, 976–989.
8. Perkins, D.N., Pappin, D.J.C., Creasy, D.M., Cottrell, J.S. (1999). Probability-based protein identification by searching sequence databases using mass spectrometry data. *Electrophoresis* **20**, 3551–3567.
9. Frank, A.M., Savitski, M.M., Nielsen, M.L., Zubarev, R.A., Pevzner, P.A. (2007) De novo peptide sequencing and identification with precision mass spectrometry. *J Proteome Res* **6**, 114–123.
10. Ma, B., Zhang, K., Hendrie, C., Liang, C., Li, M., Doherty-Kirby, A., Lajoie, G. (2003) PEAKS: powerful software for peptide de novo sequencing by MS/MS. *Rapid Commun Mass Spectrom* **17**, 2337–2342.
11. Tabb, D.L., Saraf, A., Yates, J.R. 3rd (2003) GutenTag: high-throughput sequence tagging via an empirically derived fragmentation model.*Anal Chem* **75**, 6415–6421.
12. Tanner, S., Shu, H., Frank, A., Wang, L.C., Zandi, E., Mumby, M., Pevzner, P.A., Bafna, V. (2005) InsPecT: identification of posttranslationally modified peptides from tandem mass spectra. *Anal Chem* **77**, 4626–4639.
13. Yates, J.R. 3rd, Morgan, S.F., Gatlin, C.L., Griffin, P.R., Eng, J.K. (1998) Method to compare collision-induced dissociation spectra of peptides: potential for library searching and subtractive analysis. *Anal Chem* **70**, 3557–3565.
14. Frewen, B.E., Merrihew, G.E., Wu, C.C., Noble, W.S., MacCoss, M.J. (2006) Analysis of peptide MS/MS spectra from large-scale proteomics experiments using spectrum libraries. *Anal Chem* **78**, 5678–5684.
15. Tsur, D., Tanner, S., Zandi, E., Bafna, V., Pevzner, P.A. (2005) Identification of post-translational modifications by blind search of mass spectra. *Nat Biotechnol* **23**, 1562–1567.
16. Havilio, M., Wool, A. (2007) Large-scale unrestricted identification of post-translation modifications using tandem mass spectrometry. *Anal Chem* **79**, 1362–1368.
17. Sadygov, R.G., Cociorva, D., Yates, J.R. 3rd (2004) Large-scale database searching using tandem mass spectra: looking up the answer in the back of the book. *Nat Method* 1, 195–202.
18. Nesvizhskii, A.I. (2007) Protein identification by tandem mass spectrometry and sequence database searching. *Method Mol Biol* **367**, 87–119.
19. Sadygov, R.G., Yates, J.R., 3rd (2003) A hypergeometric probability model for protein identification and validation using tandem mass spectral data and protein sequence databases. *Anal Chem* **75**, (15), 3792–3798.
20. Geer, L.Y., Markey, S.P., Kowalak, J.A., Wagner, L., Xu, M., Maynard, D.M., Yang, X., Shi, W., and Bryant, S.H. (2004) Open mass spectrometry search algorithm. *J Proteome Res* **3**, 958–964.
21. Craig, R., Beavis, R.C. (2004) TANDEM: matching proteins with tandem mass spectra. *Bioinformatics* **20**, 1466–1467.
22. Xu, T., Venable, J.D., Kyu Park, S., Cociorva, D., Lu, B., Liao, L., Wohlschlegel, J., Hewel, J., Yates, J.R. 3rd (2006) ProLuCID, a fast and sensitive tandem mass spectra-based protein identification program. *Mol Cell Proteomics* **5**(10) Supplement, 174.
23. Field, H.I., Fenyo, D., Beavis, R.C. (2002) RADARS, a bioinformatics solution that

automates proteome mass spectral analysis, optimises protein identification, and archives data in a relational database. *Proteomics* **2**, 36–47.

24. Grubbs, F.E., Procedures for detecting outlying observations in samples. *Technometrics* **1969**, 11(1), 1–21.
25. Zubarev, R.A., Horn, D.M., Fridriksson, E.K., Kelleher, N.L., Kruger, N.A., Lewis, M.A., Carpenter, B.K., McLafferty, F.W. (2000) Electron capture dissociation for structural characterization of multiply charged protein cations. *Anal Chem* **72**, 563–573.
26. Syka, J.E., Coon, J.J., Schroeder, M.J., Shabanowitz, J., Hunt, D.F. (2004) Peptide and protein sequence analysis by electron transfer dissociation mass spectrometry. *Proc Natl Acad Sci U S A* **101**, 9528–9533.
27. Tabb, D.L., McDonald, W.H., Yates, J.R. 3rd (2002) DTASelect and contrast: tools for assembling and comparing protein identifications from shotgun proteomics. *J Proteome Res* **1**, 21–26.
28. Keller, A., Nesvizhskii, A.I., Kolker, E., Aebersold, R. (2002) Empirical statistical model to estimate the accuracy of peptide identifications made by MS/MS and database search. *Anal Chem* **74**(20), 5383–5392.
29. Cociorva, D., Tabb, D., Yates, J.R. 3rd (2006) Validation of tandem mass spectrometry database search results using DTASelect. *Curr Protoc Bioinformatics* supplement **16**, 13.4.1–13.4.14.
30. Savitski, M.M., Nielsen, M.L., Kjeldsen, F., Zubarev, R.A. (2005) Proteomics-grade de novo sequencing approach. *J Proteome Res* **4**, 2348–2354.
31. Horn, D.M., Zubarev, R.A., McLafferty, F.W. (2000) Automated reduction and interpretation of high resolution electrospray mass spectra of large molecules. *J Am Soc Mass Spectrom* **11**, 320–332.
32. Lu, B., Chen, T. (2004) Algorithms for de novo peptide sequencing via tandem mass spectrometry. *Drug Discov Today: BioSilico* **2**, 85–90.
33. Mann, M., Wilm, M. (1994) Error-tolerant identification of peptides in sequence databases by peptide sequence tags. *Anal Chem* **66**, 4390–4399.
34. Han, Y., Ma, B., Zhang, K. (2005) SPIDER: software for protein identification from sequence tags containing de novo sequencing error. *J Bioinformatics Comput Biol* **3**, 697–716.
35. Searle, B.C., Dasari, S., Wilmarth, P.A., Turner, M., Reddy, A.P., David, L.L., Nagalla, S.R. (2005) Identification of protein modifications using MS/MS de novo sequencing and the OpenSea alignment algorithm. *J Proteome Res* **4**, 546–554.
36. Altschul, S.F., Gish, W., Miller, W., Myers, E.W., Lipman, D.J. (1990) Basic local alignment search tool. *J Mol Biol* **215**, 403–410.
37. Pearson, W.R., Lipman, D.J. (1988) Improved tools for biological sequence comparison. *Proc Natl Acad Sci U S A* **85**, 2444–2448.
38. Taylor, J.A., Johnson, R.S. (1997) Sequence database searches via *de novo* peptide sequencing by tandem mass spectrometry. *Rapid Commun Mass Spectrom* **11**, 1067–1075.
39. Mackey, A.J., Haystead, T.A., Pearson, W.R. (2002) Getting more from less: algorithms for rapid protein identification with multiple short peptide sequences. *Mol Cell Proteomics* **1**, 139–147.
40. Shevchenko, A., Sunyaev, S., Loboda, A., Shevchenko, A., Bork, P., Ens, W., Standing, K.G. (2001) Charting the proteomes of organisms with unsequenced genomes by MALDI-quadrupole time-of-flight mass spectrometry and BLAST homology searching. *Anal Chem* **73**, 1917–1926.
41. Heller, S. (1999) The history of the NIST/EPA/NIH mass spectral database. *Today's Chemist Work* **8**, 45–50.
42. Craig, R., Cortens, J.C., Fenyo, D., Beavis, R.C. (2006) Using annotated peptide mass spectrum libraries for protein identification. *J Proteome Res* **5**, 1843–1849.
43. Lam, H., Deutsch, E.W., Eddes, J.S., Eng, J.K., King, N., Stein, S.E., Aebersold, R. (2007) Development and validation of a spectral library searching method for peptide identification from MS/MS. *Proteomics* **7**, 655–667.
44. Liu, J., Bell, A.W., Bergeron, J.J., Yanofsky, C.M., Carrillo, B., Beaudrie, C.E., Kearney, R.E. (2007) Methods for peptide identification by spectral comparison. *Proteome Sci* **5**, 3.
45. Zhang, Z. (2004) Prediction of low-energy collision-induced dissociation spectra of peptides. *Anal Chem* **76**, 3908–3922.
46. DeGnore, J.P., Qin, J. (1998) Fragmentation of phosphopeptides in an ion trap mass spectrometer. *J Am Soc Mass Spectrom* **9**, 1175–1188.
47. Yates, J.R. 3rd, Eng, J.K., McCormack, A.L., Schieltz, D. (1995) Method to correlate tandem mass spectra of modified peptides to amino acid sequences in the protein database. *Anal Chem* **67**, 1426–1436.
48. Nesvizhskii, A.I., Keller, A., Kolker, E., Aebersold, R. (2003) A statistical model for identifying

proteins by tandem mass spectrometry. *Anal Chem* 75, 4646–4658.

49. Julka, S., Regnier, F. (2004) Quantification in proteomics through stable isotope coding: a review. *J Proteome Res* 3, 350–363.
50. Ong, S.E., Blagoev, B., Kratchmarova, I., Kristensen, D.B., Steen, H., Pandey, A., Mann, M. (2002) Stable isotope labeling by amino acids in cell culture, SILAC, as a simple and accurate approach to expression proteomics. *Mol Cell Proteomics* 1, 376–386.
51. Gygi, S.P., Rist, B., Gerber, S.A., Turecek, F., Gelb, M.H., Aebersold, R. (1999) Quantitative analysis of complex protein mixtures using isotope-coded affinity tags. *Nat Biotechnol* 17, 994–999.
52. Ross, P.L., Huang, Y.N., Marchese, J.N., Williamson, B., Parker, K. et al. (2004) Multiplexed protein quantitation in Saccharomyces cerevisiae using amine-reactive isobaric tagging reagents. *Mol Cell Proteomics* 3, 1154–1169.
53. Mirgorodskaya, O.A., Kozmin, Y.P., Titov, M.I., Korner, R., Sonksen, C.P., Roepstorff, P. (2000) Quantitation of peptides and proteins by matrix-assisted laser desorption/ionization mass spectrometry using (18)O-labeled internal standards. *Rapid Commun Mass Spectrom* 14, 1226–1232.
54. Liu, H., Sadygov, R.G., Yates, J.R. (2004) A model for random sampling and estimation of relative protein abundance in shotgun proteomics. *Anal Chem* 76, 4193–4201.
55. Blondeau, F., Ritter, B., Allaire, P.D., Wasiak, S., Girard, M., Hussain, N.K., Angers, A., Legendre-Guillemin, V., Roy, L., Boismenu, D., Kearney, R.E., Bell, A.W., Bergeron, J.J., McPherson, P.S. (2004) Tandem MS analysis of brain clathrin-coated vesicles reveals their critical involvement in synaptic vesicle recycling. *Proc Natl Acad Sci U S A* 101, 3833–3838.
56. Bondarenko, P.V., Chelius, D., Shaler, T.A. (2002) Identification and relative quantitation of protein mixtures by enzymatic digestion followed by capillary reversed-phase liquid chromatography–tandem mass spectrometry. *Anal Chem* 74, 4741–4749.
57. Chelius, D., Bondarenko, P.V. (2002) Quantitative profiling of proteins in complex mixtures using liquid chromatography and mass spectrometry. *J Proteome Res* 1, 317–323.
58. Chelius, D., Zhang, T., Wang, G., Shen, R.F. (2003) Global protein identification and quantification technology using two-dimensional liquid chromatography nanospray mass spectrometry. *Anal Chem* 75, 6658–6665.
59. Higgs, R.E., Knierman, M.D., Gelfanova, V., Butler, J.P., Hale, J.E. (2005) Comprehensive label-free method for the relative quantification of proteins from biological samples. *J Proteome Res* 4, 1442–1450.
60. Wang, W., Zhou, H., Lin, H., Roy, S., Shaler, T.A., Hill, L.R., Norton, S., Kumar, P., Anderle, M., Becker, C.H. (2003) Quantification of proteins and metabolites by mass spectrometry without isotopic labeling or spiked standards. *Anal Chem* 75, 4818–4826.
61. Li, X.J., Yi, E.C., Kemp, C.J., Zhang, H., Aebersold, R. (2005) A software suite for the generation and comparison of peptide arrays from sets of data collected by liquid chromatography-mass spectrometry. *Mol Cell Proteomics* 4, 1328–1340.
62. Wiener, M.C., Sachs, J.R., Deyanova, E.G., Yates, N.A. (2004) Differential mass spectrometry: a label-free LC-MS method for finding significant differences in complex peptide and protein mixtures. *Anal Chem* 76, 6085–6096.
63. Pan, C., Kora, G., McDonald, W.H., Tabb, D.L., VerBerkmoes, N.C., Hurst, G.B., Pelletier, D.A., Samatova, N.F., Hettich, R.L. (2006) ProRata: a quantitative proteomics program for accurate protein abundance ratio estimation with confidence interval evaluation. *Anal Chem* 15, 7121–7131.
64. Park, S.K., Venable, J.D., Xu, T., Yates, J.R. 3rd (2008) A quantitative analysis software tool for mass spectrometry-based proteomics. *Nat Methods* 5(4), 319–22.
65. Schulze, W.X., Mann, M. (2004) A novel proteomic screen for peptide-protein interactions. *J Biol Chem* 279, 10756–10764.
66. Han, D.K., Eng, J., Zhou, H., Aebersold, R. (2001) Quantitative profiling of differentiation-induced microsomal proteins using isotope-coded affinity tags and mass spectrometry. *Nat Biotechnol* 19, 946–951.
67. Li, X.J., Zhang, H., Ranish, J.A., Aebersold, R. (2003) Automated statistical analysis of protein abundance ratios from data generated by stable-isotope dilution and tandem mass spectrometry. *Anal Chem* 75, 6648–6657.
68. MacCoss, M.J., Wu, C.C., III, Yates, J.R. (2003) A correlation algorithm for the automated analysis of quantitative "shotgun" proteomics data. *Anal Chem* 75, 6912–6921.
69. Pang, J.X., Ginanni, N., Dongre, A.R., Hefta, S.A., Opiteck, G.J.J. (2002) Biomarker discovery in urine by proteomics. *J Proteome Res* 1, 161–169.
70. Gao, J., Opiteck, G.J., Friedrichs, M.S., Dongre, A.R., Hefta, S.A.J. (2003) Changes in the protein expression of yeast as a function of carbon source. *J Proteome Res* 2, 643–649.
71. Zybailov, B.L., Florens, L., Washburn, M.P. (2007) Quantitative shotgun proteomics using a protease

with broad specificity and normalized spectral abundance factors. *Mol Biosyst* **3**, 354–360.

72. Clauser, K.R., Baker, P., Burlingame, A.L. (1999) Role of accurate mass measurement (+/- 10 ppm) in protein identification strategies employing MS or MS/MS and database searching. *Anal Chem* **71**, 2871–82.
73. Mo, L., Dutta, D., Wan, Y., Chen, T. (2007) MSNovo: a dynamic programming algorithm for de novo peptide sequencing via tandem mass spectrometry. *Anal Chem* **79**, 4870–4878.
74. Fischer, B., Roth, V., Roos, F., Grossmann, J., Baginsky, S., Widmayer, P., Gruissem, W., Buhmann, J.M. (2005) NovoHMM: a hidden Markov model for de novo peptide sequencing. *Anal Chem* **77**, 7265–7273.
75. Frank, A., Pevzner, P. (2005) PepNovo: de novo peptide sequencing via probabilistic network modeling. *Anal Chem* **77**, 964–973.
76. Bern, M., Goldberg, D. (2006) De novo analysis of peptide tandem mass spectra by spectral graph partitioning. *J Comput Biol* **13**, 364–378.
77. DiMaggio, P.A. Jr, Floudas, C.A. (2007) De novo peptide identification via tandem mass spectrometry and integer linear optimization. *Anal Chem* **79**, 1433–1446.
78. Bern, M., Cai, Y., Goldberg, D. (2007) Lookup peaks: a hybrid of de novo sequencing and database search for protein identification by tandem mass spectrometry. *Anal Chem* **79**, 1393–1400.

# Part VI

## Analysis of Protein Modifications

# Chapter 16

# Proteomics Identification of Oxidatively Modified Proteins in Brain

Rukhsana Sultana, Marzia Perluigi, and D. Allan Butterfield

## Summary

Several studies demonstrated the involvement of free radicals in the pathophysiology of neurodegenerative diseases. Once formed, reactive oxygen species (ROS) can promote multiple forms of oxidative damage, including protein oxidation, and thereby influence the function of a diverse array of cellular processes leading inevitably to neuronal dysfunctions. Protein oxidation can therefore rapidly contribute to oxidative stress by directly affecting cell signaling, cell structure, and enzymatic processes such as metabolism. There are many different modes of inducing protein oxidation including metal-catalyzed oxidation, oxidation-induced cleavage of peptide chain, amino acid oxidation, and the covalent binding of lipid peroxidation products or advanced glycation end proteomics.

In this paper we describe the protocol of redox proteomics, a tool to identify post-translational modifications of proteins. We focus our attention on the identification of carbonylated and 4-hydroxy-2-*trans*-nonenal-bound proteins. In redox proteomics, samples for the identification of protein carbonyls are first derivatized with 2,4-dinitrophenolhydrazine (DNPH) followed by two-dimensional (2D) separation of these proteins based on their isoelectric point and rate of migration. The carbonylated proteins are then detected using 2D Western blot techniques. Similarly, HNE-bound proteins can be detected using the above-mentioned strategy except that the sample does not need to be derivatized. Separated proteins are identified following tryptic digestion, mass spectrometry, and interrogation of appropriate databases.

**Key words:** Oxidative stress, Redox proteomics, Neurodegeneration, Protein oxidation, Protein carbonyls, Lipid peroxidation, 4-Hydroxy-2-*trans*-nonenal, Post-translational modifications, Two-dimensional electrophoresis, Isoelectric focusing

## 1. Introduction

Oxidative stress has been described as a component of numerous neurological diseases, among which are Alzheimer's disease (AD), Parkinson disease (PD), and amyotrophic lateral sclerosis (ALS) *(1–3)*. Oxidative stress is the condition arising upon imbalance

Jörg Reinders and Albert Sickmann (eds.), *Proteomics*, Methods in Molecular Biology, Vol. 564
DOI: 10.1007/978-1-60761-157-8_16, © Humana Press, a part of Springer Science+Business Media, LLC 2009

between the (physiological) production of potentially toxic reactive oxygen species (ROS) such as superoxide, hydrogen peroxide, and hydroxyl radical and (physiological) scavenging activities. Different tissues have different susceptibilities to oxidative stress. The central nervous system is particularly sensitive to this kind of damage for a number of reasons, including a low level of some antioxidant enzymes, a high content of easily oxidized substrates (e.g., membrane polyunsaturated lipids), and an inherently high flux of ROS generated during neurochemical reactions such as dopamine oxidation *(4)*. Furthermore, in nervous tissue, oxidative stress can be detrimental by several interacting mechanisms, including direct damage to crucial molecular species, increase in intracellular free $Ca^{2+}$, and release of excitatory amino acids.

Oxidative stress may cause reversible and/or irreversible modifications on sensitive proteins leading to structural, functional, and stability modulations *(5, 6)*. The importance of protein oxidation toward cellular homeostasis derives from the fact that proteins serve vital roles in regulating cell structure, cell signaling, and the various enzymatic processes of the cell. There are many different modes of inducing protein oxidation including metal-catalyzed oxidation, oxidation-induced cleavage of the peptide backbone, amino acid oxidation, and the conjugation of lipid peroxidation products or advanced glycation end products. Protein modifications such as carbonylation, nitration, and protein-protein cross-linking are generally associated with loss of function and may lead to either the unfolding and degradation of the damaged proteins, or aggregation leading to accumulation as cytoplasmic inclusions, as observed in age-related neurodegenerative disorders *(7)*.

Chemically, the oxidative modification of proteins generally involves the introduction of carbonyl groups on amino acid side chains *(8, 9)*. First of all, this can occur either by direct hydroxyl radical attack or by iron-catalyzed reactions with hydrogen peroxide that generate hydroxyl radical. In the case of direct reactivity with hydroxyl radical, several amino acids can form carbonyl moieties. Second, carbonylation can take place as a consequence of lipid peroxidation. This can either be due to reaction of amino acids with lipid radicals, e.g., peroxyl and alkoxyl, or occur by Michael-type addition reactions between lipid peroxidation-derived aldehydic end products such as hydroxyalkenals primarily affecting cysteine, lysine, or histidine residues *(10–14)*. HNE can rapidly diffuse from the lipid bilayer to modify extracellular or intracellular membrane proteins that are necessary for normal cellular function.

In our laboratory we have used redox proteomics, the branch of proteomics that allows the identification of oxidatively modified proteins *(15–18)*, to identify the protein targets of oxidation in neurodegenerative disorders. Post-translational modifications (PTM) of proteins determine their tertiary and quaternary structures and regulate their activities and functions, by causing changes in protein activity, their cellular locations, and dynamic interactions with other proteins. Redox-dependent PTM of proteins is emerging as a key signaling system and influences many aspects of cellular homeostasis. By following this approach, our laboratory has extensively investigated, in different models of neurodegenerative diseases, oxidative modifications of proteins that lead to their altered function thus contributing to neurodegenerative process.

Our proteomic analysis to identify specifically oxidized proteins in animal models of neurodegenerative diseases couples 2D-PAGE with immunochemical detection of protein carbonyls that are derivatized by 2,4-dinitrophenylhydrazine (DNPH) and HNE-bound proteins followed by MS analysis (**Fig. 1**) *(19–25)*. A 2D Western blot map is achieved by using specific antibodies, e.g., anti-DNP, anti-3-NT, or anti-HNE, that react with those proteins containing reactive carbonyl groups/HNE in experimental and control brain. Two-dimensional gel images, used to obtain the protein

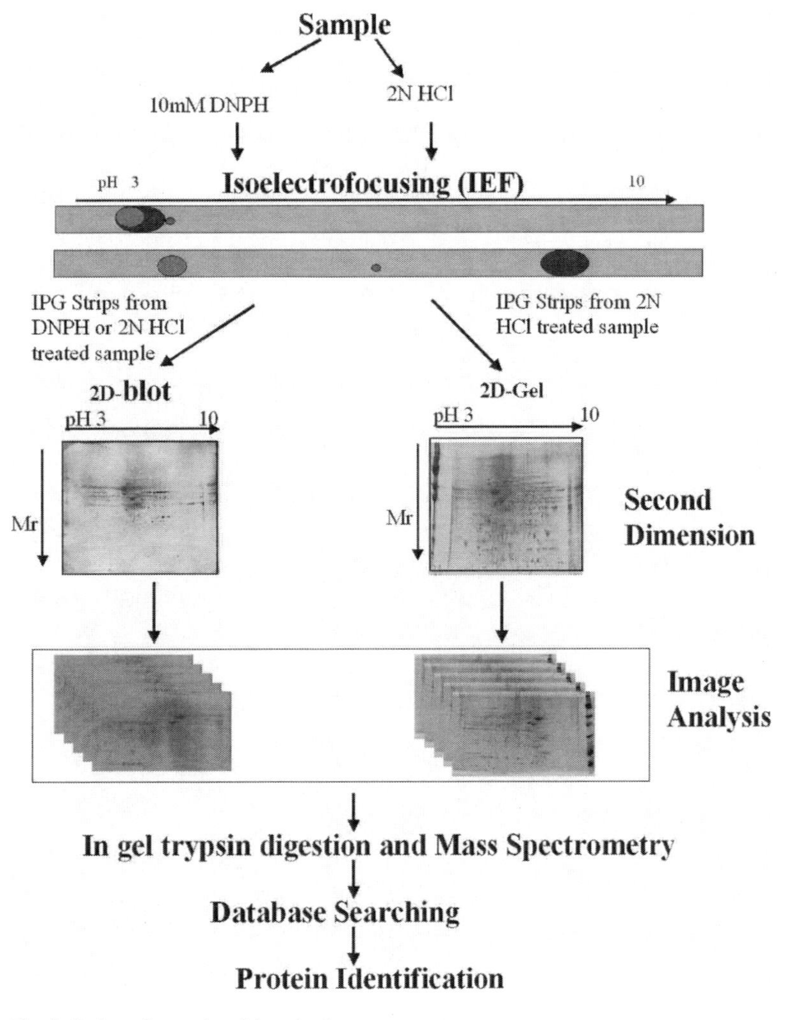


Fig. 1. Proteomics protocol (see text).

expression profile, and the 2D Western blots are analyzed by image software (PD Quest, BioRad). This sophisticated software offers powerful comparative analysis and is specifically designed to analyze immediately many gels or blots that were performed under identical experimental conditions. Powerful automatching algorithms quickly and accurately match gels or blots, and sophisticated statistical analysis tools identify experimentally significant spots. The principles of measuring intensity values by 2D analysis software are similar to those of densitometric measurements. After completion of spot matching, the normalized intensity of each protein spot from individual gels (or membranes) is compared between groups using statistical analysis.

Using redox proteomics we identified the proteins that demonstrate increased carbonyl levels and HNE-bound proteins in the spinal cord of G93A-SOD1 transgenic mouse compared to those of the nontransgenic mice. Superoxide dismutase 1 (SOD1), translationally controlled tumor protein (TCTP), ubiquitin carboxyl-terminal hydrolase-L1 (UCH-L1), and αB-crystallin *(23)* were identified as excessively carbonylated proteins, whereas, dihydropyrimidinase-related protein 2 (DRP-2), the heat-shock protein 70 (Hsp70) and possibly α-enolase *(22)* were shown to have increased amounts of HNE-bound proteins. Taken together, the redox proteomics results obtained from both protein carbonyls and HNE-bound proteins provide insight into the mechanism of G93A-SOD1 neurotoxicity in vivo, which involves oxidative modification of a $Ca^{2+}$ regulating protein (TCTP) and proteins involved in inclusion formation (SOD1, UCH-L1, Hsp70, and αB-crystallin), suggesting a potential relationship between protein oxidation, protein aggregation, and $Ca^{2+}$ regulation in ALS. Moreover, one can speculate that the oxidative modification of these proteins impairs protein stability (αB-crystallin), $Ca^{2+}$ binding (TCTP), protein degradation (UCH-L1, Hsp70), neuronal development and repair (DRP-2), energy metabolism (α-enolase), and antioxidant capacity (SOD1). Moreover, our findings unify the two prominent hypothesis of the pathogenesis of ALS: oxidative stress and protein aggregation.

## 2. Materials

### *2.1. Sample Preparation for Detection of Protein Carbonyls and HNE-Bound Proteins*

1. Homogenization buffer (pH 7.4): 10 mM HEPES, 137 mM NaCl, 4.6 mM KCl, 1.1 mM $KH_2PO_4$, 0.6 mM $MgSO_4$, 0.5 μg/mL leupeptin (stored as an aliquot at –20°C), 0.7 μg/mL pepstatin (stored as an aliquot at –20°C), 0.5 μg/mL trypsin inhibitor, 40 μg/mL PMSF dissolved in deionized water. This solution should be prepared fresh.

2. DNPH solution: 10 mM 2,4-dinitrophenylhydrazine dissolved in 2 N HCl. Store this solution in a brown bottle at room temperature. This solution is stable for months at room temperature.
3. Protein precipitation: 100% solution of trichloroacetic acid. Store at 4°C.
4. Lipid removal wash: Ethanol:ethyl acetate (1:1). This solution should be prepared fresh.

### *2.2. Isoelectric Focusing of Samples or First Dimension Separation of Proteins*

Rehydration buffer: 8 M urea, 2 M thiourea, 2% CHAPS, 0.2% biolytes, 50 mM dithiothreitol (DTT), bromophenol blue dissolved in deionized water. This solution should be prepared fresh.

### *2.3. Second Dimension Electrophoresis*

1. DTT equilibration buffer (pH 6.8): 50 mM Tris-HCl, 6 M urea, 1% (m/v) SDS, 30% (v/v) glycerol, 0.5% DTT dissolved in deionized water. This solution should be prepared fresh.
2. Iodoacetamide (IA) equilibration buffer (pH 6.8): 50 mM Tris-HCl, 6 M urea, 1% (m/v) SDS, 30% (v/v) glycerol, 4.5% IA dissolved in deionized water. This solution should be prepared fresh.
3. A 10X Tris-glycine-SDS (TGS) running buffer is purchased from Bio-Rad (Hercules, CA). Store at room temperature.
4. Fixing solution: 10% (v/v) methanol, 7% (v/v) acetic acid dissolved in deionized water. Store at room temperature.
5. SYPRO Ruby stain: Purchased from Bio-Rad, (Hercules, CA). Store at room temperature.

### *2.4. Immunochemical Detection of Protein Carbonyls and HNE-Bound Proteins*

1. Transfer buffer: 20% (w/v) glycine, 5% (w/v) Tris(hydroxymethyl) aminomethane, 10% (v/v) methanol dissolved in deionized water. This solution can be reused until the color of the solution changes to yellow. Store at 4°C.
2. Wash buffer (PBST): 0.01% (w/v) sodium azide and 0.2% (v/v) Tween 20 dissolved in phosphate-buffered saline (PBS). Store at room temperature.
3. Blocking buffer: 2% bovine serum albumin (BSA) in PBST. Make fresh.
4. Primary antibody solution: Anti-HNE antibody (Alpha Diagnostic, San Antonio, TX) (1:5,000) or 1:100 of anti-DNPH antibody (Chemicon International, Temecula, CA, USA), diluted in PBST directly before use.
5. Secondary antibody solution: Anti-rabbit conjugated to alkaline phosphatase antibody (Sigma Aldrich, St Louis, MO) diluted in PBST (1:4,000) directly before use.

6. Developing solution: Sigma Fast tablet [5-bromo-4-chloro-3-indolyl phosphate/nitro blue tetrazolium (BCIP/NBT)] (Sigma Aldrich, St Louis, MO) dissolved in 10 mL DI water. Make fresh.

## 3. Methods

### 3.1. Sample Preparation

Homogenize the brain tissue in sample homogenization buffer (20% w/v). Centrifuge the sample at 1,500 *g*. Decant the pellet and save the supernatant for proteomics.

### 3.2. Sample Derivatization for Detection of Protein Carbonyls

1. To 100–150 μg of the protein add DNPH solution or 2 N HCl four times the volume of samples, vortex the samples and incubate at room temperature for 20 min without shaking.
2. After 20 min of incubation add 100% TCA to the samples to get a final concentration of 30% TCA, and incubate the samples on ice for 10 min. This step precipitates the protein (*see* **Note 1**).
3. Centrifuge the samples at 10,000 *g* for 5 min at 4°C.
4. Decant the supernatant and wash the pellet four times with ice-cold ethanol: ethyl acetate (1:1) mixture. All the washing steps should be carried out at 10,000 *g* for 5 min at 4°C (*see* **Note 2**).
5. Dry the protein pellets.

### 3.3. Sample Preparation for IEF

1. Suspend the above protein pellets obtained from DNPH derivatization and 100–150 μg of the underivatized protein sample in 200 μL of IEF rehydration buffer. Vortex the samples at room temperature from 1–2 h (*see* **Note 3**).
2. Sonicate the samples for 10 s before loading onto IEF tray (Bio-Rad, Hercules, CA).

### 3.4. Isoelectrofocusing of Samples or First Dimension

1. Isoelectric focusing is performed with a Bio-Rad (Hercules, CA, USA) protein IEF cell apparatus using 110 mm, pH 3–10 immobilized pH gradient (IPG) strips.
2. Transfer 200 μL of the samples into the bottom of the well in the IEF tray carefully by using a micropipette, avoiding bubbles (*see* **Note 4**).
3. Place IPG strips on top of the sample with gel side facing down. Make sure that the positive (+) end of strip is placed toward the positive (+) end of the IEF tray. This is essential for proper isoelectrofocusing of the proteins.

4. Cover the IEF tray and place it in the Protean IEF cell (Bio-Rad, Hercules, CA), and the IPG strips are actively rehydrated at 50 V (20°C) overnight.
5. After 1 h of active rehydration, pause the active rehydration program and add 2 mL of mineral oil to each well to cover the IPG strips and resume the active rehydration (*see* **Note 5**).
6. After 16-18 h of active rehydration, wet paper wicks (Bio-Rad, Hercules, CA) with 8 µL of nanopure water and place a paper wick between the electrodes and IPG strips (*see* **Note 6**).
7. Isoelectrofocusing is carried out at 20°C as follows: 300 V for 2 h linear gradient, 500 V 2 h linear gradient, 1,000 V 2 h linear gradient, 8,000 V 8 h linear gradient, 8,000 V 10 h rapid gradient.
8. After completion of IEF, the IPG strips are transferred to an equilibration tray (Bio-Rad, Hercules, CA) and either processed directly for second dimension or stored at –80°C until use (*see* **Note 7**).

### *3.5. Second Dimensional Electrophoresis (Separation Based on Relative Mobility of Proteins)*

1. Incubate the IPG strips in 4 mL of freshly prepared DTT equilibration buffer in a disposable equilibration tray with lid (Bio-Rad, Hercules, CA, USA) with the gel side facing up for 10 min at room temperature. This step needs to be carried out in dark (*see* **Note 8**).
2. After 10 min transfer the IPG strips into a new well in the equilibration tray and add 4 mL of IA equilibration buffer, again with the gel side facing up for another 10 min. This step needs to be carried out in dark.
3. After 10 min of incubation in IA solution, excess equilibration is removed by rinsing the IPG strips with 1X running buffer.
4. Place the IPG strips with gel side facing up into Criterion gels (Bio-Rad, Hercules, CA , USA). Load unstained molecular weight marker (Bio-Rad, Hercules, CA) into the standard well adjacent to the IPG strip well to stain the gels and load precision stained molecular weight markers (Bio-Rad, Hercules, CA) on the gel that will be used for Western blotting.
5. Add warm overlay agarose solution into the wells of Criterion gels, avoiding bubbles, and then slowly push the IPG strip on either end until a contact is established between the gel and IPG strip.
6. After 5–10 min place the gels in the tank filled with 1X running buffer and fill the upper tank of the gel with running buffer.
7. Run gels at 200 V for 65 min at room temperature until the dye front (bromophenol blue) exits the gel into the lower tank. The gels are removed and one end of the gel is cut to allow its orientation to be tracked.

### 3.6. Protein Staining

1. Fix the gels in 50 mL of fixative solution at room temperature for 60 min with gentle agitation.
2. Remove the fixative solution and add 50 mL of SYPRO Ruby gel stain (Bio-Rad, Hercules, CA, USA). Incubate overnight at room temperature on a rocking platform.

### 3.7. Immunochemical Detection of Carbonylated or HNE-Modified Proteins

1. Transfer the proteins in the gels to nitrocellulose membranes using a semidry transfer unit (Bio-Rad, Hercules, CA) for the immunochemical detection of carbonylated or HNE-modified proteins.
2. Soak the gels in cold transfer buffer for 10 min. Prepare a setup of transfer in the following order: place one soaked filter paper on the transfer unit platform, followed by the nitrocellulose membrane, gel, and one more filter paper. Remove the trapped air bubbles between the gel and nitrocellulose by rolling a glass rod over the transfer sandwich.
3. Carry out transfer of 15 V for 2 h at room temperature. The semitransfer method of proteins saves time and requires less transfer buffer compared to wet transfer.
4. Incubate the nitrocellulose membrane after protein transfer in 50 mL of blocking buffer for 1 h at room temperature on a rocking platform, *see* **Note 9**.
5. After 1 h incubate the membrane for 1 h (or overnight) in the primary antibody solution at room temperature (or 4°C) on a rocking platform.
6. Remove the primary antibody and wash the membrane three times for 5 min each with 50 mL of wash blot.
7. Incubate the membrane in 50 mL of secondary antibody solution for 1 h on a rocking platform.
8. Remove the secondary antibody and wash the membrane three times for 5 min each with 50 mL of wash blot.
9. Add 15 mL of the developing solution to each nitrocellulose membrane. It normally takes 30–40 min for the development of optimal color intensity.
10. After color development, decant the developer and wash the membrane with tap water and dry the blot between Kim-Wipes. An example of the results produced is shown in **Fig. 2**.

### 3.8. Image Analysis and Detection of Specifically Carbonylated or HNE-Modified Proteins

1. Scan the SYPRO Ruby stained gels using a UV transilluminator ($\lambda_{ex}$ = 470 nm, $\lambda_{em}$ = 618 nm, Molecular Dynamics, Sunnyvale, CA, USA).
2. Scan the Western blot membranes with Adobe Photoshop on a Microtek Scanmaker 4900.

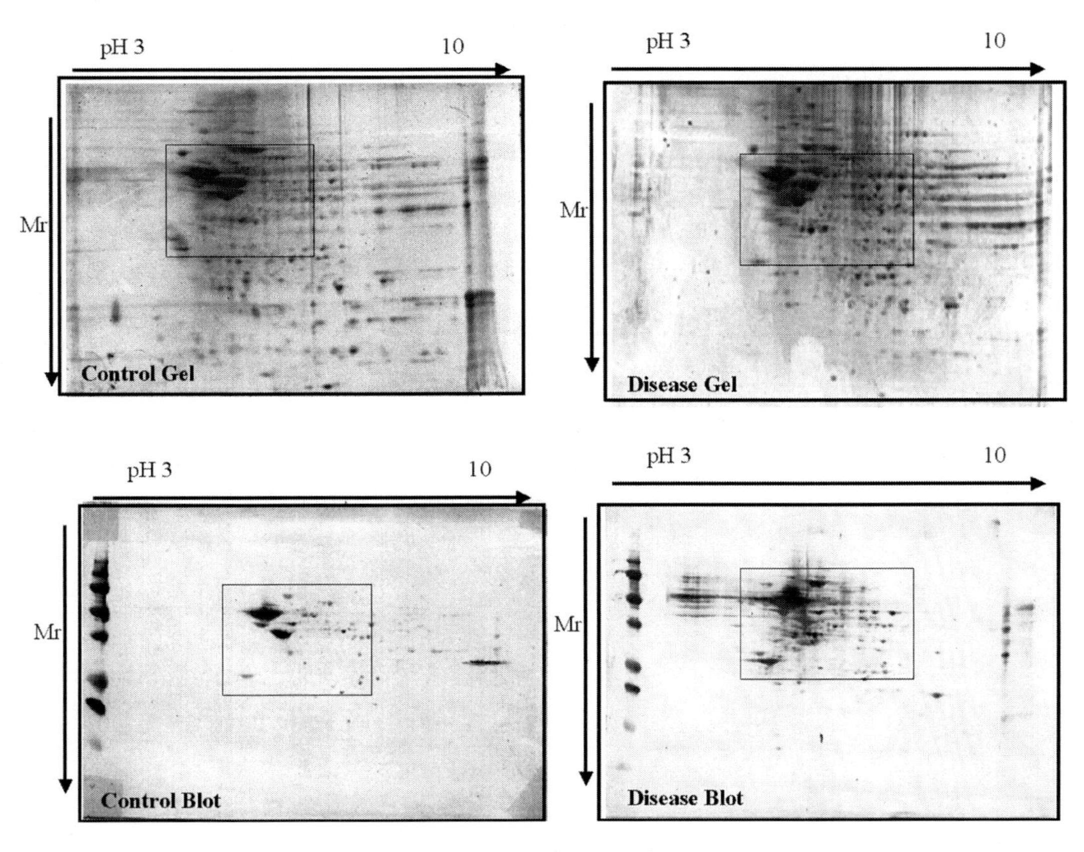


Fig. 2. A sample gel and blot developed using redox proteomics protocol.

3. The proteins that are significantly modified are determined using PDQuest image analysis software (Bio-Rad, Hercules, CA, USA) that allows us to match Western blots and 2D gel maps. The data obtained from PDQuest analysis can be used to calculate the amount of specific oxidation of a protein.
4. Excise and digest the protein spot showing a significant specific oxidation in disease vs. control with trypsin. The protein digest is then subjected to mass spectrometry analysis. Finally, proteins are identified using protein databases such as Mascot, etc.

## 4. Notes

1. TCA should be ice-cold and incubation should be done for 10 min in order to avoid the salt precipitation.
2. This step will remove excess TCA and also lipids from the samples.

3. This step ensures proper better solubilization of the proteins.
4. Bubbles will interfere with the current flow thereby interfering with the IEF.
5. Addition of oil will prevent the evaporation of the rehydration buffer, and thereby drying of the strips due to heat generation during the process of active rehydration.
6. This step will prevent burning of the IPG strip.
7. Strips turn milky white when the strips are frozen due to crystallization of urea, but once the strip comes to room temperature urea dissolves making the IPG strips appear normal.
8. DTT is light sensitive.
9. This step allows the binding of BSA to the remaining unoccupied nitro sites on the membrane that will prevent the nonspecific binding of primary antibody to the membrane.

## Acknowledgement

This work was supported in part by grants from NIH [AG-10836; AG-05119].

### References

1. Butterfield DA, Castegna A, Lauderback CM, Drake J. Evidence that amyloid beta-peptide-induced lipid peroxidation and its sequelae in Alzheimer's disease brain contribute to neuronal death. *Neurobiol Aging* 2002; 23(5):655–64.
2. Migliore L, Petrozzi L, Lucetti C, Gambaccini G, Bernardini S, Scarpato R, Trippi F, Barale R, Frenzilli G, Rodilla V, Bonuccelli U. Oxidative damage and cytogenetic analysis in leukocytes of Parkinson's disease patients. *Neurology* 2002; 58(12):1809–15.
3. Pedersen WA, Cashman NR, Mattson MP. The lipid peroxidation product 4-hydroxynonenal impairs glutamate and glucose transport and choline acetyltransferase activity in NSC-19 motor neuron cells. *Exp Neurol* 1999; 155(1):1–10.
4. Halliwell B. Reactive oxygen species and the central nervous system. *J Neurochem* 1992; 59(5):1609–23.
5. Nakamura K, Hori T, Sato N, Sugie K, Kawakami T, Yodoi J. Redox regulation of a src family protein tyrosine kinase p56lck in T cells. *Oncogene* 1993; 8(11):3133–9.
6. Staal FJ, Anderson MT, Staal GE, Herzenberg LA, Gitler C. Redox regulation of signal transduction: tyrosine phosphorylation and calcium influx. *Proc Natl Acad Sci U S A* 1994; 91(9):3619–22.
7. Dalle-Donne I, Scaloni A, Giustarini D, Cavarra E, Tell G, Lungarella G, Colombo R, Rossi R, Milzani A. Proteins as biomarkers of oxidative/nitrosative stress in diseases: the contribution of redox proteomics. *Mass Spectrom Rev* 2005; 24(1):55–99.
8. Stadtman ER. Protein oxidation and aging. *Science* 1992; 257(5074):1220–4.
9. Stadtman ER, Berlett BS. Reactive oxygen-mediated protein oxidation in aging and disease. *Chem Res Toxicol* 1997; 10(5):485–94.
10. Esterbauer H, Zollner H. Methods for determination of aldehydic lipid peroxidation products. *Free Radic Biol Med* 1989; 7(2):197–203.
11. Uchida K, Stadtman ER. Covalent attachment of 4-hydroxynonenal to glyceraldehyde-3-phosphate dehydrogenase. A possible involvement of intra- and intermolecular cross-linking reaction. *J Biol Chem* 1993; 268(9):6388–93.

12. Blakeman DP, Ryan TP, Jolly RA, Petry TW. Diquat-dependent protein carbonyl formation. Identification of lipid-dependent and lipid-independent pathways. *Biochem Pharmacol* 1995; 50(7):929–35.
13. Esterbauer H, Schaur RJ, Zollner H. Chemistry and biochemistry of 4-hydroxynonenal, malonaldehyde and related aldehydes. *Free Radic Biol Med* 1991; 11(1):81–128.
14. Malecki A, Garrido R, Mattson MP, Hennig B, Toborek M. 4-Hydroxynonenal induces oxidative stress and death of cultured spinal cord neurons. *J Neurochem* 2000; 74(6):2278–87.
15. Sultana R, Perluigi M, Butterfield DA. Protein oxidation and lipid peroxidation in brain of subjects with Alzheimer's disease: Insights into mechanism of neurodegeneration from redox proteomics. *Antioxid Redox Signal* 2006; 8(11-12):2021–37.
16. Butterfield DA, Reed T, Newman S, Sultana R. Roles of amyloid b-peptide-associated oxidative stress and brain protein modifications in the pathogenesis of Alzheimer's disease and mild cognitive impairment. *Free Radic Biol Med* 2007; 43:658–77.
17. Butterfield DA. Proteomics: a new approach to investigate oxidative stress in Alzheimer's disease brain. *Brain Res* 2004; 1000(1–2):1–7.
18. Butterfield DA, Perluigi M, Sultana R. Oxidative stress in Alzheimer's disease brain: new insights from redox proteomics. *Eur J Pharmacol* 2006; 545(1):39–50.
19. Castegna A, Aksenov M, Aksenova M, Thongboonkerd V, Klein JB, Pierce WM, Booze R, Markesbery WR, Butterfield DA. Proteomic identification of oxidatively modified proteins in Alzheimer's disease brain. Part I: creatine kinase BB, glutamine synthase, and ubiquitin carboxy-terminal hydrolase L-1. *Free Radic Biol Med* 2002; 33(4):562–71.
20. Castegna A, Aksenov M, Thongboonkerd V, Klein JB, Pierce WM, Booze R, Markesbery WR, Butterfield DA. Proteomic identification of oxidatively modified proteins in Alzheimer's disease brain. Part II: dihydropyrimidinase-related protein 2, alpha-enolase and heat shock cognate 71. *J Neurochem* 2002; 82(6):1524–32.
21. Castegna A, Thongboonkerd V, Klein J, Lynn BC, Wang YL, Osaka H, Wada K, Butterfield DA. Proteomic analysis of brain proteins in the gracile axonal dystrophy (gad) mouse, a syndrome that emanates from dysfunctional ubiquitin carboxyl-terminal hydrolase L-1, reveals oxidation of key proteins. *J Neurochem* 2004; 88(6):1540–6.
22. Perluigi M, Fai Poon H, Hensley K, Pierce WM, Klein JB, Calabrese V, De Marco C, Butterfield DA. Proteomic analysis of 4-hydroxy-2-nonenal-modified proteins in G93A-SOD1 transgenic mice – a model of familial amyotrophic lateral sclerosis. *Free Radic Biol Med* 2005; 38(7):960–8.
23. Poon HF, Hensley K, Thongboonkerd V, Merchant ML, Lynn BC, Pierce WM, Klein JB, Calabrese V, Butterfield DA. Redox proteomics analysis of oxidatively modified proteins in G93A-SOD1 transgenic mice – a model of familial amyotrophic lateral sclerosis. *Free Radic Biol Med* 2005; 39(4):453–62.
24. Sultana R, Boyd-Kimball D, Poon HF, Cai J, Pierce WM, Klein JB, Markesbery WR, Zhou XZ, Lu KP, Butterfield DA. Oxidative modification and down-regulation of Pin1 in Alzheimer's disease hippocampus: A redox proteomics analysis. *Neurobiol Aging* 2006; 27(7):918–25.
25. Sultana R, Boyd-Kimball D, Poon HF, Cai J, Pierce WM, Klein JB, Merchant M, Markesbery WR, Butterfield DA. Redox proteomics identification of oxidized proteins in Alzheimer's disease hippocampus and cerebellum: An approach to understand pathological and biochemical alterations in AD. *Neurobiol Aging* 2006; 27:1564–76.

# Chapter 17

# Isotope-Labeling and Affinity Enrichment of Phosphopeptides for Proteomic Analysis Using Liquid Chromatography–Tandem Mass Spectrometry

Uma Kota, Ko-yi Chien, and Michael B. Goshe


## Summary

The reversible phosphorylation of proteins is a dynamic process that plays a major role in many vital physiological processes by transmitting signals within cellular pathways and networks. Proteomic measurements using mass spectrometry are capable of characterizing the sites of protein phosphorylation and to quantify their change in abundance. However, the low stoichiometry of protein phosphorylation events often preclude mass spectrometry detection and require additional sample preparation steps to facilitate their characterization. Many analytical methods have been used to map and quantify changes in phosphorylation, and this chapter will present two methods that can be used for extraction of phosphopeptides from protein and proteome digests to map phosphorylation sites using liquid chromatography–tandem mass spectrometry (LC/MS/MS). The first method describes an immobilized metal affinity chromatography (IMAC) technique using $Ga^{3+}$ to enrich for phosphopeptides from protein digests. The second method describes the utilization of phosphoprotein isotope-coded solid-phase tags (PhIST) to label and enrich phosphopeptides from complex mixtures to both identify and quantify changes in protein phosphorylation. The IMAC and PhIST protocols can be applied to any isolated protein sample and is amenable to additional fractionation using strong cation/anion exchange chromatography prior to reversed-phase LC/MS/MS analysis.


**Key words:** Phosphorylation, Phosphopeptides, Liquid chromatography–tandem mass spectrometry, IMAC, Chemical labeling, Isotope-coding, Affinity labeling

## 1. Introduction

An important aspect of protein analysis is characterizing their post-translational modifications (PTMs) that cannot be predicted from genomic data. Proteins can undergo a number of modifications by the covalent atttachment of chemical moieties by various

Jörg Reinders and Albert Sickmann (eds.), *Proteomics,* Methods in Molecular Biology, Vol. 564
DOI: 10.1007/978-1-60761-157-8_17, © Humana Press, a part of Springer Science+Business Media, LLC 2009

enzymes. Examples of these include phosphorylation, sulfation, nitrosylation, *O*-linked and *N*-linked glycosylation, and ubiquitination *(1)*. These modifications have been used to predict protein function *(2)*, their cellular localization, and their fate *(3,4)*. Reversible protein phosphorylation is a particularly versatile modification that, in addition to being an essential component of signal transduction, is involved with cell growth, cell division and differentiation, apoptosis, gene expression, and cytoskeletal regulation *(5)*. Protein phosphorylation on serine, threonine, and tyrosine residues are of common occurrence; however, the phosphate group can also modify other amino acid side chains *(6)*.

To better understand the role of phosphorylation, multiple techniques have been developed and successfully applied at both the single protein and proteome level *(7–9)*. Recent years have witnessed the widespread application of mass spectrometry (MS) as a technique in protein characterization *(8)*. The technological advances in mass spectrometry instrumentation, improved multidimensional separation techniques, and powerful bioinformatics tools for data analysis have allowed the rapid development of the MS-based proteomics *(10)*. Mass spectrometry has also been used to identify phosphorylation and other PTMs while studying their dynamics *(1, 11–13)*. Analysis of protein phosphorylation, especially on a proteome-wide scale with high reproducibility and specificity, poses an analytical challenge *(14)*. This is partly because the phosphorylation reaction is substoichiometric and phosphoproteins are present in low abundance *(15)*. In addition, the ionization efficiency of a phosphopeptide is lower than its non-phosphorylated counterpart under positive-mode MS analysis, which will require more concentrated starting material for it to be detected *(16)*. Hence, selective enrichment of phosphoproteins/phosphopeptides from complex mixtures tends to be a necessary prerequisite to enhance their MS detection.

This chapter describes two approaches we have used to identify phosphorylation sites for both phosphopeptide mapping and quantitative analysis. First is the noncovalent enrichment of phosphopeptides by immobilized metal affinity chromatography (IMAC) used to globally capture pSer, pThr, and pTyr residue-containing components within change peptide mixtures by a procedure that was implemented in our previous published studies *(17)*. Our second approach involves the use of phosphoprotein isotope-coded solid-phase tags (PhIST) to chemically label phosphate groups and enable enrichment of phosphopeptides through covalent modification via solid-phase supports *(18–20)*. In the PhIST method, base-catalyzed β-elimination of the phosphate groups from pSer and pThr residues converts them to dehydroalanyl and dehydroaminobutyric acid residues, respectively. The newly formed double bonds serve as Michael acceptor sites that are reactive towards various nucleophilic reagents. Thus, the labile phosphate group is replaced with a stable isotope-coded chemical modification which is more stable under collision-induced dissociation (CID) conditions used during LC/MS/MS and leads to enhance phosphorylation site analysis.

## 2. Materials

### 2.1. Proteolysis

1. Siliconized polypropylene 1.5 or 2.0 mL Eppendorf tubes (Thermo Fisher Scientific).
2. Prevail $C_{18}$ Extract-clean columns (particle size 50 μm, pore size 60 Å, 1.5 mL) from Alltech Associates Inc. (Thermo Fisher Scientific).
3. Speed vacuum concentrator from Savant (Thermo Fisher Scientific).
4. Fifty millimolar ammonium bicarbonate, pH 8.2 (BioChemika Ultra ≥ 99.5%, Fluka).
5. β-Casein (from bovine milk, minimum 90% purity, Sigma-Aldrich).
6. Guanidine hydrochloride (Sigma-Aldrich).
7. Tris(2-carboxyethyl)phosphine hydrochloride (TCEP·HCl) from Pierce (Thermo Fisher Scientific).
8. Sequencing grade-modified trypsin (Promega).
9. Acetonitrile (HPLC grade) and formic acid (ACS regent grade) from Sigma-Aldrich.

### 2.2. Immobilized Metal Affinity Chromatography

1. SwellGel Gallium-chelated disk spin column from Pierce (Thermo Fisher Scientific).
2. Siliconized polypropylene 1.5 mL Eppendorf tubes (Thermo Fisher Scientific).
3. Loading buffer: 5% acetic acid (glacial 99.99+%, Aldrich).
4. Washing buffer 1: 0.1% acetic acid (glacial 99.99+%, Aldrich).
5. Washing buffer 2: 0.1% acetic acid (glacial 99.99+%, Aldrich) and 10% acetonitrile (HPLC grade, Sigma-Aldrich).
6. Elution buffer: 100 mM ammonium bicarbonate, pH 8.5 (BioChemika Ultra ≥ 99.5%, Fluka).
7. Water (18 MΩ) was distilled and purified using a High-Q 103S water purification system (Wilmette, IL).
8. Variable speed microcentrifuge (Spectrafuge 16 M Microcentrifuge, Labnet International, Inc.).
9. Speed vacuum concentrator from Savant (Thermo Fisher Scientific).

#### 2.3. Synthesis of the Solid-Phase Isotope-Coded Reagent

1. Aminopropyl-modified controlled pore glass beads (amine content: 91 μmol/g) 550 Å pore size (Sigma-Aldrich).
2. 1-Hydroxybenzotriazole (HOBt) (Sigma-Aldrich).
3. 4-[4-(1-(Fmoc-amino)ethyl)-2-methoxy-5-nitrophenoxy) butanoic acid] (Fmoc-Photolinker) from Novabiochem (EMD Biosciences).
4. Acetic anhydride (≥98.0%, Sigma-Aldrich).
5. Pyridine (≥99.9%, Sigma-Aldrich).
6. Piperidine (≥99.5%, Sigma-Aldrich).
7. Iodoacetic anhydride (Sigma-Aldrich).
8. N-α-Fmoc-L-leucine (Fmoc-leucine) from Novabiochem (EMD Biosciences).
9. N-α-Fmoc-L-leucine ($^{13}C_6$, 98%; $^{15}N$, 98%) from Cambridge Isotope Laboratories, Inc.
10. *N,N'*-Diisopropylcarbodiimide (DIC) (Sigma-Aldrich).
11. *N,N*-Dimethylformamide (DMF) (Sigma-Aldrich).
12. Dichloromethane ($CH_2Cl_2$) (≥99.9%, Sigma-Aldrich).
13. *N,N*-Diisopropylethylamine (≥99.5%, Sigma-Aldrich).

#### 2.4. Performic Acid Oxidation

1. Speed vacuum concentrator from Savant (Thermo Fisher Scientific).
2. Formic acid (98%, ACS reagent grade) from Sigma-Aldrich.
3. Hydrogen peroxide (30%, Fluka).
4. Water (18 MΩ) was distilled and purified using a High-Q 103S water purification system (Wilmette, IL).

#### 2.5. β-Elimination and Michael Addition

1. Dispo-Biodialyzer, MWCO of 1,000 (The Nest Group Inc., Southborough, MA).
2. Barium hydroxide (Aldrich, 99.99%).
3. Sodium hydroxide (Sigma-Aldrich, 99.99%).
4. (*R*,*R*)-Dithiothreitol (DTT) (Acros Organics).
5. Formic acid (98%, ACS reagent grade from Sigma-Aldrich).
6. Twenty millimolar Tris·HCl pH 8.0 (Sigma).

#### 2.6. PhIST Labeling and Release of Labeled Peptides

1. Glass vial (4 mL clear vial with Teflon cap, National Scientific).
2. Supelco 3 mL filter columns (Sigma-Aldrich).
3. Long-wave UV lamp (Model EN-160L, Spectronics Corp., Westbury, NY).
4. Disposable methacrylate cuvettes (4.5 mL) (Thermo Fischer Scientific).
5. Dispo-Biodialyzer, MWCO of 1,000 (The Nest Group Inc., Southborough, MA).

6. Speed vacuum concentrator from Savant (Thermo Fisher Scientific).
7. Ammonium bicarbonate (BioChemika Ultra ≥99.5%, Fluka).
8. Ammonium hydroxide (Sigma-Aldrich).
9. Copper(II) sulfate (Sigma).
10. (*R*,*R*)-dithiothreitol (DTT) (Acros Organics).
11. Ethylenediaminetetraacetic acid (EDTA) (Sigma-Aldrich).
12. Sodium chloride (Sigma-Aldrich).
13. Tris(2-carboxyethyl)phosphine hydrochloride (TCEP·HCl) from Pierce).
14. Tris·HCl (Sigma-Aldrich).
15. Acetonitrile (HPLC grade) from Sigma-Aldrich.
16. Dimethyl sulfoxide (DMSO) (≥99.9%, Sigma).
17. Formic acid (98%, ACS regent grade) from Sigma-Aldrich.
18. Hydrochloric acid (Sigma-Aldrich).
19. 2-Mercaptoethanol (BME) (≥99.0%, Sigma-Aldrich).
20. Methanol (Optima Grade) from Thermo Fisher Scientific.
21. 2-Propanol (CHROMASOLV Plus for HPLC, 99.9%, Sigma-Aldrich).
22. Water (18 MΩ) was distilled and purified using a High-Q 103S water purification system (Wilmette, IL).

### 2.7. Liquid Chromatography-Tandem Mass Spectrometry Analysis

1. A reversed-phase HPLC column containing $C_{18}$ stationary phase.
2. Acetonitrile (HPLC grade) and formic acid (ACS regent grade) (Sigma-Aldrich).
3. Mobile phases: (A) 0.1% formic acid in water and (B) 0.1% formic acid in acetonitrile.
4. Water (18 MΩ) was distilled and purified using a High-Q 103S water purification system (Wilmette, IL).

## 3. Methods

The IMAC *(17)* and PhIST *(18–20)* protocols presented in detail here are based on those previously described. The synthesis of the solid-phase isotope-coded reagents is a modified version as reported by Zhou et al. *(21)*. The β-elimination and Michael addition protocol is a modified version of McLachlin and Chait *(22)*. The overall PhIST labeling protocol is outlined in **Fig. 1** (*see* **Note 1**).

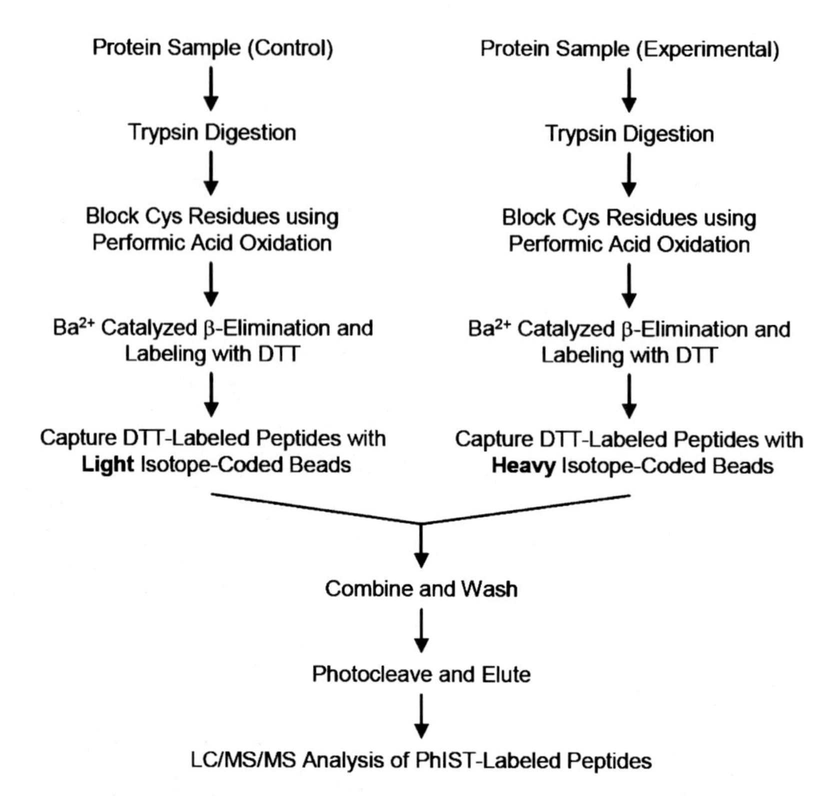


Fig. 1. PhIST labeling strategy for isolation and relative quantification of phosphopeptides from two distinct samples.

### 3.1. β-Casein Sample Preparation

1. Prepare a stock solution of 5.0 mg/mL β-casein by dissolving the lyophilized protein into 50 mM ammonium bicarbonate, pH 8.2. Divide the sample into 250 μg aliquots and transfer each to a siliconized Eppendorf tube.
2. Denature the protein sample by adding guanidine hydrochloride to a concentration of 6 M and thermally denature by boiling for 5 min in a water bath.
3. After cooling the sample to room temperature, reduce disulfide bonds by adding 100 mM of TCEP·HCl to produce a final concentration of 10 mM and incubate the sample at 37°C for 30 min (*see* **Note 2**).
4. Dilute the guanidine concentration to 1 M with 50 mM ammonium bicarbonate (pH 8.2) and perform proteolysis using a 1:20 to 1:50 (w/w) trypsin-to-protein ratio at 37°C overnight (*see* **Note 3**).
5. Following overnight digestion, desalt the sample via solid-phase extraction using a Prevail $C_{18}$ Extract-clean column *(23)*. Using a flow rate of approximately 1.0 mL/min, equilibrate the column with 5 mL of water followed by 2 mL of 10 mM ammonium bicarbonate (pH 8.0). Load the sample and then

wash the column using 3 mL of each solution: 10 mM ammonium bicarbonate (pH 8.0), water, and 20:1 (v/v) of water/acetonitrile with 0.1% formic acid. Elute the peptides with 1 mL of 20:1 (v/v) of acetonitrile/water with 0.1% formic acid and collect into a siliconized polypropylene Eppendorf tube.

6. Lyophilize the sample using vacuum centrifugation and store at –80°C. These preparations can be used as model phosphopeptide samples to verify enrichment and labeling reactions and serve as internal standards for the protocols described below.

### 3.2. Immobilized Metal Affinity Chromatography

1. Resuspend tryptic peptides in an equal volume (*see* **Note 4**) of loading buffer (*see* **Note 5**).
2. Load each sample into the spin column containing dehydrated agarose media preloaded with gallium. Incubate for 5 min to allow the samples to be absorbed completely (*see* **Note 6**). Gently tap the spin column every 2 min to facilitate sample distribution and absorption by the agarose resin.
3. Remove the bottom seal of the spin column and place the column on a 1.5 mL siliconized Eppendorf tube (*see* **Note 7**). Spin the column at 1,000 rpm to remove the unbound peptides.
4. Wash the resin with 50 μL of Washing Buffer 1. Spin the column at 1,000 rpm to remove the unbound peptides and pool them into the same tube used in **step 3**. Repeat this washing step.
5. Wash the resin with 50 μL of Washing Buffer 2 (*see* **Note 8**). Spin the column at 1,000 rpm to remove the unbound peptides and pool them into the same tube used in **step 3**. Repeat this washing step.
6. Wash the resin with 75 μL of water. Spin the column at 1,000 rpm and collect the flow-through to the same tube used in **step 3** (*see* **Note 9**).
7. Transfer the spin column to a new 1.5 mL siliconized Eppendorf tube.
8. Elute the phosphopeptides with 20 μL of Elution Buffer (*see* **Note 10**). Incubate for 3–5 min at room temperature before centrifuging at 1,000 rpm for 1 min.
9. Add an additional 20 μL of Elution Buffer into the spin column and incubate for another 3–5 min at room temperature. Spin the column and collect the eluate into the same tube used in **step 8**.
10. Remove the solvent from the flow-through (generated after **step 6**) eluate (generated after **step 9**) using vacuum centrifugation and store the samples at –80°C until LC/MS/MS analysis.

### 3.3. Synthesis of the Solid-Phase Isotope-Coded Reagent

1. Mix 600 μmol each of HOBt (91.8 mg), Fmoc-Photolinker (311.7 mg), DIC (75.8 mg) with 2 mL of anhydrous DMF in a 4 mL glass vial for 30 min (*see* **Note 11**).
2. Weigh approximately 1.09 g aminopropyl-coated controlled-pore glass (CPG) beads and wash with DMF (*see* **Note 12**).
3. Add the solution from **step 1** into the vial containing CPG beads. The reaction is carried out for 90 min at room temperature with gentle stirring.
4. Wash the beads with 5 mL of DMF and then 5 mL of $CH_2Cl_2$.
5. Acetylate the unmodified amines with 2 mL of 40% acetic anhydride, 60% pyridine in $CH_2Cl_2$ for 30 min at room temperature with gentle stirring (*see* **Note 13**).
6. Wash the beads with 5 mL of DMF.
7. Expose the beads twice with 2 mL of 20% piperidine in DMF for 20 min each to remove the Fmoc protection group and expose the amine group.
8. Make another solution containing 600 μmol each of HOBt, Fmoc-leucine, DIC in DMF as described in **step 1** (*see* **Note 14**).
9. Add the solution from **step 8** into the vial containing the photolinker-containing beads. The reaction is carried out for 90 min at room temperature with gentle stirring.
10. Repeat **steps 4–7**.
11. Wash the beads with 5 mL of DMF, then add 650 μmol of diisopropyethylamine and 600 μmol of iodoacetic anhydride in DMF, and react for 2 h at room temperature with gentle stirring.
12. Wash the beads sequentially with 5 mL of DMF and 5 mL of methanol and then dry them under reduced pressure using a vacuum desiccator. These beads are the solid-phase isotope-coded reagents (**Fig. 2**), which are stored in the dark under desiccation at 4°C.

### 3.4. Oxidation of Cysteinyl Residues with Performic Acid

1. Prepare the performic acid solution by combining 0.1 mL of 30% hydrogen peroxide and 0.9 mL of formic acid and incubate the solution while covered with foil at ambient temperature for 2 h.
2. Cool the performic acid solution and peptide samples in a –20°C freezer for 10 min (*see* **Note 15**).
3. Add an aliquot of the performic acid solution to the peptide sample to achieve a 10% (v/w) solution and incubate the samples at –20°C for 15 min (*see* **Note 16**).

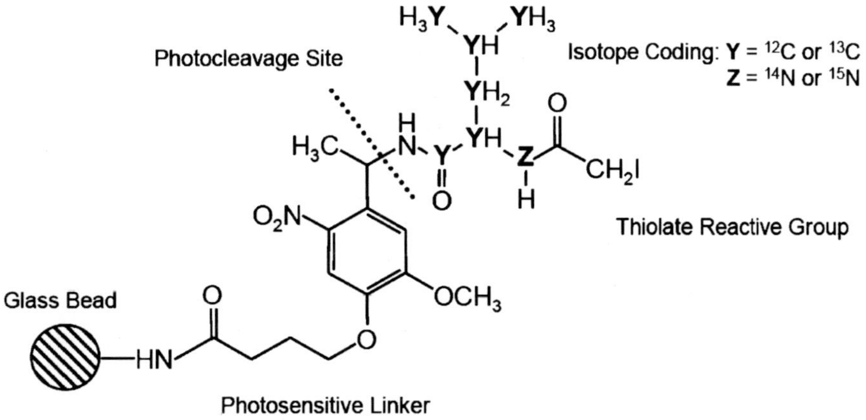

Fig. 2. Isotope-coded solid-phase reagent. Aminopropyl glass beads are covalently linked with a photosensitive linker, a stable isotope-coded leucine moiety, and a thiolate-reactive group *(18)*. The photosensitive linker can be cleaved by UV illumination (360 nm) as indicated with the dashed line. The use of light (Y = $^{12}C$, Z = $^{14}N$) and heavy (Y = $^{13}C$, Z = $^{15}N$) isotopic versions of the solid-phase tag enables the relative quantitation of phosphorylation states between two different protein samples by providing a mass difference of 7.0 Da for each labeled residue.

4. After oxidation, dilute the sample 10-fold by the addition of water, freeze at –80°C, lyophilize the sample, and store at –80°C.

#### 3.5. β-Elimination and Michael Addition

1. The β-elimination and Michael addition reactions to be performed are presented in **Fig. 3**. Prepare stock solutions of 1.0 M DTT, 2.17 M sodium hydroxide, and 133 mM barium hydroxide (*see* **Note 17**).
2. Prepare the β-elimination reaction mixture by combining appropriate volumes of DTT, sodium hydroxide, and barium hydroxide stock solutions to produce a final concentration of 20, 65, and 100 mM, respectively. The dilution is performed with acetonitrile.
3. Add the β-elimination reaction mixture to the oxidized peptides to produce a final concentration of no greater than 1.0 µg/µL.
4. Allow the reaction to proceed for 1 h at 37°C facilitated with gentle agitation (*see* **Note 18**).
5. Quench the reaction by adding 1 µL of 5% formic acid.
6. Dialyze the peptides at 4°C against 20 mM Tris·HCl (pH 8.0) using a Dispo-Biodialyzer (cellulose membrane, MWCO of 1,000). To increase the sample volume to facilitate transfer from the reaction tube, dilute the sample with two or three 20 µL aliquots of 0.1% formic acid. Carefully transfer the

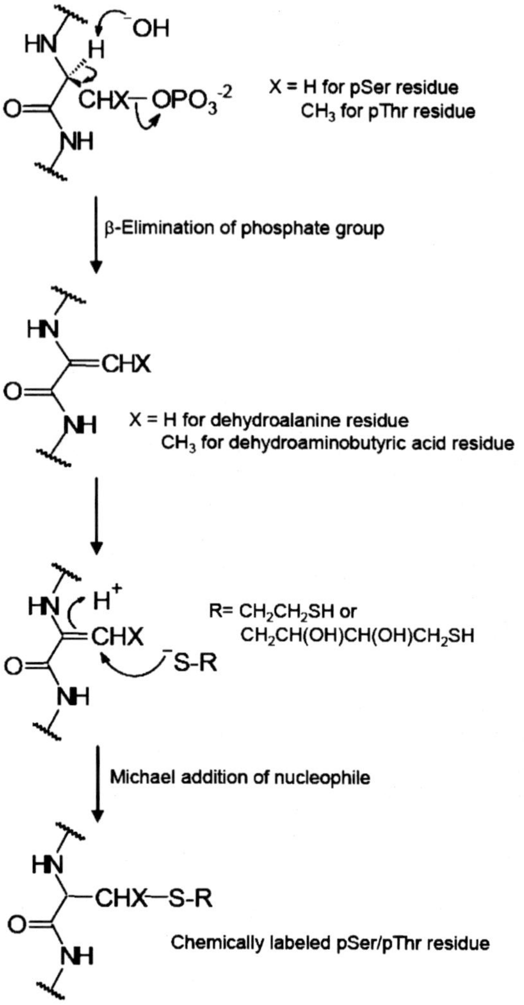

Fig. 3. Chemical labeling of phosphopeptides using β-elimination/Michael addition. Base-catalyzed β-elimination of phosphate groups from pSer and pThr residues converts them to dehydroalanine and dehydroaminobutyric acid residues, respectively. These modified residues are labeled with a nucleophile such as ethanedithiol (EDT) or dithiothreitol (DTT) via Michael addition to produce EDT-labeled peptides *(18)* or DTT-labeled peptides *(19, 20)*, which can be isotope-coded using the solid-phase reagents.

sample into the dialyzer and place the unit into a liter of 20 mM Tris·HCl (pH 8.0). Exchange the Tris buffer every 8 h. Dialysis can take 24–36 h to remove the excess DTT.

7. Once dialysis is complete, place the Dispo-Biodialyzer in the collection tube and spin at low speed (approximately 500 rpm) for 10 s.
8. Store the collected sample at –80°C until further use (*see* **Note 19**).

***3.6. Capture and Isotope Coding of DTT-Labeled Peptides Using Solid-Phase Reagents***

1. The DTT-labeled peptides are captured and isotope-coded as outlined in **Fig. 4**. Reduce the peptide samples with 5 mM TCEP·HCl for 1 h at 37°C.
2. Weigh 5 mg of either the light ($^{12}C_6$, $^{14}N$) and heavy ($^{13}C_6$, $^{15}N$) isotope-coded beads to the sample generated in **Subheading 3.5** and incubate the reaction for 2.5 h in the dark (wrap the sample tubes with aluminum foil) with gentle mixing (*see* **Note 20**).

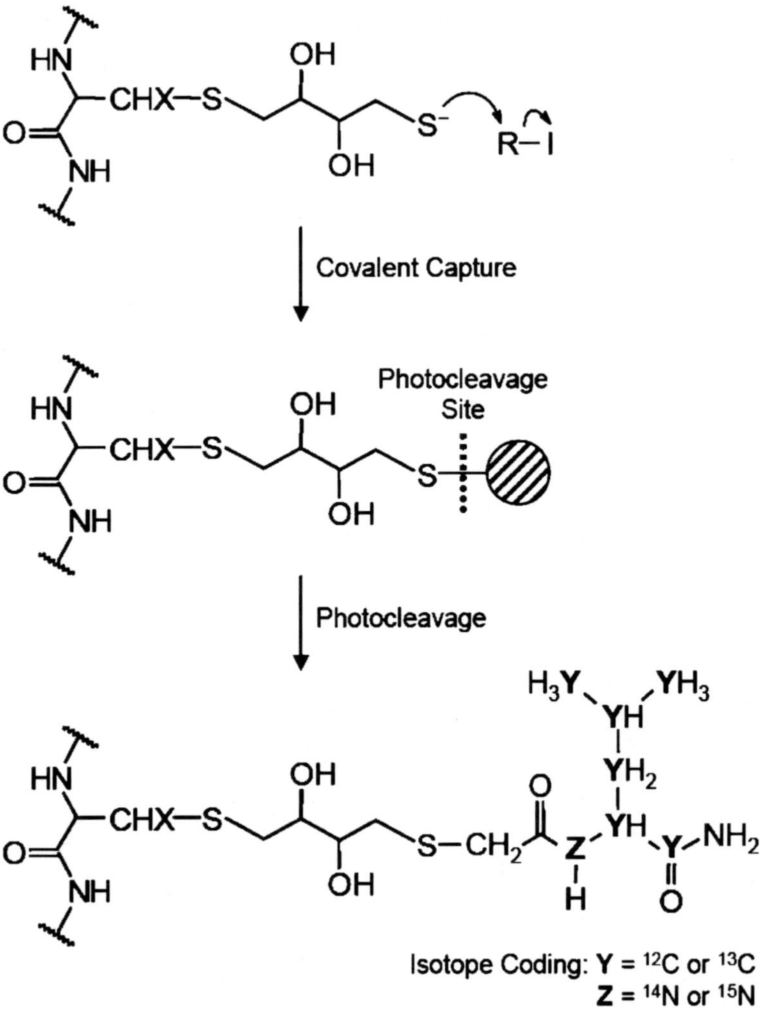

Fig. 4. PhIST labeling using solid-phase reagents. DTT-labeled peptides containing phosphoseryl (X = H) or phosphothreonyl (X = $CH_3$) residues are isolated from non-phosphorylated peptides via covalent modification to the isotope-coded solid-phase reagents. UV illumination is used to photocleave the captured peptides and generate the PhIST-labeled peptides. An additional light (306.13 Da) or heavy (313.13 Da) PhIST tag is added to phosphoseryl and phosphothreonyl residues when DTT is used during Michael addition.

3. After labeling, quench any remaining unreacted thiolate-reactive sites by adding 2-mercaptoethanol to produce a concentration of 1% (v/v) and incubate the reaction at room temperature for 10 min.
4. Wash samples twice with 1 mL of 2.0 M sodium chloride. This is accomplished by adding 1 mL of the wash solution to the sample tube, gently mixing the beads and solution, centrifuging the sample for 30 s at 9,000 rpm to pellet the beads, and removing the supernatant.
5. Next, a series of washing steps is performed to remove non-specifically absorbed peptides. These washes are best performed using the Supelco 3 mL filter columns and an SPE vacuum manifold rather than the centrifugation method used in **step 4** due to incomplete removal of the solvent from each wash during centrifugation and the time it will take to perform each wash. Resuspend the beads in 1 mL of 0.1 N HCl and transfer the samples into the filter column, combining the light and heavy isotope-coded beads. Perform the washes using the following amounts and solution/solvent order: 5–10 mL of 0.1 N HCl, 5 mL of 1% ammonium bicarbonate (pH 8.2), 5–10 mL of 2.0 M sodium chloride, 5–10 mL of methanol, 5–10 mL of acetonitrile, 5 mL of isopropanol, and 5–10 mL of water.
6. Resuspend the washed beads in 100 mM Tris.HC1/10 mM EDTA (pH 6.7) and transfer the beads into a 2 mL autosampler glass vial. After the beads settle to the bottom of the vial, remove as much of the Tris.HC1/EDTA solution. Samples could be stored (wrapped with foil) at –20°C until release of the isotopically labeled peptide can be performed.

#### 3.7. Release of Isotopically Labeled Peptides by Photocleavage

1. Freshly prepare the photolysis buffer 100 mM Tris.HC1/10 mM EDTA (pH 6.7), 3% DTT (w/v), and 5% DMSO (v/v). Degas the solution with nitrogen for 5 min and then suspend the beads with 100 μL of buffer.
2. Fill a 4.5 mL methacrylate cuvette with 25% copper(II) sulfate. Make a hole in the cuvette lid to accommodate the glass vial containing the sample (*see* **Note 21**).
3. Resuspend the beads in 100 μL of the photolysis buffer and place the glass vial into the cuvette; make sure that the glass vial is submerged into the copper(II) sulfate solution.
4. Place the sample under UV illumination (360 nm) using a long-wave UV lamp for 6 h to photocleave the isotope-coded peptides. Gently shake the samples occasionally to facilitate uniform UV illumination of the beads.
5. After the cleavage reaction, collect the supernatant into a 1.5 mL siliconized Eppendorf tube.

6. Wash the beads twice with 100 μL of 30% methanol and add each supernatant to the 1.5 mL Eppendorf tube. This sample contains both the light and heavy PhIST-labeled peptides.
7. Concentrate the pooled sample to a volume of approximately 20 μL by vacuum centrifugation or using a stream of nitrogen to evaporate the solvent (do not reduce the sample to complete dryness to prevent irreversible absorption of sample to the walls of the tube). If to be analyzed later, samples could be stored at –20°C.
8. Dilute the sample with 80 μL of 10% acetonitrile/0.1% formic acid solution and dialyze against 100 mM ammonium bicarbonate (pH 8.0) using a 1 kDa MWCO dialysis membrane (Dispo-Biodialyzer) at 4°C in the same manner as described in **step 6** of **Subheading 3.5**.
9. Once dialysis is complete, place the Dispo-Biodialyzer in the collection tube and spin at low speed (approximately 500 rpm) for 10 s. The PhIST-labeled sample can be stored at –80°C until LC/MS/MS analysis can be performed.
10. Based on the LC/MS/MS system used in the analysis, the sample may have to be reduced in volume using vacuum centrifugation or a stream of nitrogen or lyophilized to complete dryness and then solubilized at the time of analysis (*see* **Note 22**).

### 3.8. Liquid Chromatography–Tandem Mass Spectrometry Analysis

1. The analysis of the IMAC-enriched or PhIST-labeled peptides can be performed using a micro or nanoflow high-performance liquid chromatograph coupled to a tandem mass spectrometer.
2. To prepare the sample for analysis, solubilize the peptides in 5% Mobile Phase B/95% Mobile Phase A to produce a concentration suitable for loading onto the reversed-phase column. To facilitate solubilization of the peptides, the sample can be vortexed with intermittent sonication using a sonicating water bath (*see* **Note 23**).
3. For the initial LC/MS/MS analysis, use a 1% per min linear gradient of Mobile Phase B to determine the separation efficiency of the peptides contained in the sample. Based on this initial analysis, a more complex gradient can be utilized to achieve better peptide separation (*see* **Note 24**).
4. To analyze the acquired product ion spectra, a number of mass modifications need to be used in the database searching algorithm based on the method used. For the IMAC method, mass modifications need to include a variable modification (i.e., both modified and unmodified forms of the residues) of 16.0 Da on Met residues due to oxidation (Met sulfoxide) and a variable phosphorylation modification of 80.0 Da on

Ser, Thr, and Tyr residues. For the PhIST method, static modifications of 32.0 Da on Met and Trp residues and 48.0 Da on Cys residues are needed to account for oxidation via performic acid treatment *(19)* with all product ion spectra analyzed twice: once using a variable mass modification on Ser and Thr residues equal to the additional mass of the light isotopic version of the PhIST label (306.13 Da) and then performed a second time using the mass for the heavy version (313.13 Da) *(20)* (*see* **Note 25**).

## 4. Notes

1. The PhIST labeling strategy cannot be applied to phosphotyrosine residues because of their inability to undergo β-elimination. This will reveal a bias of phosphoproteome studies towards pSer and pThr proteins; however, the chemical labeling strategy allows for simultaneous enrichment of phosphopeptides and site-specific mapping of phosphorylation sites while utilizing stable isotopes for relative quantification between two distinct samples *(14, 24)*. The use of this approach was first demonstrated by Goshe et al. with the phosphoprotein isotope-coded affinity tag (PhIAT) approach using ethanedithiol (EDT) and a biotin affinity label *(25, 26)* and later expanded to PhIST using EDT *(18)* or DTT *(20)* with an isotope-coded label attached via a solid-phase reagent.

   It should be noted that several other chemical labeling techniques have been reported in the literature *(22, 27–36)*. These labeling strategies have been developed on model proteins or peptides and few have been applied to complex proteome samples. It must be emphasized that there is no single technique that will yield the best results. The success of any labeling protocol depends on a number of factors such as sample quantity, preparation, enrichment techniques, and the type of mass spectrometry analysis. Usually, a combination of different enrichment and/or labeling strategies and different LC/MS/MS methods has to be employed for the complete characterization of a phosphoprotein or phosphoproteome. This chapter makes an attempt to address some of these issues using our described protocols.
2. The 100 mM TCEP·HCl (made using water) can be repeatedly used for up to one or two months.
3. Prepare a stock solution of 1.0 μg/μL trypsin by adding the required amount of enzyme resuspension buffer (provided by the manufacturer) to the vial containing the protease. Add the appropriate amount of this trypsin solution to the

sample and store the remaining stock solution at –20°C for future use.

4. If the sample has been dried, add between 25 and 50 μL of loading buffer to the tube and slightly vortex and/or sonicate the sample for 3–5 min using a sonicating water bath to facilitate dissolution of the peptides.

5. The phosphate group can bind to metal ions at a pH between 2.0 and 5.5. However, the carboxylic acid group on aspartyl and glutamyl residues (Asp and Glu) will cause non-specific binding mediated by similar metal coordination as the phosphate group. To reduce this non-specific binding, it is essential to maintain the buffer pH at or below 3.5 so that the acidic residues will be protonated and thus reduce their affinity toward metal ions. To achieve the proper pH, the addition of acetic acid stock solutions (0.2% to 20% stock solutions) can be used to achieve optimal loading pH. Another option to avoid significant non-specific binding is to etherify the carboxylic acid groups on Asp and Glu as well as peptide C-termini *(37)*.

6. If the sample volume is more than 50 μL, extend the incubation time to 4-8 min to facilitate peptide binding.

7. The solution in the column may leak out while removing the bottom seal. To minimize sample loss from this step, pull out the seal carefully and then immediately place the column on a collecting tube.

8. If the samples need more stringent washing to remove the nonphosphorylated peptides, the percentage of acetonitrile in Washing Buffer 2 can be increased to 30%.

9. The collected peptides can be retained and analyzed by LC/MS/MS to determine whether or not the phosphorylated peptides were captured by the IMAC column.

10. The content of Elution Buffer can be modified depending on the subsequent analysis strategy. Theoretically, phosphopeptides will elute at a pH > 6, so 400 mM ammonium hydroxide can be used during elution since it is also compatible with LC/MS/MS analysis. If desalting via a $C_{18}$ online trap column or offline SPE, 10–50 mM sodium phosphate at pH 8.5 can also be used during elution. To increase the elution efficiency, up to 30% of acetonitrile can also be added into the elution solution.

11. To dry DMF, plug a Pasteur glass pipette with glass wool and fill it with a few grams of aluminum oxide. Pass the DMF through the aluminum oxide and collect the solvent in a glass test tube. Other organics like piperidine, pyridine, and $CH_2Cl_2$ must also be dried prior to use in the subsequent reaction steps.

12. The amount of beads used for solid-phase synthesis will depend on the number of amine groups present on the beads as determined by the manufacturer. Our CPG beads contained an amine content of 91 μmol/g, but this value will depend on the processing and pore size and will be specified by the manufacturer. Although smaller pore size glass beads have been used for synthesis (300 Å), the larger pore size (550 Å) facilitates better peptide accessibility and thus is highly recommended.
13. The reaction with acetic anhydride can be carried out for longer (60-90 min) to ensure acetylation of all non-photolinker-modified amine groups.
14. To synthesize the heavy isotope-coded version of the PhIST reagent, use N-α-Fmoc-L-leucine ($^{13}C_6$, 98%; $^{15}N$, 98%). Owing to the high costs of isotope-coded amino acids, it is advisable to perform the synthesis using a smaller molar excess than used with the light version ($^{12}C_6$, $^{14}N$) of the reagent. In this case, synthesis of both light and heavy versions of the solid phase reagent must be carried out using the same molar excess ratios to preserve subsequent labeling of peptides using the same amount of each PhIST reagent.
15. The performic acid solution and the peptide samples must be kept at –20°C for the remainder of the procedure as previously described *(19)*. Performic acid oxidation is a necessary step to prevent β-elimination of the Cys side chain and subsequent PhIST-labeling which complicates LC/MS/MS analysis.
16. Incubation time of 15 min is sufficient for Cys and Met oxidation of peptides. To minimize sample loss, this oxidation step is performed in the same siliconized polypropylene Eppendorf tube that contains the dried peptides.
17. For optimum results, all stock solutions must be freshly prepared before use.
18. Although the β-elimination of phosphate groups at high temperatures (at least 50°C) can be performed to increase the rate of the reaction, the occurrence of side reactions (i.e., β-elimination of Ser residues) also increases *(38)*.
19. The quantitation of the sulfhydryl groups of the DTT-labeled sample can be performed using 5,5′-dithiobis (2-nitrobenzoic acid) assay (DTNB assay) *(39)*. The absorption coefficient $\varepsilon_{412nm} = 13{,}600\ M^{-1}cm^{-1}$ can be used to calculate the molar concentration of sulfhydryl groups.
20. The amount of the isotope-coded reagent added should be in a five-fold molar excess over the amount of DTT-labeled

peptides used during the reduction step. Typically, 5 mg of beads is appropriate for most applications. For this amount of beads, a solution volume of 100 μL promotes sufficient hydration of the beads while facilitating gentle mixing during the labeling reaction.

21. Copper(II) sulfate serves like a filter to block visible light during the photocleavage reaction and helps to reduce photocleavage by-products ***(18)***.

22. In some cases, it may be necessary to reduce the sample to complete dryness. The sample can be stored as a solution after **step 9**, but LC/MS/MS analysis should be performed as soon as possible to avoid unwanted degradation (i.e., oxidation) by-products that may form during long-term storage of the peptide solution. For long-term storage (greater than one month), it is better to lyophilize the sample to complete dryness. In this manner, the peptides can be stored in an air-tight container containing desiccant for months at –80°C and can then be solubilized at the time of LC/MS/MS analysis using the appropriate mobile phase of the LC system.

23. If the dried PhIST-labeled peptides are difficult to solubilize and additional sonication is not helpful, the solution can be adjusted to contain 10% of the organic phase. Higher concentrations of organic mobile phase (up to 25%) may be used to increase solubilization of the PhIST-labeled peptides, but higher percentages of organic mobile phase may compromise column loading such that more polar peptides are not retained or will tend to elute in a narrow retention time window. If any components remain insoluble, centrifuge the sample at 15,000 rpm using a microcentrifuge and transfer the supernatant into an autosampler vial. Alternatively, the sample could be filtered using a pipette tip containing a porous filter.

24. Reversed-phase separation of complex peptide mixtures can be accomplished using a wide variety of stationary phases that contain various imbedded polar groups and proprietary chemical functionalities that can alter peptide separation using a given mobile phase gradient. Consequently, each LC system needs to be standardized for a particular stationary and mobile phase, but for the PhIST-labeled peptides, separations using a mobile phase composition from low organic (1% to 5%) to high organic (75% to 95%) are usually achieved with a single linear gradient at a defined rate (1% to 3% per min), although multiple linear gradients can be introduced to achieve more effective separations.

25. For more details on analyzing product ion spectra for phosphopeptides, please refer the articles utilizing the IMAC and PhIST methods *(14, 17–20, 24)*.

## Acknowledgements

Portions of this work were supported by a grant from United States Department of Agriculture (NRI 2005-35604-15420). We also thank the research agencies of North Carolina State University and the North Carolina Agricultural Research Service for continued support to our biological mass spectrometry research.

### References

1. Larsen, M. R., Trelle, M. B., Thingholm, T. E., and Jensen, O. N. (2006) Analysis of posttranslational modifications of proteins by tandem mass spectrometry. *Biotechniques* **40**, 790–8.
2. Jensen, L. J., Gupta, R., Blom, N., Devos, D., Tamames, J., Kesmir, C., Nielsen, H., Staerfeldt, H. H., Rapacki, K., Workman, C., Andersen, C. A., Knudsen, S., Krogh, A., Valencia, A., and Brunak, S. (2002) Prediction of human protein function from post-translational modifications and localization features. *J Mol Biol* **319**, 1257–65.
3. Nakai, K. (2001) Review: prediction of in vivo fates of proteins in the era of genomics and proteomics. *J Struct Biol* **134**, 103–16.
4. Garavelli, J. S. (2004) The RESID Database of Protein Modifications as a resource and annotation tool. *Proteomics* 4, 1527–33.
5. McLachlin, D. T. and Chait, B. T. (2001) Analysis of phosphorylated proteins and peptides by mass spectrometry. *Curr Opin Chem Biol* **5**, 591–602.
6. Aebersold, R. and Goodlett, D. R. (2001) Mass spectrometry in proteomics. *Chem Rev* **101**, 269–95.
7. Salih, E. (2005) Phosphoproteomics by mass spectrometry and classical protein chemistry approaches. *Mass Spectrom Rev* **24**, 828–46.
8. Domon, B. and Aebersold, R. (2006) Mass spectrometry and protein analysis. *Science* **312**, 212–7.
9. Collins, M. O., Yu, L., and Choudhary, J. S. (2007) Analysis of protein phosphorylation on a proteome-scale. *Proteomics* 7, 2751–68.
10. Leitner, A. and Lindner, W. (2006) Chemistry meets proteomics: the use of chemical tagging reactions for MS-based proteomics. *Proteomics* **6**, 5418–34.
11. Yates, J. R., 3rd, Eng, J. K., McCormack, A. L., and Schieltz, D. (1995) Method to correlate tandem mass spectra of modified peptides to amino acid sequences in the protein database. *Anal Chem* **67**, 1426–36.
12. Wilkins, M. R., Gasteiger, E., Gooley, A. A., Herbert, B. R., Molloy, M. P., Binz, P. A., Ou, K., Sanchez, J. C., Bairoch, A., Williams, K. L., and Hochstrasser, D. F. (1999) High-throughput mass spectrometric discovery of protein post-translational modifications. *J Mol Biol* **289**, 645–57.
13. Loyet, K. M., Stults, J. T., and Arnott, D. (2005) Mass spectrometric contributions to the practice of phosphorylation site mapping through 2003: a literature review. *Mol Cell Proteomics* **4**,235–45. Epub 2005 Jan 7.
14. Goshe, M. B. (2006) Characterizing phosphoproteins and phosphoproteomes using mass spectrometry. *Brief Funct Genomic Proteomic* **4**, 363–76. Epub 2006 Feb 7.
15. Bodenmiller, B., Mueller, L. N., Mueller, M., Domon, B., and Aebersold, R. (2007) Reproducible isolation of distinct, overlapping segments of the phosphoproteome. *Nat Methods* **4**, 231–7. Epub 2007 Feb 11.
16. Mann, M., Ong, S. E., Gronborg, M., Steen, H., Jensen, O. N., and Pandey, A. (2002) Analysis of protein phosphorylation using mass spectrometry: deciphering the phosphoproteome. *Trends Biotechnol* **20**, 261–8.
17. Wang, X., Goshe, M. B., Soderblom, E. J., Phinney, B. S., Kuchar, J. A., Li, J., Asami, T., Yoshida, S., Huber, S. C., and Clouse, S. D. (2005) Identification and functional analysis of in vivo phosphorylation sites of the Arabidopsis BRASSINOSTEROID-INSENSITIVE1

receptor kinase. *Plant Cell* **17**, 1685–703. Epub 2005 May 13.

18. Qian, W. J., Goshe, M. B., Camp, D. G., 2nd, Yu, L. R., Tang, K., and Smith, R. D. (2003) Phosphoprotein isotope-coded solid-phase tag approach for enrichment and quantitative analysis of phosphopeptides from complex mixtures. *Anal Chem* **75**, 5441–50.
19. Soderblom, E. J., Cawthon, D., Duhart, H., Xu, Z. A., Slikker, Jr., W., Ali, S. F., and Goshe, M. B. (2005) An improved labeling method using phosphoprotein isotope-coded solid-phase tags for neuronal cell applications. *Int J Neuroprotec Neuroregen* **1**, 91–7.
20. Cawthon, D., Soderblom, E. J., Xu, Z. A., Duhart, H., Slikker W, JrAli, S. F., and Goshe, M. B. (2005) Quantitative analysis of phosphoproteins in a Parkinson's disease model using phosphoprotein isotope-coded solid-phase tags. *Int J Neuroprotec Neuroregen* **1**, 98–106.
21. Zhou, H., Ranish, J. A., Watts, J. D., and Aebersold, R. (2002) Quantitative proteome analysis by solid-phase isotope tagging and mass spectrometry. *Nat Biotechnol* **20**, 512–5.
22. McLachlin, D. T. and Chait, B. T. (2003) Improved beta-elimination-based affinity purification strategy for enrichment of phosphopeptides. *Anal Chem* **75**, 6826–36.
23. Liu, H. C., Soderblom, E. J., and Goshe, M. B. (2006) A mass spectrometry-based proteomic approach to study Marek's disease virus gene expression. *J Virol Methods* **135**, 66–75.
24. Goshe, M. B. and Smith, R. D. (2003) Stable isotope-coded proteomic mass spectrometry. *Curr Opin Biotechnol* **14**, 101–9.
25. Goshe, M. B., Conrads, T. P., Panisko, E. A., Angell, N. H., Veenstra, T. D., and Smith, R. D. (2001) Phosphoprotein isotope-coded affinity tag approach for isolating and quantitating phosphopeptides in proteome-wide analyses. *Anal Chem* **73**, 2578–86.
26. Goshe, M. B., Veenstra, T. D., Panisko, E. A., Conrads, T. P., Angell, N. H., and Smith, R. D. (2002) Phosphoprotein isotope-coded affinity tags: application to the enrichment and identification of low-abundance phosphoproteins. *Anal Chem* **74**, 607–16.
27. Jaffe, H., Veeranna, and Pant, H. C. (1998) Characterization of serine and threonine phosphorylation sites in beta-elimination/ethanethiol addition-modified proteins by electrospray tandem mass spectrometry and database searching. *Biochemistry* **37**, 16211–24.
28. Weckwerth, W., Willmitzer, L., and Fiehn, O. (2000) Comparative quantification and identification of phosphoproteins using stable isotope labeling and liquid chromatography/mass spectrometry [in process citation]. *Rapid Commun Mass Spectrom* **14**, 1677–81.
29. Oda, Y., Nagasu, T., and Chait, B. (2001) Enrichment analysis of phosphorylated proteins as a tool for probing the phosphoproteome. *Nat Biotechnol* **19**, 379–82.
30. Zhou, H., Watts, J. D., and Aebersold, R. (2001) A systematic approach to the analysis of protein phosphorylation. *Nat Biotechnol* **19**, 375–8.
31. Molloy, M. P. and Andrews, P. C. (2001) Phosphopeptide derivatization signatures to identify serine and threonine phosphorylated peptides by mass spectrometry. *Anal Chem* **73**, 5387–94.
32. Li, W., Boykins, R. A., Backlund, P. S., Wang, G., and Chen, H. C. (2002) Identification of phosphoserine and phosphothreonine as cysteic acid and beta-methylcysteic acid residues in peptides by tandem mass spectrometric sequencing. *Anal Chem* **74**, 5701–10.
33. Adamczyk, M., Gebler, J. C., and Wu, J. (2002) Identification of phosphopeptides by chemical modification with an isotopic tag and ion trap mass spectrometry. *Rapid Commun Mass Spectrom* **16**, 999–1001.
34. Knight, Z. A., Schilling, B., Row, R. H., Kenski, D. M., Gibson, B. W., and Shokat, K. M. (2003) Phosphospecific proteolysis for mapping sites of protein phosphorylation. *Nat Biotechnol* **21**, 1047–54. Epub 2003 Aug 17.
35. Amoresano, A., Marino, G., Cirulli, C., and Quemeneur, E. (2004) Mapping phosphorylation sites: a new strategy based on the use of isotopically labelled DTT and mass spectrometry. *Eur J Mass Spectrom (Chichester, Eng)* **10**, 401–12.
36. Tseng, H. C., Ovaa, H., Wei, N. J., Ploegh, H., and Tsai, L. H. (2005) Phosphoproteomic analysis with a solid-phase capture-release-tag approach. *Chem Biol* **12**, 769–77.
37. Ficarro, S. B., McCleland, M. L., Stukenberg, P. T., Burke, D. J., Ross, M. M., Shabanowitz, J., Hunt, D. F., and White, F. M. (2002) Phosphoproteome analysis by mass spectrometry and its application to Saccharomyces cerevisiae. *Nat Biotechnol* **20**, 301–5.
38. Li, W., Backlund, P. S., Boykins, R. A., Wang, G., and Chen, H.-C. (2003) Susceptibility of the hydroxyl groups in serine and threonine to [beta]-elimination/Michael addition under commonly used moderately high-temperature conditions. *Analytl Biochem* **323**, 94–102.
39. Ellman, G. L. (1959) Tissue sulfhydryl groups. *Arch Biochem Biophys* **82**, 70–7.

# Part VII

## Subcellular Proteomics

# Chapter 18

# Organelle Proteomics: Reduction of Sample Complexity by Enzymatic In-Gel Selection of Native Proteins


## Veronika Reisinger and Lutz A. Eichacker



## Summary

One major problem in proteomics is the biochemical complexity of living cells. Therefore, strategies are needed to reduce the number of proteins to a manageable amount, enabling researchers to make a statement concerning protein functions. One possibility is the isolation of organelles, which reduces the protein complexity, e.g., for the chloroplast to an estimated number of 2,700 different proteins. For further limitation of the protein number, proteins can be divided into membrane and soluble proteins, which can be analyzed separately in a subsequent step. For membrane proteins, blue native polyacrylamide gel electrophoresis (BN-PAGE) in combination with enzymatic in-gel assays (e.g. detection of NADPH dehydrogenases) is a suitable method for a fast and easy visualization and identification of only one class of membrane proteins.



**Key words:** Organelle proteomics, Chloroplast isolation, Native electrophoresis, Enzymatic assay, NADPH dehydrogenase, Protein identification


## 1. Introduction

The challenge of proteomic research is the biochemical complexity, the number, and the dynamic range of proteins. One approach for reducing the complexity of a sample is cell fractionation to an organellar level. In the field of plant research, the isolation of chloroplasts is a standard technique. On the level of the chloroplast, around 2,700 different proteins are predicted *(1)*. To cope with the high number of proteins, two-dimensional IEF/SDS-PAGE, a technique combining a separation according to net charge and molecular mass, is used *(2)*. It allows the separation


Jörg Reinders and Albert Sickmann (eds.), *Proteomics,* Methods in Molecular Biology, Vol. 564
DOI: 10.1007/978-1-60761-157-8_18, © Humana Press, a part of Springer Science+Business Media, LLC 2009

of up to 2,000 different protein spots per gel down to 1 ng *(3–6)*. Despite its outstanding properties, IEF/SDS-PAGE has clear limitations with respect to hydrophobic membrane proteins *(7, 8)*. Therefore, alternative strategies for the separation of membrane proteins were developed *(9)*. But, all techniques struggle with the high dynamic range of proteins in the organelles. In the chloroplast, just 10–15 protein species make up 90% of the protein biomass *(10)*. For visualization and identification of low abundant proteins, alternative strategies have to be applied. One possibility is detection of native proteins by their specific enzymatic activity, which can be directly performed after electrophoresis *(11)*. The following protocol describes the detection of NADPH oxidizing enzyme complexes of the thylakoid membrane after separation by blue native electrophoresis *(12)*.

## 2. Materials

### 2.1. Plastid Isolation

1. Isolation medium: 400 mM sorbitol, 50 mM Hepes-KOH, pH 8.0, 2 mM EDTA. Store at 4°C.
2. 40% Percoll: 400 mM sorbitol, 50 mM Hepes-KOH, pH 8.0, 2 mM EDTA, 40% (v/v) Percoll™ (GE Healthcare). Store at 4°C.
3. 80% Percoll: 400 mM sorbitol, 50 mM Hepes-KOH, pH 8.0, 2 mM EDTA, 80% (v/v) Percoll™ (GE Healthcare). Store at 4°C.
4. Wash medium: 400 mM sorbitol, 50 mM Hepes-KOH, pH 8.0. Store at 4°C.
5. Frequency adjustable Ultrathurax (e.g. Polytron, Kinematica).

### 2.2. Membrane Complex Isolation

1. TMK buffer: 10 mM Tris–HCl pH 6.8, 10 mM $MgCl_2$, 20 mM KCl.
2. Sample buffer: 750 mM ε-amino caproic acid, 50 mM Bis-Tris–HCl pH 7.0, 0.5 mM EDTA-$Na_2$. Store at –20°C.
3. Detergent solution: 10% (w/v) n-Dodecyl-β-D-maltoside. Store at –20°C.
4. Loading buffer: 750 mM ε-amino caproic acid, 5% (w/v) Coomassie G 250. Store at –20°C.

### 2.3. Blue Native Polyacrylamide Gel Electrophoresis

1. Gel buffer (6×): 3 M ε-amino caproic acid, 0.3 M Bis–Tris–HCl pH 7.0. Store at 4°C.
2. Acrylamide solution: 30% (w/v) acrylamide/bisacrylamide solution (37.5:1, 2.6% C), acts in unpolymerized state as a neurotoxin. Store at 4°C.

3. Glycerol (100%). Store at room temperature.
4. TEMED: N,N,N,N′-tetramethyl-ethylenediamine. Store at room temperature.
5. APS: ammonium persulfate: 10% (w/v) solution. Stable at 4°C for up to 2 weeks (*see* **Note 1**).
6. Water-saturated isobutanol: 50% (v/v) isobutanol. Store at room temperature.
7. Blue cathode buffer: 50 mM Tricine, 15 mM Bis-Tris-HCl pH 7.0, 0.02% Coomassie 250 G (Serva) (*see* **Note 2**). Store at 4°C (*see* **Note 3**).
8. Colorless cathode buffer: 50 mM Tricine, 15 mM Bis–Tris–HCl pH 7.0. Store at 4°C (*see* **Note 3**).
9. Anode buffer: 50 mM Bis–Tris–HCl pH 7.0. Store at 4°C (*see* **Note 3**).

### 2.4. Enzymatic Assay

1. Staining solution: 150 mM Tris-HCl pH 8.2, 0.02% (w/v) NADPH, 0.004% (w/v) 2,6-dichlorophenol indophenol (DCIP), 0.014% (w/v) methyl thiazolyl tetrazolium (MTT). MTT does not dissolve completely. Prepare freshly before use.

## 3. Methods

The protocol given here deals with the visualization of thylakoid membrane enzymes catalyzing the oxidation of NADPH in barley chloroplasts. As the whole experiment can be carried out in about two and a half hours, it represents a fast possibility to identify one suborganellar protein fraction in plastids. In general, this method can be assigned to all other kinds of organelles and protein complexes of the cell.

As the proteins are visualized by their enzymatic activity in the terminal step, it is important to keep the proteins in a native state during the whole isolation procedure. Therefore, the equipment and all buffers should be precooled to 4°C before use. Also, all work steps should be performed on ice as far as possible.

### 3.1. Plastid Isolation

1. Cut a tray (21 × 35 cm) of barley leaves about 1 cm above the seed by scissors and collect them in 250 mL ice cold isolation medium. Barley seedlings were grown on Vermiculit for 4, 5 days in a light/dark (16/8 h) regime at 25°C before.
2. Dissect leaves with the scissors roughly in 1-cm long pieces. Subsequently, grind leave pieces with a frequency adjustable Ultrathurax (e.g. Polytron) (*see* **Note 4**).

3. Filtrate the leave pieces through three layers of gauze bandage and one layer of nylon gauze (mesh aperture 0.22 μm) in a centrifuge beaker.
4. Take up the leave residue in another 250 mL of ice cold isolation medium and repeat grinding with the Ultrathurax and filtering.
5. Spin both centrifuge beaker at 3800 × *g* and 4°C for 3 min.
6. Discard the supernatant and resuspend each pellet in 3 mL isolation medium.
7. Prepare an 80%/40% Percoll gradient: Fill 12 mL 40% Percoll in a 30 mL Corex tube. Underlay the 40% Percoll by 7 mL of 80% Percoll (*see* **Note 5**).
8. Filtrate the resupended pellet carefully through the nylon gauze on top of the Percoll gradient.
9. Spin the Percoll gradient in swing out buckets at 4100 × *g* and 4°C for 8 min. Broken chloroplasts will form a band at the top of the Percoll gradient and the intact chloroplasts will form a band at the interface between the 40% and 80% Percoll layers.
10. Discard the broken plastids by pipetting off the top percoll layer. Transfer intact chloroplasts from the 40%/80% interphase with a 1 mL pipette to a new Corex tube.
11. After removal of all intact chloroplasts from the interphase, add at least a twofold volume of wash medium to the intact chloroplasts.
12. Spin the intact chloroplasts down at 4100 × *g* and 4°C for 3 min.
13. Discard the supernatant and resuspend the pellet in about 300 μL wash medium.
14. Transfer the resuspended pellet to a 1.5 mL reaction tube.
15. Determine the chloroplast concentration (*see* **Note 6**): Add 2 μL of chloroplast suspension to 998 μL 80% acetone (1:500 dilution) and mix.
16. Centrifuge for 2 min at 3,000 × *g*. Retain the supernatant.
17. Measure the absorbance of the supernatant at 652 nm. Use the 80% acetone solution as the reference (*see* **Note 7**).
18. Multiply the absorbance by the dilution factor 500 and divide by the extinction coefficient of 36 to obtain the μg of chlorophyll per μL of chloroplast suspension.
19. Keep the chloroplast suspension on ice.

### 3.2. Sample Preparation

1. Take an aliquot of the chloroplast suspension corresponding to 50 μg chlorophyll (*see* **Note 8**) and add 200 μL TMK buffer.

2. Incubate the sample for 10 min on ice. During this incubation step, chloroplasts are lysed.
3. Spinlysed plastids at 7,500 × g in a microfuge for 3 min at 4°C.
4. Discard the supernatant containing stromal proteins and resuspend the thylakoid membrane pellet in 200 μL TMK buffer. Repeat this step one more time to remove all peripheral proteins.
5. Resuspend the pellet in 70 μL sample buffer.
6. Add 10 μL detergent solution and mix gently (*see* **Note 9**). Avoid foaming. This eases underlaying the complete sample in the wells of the acrylamide gel in the subsequent step.
7. Incubate the sample on ice for at least 10 min to solubilize the thylakoid membrane protein complexes.
8. Centrifuge for 10 min at 16.000 × *g* at 4°C to pellet the unsolubilized material. Unsolubilized material can affect the subsequent electrophoretic separation of the protein complexes in a negative way.
9. Transfer the supernatant consisting of the solubilized membrane protein complexes to a new reaction tube containing 5 μL loading buffer. Sample is now ready for blue native electrophoresis.

### 3.3. Blue Native Polyacrylamide Gel Electrophoresis

Electrophoresis is carried out in mini gel format (e.g. SE 250, Hoefer) (*see* **Note 10**). The amounts for casting the separating gel stated here are sufficient to cast 5 mini-gels (10 × 8 × 0.075 cm) in a multiple gel caster.

1. Clean glass plates, notched alumina plates, and spacers with denatured ethanol and assemble the multiple gel caster.
2. Prepare the separating gel solution (7.5% PAG): mix 7.87 mL acrylamide solution with 5.25 mL gel buffer and 18.38 mL $ddH_2O$. Add 30 μL TEMED and 120 μL APS and mix. Immediately pour the gel with a 5 mL pipette, leaving about 1 cm for the stacking gel.
3. Overlay the cast gels with water-saturated isobutanol. Polymerization of the separating gel takes place in about 1 hour.
4. After polymerization, remove the isobutanol by rinsing the gels with $ddH_2O$ (*see* **Note 11**). Dry the area above the separating gel completely with 3MM Whatman paper.
5. Prepare the stacking gel solution (for 1 mini-gel): mix 266 μL acrylamide solution with 333 μL gel buffer and 1378 μL $ddH_2O$ in a 2 mL reaction tube.
6. Fix a gel sandwich with four spring steel clips on the bench. Clean a 10-well comb with denatured ethanol and insert it in the gel sandwich.

7. Add 2 μL TEMED and 20 μL APS to the stacking gel solution and pipette the stacking gel solution between the plates with a 1 mL pipette. The stacking gel polymerizes within 20 min.
8. After the stacking gel has set, remove the comb carefully from the gel sandwich and assemble the buffer chamber.
9. Pour the blue cathode buffer in the upper buffer chamber and rinse the single wells with a 25 μL microsyringe to remove residues of polyacrylamide.
10. Underlay 20 μL of the sample in each well using the microsyringe. If desired, use one well for native molecular weight marker (e.g. HMW native marker kit, GE Healthcare).
11. Add anode buffer to the lower buffer chamber and complete the assembly of the electrophoretic apparatus. Connect the electrophoretic apparatus to the power supply.
12. Set the power supply to 6 mA, 1,000 V, and 24 W and start the electrophoresis. Perform the electrophoretic separation at 4°C either by using a thermostatic circulator or performing electrophoresis in a cold room.
13. After the blue Coomassie dye front has reached half of the separating gel, pause the electrophoretic run and replace the blue cathode buffer by colorless cathode buffer.
14. Continue the electrophoresis until the blue Coomassie dye front reaches the bottom of the separating gel. After the end of electrophoresis, stop cooling, disconnect the electrophoretic unit from the power supply, and disassemble it as well as the gel sandwich. Remove the stacking gel and discard it.
15. The separating gel can now be used for the enzymatic in-gel assay (*see* **Note 12**).

#### 3.4. Enzymatic Assay

1. Incubate the gel in staining solution in the dark at 37°C until purple bands appear.
2. After purple bands are clearly visible, scan the gel immediately by a flat bed scanner and fix the gel in 25% ethanol (**Fig. 1**)
3. After fixation, identification of the purple-stained bands can be performed by mass spectrometry. Therefore, wash the gel two times in dd$H_2O$. Cut out the lanes, digest them enzymatically (e.g. OMX-S, OMX GmbH) and analyze the generated peptides by online mass spectrometry (*see* **Note 13**).

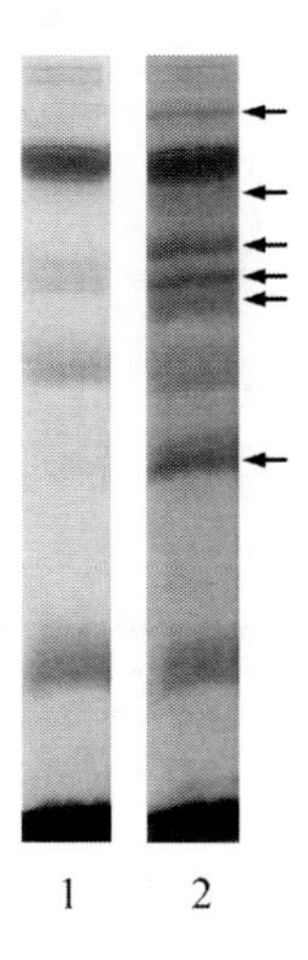


Fig. 1. Detection of NADPH dehydrogenases after BN-PAGE. Native thylakoid membrane complexes were isolated from intact chloroplasts and separated by BN-PAGE *(1)*. After electrophoresis, protein complexes possessing NADPH dehydrogenase activity were visualized as purple bands by an in-gel activity assay (**2**, *black arrows*).

## 4. Notes

1. 10% (w/v) ammonium persulfate solution can be prepared in larger amounts, aliquoted in 2 mL reaction tubes, and stored at –20°C for more than 2 months.
2. It is crucial to use Coomassie 250 G for all buffers used in BN-PAGE owing to the better water solubility of Coomassie G in comparison with Coomassie R.
3. Cathode and anode buffers can be prepared as 10 x stock solutions and stored at 4°C for 1 month or at –20°C for longer time.
4. The use of an ultrathurax is mainly recommended if the leave material is made up primarily of bast fibers. In case of herbaceous leaves, the use of a hand blender (normally used in household) is also fine.
5. Use a 5 mL automatic pipette for underlaying the 80% Percoll solution. Avoid air bubbles releasing from the pipette during underlaying the 40% Percoll by the 80% Percoll solution. Air bubbles can disturb the two phases of the gradient. It is also possible to overlay the 80% solution by the 40% solution, but this is the more time-consuming way.
6. Instead of chlorophyll determination, the amount of protein can also be determined by the number of organelles. Therefore, organelles can be counted in an Abbe-Zeiss counting cell chamber. $1 \times 10^8$ chloroplasts correspond approximately to 50 µg chlorophyll.

7. For absorbance spectroscopy in 80% acetone, the use of optical glass cuvettes is recommended. Disposable cuvettes are incompatible with acetone.
8. 50 μg of chlorophyll correspond approximately to 400 μg membrane proteins. One lane of a mini-gel can be loaded with about 100 μg protein for native electrophoresis.
9. The amount of detergent was proven to be optimal for the solubilization of thylakoid membranes. In case of other types of membranes, the optimal type and concentration of detergent has to be tested individually by dilution series.
10. The use of mini-gels is a fast and small-scale method for the analysis of protein samples, but resolution of minigels is limited. Increase in resolution can be achieved by the combination of gradient gels and large format gels (20 × 20 cm).
11. Polymerization of the separating gel is completed when a layer of water is formed between the upper edge of the gel and the isobutanol layer.
12. As the proteins are fixed in the gel after the enzymatic assay, the three additional lanes (which should be all identical) can be used for Western blotting or analysis of complex composition by SDS-PAGE.
13. On the level of protein complexes, the use of online LC-MS approaches is advisable owing to the possible high number of different proteins contained in the complex.

## References

1. Millar, A. H., Whelan, J., and Small, I. (2006) Recent surprises in protein targeting to mitochondria and plastids. *Curr. Opin. Plant Biol.* **9**, 610–5
2. O'Farrell, P. H. (1975) High resolution two-dimensional electrophoresis of proteins. *J. Biol. Chem.* **250**, 4007–21
3. Peltier, J. B., Friso, G., Kalume, D. E., Roepstorff, P., Nilsson, F., Adamska, I.et al. (2000) Proteomics of the chloroplast: systematic identification and targeting analysis of lumenal and peripheral thylakoid proteins. *Plant Cell* **12**, 319–41
4. Schubert, M., Petersson, U. A., Haas, B. J., Funk, C., Schroder, W. P., and Kieselbach, T. (2002) Proteome map of the chloroplast lumen of Arabidopsis thaliana. *J. Biol. Chem.* **277**, 8354–65
5. Friso, G., Giacomelli, L., Ytterberg, A. J., Peltier, J. B., Rudella, A., Sun, Q.et al. (2004) In-depth analysis of the thylakoid membrane proteome of Arabidopsis thaliana chloroplasts: new proteins, new functions, and a plastid proteome database. *Plant Cell* **16**, 478–99
6. Lopez, J. L. (2007) Two-dimensional electrophoresis in proteome expression analysis. *J. Chromatogr. B Analyt. Technol. Biomed. Life Sci.* **849**, 190–202
7. Canas, B., Pineiro, C., Calvo, E., Lopez-Ferrer, D., and Gallardo, J. M. (2007) Trends in sample preparation for classical and second generation proteomics. *J. Chromatogr. A* **1153**, 235–58
8. Kashino, Y. (2003) Separation methods in the analysis of protein membrane complexes. *J. Chromatogr. B Analyt. Technol. Biomed. Life Sci.* **797**, 191–216
9. Braun, R. J., Kinkl, N., Beer, M., and Ueffing, M. (2007) Two-dimensional electrophoresis of membrane proteins. *Anal. Bioanal. Chem.* **389**, 1033–45

10. van Wijk, K. J. (2004) Plastid proteomics. *Plant Physiol. Biochem.* **42**, 963–77
11. Manchenko, G. (ed.) (2003) *Handbook of Detection of Enzymes on Electrophoretic Gels.* CRC Press LLC, Florida
12. Schagger, H. and von Jagow, G. (1991) Blue native electrophoresis for isolation of membrane protein complexes in enzymatically active form. *Anal. Biochem.* **199**, 223–31

# Chapter 19

# Isolation of Plasma Membranes from the Nervous System by Countercurrent Distribution in Aqueous Polymer Two-Phase Systems

**Jens Schindler and Hans Gerd Nothwang**

## Summary

The plasma membrane separates the cell-interior from the cell's environment. To maintain homeostatic conditions and to enable transfer of information, the plasma membrane is equipped with a variety of different proteins such as transporters, channels, and receptors. The kind and number of plasma membrane proteins are a characteristic of each cell type. Owing to their location, plasma membrane proteins also represent a plethora of drug targets. Their importance has entailed many studies aiming at their proteomic identification and characterization. Therefore, protocols are required that enable their purification in high purity and quantity. Here, we report a protocol, based on aqueous polymer two-phase systems, which fulfils these demands. Furthermore, the protocol is time-saving and protects protein structure and function.

**Key words:** Brain, Plasma membrane, Countercurrent distribution, Two-phase system, Enrichment

## 1. Introduction

Plasma membrane (PM) proteins mediate signal transduction, solute transport, secretion, and cell-cell contact. They are also the central players in the propagation and transmission of action potentials, which are the lingua franca in the nervous system. Finally, ~70% of all known drug targets act on them *(1)*. PM proteins are thus of prime interest in many areas of both basic and biomedical research. However, their proteome analysis is rather difficult, as they encompass only 0.4–2.5% of the total cellular protein amount *(2)*. This renders their identification difficult in the bulk of other, more abundant proteins of the cytoskeleton,

Jörg Reinders and Albert Sickmann (eds.), *Proteomics,* Methods in Molecular Biology, Vol. 564
DOI: 10.1007/978-1-60761-157-8_19, © Humana Press, a part of Springer Science+Business Media, LLC 2009

the energy metabolism, and alike. Furthermore, many properties of PMs such as density overlap with those of other membranous compartments, mainly the endoplasmic reticulum. Classical purification protocols of PM proteins are therefore rather cumbersome and material-consuming. This is for instance the case with the most popular method, consisting of a combination of differential and density gradient centrifugation steps *(3, 4)*. Other methods are either quite expensive and contamination-prone (e.g. immunoprecipitation) or apply only to cultured cells such as surface labeling, which cannot be applied to bulky tissue.

Interest in novel, more efficient subcellular purification protocols has recently emerged from the impressive progress in the analytical part of proteomics. Current mass spectrometry can identify proteins at concentrations of less than 1 pM. This allows for detection of proteins such as the neuronal PSD-95, which is present at 300 copies/postsynaptic density *(5)* in as few as $2 \times 10^6$ neurons, based on 1,000 postsynaptic densities/cells. Hence, proteomics studies on functional or anatomically well-characterized small tissue samples or scarce biopsy material come into reach. However, protocols for the isolation of defined subcellular compartments did not keep pace, despite the recognized need to analyze the subproteome of the various compartments separately. This cellular dissection is mandatory to detect low abundant proteins and to identify compartment-specific post-translational modifications or significant changes in protein localization. Changes therein often underlie physiological and pathophysiological processes *(6)*.

A highly selective and efficient method to separate membranes of different subcellular origin was developed more than 30 years ago by Albertsson and colleagues and was based on the use of aqueous polymer two-phase systems *(7, 8)*. These systems often form, when aqueous solutions of two structural different water-soluble polymers are mixed above a defined concentration. Most often, poly(ethylene glycol) (PEG) and dextran are used as polymers, as they are cheap, require only moderate concentrations, separate easily, and preserve protein structure and function well. When mixed, phases will form, and the upper phase will be enriched in PEG, whereas the bottom phase will mainly contain dextran. Interestingly, the various cellular membranes have different affinities to partition in either of the two phases. PMs prefer the upper phase, whereas mitochondria partition rather to the bottom phase. This different behavior can be attributed to differences in hydrophobic and hydrophilic surface properties of membranes, most likely arising from differences in their phospholipid composition.

Membranes differ only subtly in their surface properties and the isolation of PMs by aqueous polymer two-phase systems cannot be achieved in a single step. One possibility to increase the purity of the PM fraction is multistep extraction procedures such as countercurrent distribution (CD) (**Fig. 1**). CD is based on the Nernst distribution law. Membranes are separated by this method on the basis of their different solubilities in two immiscible aqueous solutions of structurally different polymers. These two phases, flowing into opposite directions, are brought into contact, mixed, and allowed to separate. PMs preferentially partition to the top phase and will reside there throughout the multiple extractions. Intracellular membranes such as mitochondria will preferentially partition to the bottom phase throughout the procedure. We recently adapted this principle to the isolation of

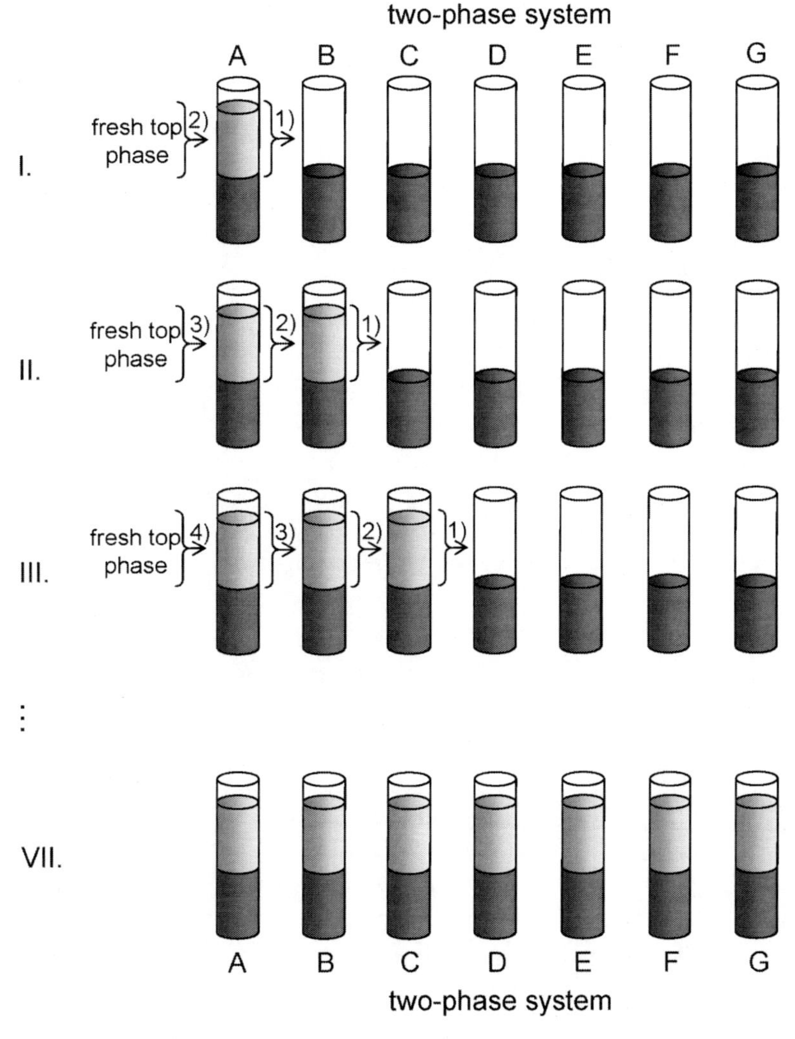


Fig. 1. Scheme of countercurrent distribution. In CD experiments, the top-phase of the first two-phase system A is transferred to a fresh bottom-phase B and the bottom-phase of two-phase system A is re-extracted with a fresh top-phase. After six iterations, biomaterial is efficiently separated.

PMs from the nervous system *(9)*. The protocol is fast, easy to perform, and yields up to 30% of the initial PMs with high purity. Contaminations by endoplasmic reticulum and mitochondria, the major contaminations in standard protocols, were low.

## 2. Materials

Owing to the strong influence of ions on membrane partitioning in the two-phase systems, double distilled water should be used throughout the experiments.

### 2.1. Two-Phase Systems

1. Glass-Teflon homogenizer.
2. Dextran stock solution: Dextran T500 (20%, w/w) (*see* **Note 1**).
3. PEG stock solution: PEG 3350 (40%, w/w).
4. Tris–$H_2SO_4$: Tris (200 mM), pH 7.8 adjusted with $H_2SO_4$.

## 3. Methods

### 3.1. Two-Phase Partitioning

All steps of the affinity two-phase partitioning protocol should be performed at 4°C. Working at room temperature prevents phase separation. The procedure is illustrated in **Fig. 1**. The numbers in **Fig. 1** correspond to the numbered two-phase systems given in **Table 1**, and the letters refer to the phases as indicated in the protocol given below.

1. Prepare seven two-phase systems with the compositions indicated in **Table 1**, 1 day prior to use. Mix them by 20 invertations, vortexing for 10 s, another 20 invertations, and store the mixtures at 4°C overnight. Two-phase systems will form

**Table 1**
**Composition of two-phase systems**

| | Two-phase system "A" (g) | Two-phase systems "B–G" (g) |
|---|---|---|
| Dextran stock solution | 1.035 | 1.035 |
| PEG stock solution | 0.518 | 0.518 |
| Tris–$H_2SO_4$ | 0.750 | 0.750 |
| Water | 0.598 | 0.698 |

overnight with the top phase enriched in PEG and the bottom phase enriched in dextran.

2. On the next day, remove all top phases from two-phase systems "B–G" and store them separately.
3. Homogenize 0.1 g brain tissue in two-phase system "A" using a glass-Teflon homogenizer followed by 45 s of sonication. Centrifuge at 700 × *g* for 5 min to accelerate phase separation.
4. Transfer the top phase (Top) of two-phase system "A" onto the bottom-phase (Bot) of two-phase system "B" (TopA → BotB) (*see* **Note 2**). Add an equal amount of fresh top-phase (stored in **step 2**) onto bottom-phase "A" (I. in **Fig. 1**). Mix both two-phase systems by 20 invertations, vortex for 10 s, then mix again by another 20 invertations. Centrifuge at 700 × *g* for 5 min to accelerate phase separation.
5. Transfer top phases in the following order (II. in **Fig. 1**): (1) TopB → BotC; (2) TopA → BotB. Add an equal amount of fresh top-phase (stored in **step 2**) onto bottom-phase "A." Mix all two-phase systems by 20 invertations, vortexing for 10 s, and another 20 invertations. Centrifuge at 700 × *g* for 5 min to accelerate phase separation.
6. Transfer top phases in the following order (III. in Fig. 1): (1) TopC → BotD; (2) TopB → BotC; (3) Top A → BotB. Add an equal amount of fresh top-phase (stored in **step 2**) onto bottom-phase "A."

   Mix all two-phase systems by 20 invertations, vortexing for 10 s, and another 20 invertations. Centrifuge at 700 × *g* for 5 min to accelerate phase separation.
7. Transfer top phases in the following order: (1) Top D → BotE; (2) TopC → BotD; (3) TopB → BotC; (4) Top A → BotB. Add an equal amount of fresh top-phase (stored in **step 2**) onto bottom-phase "A." Mix all two-phase systems by 20 invertations, vortexing for 10 s, and another 20 invertations. Centrifuge at 700 × *g* for 5 min to accelerate phase separation.
8. Transfer top phases in the following order: (1) TopE → BotF; (2) Top D → BotE; (3) TopC → BotD; (4) TopB → BotC; (5) Top A → BotB. Add an equal amount of fresh top-phase (stored in **step 2**) onto bottom-phase "A." Mix all two-phase systems by 20 invertations, vortexing for 10 s, and another 20 invertations. Centrifuge at 700 × *g* for 5 min to accelerate phase separation.
9. Transfer top phases in the following order: (1) TopF → BotG: (2) TopE → BotF; (3) Top D → BotE; (4) TopC → BotD; (5) TopB → BotC; (6) Top A → BotB. Add an equal amount of fresh top-phase (stored in **step 2**) onto bottom-phase "A."

   Mix all two-phase systems by 20 invertations, vortexing for 10 s, and another 20 invertations. Centrifuge at 700 × *g* for 5 min to accelerate phase separation. After phase separation, you end up with seven two-phase systems (VII. in **Fig. 1**). PMs are enriched in TopF and TopG.

10. PMs can be recovered from TopG or combined TopF + G (*see* **Note 3**) by diluting the phases 1:10 with water followed by ultracentrifugation at 1,50,000 × *g* and 4°C for 1 h.

## 4. Notes

1. Dextran can contain up to 10% water and for that reason has to be freeze-dried. For freeze-drying, dissolve dextran in distilled water in a plastic dish with a large surface (e.g. Petri dish), freeze it at –80°C, and dry it by sublimating the water under vacuum. Store the freeze-dried dextran in closed plastic tubes sealed tightly with parafilm at –20°C. Let it come to room temperature before opening to protect it from humidity.
2. In two-phase systems, interphases are always considered as part of the bottom phase.
3. The purity of PMs from TopG alone is slightly higher but combining TopF and TopG nearly doubles the yield.

## Acknowledgements

This work was funded in part by the Nano+Bio-Center of the University of Kaiserslautern.

### References

1. Hopkins AL, Groom CR. The druggable genome. *Nat Rev Drug Discov* 2002 Sep; 1:727–30.
2. Evans WH. Isolation and characterization of membranes and cell organelles. In: Rickwood D, editor. Preparative Centrifugation. 1st ed. Oxford: IRL Press; 1991. p. 233–70.
3. DePierre JW, Karnovsky ML. Isolation of a nuclear fraction from guinea pig polymorphonuclear leukocytes after controlled hypotonic homogenization. *Biochim Biophys Acta* 1973 Aug 17; 320(1):205–9.
4. Cao R, Li X, Liu Z, Peng X, Hu W, Wang X, et al. Integration of a two-phase partition method into proteomics research on rat liver plasma membrane proteins. *J Proteome Res* 2006 Mar; 5:634–42.
5. Chen X, Vinade L, Leapman RD, Petersen JD, Nakagawa T, Phillips TM, Sheng M, Reese TS (2005) Mass of the postsynaptic density and enumeration of three key molecules. Proc Natl Acad Sci U S A 2005 Aug; 102(32): 11551–56.
6. Dreger M. Subcellular proteomics. *Mass Spectrom Rev* 2003 Jan; 22(1):27–56.
7. Albertsson PA. Partition of Cell Particles and Macromolecules. 3rd ed. New York: Wiley; 1986.
8. Schindler J, Nothwang HG. Aqueous polymer two-phase systems: Effective tools for plasma membrane proteomics. *Proteomics* 2006 Oct; 6:5409–17.
9. Schindler J, Lewandrowski U, Sickmann A, Friauf E. Aqueous polymer two-phase systems for the proteomic analysis of plasma membranes from minute brain samples. J Proteome Res. 2008 Jan; 7(1):432–42.

# Chapter 20

# Enrichment and Preparation of Plasma Membrane Proteins from *Arabidopsis thaliana* for Global Proteomic Analysis Using Liquid Chromatography–Tandem Mass Spectrometry

**Srijeet K. Mitra, Steven D. Clouse, and Michael B. Goshe**

## Summary

The plasma membrane proteins are critical components in cellular control and differentiation and thus are of special interest to those studying signal transduction mechanisms in all organisms. When conducting proteomic studies on membrane components of cells and tissues, the complexity is not simply confined to the large number of proteins present in the sample but also to the highly hydrophobic nature of membrane proteins containing multiple transmembrane domains. Consequently, these proteins are more difficult to analyze by mass spectrometry, particularly if protein sequence coverage is to be established. This chapter contains a method for extraction, solubilization, alkylation, proteolysis, and identification of hydrophobic integral plasma membrane proteins for large-scale proteomic analysis using strong cation exchange chromatography (SCXC) and liquid chromatography–tandem mass spectrometry (LC/MS/MS). In our approach, microsomes are isolated from plant tissue and then subjected to a two-phase extraction procedure to enrich for plasma membranes. Proteins are extracted and solubilized from the membrane using a methanol-aqueous buffer system that allows for effective reduction, cysteinyl alkylation, and tryptic digestion for subsequent SCXC–LC/MS/MS analysis. Our protocol is also amenable to isotope labeling methods to quantify integral membrane protein expression and post-translational modifications. In addition to plants, the method can be applied to other systems quite readily; thus, we anticipate that it will be of general interest to those characterizing plasma membrane proteins of any organism.

**Key words:** Membrane proteins, Mass spectrometry, Liquid chromatography, Enrichment, Phase partitioning, Proteomics, Plant, Arabidopsis

## 1. Introduction

Proteomic analysis of complex mixtures involves mass spectrometry analysis to identify large numbers of proteins, quantify their abundances, and characterize post-translational modifications *(1–4)*. Many researchers are implementing discovery-based

Jörg Reinders and Albert Sickmann (eds.), *Proteomics,* Methods in Molecular Biology, Vol. 564
DOI: 10.1007/978-1-60761-157-8_20, © Humana Press, a part of Springer Science+Business Media, LLC 2009

proteomic measurements that comprise many different species and cell types, including plants such as *Arabidopsis thaliana* (*5–14*). Because of sample complexity, extensive fractionation is often needed to increase proteomic coverage, including isolation of cellular organelles and microsomal fractions before mass spectrometry analysis (*12, 15*). Membrane proteins residing in the lipid bilayer have an additional layer of analytical complexity when compared with aqueous proteins because of their larger size, considerable hydrophobicity, and differing physicochemical properties. Integral membrane proteins containing multiple helical regions or transmembrane domains composed of highly hydrophobic amino acid residues tend to be extremely hydrophobic, whereas β-barrel porins and the peripheral membrane proteins tend to be more hydrophilic. Thus, the ability to successfully conduct membrane proteomic studies requires both effective enrichment of membrane components from cells and tissue samples and subsequent solubilization of both hydrophobic and hydrophilic proteins that are amenable to further processing and mass spectrometry analysis.

Proteomic studies are typically accomplished using either an in-gel or in-solution-based approach that can be coupled to further analysis using liquid chromatography–tandem mass spectrometry (LC/MS/MS). Two-dimensional polyacrylamide gel electrophoresis (2D-PAGE) is the most widely used separation method in proteomics because of its amenability to a variety of protein sample preparation methods. Isoelectric focusing (IEF) used in the first dimension of separation is performed on proteins solublized in nonionic detergents (e.g., Triton X-100 and Brij-58), but these separations usually result in the precipitation of integral membrane proteins at their corresponding IP, thus limiting their detection in 2D-PAGE analysis (*16*). Although these detergents effectively solubilize many types of membrane proteins, they can impede direct analysis using reversed-phase LC/MS/MS by compromising chromatographic separations and electrospray ionization, and thus their use necessitates additional purification procedures that lead to sample losses (*17–19*). The use of MS-compatible organic solvents such as methanol that promote membrane protein solubilization and direct in-solution tryptic digestion for effective reversed-phase LC/MS/MS analysis was first established for proteomic studies of bacterial membranes (*20*) and has been successfully used in other membrane-enriched samples (*18, 20–24*), including our study using 11-day-old Arabidopsis seedlings (*25*).

This chapter describes our detailed methodology of incorporating methanol as a chemical denaturant and solubilization reagent for effectively identifying membrane proteins using LC/MS/MS analysis from both microsomal and plasma membrane enriched fractions isolated from Arabidopsis (*25*). First, we address the method of homogenization and enrichment of microsomes from plant tissue. Second, we describe the preparation of samples enriched in plasma membranes using phase partitioning *(26, 27)*. Third, we present methods using methanol to extract and solubilize membrane proteins from the isolated lipid components to facilitate tryptic digestion that is directly amenable to mass spectrometry analysis. In addition, we reveal several important distinctions that can be applied to other organelle fractionation techniques to enhance membrane protein solubilization and identification using multidimensional LC-based proteomic analysis. Based on its design, the method as described can be applied to other plasma membrane protein systems and is easily coupled to 2D LC/MS/MS analysis using SCXC and reversed-phase chromatography.

## 2. Materials

The protocols described relate to those previously used to obtain membrane proteins from 11-day-old Arabidopsis seedlings *(25)* as outlined in **Fig. 1**. For those interested in applying this method to different cell or tissue types, please refer to **Note 1**.

### 2.1. Enriching Microsomal Fractions from Plant Tissue

1. *Arabidopsis thaliana* ecotype Columbia-0 seeds were obtained from Arabidopsis Stock Center at the Ohio State University, Columbus, OH.
2. Knife blender (Oster mini-blender, Model Galaxie).
3. Sonicating water bath (Branson).
4. Centrifuge capable of 6,000 × *g* at 4°C.
5. Ultracentrifuge capable of 115,000 × *g* at 4°C.
6. Lysis buffer: 50 mM Tris.HC1 pH 7.5 (Thermo Fisher Scientific), 5 mM EDTA (Thermo Fisher Scientific), 5 mM ascorbate (Sigma), 5 mM dithiothreitol (DTT) (Sigma), 0.2% casein hydrolysate (Sigma), 0.6% polyvinylpyrrolidone (PVP) (Sigma), 50 nM microcystin (Calbiochem), 20 mM sodium fluoride (Sigma), 1 mM phenylmethanesulfonyl fluoride (PMSF) (Sigma), and one protease inhibitor cocktail tablet (Roche Diagnostics, Indianapolis, IN).
7. Water (18 MΩ) was distilled and purified using a Barnstead NANOpure Infinity water purification system (Dubuque, IA).

### 2.2. Phase Separation of Plasma Membranes from Microsomal Fractions

1. Sonicating probe (sonic dismembrator model F60, Thermo Fisher Scientific).
2. Siliconized polypropylene Eppendorf tubes (Thermo Fisher Scientific).

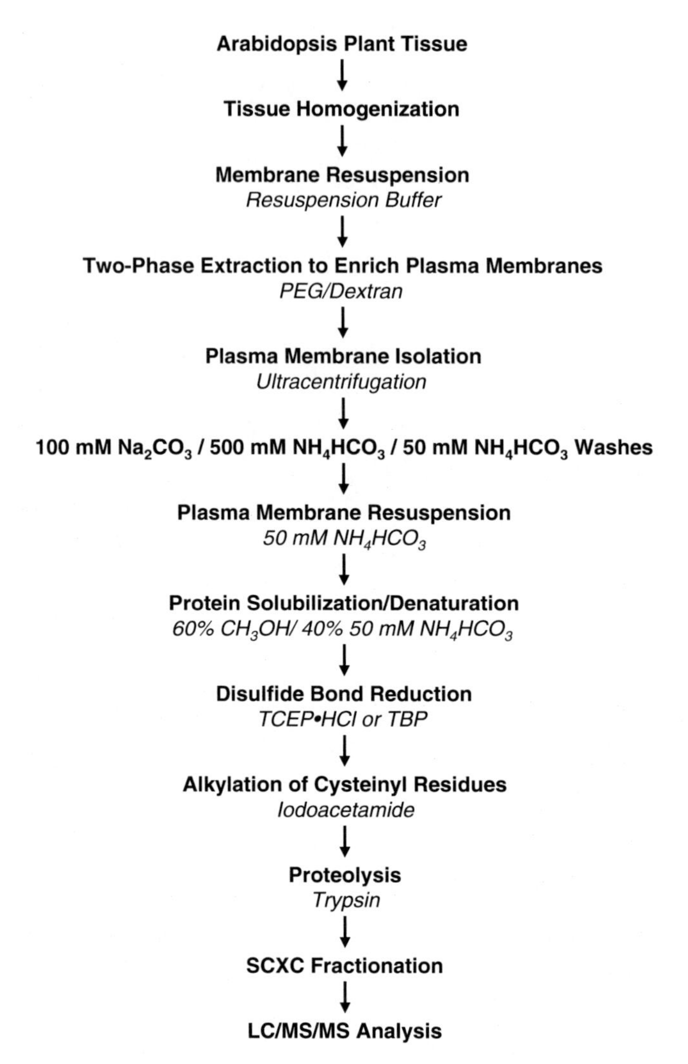


Fig. 1. Flow-diagram of the protocol for isolating and processing plasma membrane proteins from Arabidopsis plants for LC/MS/MS analysis.

3. BCA assay kit from Pierce (Thermo Fisher Scientific).
4. Polypropylene culture tube (16 mL) (Genesee Scientific).
5. Resuspension buffer: 330 mM sucrose (Thermo Fisher Scientific), 4 mM potassium phosphate, pH 7.8 (Thermo Fisher Scientific), 2 mM potassium chloride (Thermo Fisher Scientific), 2 mM EDTA (Sigma), 50 nM microcystin (Calbiochem), 20 mM sodium fluoride (Sigma), 1 mM PMSF (Sigma), and one protease inhibitor cocktail tablet (Roche Diagnostics, Indianapolis, IN).

6. Phase dilution buffer: 330 mM sucrose (Thermo Fisher Scientific), 4 mM potassium phosphate, pH 7.8 (Thermo Fisher Scientific), 2 mM potassium chloride (Thermo Fisher Scientific), 2 mM EDTA (Sigma).
7. Twenty percent polyethylene glycol 3,350 (PEG) (Sigma).
8. Forty percent dextran T500 (Sigma).
9. One hundred millimolar sodium carbonate, pH 11 (Thermo Fisher Scientific).
10. Water (18 MΩ) was distilled and purified using a Barnstead NANOpure Infinity water purification system (Dubuque, IA).

#### 2.3. Methanol-Assisted Solubilization, Denaturation, Reduction, Alkylation, and Proteolytic Digestion of Membrane Proteins

1. Shaker/rotisserie (Labquake).
2. Fifty millimolar and five hundred millimolar ammonium bicarbonate, pH 8.0 (BioChemika Ultra ≥99.5%, Fluka).
3. Methanol (HPLC grade, Thermo Fisher Scientific).
4. Tris-(2-carboxyethyl)phosphine hydrochloride (TCEP·HCl) from Pierce (Thermo Fisher Scientific).
5. Iodoacetamide (IA) (Sigma).
6. Sequencing grade-modified trypsin (Promega).
7. Water (18 MΩ) was distilled and purified using a High-Q 103S water purification system (Wilmette, IL).

#### 2.4. Strong Cation Exchange Chromatography

1. polySULFOETHYL aspartamide SCX column (PolyLC Inc., Columbia, MD).
2. Ammonium bicarbonate and ammonium formate (BioChemika Ultra ≥99.5%, Fluka).
3. Acetonitrile (HPLC grade) and formic acid (ACS regent grade) (Sigma-Aldrich).
4. Mobile phases: (A) 75% 10 mM ammonium formate, pH 3.0/25% acetonitrile (v/v) and (B) 75% 200 mM ammonium formate, pH 8.0/25% acetonitrile (v/v).
5. Water (18 MΩ) was distilled and purified using a High-Q 103S water purification system (Wilmette, IL).

#### 2.5. Liquid Chromatography–Tandem Mass Spectrometry Analysis

1. A reversed-phase HPLC column containing $C_{18}$ stationary phase.
2. Acetonitrile (HPLC grade) and formic acid (ACS regent grade) (Sigma-Aldrich).
3. Mobile phases: (A) 0.1% formic acid in water and (B) 0.1% formic acid in acetonitrile.
4. Water (18 MΩ) was distilled and purified using a High-Q 103S water purification system (Wilmette, IL).

## 3. Methods

### 3.1. Enriching Microsomal Fractions from Plant Tissue

1. Grow Arabidopsis under a given experimental condition and harvest the plants (*see* **Note 2**).
2. Homogenize approximately 20 g of plant material with a blender using 50 mL of ice-cold lysis buffer for up to 10 min until a homogeneous mixture is obtained (*see* **Note 3**).
3. After homogenization, evenly divide the sample into two 25 mL polypropylene centrifuge tubes and place the sample in a sonicating water bath for 20 min.
4. Remove cellular debris by centrifuging at 6,000 × *g* for 15 min at 4°C.
5. Transfer the supernatants into one ultracentrifuge tube (*see* **Note 4**).
6. Perform ultracentrifugation at 100,000 × *g* for 2 h at 4°C to enrich for the membranes (*see* **Note 5**).

### 3.2. Phase Separation of Plasma Membranes from Microsomal Fractions

1. To conduct the PEG/dextran phase separation to enrich for plasma membranes from the microsomal fraction isolated in **Subheading 3.1**, a 6.5% polymer concentration is used *(27)* (*see* **Note 6**).
2. To enrich plasma membranes from the microsomal fractions, a series of three consecutive extractions are performed on ice (**Fig. 2**), each containing a prepared PEG/dextran aqueous medium in 16 mL polypropylene culture tubes using ice-cold (4°C) solutions. To prepare the two-phase system, add 0.65 g of 6.5% PEG, 1.2 g of 6.5% dextran, and 1.15 g of phase dilution buffer to each culture tube. Once prepared, keep all solutions on ice (*see* **Note 7**). For each microsomal preparation described in **Subheading 3.1**, prepare two sets of 3 tubes (referred to as P1, P2, and P3) containing the two-phase system (**Fig. 2a**).
3. Resuspend the microsomal fraction in 2.0 mL of resuspension buffer using a sonicating probe to disrupt the endomembranes (**Fig. 2a**) (*see* **Note 8**).
4. To perform the separation, add 1.0 mL of the resuspended microsomes to the two-phase system P1 (**Fig. 2b**) and mix thoroughly by gently inverting the capped tube 20 times. Place the tube on ice and allow the phase separation to occur (*see* **Note 9**).
5. Remove the upper layer ($TL_1$ of **Fig. 2b**) and transfer it to the pretreated dextran layer $P2_B$. Mix thoroughly and allow the phase separation to occur as described in **step 4**.

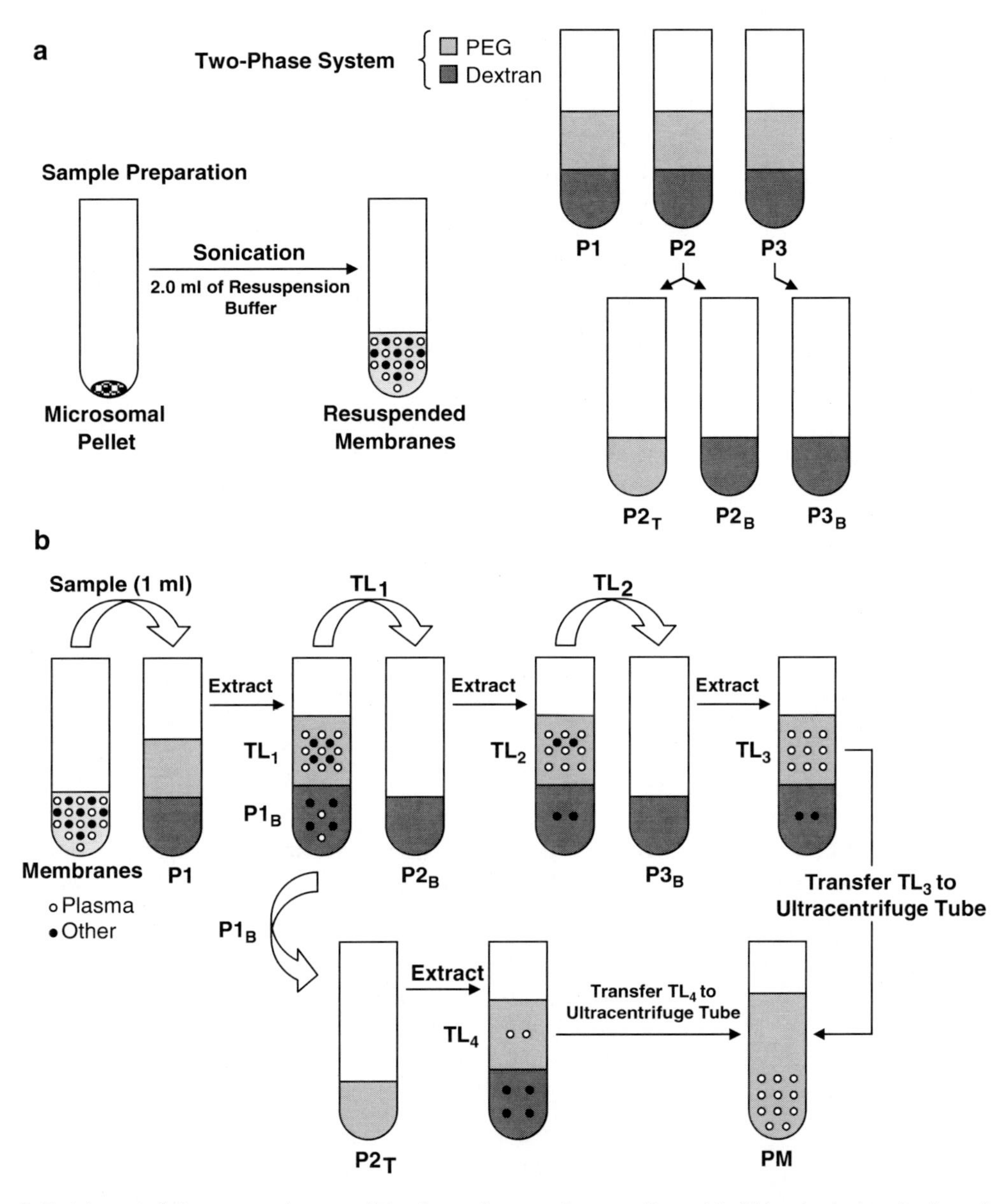


Fig. 2. Enrichment of plasma membrane proteins from microsomal preparations of Arabidopsis plants using two-phase extraction. (**a**) Preparation of the sample and the two-phase system. The membrane components of the microsomal fraction are prepared by adding resuspension buffer and sonicating the sample as described in **Subheading 3.2**. The two-phase system consisting of PEG and dextran is prepared as described in **Subheading 3.2** and then is divided into P1 (the two-phase system), $P2_T$ (pretreated PEG layer), $P2_B$ and $P3_B$ (pretreated dextran layers) for use in the extraction. (**b**) Extraction of plasma membranes. The solutions made in (**a**) are used to enrich for the plasma membranes from the resuspended microsomal fraction. In our protocol, the resuspended membranes are transferred to P1, mixed, and allowed to equilibrate, thus producing the upper PEG layer ($TL_1$) enriched in plasma membranes and a lower dextran layer ($P1_B$) primarily containing the other membrane components. $TL_1$ is further refined with subsequent extractions with the pretreated dextran solutions ($P2_B$ and $P3_B$) to produce a sample highly enriched in plasma membranes ($TL_3$). The plasma membranes remaining in the lower dextran layer from the first extraction step ($P1_B$) are subsequently extracted by a pretreated PEG layer ($P2_T$) and the resulting upper layer ($TL_4$) is transferred to the pooled plasma membrane sample (PM).

6. Remove the upper layer ($TL_2$ of **Fig. 2b**) and transfer it to the pretreated dextran layer $P3_B$. Mix thoroughly and allow the phase separation to occur as described in **step 4**. Transfer the upper layer ($TL_3$ of **Fig. 2b**) to a new polypropylene ultracentrifuge tube. This is the PEG extraction layer enriched in plasma membranes (PM in **Fig. 2b**).
7. To extract any residual plasma membrane that may remain in the dextran layer from the first extraction ($P1_B$ of **Fig. 2b**), transfer this layer to the pretreated PEG solution $P2_T$ (*see* **Note 10**). Mix thoroughly and allow the phase separation to occur as described in **step 4**. Transfer the upper layer ($TL_4$ of **Fig. 2b**) to the PEG extraction layer PM obtained in **step 6**.
8. Repeat **steps 4–7** for the remaining 1.0 mL of resuspended membranes and add the extracted PEG layers to the pooled PM sample of **step 7**.
9. Centrifuge the pooled PM sample at 100,000 × *g* for 2 h at 4°C to pellet the plasma membranes (*see* **Note 11**).
10. Resuspend the plasma membrane pellet in 100 mM sodium carbonate (pH 11) by intermittent vortexing and sonication using a sonicating bath (*see* **Note 12**).
11. Centrifuge at 100,000 × *g* for 1 h at 4°C. (*see* **Note 13**).

### 3.3. Methanol-Assisted Solubilization, Denaturation, Reduction, Alkylation, and Proteolytic Digestion of Membrane Proteins

1. Resuspend the membranes in 500 mM ammonium bicarbonate (pH 8.0) by intermittent vortexing and sonication using a sonicating bath. Centrifuge at 100,000 × *g* for 1 h at 4°C and discard the supernatant. Repeat this procedure using 50 mM ammonium bicarbonate (pH 8.0).
2. Resuspend the membranes using 0.4 mL of 50 mM ammonium bicarbonate (pH 8.0) and transfer the sample into a siliconized polypropylene Eppendorf tube (1.5 mL or 2.0 mL) (*see* **Note 14**).
3. Determine the protein concentration using the BCA assay *(28)* (*see* **Note 15**).
4. To the resuspended membrane sample, add enough methanol to achieve a final composition of 60% methanol/40% 50 mM ammonium bicarbonate (v/v). Solubilize the membrane proteins by intermittent mixing and sonication using a sonicating bath.
5. Incubate the sample in a boiling water bath for 5 min to promote thermal denaturation of the proteins.
6. Once the sample reaches room temperature, add a 30-molar excess of TCEP·HCl to protein and incubate the sample at 37°C for 30 min to reduce disulfide bonds (*see* **Note 16**).

7. Alkylate the thiolate groups of the cysteinyl residues by adding a 30-molar excess of IA to protein and incubate the solution for 90 min in the dark at room temperature while using gentle agitation performed with a shaker/rotisserie (*see* **Note 17**).
8. After alkylation, proteolysis is performed using a 1:20 (w/w) trypsin-to-protein ratio overnight at 37°C (*see* **Note 18**).
9. Once digestion is complete, the sample is frozen by submerging the tube into liquid nitrogen. At this point, the sample can be stored at –80°C until SCXC or LC/MS/MS analysis.

### 3.4. Fractionation of Peptides Using Strong Cation Exchange Chromatography

1. After tryptic digestion, the peptides can be fractionated using SCXC (*see* **Note 19**).
2. To prepare the sample for analysis, solubilize the peptides in Mobile Phase A to produce a concentration suitable for injection onto the SCXC column in use. To facilitate solubilization of the peptides, the sample can be vortexed with intermittent sonication using a sonicating water bath.
3. When using a 5 µm polySULFOETHYL aspartamide SCX column (4.6 mm × 200 mm), load at least 0.2 mg of peptides using a flow rate of 1.0 mL/min. To separate the peptides, use a linear gradient from 0% to 100% Mobile Phase B over 30 min and continue the separation isocratically at 100% Mobile Phase B for 25 min (*see* **Note 20**).
4. Monitor the peptide elution at 215 nm (for all peptides) and 280 nm (for all aromatic-containing peptides) and collect fractions at every 0.5 or 1.0 min.
5. Lyophilize the collected fractions and store at –80°C until LC/MS/MS analysis.

### 3.5. Liquid Chromatography–Tandem Mass Spectrometry Analysis

1. The analysis of the peptides obtained from SCXC fractionation can be performed using a micro- or nanoflow high-performance liquid chromatograph coupled to a tandem mass spectrometer.
2. To prepare the sample for analysis, solubilize the peptides in 5% Mobile Phase B/95% Mobile Phase A to produce a concentration suitable for loading onto the reversed-phase column. To facilitate solubilization of the peptides, the sample can be vortexed with intermittent sonication using a sonicating water bath (*see* **Note 21**).
3. For the initial LC/MS/MS analysis, use a 1% per min linear gradient of Mobile Phase B to determine the separation efficiency of the peptides contained in the sample. Based on this initial analysis, a more complex gradient can be used to achieve better peptide separation (*see* **Note 22**).

## 4. Notes

1. The described protocol for membrane protein enrichment and methanol-assisted, extraction, solubilization, and proteolytic digestion should be applicable to any isolated cells or tissues once the sample has been effectively lysed or homogenized.
2. Plant material can be obtained from a variety of sources. In our study *(25)*, Arabidopsis seedlings were obtained by sterilizing 60 mg of Arabidopsis seeds in ethanol followed by a washing with a 30% bleach solution for 20 min and then extensive washing using copious amounts of water. The seeds were stored at 4°C for 48 h (vernalization) and then transferred into a 250 mL Erlenmeyer flask containing 40 mL of sterile MS medium, which is subjected to constant white light with shaking for 11 days. The lysis buffer used during homogenation can vary, depending on the sample being analyzed and the nature of the study being conducted. For example, our lysis buffer contains the phosphatase inhibitor, microcystin, which can be omitted if phosphorylation sites are not of particular interest. At the time of harvest, the plants were rapidly frozen in liquid nitrogen and stored at –80°C until membrane isolation could be performed.
3. To homogenize the plant material, operate the blender using five consecutive pulses of 2 min each to avoid overheating of the sample.
4. When transferring the samples into the ultracentrifuge tube, use a cheese cloth to filter any remaining debris.
5. The green pellet lying at the bottom of the ultracentrifuge tube contains the membrane-enriched microsomal fraction. The sample can be immediately processed by phase partitioning (**Subheading 3.2**) or stored in the ultracentrifuge tube at –80°C until phase partitioning can be performed. It should be noted that if there is an interest in the microsomal proteome, the sample can be processed at this stage for mass spectrometry analysis according to **Subheading 3.3** *(25)*.
6. Dextran T500 with an average molecular weight of 500 kDa (Sigma) is a good first choice for the dextran component of the medium, since it easily partitions away from the layer containing the lower molecular weight PEG at the concentrations used in our extractions *(29)*.
7. Each stock solution can be stored at 4°C, and the composed two-phase system may be indefinitely stored at –20°C if tightly sealed *(30)*.

8. Microsomal pellets obtained from Arabidopsis plants are green and reside at the bottom of the centrifuge tube as a thick viscous mass. Resuspend the microsomal pellet by adding 2 mL of ice-cold resuspension buffer and use the sonicating probe in continuous mode at setting 5 and sonicate with four consecutive pulses of 1 min. The overall time of sonication depends on the thickness of the pellet and its resuspension in solution. Sonication is complete when the sample becomes translucent.
9. A typical phase separation by gravity takes about 20 min. To accelerate the separation, the tube can be centrifuged at 1,000 rpm for 5 min.
10. Instead of transferring $P1_B$ (the lower dextran layer) to the tube containing $P2_T$ (the pretreated PEG solution), $P2_T$ can be transferred to the tube containing $P1_B$.
11. For ultracentrifugation, dilute the sample with an appropriate volume of resuspension buffer to prevent collapse of the ultracentrifuge tube during the spin.
12. Sodium carbonate treatment disrupts micelles to facilitate the removal of peripheral proteins. The method used is a modified version of those previously reported *(31, 32)*.
13. The pelleted sample can be immediately processed for protease digestion (**Subheading 3.3**) or stored in the ultracentrifuge tube at –80°C until digestion can be performed.
14. Resuspend the microsomal pellet in a minimal volume of 50 mM ammonium bicarbonate (0.4 mL when starting with 20 g of Arabidopsis seedlings) to mediate transfer of the pellet to the microfuge tube. An efficient way to perform the transfer is to use a small nylon brush (Royal 9250 No. 3 round brush) to facilitate the release of the pellet from the bottom of the centrifuge tube, thus allowing the solution and pellet to be easily transferred to the microfuge tube.
15. Determine the protein concentration using the BCA assay with a standard curved generated with bovine serum albumin standards diluted with 50 mM ammonium bicarbonate. The plasma membrane protein concentration should be between 1 and 2 mg/mL so that the addition of methanol produces a protein concentration between 0.5 and 1 mg/mL. If the measured concentration is greater than 2 mg/mL, dilute the sample accordingly with 50 mM ammonium bicarbonate. In a typical preparation, between 0.50 and 0.75 mg of plasma membrane protein is obtained from 20 g of Arabidopsis seedlings.
16. Tributylphosphine (TBP) can be used as an alternative reducing reagent.

17. The addition of IA can be performed by adding the appropriate amount of the solid reagent. IA is a white solid and should be stored in a desiccator at −20°C and completely thawed in a desiccator before use to prevent degradation. If the compound starts to exhibit a noticeable yellow color, it should not be used for alkylation. For most applications, the amount of IA usually required is in the microgram range and thus cannot be accurately measured using a bench-top microbalance. To add an appropriate amount of IA, dissolve at least 5 mg into 50 μL of 60% methanol/40% 50 mM ammonium bicarbonate and then add an appropriate aliquot of this stock IA solution to the sample and discard the remaining material. Once IA has been added to the sample, incubate the sample in the dark by covering the entire tube with aluminum foil to help prevent iodination of tyrosyl residues. Further details regarding the optimization of the cysteinyl alkylation reaction in 60% methanol/40% 50 mM ammonium bicarbonate are described elsewhere *(21)*.
18. Prepare a stock solution of 1.0 μg/μL trypsin by adding the required amount of enzyme resuspension buffer (provided by the manufacturer) to the vial containing the protease. Add the appropriate amount of this trypsin solution to the sample and store the remaining stock solution at −20°C for future use.
19. SCXC is an optional fractionation step that can be bypassed for direct analysis by reversed-phase LC/MS/MS as described in **Subheading 3.5**. Also an optional solid-phase extraction step can be used prior to SCXC or LC/MS/MS as described previously *(25)*.
20. For SCXC, various stationary phases and column lengths can be used, but the mobile phase needs to be comprised of volatile compounds that can be easily removed via vacuum centrifugation. Ammonium formate and ammonium acetate are good reagents for the SCXC mobile phase components. We have used ammonium formate and the polySULFOETHYL aspartamide stationary phase to separate peptides obtained from microsomal fractions of 11-day-old Arabidopsis seedlings *(25)* in addition to other complex peptide mixtures *(33)*. The separations conducted in these studies can be modified using more complex linear gradients and other column configurations if an alternative peptide fractionation is desired.
21. If additional sonication is not helpful, the solution can be adjusted to contain 10% of the organic phase. In this case, the reversed-phase column needs to be equilibrated with the same composition. Higher concentrations of organic mobile phase (up to 25%) may be used to increase solubilization of more hydrophobic peptides, but such a high percentage

of organic mobile phase may compromise column loading such that more polar peptides are not retained or will tend to elute in a narrow retention time window. If any components remain insoluble, centrifuge the sample at 15,000 rpm using a microcentrifuge and transfer the supernatant into an autosampler vial. Alternatively, the sample could be filtered using a pipette tip containing a porous filter.

22. Reversed-phase separation of complex peptide mixtures can be accomplished using a wide variety of $C_{18}$ stationary phases that contain various embedded polar groups and proprietary chemical functionalities that can alter peptide separation using a given mobile phase gradient. Thus, each LC system needs to be standardized for a particular stationary and mobile phase, but for complex peptide mixtures such as those obtained from proteome digests and SCXC fractions, separations from low organic (1% to 5%) to high organic (75% to 95%) are usually achieved with a single linear gradient at a defined rate (1% to 3% per min), although multiple linear gradients can be introduced to achieve more effective separations. With the higher hydrophobic content of the samples generated in our approach, a final organic composition of 75% acetonitrile was not high enough to elute very hydrophobic peptides, so a gradient to 95% acetonitrile was implemented *(25)*.

## Acknowledgements

The authors thank the research agencies of North Carolina State University and the North Carolina Agricultural Research Service for continued support of our biological mass spectrometry research. This work was supported by a grant from the National Science Foundation (MCB-0419819) as part of the Arabidopsis 2010 Project.

## References

1. Goshe, M. B. (2006) Characterizing phosphoproteins and phosphoproteomes using mass spectrometry. *Brief Funct Genomic Proteomic* **4**, 363–76. Epub 2006 Feb 7.
2. Smith, J. C., Lambert, J. P., Elisma, F., and Figeys, D. (2007) Proteomics in 2005/2006: developments, applications and challenges. *Anal Chem* **79**, 4325–43. Epub 2007 May 4.
3. Issaq, H. J., Xiao, Z., and Veenstra, T. D. (2007) Serum and plasma proteomics. *Chem Rev* **107**, 3601–20. Epub 2007 Jul 18.
4. Kota, U. and Goshe, M.B. (2007) Characterizing Proteins and Proteomes using Isotope-Coded Mass Spectrometry in *Spectral Techniques in Proteomics* edited by Daniel S. Sem. CRC Press LLC, Boca Raton FL, pp. 255–285.
5. Ferro, M., Salvi, D., Brugiere, S., Miras, S., Kowalski, S., Louwagie, M., Garin, J., Joyard, J., and Rolland, N. (2003) Proteomics of the chloroplast envelope membranes from Arabidopsis thaliana. *Mol Cell Proteomics* **2**, 325–45. Epub 2003 May 23.

6. Brugiere, S., Kowalski, S., Ferro, M., Seigneurin-Berny, D., Miras, S., Salvi, D., Ravanel, S., d'Herin, P., Garin, J., Bourguignon, J., Joyard, J. and Rolland, N. (2004) The hydrophobic proteome of mitochondrial membranes from Arabidopsis cell suspensions. *Phytochemistry* **65**, 1693–707.
7. Canovas, F. M., Dumas-Gaudot, E., Recorbet, G., Jorrin, J., Mock, H. P. and Rossignol, M. (2004) Plant proteome analysis. *Proteomics* **4**, 285–98.
8. Carter, C., Pan, S., Zouhar, J., Avila, E. L., Girke, T., and Raikhel, N. V. (2004) The vegetative vacuole proteome of *Arabidopsis thaliana* reveals predicted and unexpected proteins. *Plant Cell* **16**, 3285–303. Epub 2004 Nov 11.
9. Ephritikhine, G., Ferro, M. and Rolland, N. (2004) Plant membrane proteomics. *Plant Physiol Biochem* **42**, 943–62.
10. Nuhse, T. S., Stensballe, A., Jensen, O. N. and Peck, S. C. (2004) Phosphoproteomics of the Arabidopsis plasma membrane and a new phosphorylation site database. *Plant Cell* **16**, 2394–405. Epub 2004 Aug 12.
11. Rolland, N., Ferro, M., Ephritikhine, G., Marmagne, A., Ramus, C., Brugiere, S., Salvi, D., Seigneurin-Berny, D., Bourguignon, J., Barbier-Brygoo, H., Joyard, J. and Garin, J. (2006) A versatile method for deciphering plant membrane proteomes. *J Exp Bot* **57**, 1579–89. Epub 2006 Apr 4.
12. Dunkley, T. P., Hester, S., Shadforth, I. P., Runions, J., Weimar, T., Hanton, S. L., Griffin, J. L., Bessant, C., Brandizzi, F., Hawes, C., Watson, R. B., Dupree, P. and Lilley, K. S. (2006) Mapping the Arabidopsis organelle proteome. *Proc Natl Acad Sci U S A* **103**, 6518–23. Epub 2006 Apr 17.
13. Baginsky, S. and Gruissem, W. (2006) *Arabidopsis thaliana* proteomics: from proteome to genome.*J Exp Bot* **57**, 1485–91. Epub 2006 Mar 21.
14. Komatsu, S., Konishi, H. and Hashimoto, M. (2007) The proteomics of plant cell membranes.*J Exp Bot* **58**, 103–12. Epub 2006 Jun 27.
15. Lilley, K. S. and Dupree, P. (2006) Methods of quantitative proteomics and their application to plant organelle characterization. *J Exp Bot* **57**, 1493–9.
16. Whitelegge, J. P., Laganowsky, A., Nishio, J., Souda, P., Zhang, H. and Cramer, W. A. (2006) Sequencing covalent modifications of membrane proteins. *J Exp Bot* **57**, 1515–22. Epub 2006 Mar 30.
17. Funk, J., Li, X. and Franz, T. (2005) Threshold values for detergents in protein and peptide samples for mass spectrometry. *Rapid Commun Mass Spectrom* **19**, 2986–8.
18. Blonder, J., Conrads, T. P., Yu, L. R., Terunuma, A., Janini, G. M., Issaq, H. J., Vogel, J. C. and Veenstra, T. D. (2004) A detergent- and cyanogen bromide-free method for integral membrane proteomics: application to Halobacterium purple membranes and the human epidermal membrane proteome. *Proteomics* **4**, 31–45.
19. Loo, R. R., Dales, N. and Andrews, P. C. (1996) The effect of detergents on proteins analyzed by electrospray ionization. *Methods Mol Biol* **61**, 141–60.
20. Blonder, J., Goshe, M. B., Moore, R. J., Pasa-Tolic, L., Masselon, C. D., Lipton, M. S. and Smith, R. D. (2002) Enrichment of integral membrane proteins for proteomic analysis using liquid chromatography-tandem mass spectrometry. *J Proteome Res* **1**, 351–60.
21. Goshe, M. B., Blonder, J. and Smith, R. D. (2003) Affinity labeling of highly hydrophobic integral membrane proteins for proteome-wide analysis. *J Proteome Res* **2**, 153–61.
22. Blonder, J., Goshe, M. B., Xiao, W., Camp, D. G., 2nd, Wingerd, M., Davis, R. W. and Smith, R. D. (2004) Global analysis of the membrane subproteome of *Pseudomonas aeruginosa* using liquid chromatography–tandem mass spectrometry. *J Proteome Res* **3**, 434–44.
23. Karsan, A., Blonder, J., Law, J., Yaquian, E., Lucas, D. A., Conrads, T. P. and Veenstra, T. (2005) Proteomic analysis of lipid microdomains from lipopolysaccharide-activated human endothelial cells. *J Proteome Res* **4**, 349–57.
24. Fischer, F., Wolters, D., Rogner, M. and Poetsch, A. (2006) Toward the complete membrane proteome: high coverage of integral membrane proteins through transmembrane peptide detection. *Mol Cell Proteomics* **5**, 444–53. Epub 2005 Nov 16.
25. Mitra, S. K., Gantt, J. A., Ruby, J. F., Clouse, S. D. and Goshe, M. B. (2007) Membrane proteomic analysis of *Arabidopsis thaliana* using alternative solubilization techniques. *J Proteome Res* **6**, 1933–50. Epub 2007 Apr 14.
26. Kjellbom, P. and Larsson, C. (1984) Preparation and polypeptide composition of chlorophyll-free plasma membranes from leaves of light-grown spinach and barley. *Physiol Plant* **62**, 501–9.
27. Larsson, C., Sommarin, M. and Widell, S. (1994) Isolation of highly purified plant plasma membranes and separation of inside-out and right-side-out vesicles. *Methods Enzymol* **228**, 451–69.
28. Smith, P. K., Krohn, R. I., Hermanson, G. T., Mallia, A. K., Gartner, F. H., Provenzano, M.

D., Fujimoto, E. K., Goeke, N. M., Olson, B. J. and Klenk, D. C. (1985) Measurement of protein using bicinchoninic acid. *Anal Biochem* **150**, 76–85.

29. Morre, D. J. and Andersson, B. (1994) Isolation of all major organelles and membranous cell components from a single homogenate of green leaves. *Methods Enzymol* **228**, 412–9.
30. Brooks, D. E. and Norris-Jones, R. (1994) Preparation and analysis of two-phase systems. *Methods Enzymol* **228**, 14–27.
31. Fujiki, Y., Hubbard, A. L., Fowler, S. and Lazarow, P. B. (1982) Isolation of intracellular membranes by means of sodium carbonate treatment: application to endoplasmic reticulum. *J Cell Biol* **93**, 97–102.
32. Molloy, M. P., Herbert, B. R., Slade, M. B., Rabilloud, T., Nouwens, A. S., Williams, K. L. and Gooley, A. A. (2000) Proteomic analysis of the Escherichia coli outer membrane. *Eur J Biochem* **267**, 2871–81.
33. Liu, H. C., Soderblom, E. J. and Goshe, M. B. (2006) A mass spectrometry-based proteomic approach to study Marek's Disease Virus gene expression. *J Virol Methods* **135**, 66–75. Epub 2006 Mar 24.

# Part VIII

## Analysis of Protein Interactions

# Chapter 21

## Tandem Affinity Purification of Protein Complexes from Mammalian Cells by the Strep/FLAG (SF)-TAP Tag

Christian Johannes Gloeckner, Karsten Boldt, Annette Schumacher, and Marius Ueffing

### Summary

Isolation and dissection of native multiprotein complexes is a central theme in functional genomics. The development of the tandem affinity purification (TAP) tag has enabled efficient and large-scale purification of native protein complexes. The SF-TAP tag, a modified version of the TAP tag, allows a fast and straightforward purification of protein complexes from mammalian cells. It consists of a tandem Strep-tag II and a FLAG epitope (SF-TAP). The SF-TAP tag allows a native elution of protein complexes without proteolytic cleavage needed in the original TAP procedure. Besides the SF-TAP protocol, the principal idea of a pathway mapping by subsequent tagging of copurified proteins is demonstrated for the interactome of the MAPKKK Raf.

**Key words:** Protein complexes, Tandem affinity tag, TAP, SF-TAP

## 1. Introduction

Determining functional relationships between proteins in eukaryotic cells is one of the most important tasks in the postgenomic era *(1)*. Increasing attention has been directed to the analysis of functional protein complexes. Affinity-based methods enabling access to the complex molecular entities that form and coordinate signaling networks within biological systems are of critical importance. Several methods for isolating native protein complexes that allow a mass spectrometric analysis of the assembled proteins have been described for yeast, bacteria, plant, insect, and

Jörg Reinders and Albert Sickmann (eds.), *Proteomics,* Methods in Molecular Biology, Vol. 564
DOI: 10.1007/978-1-60761-157-8_21, © Humana Press, a part of Springer Science+Business Media, LLC 2009

mammalian cells *(2)*. One of them, the tandem affinity purification (TAP) method, originally developed for the analysis of protein interactions in yeast, allows systematic protein–protein interaction analysis under near-physiological conditions *(3, 4)*. It consists of two protein A IgG binding domains, a TEV (tobacco etch virus) protease cleavage site and a CBP (calmodulin binding peptide) domain. In comparison to purifications via a single affinity tag or the classical immuno-precipitation, a two-step procedure results in a marked reduction of the background caused by unspecific binding of proteins. Although the original TAP-tag has been applied to mammalian cell culture systems as well *(5, 6)*, the demand for advanced TAP-tag approaches is demonstrated by several reports describing modified TAP-tag versions like the GS-TAP tag. The latter combines a Protein G and a streptavidin-binding peptide *(7)*. Like the GS-TAP tag, most of them are dependent on proteolytic cleavage for elution after the first affinity purification step. Recently, we developed a novel TAP combination (SF-TAP), which omits the proteolytic cleavage due to the combination of medium affinity tags, which can be eluted by competition *(8)*. Additionally, the selection for rather small tags resulted in a fourfold size reduction of the SF-TAP tag (5 kDa) when compared with the original TAP tag (21 kDa). This TAP combination consists of a tandem Strep-tag II *(9, 10)* and a FLAG tag (SF-TAP). Desthiobiotin is used for the elution of tagged proteins from the Strep-Tactin resin in the first step and the FLAG octapeptide for the elution from the anti-FLAG resin in the second step. An overview of the SF-TAP procedure including the tag sequence is shown in **Fig. 1**. Owing to omission of the proteolytic step, the overall time for a TAP could be reduced to 2½ h when compared with the original TAP protocol. In the following sections, the SF-TAP protocol is demonstrated by the purification of protein complexes involved in MAPK signaling using the MAPKKKs B-Raf and Raf-1, their downstream target MEK, and the scaffolding protein 14-3-3 as baits. Other examples of successful SF-TAP approaches can be found in **refs.** ***11, 12***. The subsequent tagging of proteins coprecipitated with an SF-TAP fusion protein can be used for a systematic analysis and annotation of protein interaction networks. To allow a fast and easy cloning of SF-TAP fusion constructs, expression vectors have been generated on the basis of the pcDNA3.0 (Invitrogen) expression vector containing an N- or C-terminal version of the SF-TAP tag and a Gateway for recombination-based cloning. The vector allows the expression of SF-TAP fusion proteins in mammalian cells under the control of the viral CMV promoter.

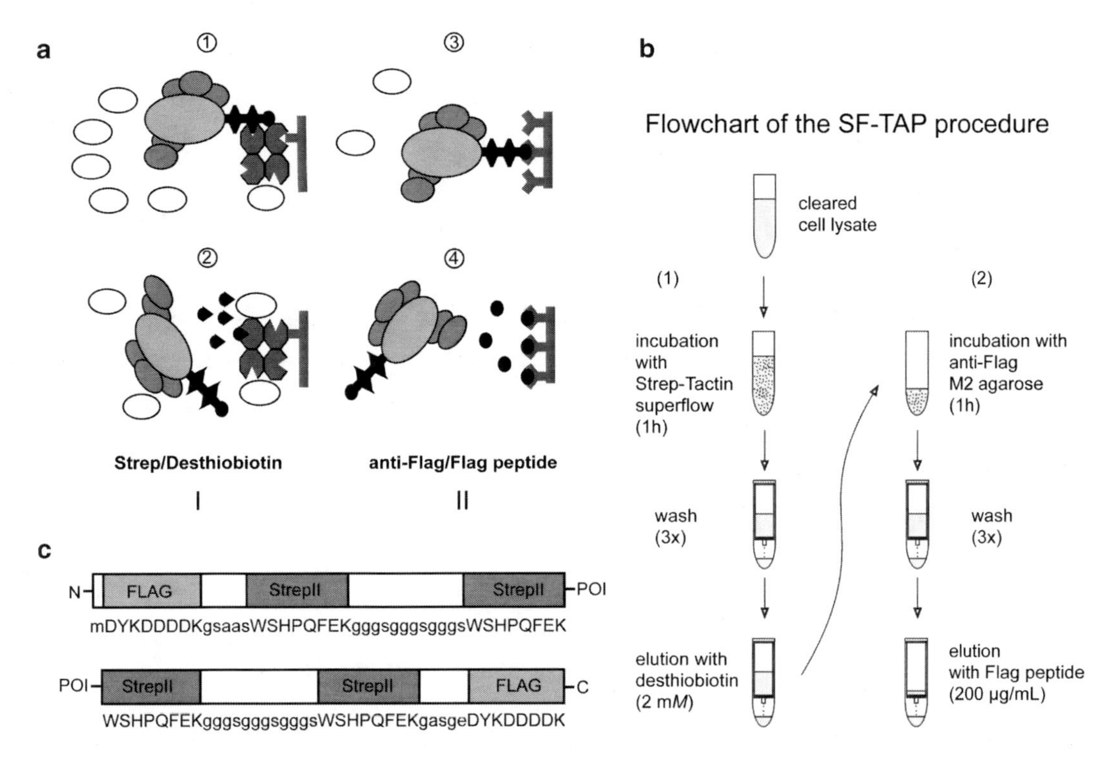


Fig. 1. Overview of the SF-TAP procedure. (**a**) In the first purification step (I), the tandem Strep-tag II is bound by Strep-Tactin matrix *(1)* and eluted by desthiobiotin *(2)*. In the second purification step (II), the FLAG moiety is bound to anit-FLAG M2 agarose and eluted by the FLAG peptide. (**b**) *(1)* First purification of the SF-fusion protein by the Strep/Strep-Tactin system. *(2)* Second purification of the SF-fusion protein by the FLAG/anti FLAG M2 system (source: ref.**8***)*. (c) Sequences of the N- and C-terminal versions of the SF-TAP tag (POI = protein of interest) (source: **ref.**8*)*.

## 2. Materials

### 2.1. Gateway Cloning

1. *Escherichia coli ccd* B-survival T1 phage resistant (T1$^{R}$) strain (Invitrogen) for amplification of donor (pDONR) and destination (pDEST/N-SF-TAP, pDEST/C-SF-TAP) plasmids.
2. pDONR201 plasmid (Invitrogen).
3. Ten micromolar stock of *att*B1 primer (**Fig. 2b** *[9]*).
4. Ten micromolar stock of *att*B2 primer (**Fig. 2b** *[10]*).
5. Ten micromolar stock of gene specific primers (**Fig. 2b** *[2, 4, 6, 8]*).
6. Phusion High Fidelity PCR Kit (NEB) including Phusion Tag, HF-Reaction Buffer, and 10 mM dNTP mix.
7. pDEST/N-SF-TAP/pDEST/C-SF-TAP.
8. Appropriate buffers and equipment for agarose gel electrophoresis for DNA separation.

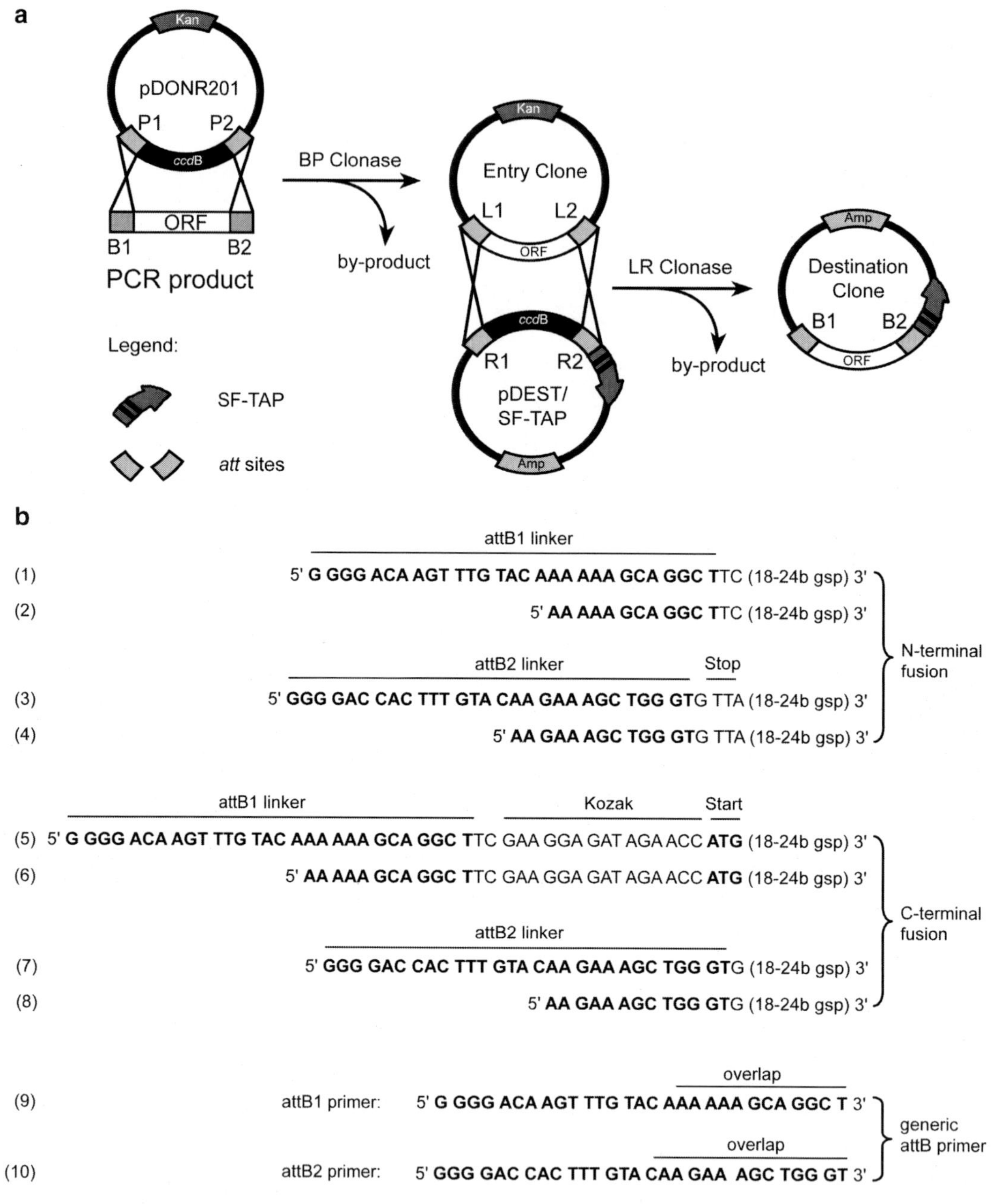


Fig. 2. Overview of the Gateway cloning strategy. (**a**) Schematic overview of the BP and LR reaction used to produce the entry clones and SF-TAP tagged destination clones, respectively. The second PCR product containing *att* B1/B2 sites is recombined with the *att* P1/P2 sites of the pDONR201 vector using the BP Clonase Mix. For selection of the entry clones, kanamycin, and the *ccd* **B** negative selection marker are used. In the LR reaction, the confirmed entry clones (containing *att* L1/L2 sites) are recombined with the SF-TAP destination vectors (containing *att* R1/R2 sites). The resulting destination clones are selected by an ampicillin resistance gene. The entry and destination vectors contain different antibiotic selection markers to select against the LR by-products. The *ccd* B marker is used for negative selection of empty destination vectors. (**b**) Primers used for the PCR steps preceding the BP reaction. *att* B1/B2 linker sequences are attached to the 5′-end of gene-specific primers (gsp). For N-terminal fusion products primers *(1)–(4)* are used. To allow in frame cloning to the N-terminal SF-TAP tag, two nucleotides have been inserted after the attB1 linker (*bold*) of the forward primer *(1)*, *(2)*. In the reverse primer *(3)*, *(4)* a stop codon has been introduced. For C-terminal fusion products, primers *(5)–(8)* are used. In the forward primer *(5)*, *(6)* a Kozak sequence including a start codon has been introduced after the *att* B1 site.

9. Optional: appropriate commercial kits for extraction of DNA from agarose gels.
10. BP Clonase Mix II (Invitrogen).
11. LR Clonase Mix II (Invitrogen).
12. Ten millimolar Tris–HCl, pH 8.0.
13. Deionized sterile water.
14. LB (Luria Bertani) agar plates and liquid media supplemented with appropriate selection antibiotics (kanamycin, ampicillin) according to standard protocols *(13)*.
15. Appropriate commercial kits for plasmid isolation from 5 mL minicultures.

### 2.2. Cell Culture of HEK293 Cells

1. Dulbeco's Modified Eagle's Medium (DMEM) (Invitrogen).
2. Heat-inactivated fetal bovine serum (Invitrogen).
3. Penicillin/Streptomycin (Invitrogen).
4. PBS (Invitrogen).
5. Trypsin/EDTA (Invitrogen).
6. Freezing buffer: 90% FBS (Invitrogen), 10% DMSO (Sigma).
7. Two milliliter cryovials (Nunc).

### 2.3. Transfection of HEK293 Cell and Generation of Cell Lines Stably Expressing the SF-TAP Fusion Protein

1. Effectene Transfection Kit (Qiagen).
2. G-418 (PAA).
3. Anti-FLAG-M2 antibody (Sigma).

### 2.4. Cell Lysis and SF-TAP Purification

1. Cell scraper (Sarstedt).
2. Millex GP 0.22 μm syringe driven filter units (Millipore).
3. Microspin columns (GE-Healthcare).
4. Microcon YM-3 centrifugal filter units (Millipore).
5. 10X TBS: to prepare 1 L of 10X TBS, dissolve 36.3 g of Tris (Merck), and 87.7 g sodium chloride in water, adjust the pH with HCl to 7.4, and fill up to 1 L with water.
6. 50X Protease Inhibitor Cocktail: dissolve 1 tablet of the protease inhibitor cocktail (Roche) in 1 mL of water.
7. Lysis buffer: to prepare 10 mL of lysis buffer, add 1 mL of 10X TBS, 200 μL of Protease Inhibitor Cocktail, 100 μL

Fig. 2. (continued) To allow in frame cloning to the C-terminal SF-TAP tag, one nucleotide has been inserted after the *att*B2 linker (*bold*) of the reverse primer *(7)*, *(8)*. To avoid inefficient PCR amplification, a two-step PCR strategy is recommended. The *att*B sites partially attached to the 5′-end of the gene-specific primers *(2)*/*(4)* and *(6)*/*(8)* used in the first amplification step. The second PCR restores full *att*B sites by generic primers *(9)*/*(10)*, which overlaps with the 5′-overhang of the gene-specific primers. The overlapping sequence is indicated.

of Phosphatase Inhibitor Cocktail I and II (Sigma-Aldrich), and 50 µL NP40 to 8.55 mL water (*see* **Note 1**).

8. Wash buffer: to prepare 10 mL of wash buffer, add 1 mL of 10X TBS, 100 µL of Phosphatase Inhibitor Cocktail I and II, and 10 µL NP40 to 8.79 mL water.
9. TBS buffer: to prepare 10 mL, dilute 1 mL of 10X TBS in 9 mL of water.
10. Desthiobiotin elution buffer: dilute 10X buffer E (IBA) 1 in 10 in water (final concentration: 2 mM desthiobiotin).
11. FLAG peptide stock solution: dissolve 1 mg of FLAG peptide (Sigma-Aldrich) in 800 µL TBS buffer. Store aliquots at –80 C.
12. FLAG elution buffer: dilute FLAG peptide stock solution in 1 in 25 in TBS buffer (final concentration: 200 µg/mL Flag peptide).
13. Strep-Tactin Superflow (IBA).
14. Anti-FLAG-M2 affinity gel (Sigma-Aldrich).

#### 2.5. Chloroform Methanol Precipitation

1. Two milliliter sample tube (Eppendorf).
2. Chlorophorm (Merck).
3. Methanol (Merck).

#### 2.6. Alkylation and Tryptic Digestion

1. RapiGest stock solution: 2% RapiGest (Waters) in water.
2. Fifty millimolar ammoniumbicarbonate: dissolve 0.39 g of ammoniumbicarbonate (Sigma-Aldrich) in 100 mL of water (*see* **Note 21**).
3. One hundred millimolar DTT: dissolve 0.154 g 1,4-dithiothreitol in 10 mL of water (*see* **Note 21**).
4. Three hundred millimolar iodacetamide: dissolve 0.277 g of 2-iodacetamide in 5 mL of water (*see* **Notes 19** and **21**).
5. Trypsin solution (0.5 mg/mL): dissolve one vial (20 µg) of sequencing grade modified trypsin (Promega) in 40 µL of trypsin resuspension dilution buffer (Promega) (*see* **Note 2**).
6. Thirty-seven percent HCl (Merck).

## 3. Methods

#### 3.1. Cloning of SF-TAP Expression Constructs by Gateway Cloning

The Gateway cloning system (Invitrogen) is based on site-specific recombination mediated by the λ integrase family of recombinases *(14)*. It allows fast and easy cloning of inserts, a prerequisite for the generation of larger numbers of expression vectors. The Gateway system consists of two classes of vectors, the donor and the destination vectors. The donor vectors are used for the generation of generic entry clones, which contain the CDS of proteins of interest (POI). The entry clones are recombined with destination vectors to obtain specific expression constructs (destination clones).

In the presented Gateway cloning strategy, the open reading frames of the POI will be first amplified by PCR. The PCR primers contain the *att*B1/2 recombination sites to allow a recombination with the pDONR201 plasmid for the generation of an Entry Clone in the LR reaction. Since the *att*B sites are quite long (29 bp), a two-step PCR strategy is advantageous. In the first PCR, half of the needed *att*B sites are attached to the primers, and in the second PCR, generic *att* B1/2 primers for all constructs are used to complete the *att*B sites (*see* **Fig. 2b**).

The entry clones are then recombined with the appropriate destination plasmid, either pDEST N-SF-TAP or pDEST C-SF-TAP, to obtain the destination expression clone. An overview of the Gateway cloning strategy is shown in **Fig. 2a**. Note: the donor vectors and destination vectors contain a *ccd*B marker for negative selection to avoid background caused by empty pDONR and pDEST vectors in the BP and LR reactions, respectively. The *ccd*B gene is toxic for *E. coli* K12 strains like DH5α. Only strains containing the *ccd*A gene, such as DB3.1 or the *ccd*B survival $T1^R$ strain (Invitrogen), can be used for amplification of these plasmids. For proper amplification of plasmids, refer standard protocols *(13)*.

1. Design gene-specific primers for the open reading frame of interest flanked by *att*B linker sequences (*see* **Fig. 2b**). For N-terminal fusion constructs, the correct open reading frame has to be considered for the design of the forward primer. Additionally, an appropriate stop codon has to be inserted into the reverse primer. For C-terminal fusion constructs, the forward primer should contain a Kozak sequence.
2. *First PCR*- 5 μL HF-Reaction Buffer (provided with the polymerase kit)
   - 0.5 μL dNTP mix (provided with the polymerase kit)
   - 1.3 μL forward/reverse (each) gene specific primer stocks (*see* **Note 3**)
   - X μL template *(5–10* ng)
   - 1 μL Phusion taq polymerase
   - Add water to a total volume of 25 μL

   Use the following PCR program for the thermocycler:

   | | |
   |---|---|
   | 1 cycle | 98° C 1 min |
   | 10 cycles | 96° C 20 s |
   | | 55° C 40s |
   | | 72° C 1 min per kB insert size |

3. *Second PCR* to 10 µL of the first PCR reaction add
   - 10 µL HF-Reaction Buffer (provided with the polymerase kit)
   - 1 µL dNTP mix (provided with the polymerase kit)
   - 1.5 µL of *att*B1 and *att*B2 primer stocks (*see* **Note 4**)
   - 1 µL Phusion taq polymerase
   - Add water to a final volume of 50 µL

   Use the following PCR program for the thermocycler:

   1 cycle 98° C 1 min

   10 cycles 96° C 20 s
   55° C 40 s
   72° C 1 min per kB insert size

   20 cycles 95° C 20 s
   55° C 40 s
   72° C 1 min per kB insert size
4. Check PCR product by agarose gel electrophoresis according to standard protocols (*see* **Note 5**).
5. *BP reaction*
   (a) To 3 µL PCR product add
      - 1 µL pDONR201 plasmid (90 ng/µL)
      - 1 µL BP Clonase Mix II (Invitrogen)

   (b) Incubate for 1–3 h at 25° C

   (c) Add 0.5 µL Proteinase K (*see* **Note 6**)

   (d) Incubate for 10 min at 37° C
6. Transform 5 µL of the BP reaction into DH5α using standard protocols *(13)* and plate cells onto LB kanamycin agar plates.
7. Collect clones and inoculate 5 mL (LB kanamycin) overnight cultures.
8. Isolate entry clone plasmids using an appropriate commercial kit.
9. Verify the inserts by sequencing (*see* **Note 7**).
10. *LR reaction:*
   (a) To 1 µL entry plasmid (90 ng/µL) add
      - 2 µL 10 mM Tris-HCl pH 8.0
      - 1 µL pDEST/(N- or C-) SF-TAP plasmid (90 ng/µL)
      - 1 µL LR Clonase Mix II (Invitrogen)

   (b) Incubate for 1–3 h at 25° C

   (c) Add 0.5 µL Proteinase K (*see* **Note 6**)

   (d) Incubate for 10 min at 37° C

11. Transform 5 μL of the LR reaction into DH5α using standard protocols and plate cells onto LB ampicillin agar plates.
12. Collect clones and inoculate 5 mL (LB ampicillin) overnight cultures.
13. Isolate plasmids using an appropriate commercial kit.
14. Confirm clones by sequencing (*see* **Note 8**).

### 3.2. Transfection of HEK293 Cells and Generation of Cell Lines Stably Expressing the SF-TAP Fusion Protein

For pilot experiments, the SF-TAP approach can be tested in HEK293 cells transiently expressing the bait protein. There are two major advantages of using HEK293 cells: (1) The transfection efficiency is usually very high. (2) They are fast growing and thereby produce high amounts of protein (10–15 mg total protein per 14 cm dish corresponding to ~7 × $10^7$ HEK293 cells). For the transfection of one 14 cm dish, 1–4 μg of plasmid DNA is usually used. Note that a strong overexpression of the bait protein leads to an increased copurification of heat shock proteins like HSP70. Stable cell lines expressing the SF-tagged protein of interest are therefore recommended for in depth analysis.

1. Grow cells in DMEM supplemented with 10% heat-inactivated FBS and penicillin/streptomycin or appropriate medium for your favorite cell line.
2. Plate cells on 3-cm dishes (1–2 × $10^6$) one day before transfection. Cells should be 40–80% confluent on the day of transfection when using Effectene transfection reagent.
3. Transfect cells using Effectene transfection reagent according to manufacturer's protocols (*see* **Note 9**).
4. Change medium after 6 h.
5. After 48 h, trypsinize cells and seed them in a low density (1 × $10^6$ cells/10-cm dish) to allow the formation of single colonies upon selection.
6. Add G-418 (500 μg/mL) for selection of the SF-TAP expression vectors, which are based on pcDNA3.0 and contain a neomycin resistance gene.
7. Grow the cells under G-418 for 2–4 weeks; change medium every second day.
8. Collect single colonies with a 20 μL pipet into 12-well dishes.
9. Keep colonies under G-418 selection until the cell density is sufficient for expanding them to 6-well dishes (2-wells per clone).
10. Grow cells to above 90% confluency and trypsinize one well of each clone for generation of a cryostock (*see* **Note 10**).

11. Lyse one well of each clone in 300 μL lysis buffer (*see* **Subheading 2.4** *[7]*) and test for expression of the bait protein by Western blot (anti-FLAG M2 antibody; 1:5,000 in 5% nonfat milk powder in TBS buffer, 1% Tween).

### 3.3. SF-TAP Purification Protocol

A flowchart of the SF-TAP purification procedure is shown in **Fig. 1b**.

1. Remove the medium from the plates.
2. Optional: rinse the cells in warm PBS.
3. Scrape of the cells in 1 mL lysis buffer per 14-cm plate on ice using a cell scraper.
4. Lyse the cell for 20 min on ice and mix the lysates during the incubation.
5. Spin down cell debris including nuclei (10 min, 10,000*g*, 4° C).
6. Clear lysates by filtration through syringe 0.22 μm filters.
7. Incubate lysates with 50 μL/plate (*see* **Note 11**) Strep-Tactin superflow resin for 1 h at 4° C (use an overhead tumbler to keep the resin evenly distributed).
8. Spin down for 30 s at 7,000*g*, remove most of the supernatant, and transfer resin to microspin columns (*see* **Note 12**).
9. Remove the remaining supernatant by centrifugation (5 s, 100*g*) and wash with 3X 500 μL wash buffer (centrifuge 5 s 100*g* each time to remove the supernatant) (*see* **Notes 13** and **14**).
10. Elute 1X with 500 μL desthiobiotin elution buffer for 10 min. During elution, the resin should be gently mixed several times.
11. After incubation, remove the plug of the spin column, transfer it to a new collection tube, and harvest the eluate by centrifugation (10 s, 5,000*g*) (*see* **Note 15**).
12. Transfer eluate to 25 μL per 14-cm plate anti-FLAG-M2-agarose in microspin columns.
13. Incubate for 1 h at 4° C (over head tumbler).
14. Wash with 1X 500 μL wash buffer and 2X 500 μL with TBS buffer (centrifuge 5 s 100*g* each time to remove the supernatant).
15. For elution, incubate with 1X 200 μL FLAG elution buffer for 10 min, gently mix the resin several times.
16. After incubation, remove the plug of the spin column, transfer it to a new collection tube, and harvest the eluate by centrifugation (10 s, 5,000*g*) (*see* **Note 15**).
17. Concentrate eluate for downstream analysis, i.e., SDS-PAGE gel electrophoresis by centrifugal filter devices (Microcon

YM-3) or chloroform-methanol precipitation for LC-MSMS analysis (*see* **Subheading 3.4**).

#### 3.4. Chloroform–Methanol Precipitation (According to Wessel and Flügge [15])

1. Transfer 200 μL SF-TAP eluate to a 2 mL sample tube.
2. Add 0.8 mL of methanol, mix, and spin for 20 s (*see* **Note 16**).
3. Add 0.2 mL chloroform, mix, and spin for 20 s.
4. Add 0.6 mL of water, vortex for 5 s, and centrifuge for 1 min.
5. Carefully remove and discard the upper layer (aqueous phase; *see* **Note 17**).
6. Add 0.6 mL of methanol, mix gently, and centrifuge for 2 min.
7. Carefully remove the supernatant and let the pellet dry on air.

#### 3.5. Preparation of Samples for an LC-MS/MS Approach (Alkylation and Tryptic Digest)

1. Dissolve the protein pellet in 24.5 μL 50 mM ammonium bicarbonate.
2. Add 3 μL of a RapiGest (*see* **Note 18**) stock solution (final concentration 0.2%).
3. Add 0.75 μL of 100 mM DTT and vortex.
4. Incubate 10 min at 60° C.
5. Cool down the samples to room temperature.
6. Add 0.75 μL of 300 mM iodoacetamide and vortex.
7. Incubate for 30 min at room temperature (*see* **Note 19**).
8. Add 2 μL trypsin stock solution and vortex.
9. Incubate at 37° C overnight.
10. For hydrolysis of the RapiGest add HCl (37%) in 1 μL steps until the pH is <2 (*see* **Note 20**).
11. Incubate for 30 min at room temperature.
12. Centrifuge for 10 min (13,000*g*).
13. Retain supernatant between pellet and the upper oily phase.
14. The sample can directly be subjected to a C-18 HPLC separation prior to MS/MS analysis (LC-MS/MS).

#### 3.6. Pathway Walking for Raf-1 and B-Raf

As an example for an application, we provide the purification of protein complexes associated with the MAP kinase kinase kinase Raf-1 and its isoform B-Raf. Raf kinases are involved in key processes of cellular signaling, including cell fate decisions like proliferation, differentiation, or prevention of apoptosis *(16)*. The Raf isoforms were C-terminally tagged with the SF-TAP tag. To demonstrate a pathway walking approach, the downstream kinase MEK1 and the scaffold protein 14-3-3ε were subsequently tagged. Both proteins have been successfully coprecipitated with Raf baits. The protein composition of the SF-TAP purifications has been analyzed in depth by direct LC-MS/MS analysis. The identified proteins are presented as a comprehensive interactome map (*see* **Fig. 3**).

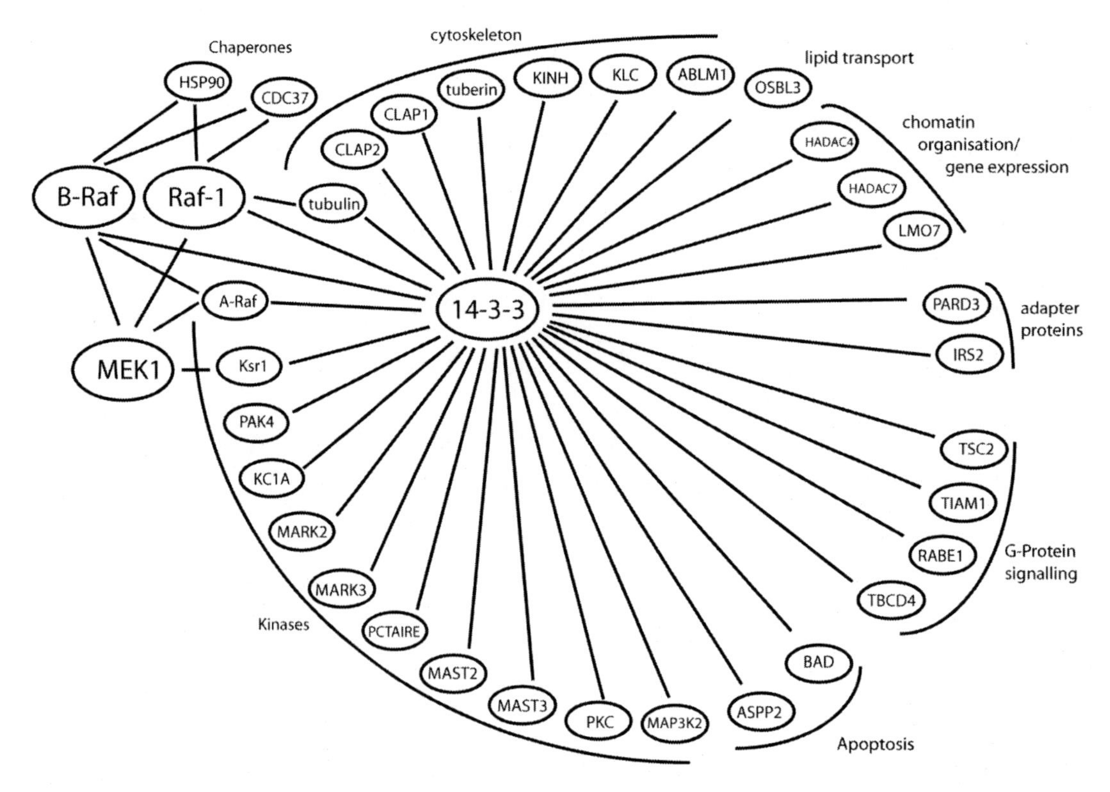


Fig. 3. Interactome map for B-Raf, Raf 1, MEK1, and 14-3-3. After in liquid digestion of SF-TAP eluates of the different bait proteins, samples were analyzed by direct LC-MS/MS analysis on a 4,700 Maldi-TOF/TOF instrument. The list of proteins and a detailed description of the MS-analysis can be found in the supplemental information of **ref.** *8*. Identified proteins were assembled to an interactome map

## 4. Notes

1. Optional: phosphatase inhibitors can be left out if no phosphorylated proteins are analyzed.
2. Store trypsin aliquots at –20° C.
3. Gene-specific primer pairs: N-terminal tagging: *(2)/(4)* or C-terminal tagging *(6)/(8)* (**Fig. 2b**).
4. Generic *att*B primer *(9)/(10)* (**Fig. 2b**).
5. 5 PCR can produce side products, which cause background in the BP reaction. If side bands appear after separation by agarose gel electrophoresis, it is recommended to isolate the band of correct size and extract the DNA using appropriate commercial kits before use in the BP reaction. For preparative agarose gel electrophoresis, TAE (tris-acetate-EDTA) running buffer should be used. Buffers and standard protocols for the agarose gel electrophoresis are found in ref. *13*.
6. Provided with the BP/LR Clonase Mix II (Invitrogen).

7. pDONR201 sequencing primers: forward: tcgcgttaacgctagcatggatctc; reverse: gtaacatcagagattttgagacac.
8. pDEST-SF-TAP sequencing primers: forward: gcggtaggcgtgtacggtggg; reverse: gggcaaacaacagatggctggc.
9. For other cell lines than HEK293, different transfection reagent could be favorable.
10. For generation of a cryostock, wash cells 1X in PBS and resuspend cells in freezing buffer. Transfer resuspended cells to cryovials. Freeze cells slowly: keep cells for 1 h at –20 C, overnight in –80°C before storage in a liquid nitrogen tank. For cultivation and expansion of confirmed clones, thaw the cryostock at 37°C, wash cells 1X with medium, and plate cells onto 10 cm culture dishes.
11. Note that a maximum of 200 µL settled resin per spin column should not be exceeded. If more than four 14 cm dishes (~$2.5 \times 10^8$ HEK293 cells) are used, reduce the volume per plate or use additional spin columns.
12. Snap of bottom closure of the spin columns before use. The maximum volume of the spin columns is 650 µL.
13. Replug spin columns with inverted bottom closure before adding the elution buffer.
14. Avoid allowing the resin to run dry. Depending on the bait-protein, this markedly reduces the yield.
15. If the spin columns were closed by the top screw cap during the incubation with elution buffer, it has to be removed before centrifugation. Spin columns must be left open (without screw cap) during centrifugation to allow pressure balance.
16. Use 9,000*g* for all centrifugation steps.
17. The protein precipitate is in the interphase.
18. RapiGest is a surfactant that can be used for better solubilization (and denaturation if no precipitation step is used) of the proteins *(17)*.
19. Samples should be light protected.
20. Check the pH with pH-indicator stripes.
21. Solution should be prepared freshly prior to use.

## Acknowledgments

The authors thank Gabriele Dütsch, Elöd Körtvély, and Pia Jores for critical reading of the manuscript. This work was supported by the German Federal Ministry for Education and Research BMBF grant 031U108A/031U208A, BMBF grant 0316865A (Quant-Pro), and EU grants INTERACTION PROTEOME (LSHG-CT-2003-505520) and ProteomeBinders (FP6-026008). The authors C.J Gloeckner and K. Boldt contributed equally to this work.

## References

1. Alberts, B. (1998) The cell as a collection of protein machines: preparing the next generation of molecular biologists. *Cell*, **92**, 291–4.
2. Chang, I.F. (2006) Mass spectrometry-based proteomic analysis of the epitope-tag affinity purified protein complexes in eukaryotes. *Proteomics*, **6**, 6158–66.
3. Gavin, A.-C., Bosche, M., Krause, R., Grandi, P., Marzioch, M., Bauer, A., Schultz, J., Rick, J.M., Michon, A.-M., Cruciat, C.-M. (2002) Functional organization of the yeast proteome by systematic analysis of protein complexes. *Nature*, **415**, 141–7.
4. Rigaut, G., Shevchenko, A., Rutz, B., Wilm, M., Mann, M. and Seraphin, B. (1999) A generic protein purification method for protein complex characterization and proteome exploration. *Nat Biotechnol*, **17**, 1030–2.
5. Gingras, A.-C., Aebersold, R. and Raught, B. (2005) Advances in protein complex analysis using mass spectrometry. *J Physiol*, **563**, 11–21.
6. Bouwmeester, T., Bauch, A., Ruffner, H., Angrand, P.-O., Bergamini, G., Croughton, K., Cruciat, C., Eberhard, D., Gagneur, J., Ghidelli, S. (2004) A physical and functional map of the human TNF-alpha/NF-kappa B signal transduction pathway. *Nat Cell Biol*, **6**, 97–105.
7. Burckstummer, T., Bennett, K.L., Preradovic, A., Schutze, G., Hantschel, O., Superti-Furga, G. and Bauch, A. (2006) An efficient tandem affinity purification procedure for interaction proteomics in mammalian cells. *Nat Methods*, **3**, 1013–9.
8. Gloeckner, J., Boldt, K., Schumacher, A., Roepman, R. and Ueffing, M. (2007) A novel tandem affinity purification strategy for the efficient isolation and characterization of native protein complexes. *Proteomics*, 7, 4228–34.
9. Junttila, M.R., Saarinen, S., Schmidt, T., Kast, J. and Westermarck, J. (2005) Single-step Strep-tag purification for the isolation and identification of protein complexes from mammalian cells. *Proteomics*, **5**, 1199–203.
10. Skerra, A. and Schmidt, T.G. (2000) Use of the Strep-Tag and streptavidin for detection and purification of recombinant proteins. *Method Enzymol*, **326**, 271–304.
11. Gloeckner, C.J., Kinkl, N., Schumacher, A., Braun, R.J., O'Neill, E., Meitinger, T., Kolch, W., Prokisch, H. and Ueffing, M. (2006) The Parkinson disease causing LRRK2 mutation I2020T is associated with increased kinase activity. *Hum Mol Genet*, **15**, 223–32.
12. den Hollander, A.I., Koenekoop, R.K., Mohamed, M.D., Arts, H.H., Boldt, K., Towns, K.V., Sedmak, T., Beer, M., Nagel-Wolfrum, K., McKibbin, M.. (2007) Mutations in LCA5, encoding the ciliary protein lebercilin, cause Leber congenital amaurosis. *Nat Genet*, **39**, 889–95.
13. Sambrook, J. and Russell, D. (2001) *Molecular Cloning: A Laboratory Manual.*3rd Cold Spring Harbor Laboratory Press, New York.
14. Hartley, J.L., Temple, G.F. and Brasch, M.A. (2000) DNA cloning using in vitro site-specific recombination. *Genome Res*, **10**, 1788–95.
15. Wessel, D. and Flugge, U.I. (1984) A method for the quantitative recovery of protein in dilute solution in the presence of detergents and lipids. *Anal Biochem*, **138**, 141–3.
16. Kolch, W. (2005) Coordinating ERK/MAPK signalling through scaffolds and inhibitors. *Nat Rev Mol Cell Biol*, **6**, 827–37.
17. Yu, Y.Q., Gilar, M., Lee, P.J., Bouvier, E.S. and Gebler, J.C. (2003) Enzyme-friendly, mass spectrometry-compatible surfactant for in-solution enzymatic digestion of proteins. *Anal Chem*, **75**, 6023–8.

# Chapter 22

# Sequential Peptide Affinity Purification System for the Systematic Isolation and Identification of Protein Complexes from *Escherichia coli*

**Mohan Babu, Gareth Butland, Oxana Pogoutse, Joyce Li, Jack F. Greenblatt, and Andrew Emili**

## Summary

Biochemical purification of affinity-tagged proteins in combination with mass spectrometry methods is increasingly seen as a cornerstone of systems biology, as it allows for the systematic genome-scale characterization of macromolecular protein complexes, representing demarcated sets of stably interacting protein partners. Accurate and sensitive identification of both the specific and shared polypeptide components of distinct complexes requires purification to near homogeneity. To this end, a sequential peptide affinity (SPA) purification system was developed to enable the rapid and efficient isolation of native *Escherichia coli* protein complexes (J Proteome Res 3:463–468, 2004). SPA purification makes use of a dual-affinity tag, consisting of three modified FLAG sequences (3X FLAG) and a calmodulin binding peptide (CBP), spaced by a cleavage site for tobacco etch virus (TEV) protease (J Proteome Res 3:463-468, 2004). Using the λ-phage Red homologous recombination system (PNAS 97:5978-5983, 2000), a DNA cassette, encoding the SPA-tag and a selectable marker flanked by gene-specific targeting sequences, is introduced into a selected locus in the *E. coli* chromosome so as to create a C-terminal fusion with the protein of interest. This procedure aims for near-endogenous levels of tagged protein production in the recombinant bacteria to avoid spurious, non-specific protein associations (J Proteome Res 3:463–468, 2004). In this chapter, we describe a detailed, optimized protocol for the tagging, purification, and subsequent mass spectrometry-based identification of the subunits of even low-abundance bacterial protein complexes isolated as part of an ongoing large-scale proteomic study in *E. coli* (Nature 433:531–537, 2005).

**Key words:** Affinity purification, Protein complex, *E. coli*, SPA-tagging, λ–Red recombination, Mass spectrometry, LC-MS, MALDI-TOF

Jörg Reinders and Albert Sickmann (eds.), *Proteomics,* Methods in Molecular Biology, Vol. 564
DOI: 10.1007/978-1-60761-157-8_22, © Humana Press, a part of Springer Science+Business Media, LLC 2009

## 1. Introduction

Since virtually all cellular processes involve physical associations between proteins, it is of immense importance to systematically map protein-protein interaction (PPI) networks as a means of investigating the mechanistic basis of biological systems. Large-scale PPI networks were initially described for *Saccharomyces cerevisiae* using yeast two-hybrid analysis *(1, 2)*, and, more recently and far more comprehensively, by purifying soluble stable protein complexes on a global scale using the tandem affinity purification (TAP) approach *(3, 4)*, wherein the tagged proteins are expressed under normal physiological conditions and then purified ~$10^6$-fold via a two-step enrichment procedure. The flexible and robust TAP method was subsequently adapted for use in flies, worms, and human cells *(5–8)*. Our group has developed the sequential peptide affinity (SPA) system, a variant of TAP in which a smaller 3X FLAG tag is substituted for the original protein A moiety and the system is optimized for creating C-terminal fusions via recombining in suitably engineered bacterial strains *(9)*. Since interacting proteins usually function in the same biological process, computational assessment of protein complexes and PPI networks provides a rational framework for systematically inferring the biological functions of uncharacterized proteins (guilt by association).

We have used SPA tags to purify hundreds of tagged proteins and their associated protein partners from *Escherichia coli*, a model microbe well suited to genomic investigations of the fundamental physiology and conserved molecular pathways of prokaryotes. As of April 2009, more than 2,000 *E. coli* open reading frames (ORFs) have been SPA-tagged using this system, with over 1,800 of these proteins purified successfully, together with their endogenous interacting partners, as judged by successful detection by mass spectrometry. An extensive PPI network, consisting of highly conserved and essential protein complexes, was reported previously *(10)*. The entire collection of SPA-tagged strains has now been made commercially available for unfettered academic use (https://www.openbiosystems.com/GeneExpression/NonMammalian/Bacteria/EcoliTaggedORFs/). Recently we performed an extensive proteomic survey using ~2000 affinity-tagged *E. coli* strains, including proteins of unknown function, and generated comprehensive genomic context inferences to derive a high-confidence putative physical interactions, most of which are novel *(11)*. Membrane-associated protein complexes are far more difficult to purify, however, and specialized methods for these proteins are currently under development.

SPA purification utilizes dual affinity tags, 3X FLAG, and a calmodulin binding peptide (CBP), separated by a tobacco etch virus (TEV) protease cleavage site (*[9]*, **Fig. 1a**), to allow for highly selective two-stage protein enrichment. We chose to use the 3X FLAG affinity tag for three reasons: first, the FLAG epitope has been widely used for tagging, monitoring, and purifying proteins from various organisms; second, there are far fewer amino acid residues (22 residues) in 3X FLAG than the 137 residues in the protein A moiety of the original TAP tag, which lessens the chance of functional perturbation and misfolding/instability of certain tagged proteins; finally, 3X FLAG is sensitively detected with a high-affinity monoclonal antibody, M2, which is commercially available and known to work well in Western blots and immunoprecipitation experiments *(12)*.

Homologous recombination is used to target the SPA-tag construct into chromosomal loci of interest so as to produce C-terminal fusions without perturbing the nearby transcriptional promoter or terminator sequences of the targeted bacterial operon. Sequence-specific linear PCR products encoding the affinity purification tag and a selectable marker are transformed into the cell and undergo recombination at the stop codon of the target gene in the chromosome of the lysogenic *E. coli* strain DY330, which harbors the highly efficient λ-phage-encoded homologous recombination enzymes *exo*, *bet*, and *gam* of the "Red" system under the control of the temperature-sensitive *CI857* repressor (*[13]*, **Fig. 1b**). Strains in which the PCR product has integrated are subjected to antibiotic selection, and tagged protein expression is confirmed by Western blotting. The tagged bait proteins are then isolated by affinity purification (**Fig. 1c**) from crude whole cell lysates prepared from cells harvested from large-scale cultures. Two complementary and highly sensitive mass spectrometry methods are used to ensure identification of co-purifying interacting proteins: gel-free liquid chromatography–tandem mass spectrometry (LC-MS), and gel-based peptide mass fingerprinting using matrix-assisted laser desorption/ionization – time-of-flight mass spectrometry (MALDI-TOF).

In this chapter, we describe detailed step-by-step procedures for effective tagging of individual *E. coli* ORFs, purification of the resulting bait proteins, and subsequent identification of interacting protein partners by mass spectrometry methods. While this method has been expressly optimized for a well-characterized *E. coli* laboratory strain, we note that the basic approach can potentially be adapted for use in any other closely related strain or other bacteria that can be cultured under standard laboratory conditions.

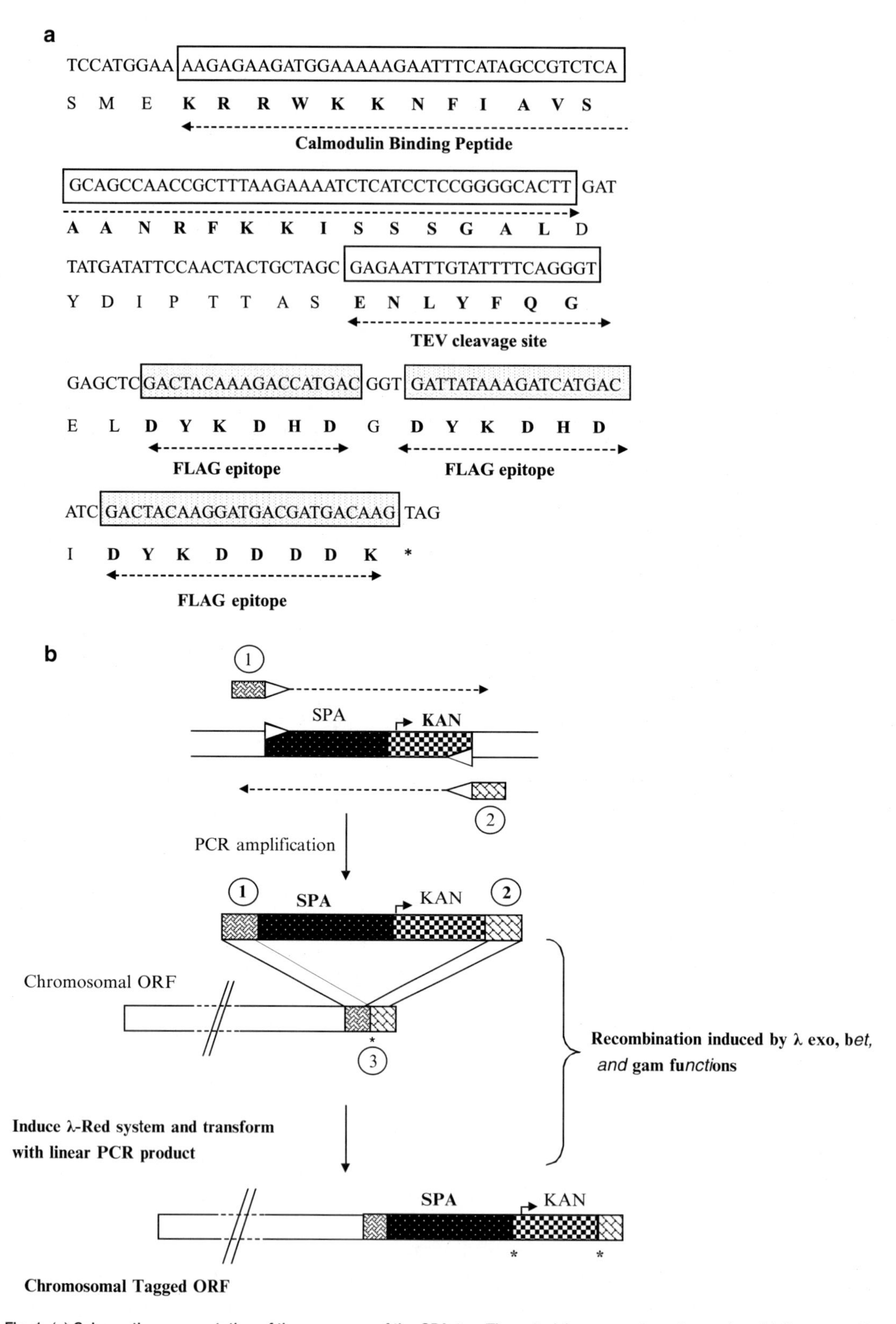


Fig. 1. (**a**) Schematic representation of the sequence of the SPA-tag. The asterisk represents a stop codon. (**b**) Gene-specific affinity-tagging cassettes produced by the polymerase chain reaction (PCR) using primers *(9,13)* homologous to the regions on either side of the target gene translational termination codon *(10)* were integrated into the *E. coli* chromosome using the λ-Red recombination system. SPA, Sequential Peptide Affinity system; Kan, Kanamycin resistance cassette; *, stop codon.

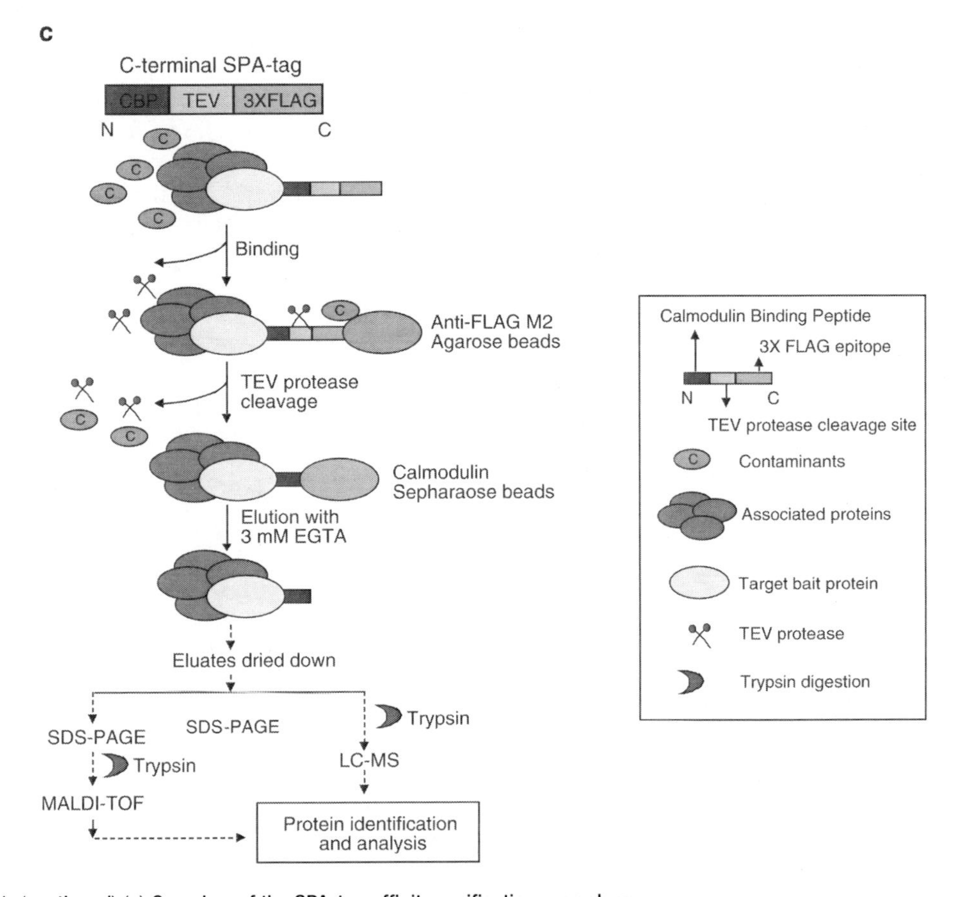


Fig. 1. (continued) (**c**) Overview of the SPA-tag affinity purification procedure.

## 2. Materials

### *2.1. SPA Vector Construction and Gene-Specific SPA-Tagging in E. coli*

1. The plasmid pJL148 was created by replacing the TAP tag from the plasmid pJL72 *(14)* with the SPA-tag from the vector pMZ13F *(14)*.
2. All the oligonucleotides used in this study were synthesized by Invitrogen Canada Inc. (Burlington, Ontario).
3. Restriction enzymes were from New England BioLabs Ltd. (Pickering, Ontario).
4. Isolation of plasmid DNA or purification of PCR products was performed using Qiagen products (Qiagen Inc., Mississauga, Ontario).
5. One molar Tris-HCl (pH 7.5) stock solution: Dissolve 121.1 g of Tris base in 800 mL of double distilled water, adjust to pH 7.5 with concentrated HCl and make up the volume to 1 L.

6. EDTA (0.2 M, pH 8.0) stock solution: Dissolve 74.44 g of EDTA in 800 mL of double distilled water, adjust the pH to 8.0 using 1 M NaOH and make up the volume to 1 L.

### 2.2. Induction of λ-Red Recombination Functions, and Preparation and Transformation of Competent E. coli Cells

1. Luria-Bertani (LB) Medium: Solid medium is prepared by dissolving 25 g of LB powder (Bioshop cat#LBL405) and 20 g of agar (Bioshop cat#AGR003) in 1,000 mL of distilled water. Liquid LB medium is prepared without agar.
2. The Red-expressing *E. coli* strain DY330 was a kind gift from Donald L.Court (National Cancer Institute, Frederick, MD 21702).
3. SOC medium was obtained from Invitrogen Inc. (Invitrogen cat#15544-034).
4. Sample stock buffer (3X): Prepare 10 mL of 3X sample stock buffer containing 1.25 mL of 0.5 M Tris–HCl (pH 6.8), 1 mL of 100% glycerol, 1 mL of 10% SDS (w/v), and 20 mg of bromophenol blue (Bio-Rad cat#161-0404).
5. To prepare 30 mL of 1X sample buffer, add 9 mL of 3X sample stock buffer, 1 mL of β-mercaptomethanol, and 20 mL of distilled water.
6. Kanamycin (Sigma cat# 60615) antibiotic stock: Dissolve 100 mg/mL of kanamycin in double distilled water and filter-sterilize using a 0.22 μm millipore filter. The filter-sterilized kanamycin stock solution is stored in single use aliquots at –20°C.

### 2.3. SDS-Polyacrylamide Gel Electrophoresis

1. SDS stock solution [10% (w/v)]: 10 g SDS is dissolved in 100 mL double distilled water. The solution is stored at room temperature.
2. Electrode running buffer (5X): [125 mM Tris-HCl, 960 mM glycine, 0.5% (w/v) SDS, pH 8.3]. Thirty gram of Tris-HCl, 144 g of glycine, and 10 g of SDS are dissolved in 2 L of double distilled water without pH adjustment and stored at 4°C.
3. Acrylamide: 30% acrylamide monomer, 0.8% *N,N'*-methylenebis-acrylamide (Bio-Rad cat#161-0125). The solution is filtered through Whatman No.1 filter paper (Fischer Scientific cat#09-806A) and stored at 4°C in the dark.
4. Separating buffer (4X): 1.5 M Tris–HCl (pH 8.7), 0.4% SDS. Store at room temperature.
5. Stacking buffer (5X): 0.5 M Tris-HCl (pH 6.8), 0.4% SDS. Store at room temperature.
6. Ammonium persulfate (APS): Prepare 20% APS (Bioshop cat#AMP001) solution by mixing 2 g of APS in 10 mL of double distilled water. Aliquot the solution into several microcentrifuge tubes in a volume of 200 μL. Store the aliquots at –20°C.

7. Water saturated *n*-Butanol: Mix equal volumes of *n*-butanol (Sigma Aldrich cat#B7906) and water in a glass bottle and leave it for a while to separate. Use topmost layer containing *n*-butanol saturated with water. Store the solution at room temperature.
8. Resolving gel (12.5%): Prepare 1.5 mm thick, 12.5% polyacrylamide resolving gel by mixing 7.2 mL of 4X separating buffer with 12 mL acrylamide solution, 9.6 mL double distilled water, 100 μL 20% APS solution, and 20 μL TEMED (Bioshop cat#TEM001).
9. Stacking gel (7.5% polyacrylamide): Prepare the stacking gel by mixing 4 mL of 5X stacking buffer with 2.4 mL acrylamide solution, 9.6 mL double distilled water, 100 μL 20% APS solution, and 10 μL TEMED.

### 2.4. Confirmation of SPA-Tagging in Transformants Using Western Blotting with Chemiluminescent Method

1. Trans-Blot® Cell system (Bio-Rad cat#170-3853) was used to transfer the protein samples from the SDS-polyacrylamide gels to nitrocellulose membranes (Bio-Rad cat#162-0115).
2. Transfer stock buffer: Prepare 10X transfer stock buffer with 120 g Tris-HCl and 576 g glycine in distilled water.
3. To prepare 1X transfer buffer, dilute 400 mL of 10X transfer stock buffer with 800 mL methanol and 2800 mL of sterile distilled water. Store the buffer at room temperature.
4. TBS buffer: Prepare 5X TBS stock buffer with 48.44 g Tris and 584.4 g NaCl.
5. Wash buffer: TBS buffer containing 0.05% Tween-20.
6. Blocking buffer: 5% (w/v) non-fat dry milk in TBS buffer containing 0.1% Tween-20.
7. Primary antibody: M2 antibody (Sigma cat#F3165) against the FLAG epitopes of the SPA-tag.
8. Primary antibody buffer: TBS buffer supplemented with 1% gelatin, 0.05% Tween-20, and 0.02% $NaN_3$.
9. Secondary antibody: Antimouse IgG conjugated to Horseradish Peroxidase (HRP; Amersham Biosciences cat#NA931V).
10. Secondary antibody buffer: 5% (w/v) non-fat dry milk in TBS buffer containing 0.1% Tween-20.
11. Chemiluminescence reagent (PIERCE cat#1856136) is prepared (0.125 mL of chemiluminescence reagent per $cm^2$ of membrane) by mixing equal volumes of the enhanced luminol reagent and the oxidizing reagent.
12. Kodak X-OMAT autoradiography film (Clonex Corp., cat#CLEC810).

13. Stripping buffer: 62.5 mM Tris-HCl (pH 6.7), 2% (w/v) SDS, and 100 mM β-mercaptoethanol. Store the buffer at room temperature.

### 2.5. SPA-Tagged Protein Purification

#### 2.5.1. Culturing SPA-Tagged E. coli Strains and Sonication

1. *E. coli* strains are used for culturing in which a bait protein is SPA-tagged.
2. Terrific Broth (TB) medium is prepared by dissolving 11.07 g of tryptone (Bioshop cat#TRP402), 22.13 g of yeast extract (Bioshop cat#YEX401), and 30 mL of 50% glycerol in 830 mL of double distilled water.
3. Potassium salt stock solution: Dissolve 18.48 g of monobasic $KH_2PO_4$ and 100.24 g of dibasic $K_2HPO_4$ in 800 mL of double distilled water.
4. Sonication stock buffer is prepared by dissolving 9.6 mL of 2 M Tris-HCl (pH 7.9), 19.2 mL of 5 M NaCl, 384 µL of 0.5 M EDTA, and 192 mL of 50% glycerol in 738.2 mL of double distilled water. Store the stock solution at 4°C.
5. Sonication working buffer is prepared by mixing 300 mL of sonication stock buffer with seven protease inhibitor tablets (Roche cat#800-363-5887) and 150 µL of 0.5 M dithiothreitol (DTT; Bioshop cat# DTT001). Aliquot the sonication working buffer into polypropylene Falcon tubes (25 mL in each) and keep them refrigerated at 4°C.

#### 2.5.2. Purification of SPA-Tagged Proteins

1. Protease inhibitor stock solution (25X): Dissolve 1 tablet of protease inhibitor (Roche cat#11 873 580 001) in 2 mL of sterile distilled water. This stock solution is stable for at least 12 weeks when stored at –15°C to –25°C.
2. Two molar Tris–HCl (pH 7.9) stock solution: Dissolve 242.2 g of Tris base in 800 mL of double distilled water, adjust to pH 7.9 with concentrated HCl, and make up the volume to 1 L.
3. Five molar NaCl stock solution: Dissolve 292.2 g of NaCl in 1 L of double distilled water.
4. Ten percent Triton X-100 solution from Sigma Aldrich (cat#93443) is stored at 4°C.
5. EDTA (0.5 M, pH 8.0) stock solution: Dissolve 186.1 g of EDTA in 800 mL of double distilled water, adjust the pH to 8.0 using 1 M NaOH, and make up the volume to 1 L.
6. AFC buffer 10X stock solution: Prepare 40 mL of 10X AFC buffer stock solution containing 2 mL of 2 M Tris-HCl (pH 7.9), 8 mL of 5 M NaCl, 4 mL of 10% Triton X-100, and 26 mL of sterile distilled water.
7. To prepare 40 mL of 1X AFC working buffer, add 4 mL of 10X AFC buffer stock solution, 200 µL of 0.2 M DTT, 1.6 mL of protease inhibitor stock solution, and 34.2 mL of sterile distilled water.

8. TEV cleavage 10X stock buffer: Prepare 40 mL of 10X TEV cleavage buffer stock solution containing 10 mL of 2 M Tris-HCl (pH 7.9), 8 mL of 5 M NaCl, 160 μL of 0.5 M EDTA, 4 mL of 10% Triton X-100, and 17.84 mL of sterile distilled water.
9. To prepare 25 mL of 1X TEV cleavage buffer, add 2.5 mL of 10X TEV cleavage stock buffer, 125 μL of 0.2 M DTT, 1 mL of protease inhibitor stock solution, and 21.4 mL of sterile distilled water.
10. Calmodulin binding 10X stock buffer: Prepare 40 mL of 10X calmodulin binding stock buffer solution containing 2 mL of 2 M Tris–HCl (pH 7.9), 8 mL of 5 M NaCl, 800 μL of 1 M $CaCl_2$, 4 mL of 10% Triton X-100, and 25.2 mL of sterile distilled water.
11. To prepare 12 mL of 1X calmodulin binding buffer, add 1.2 mL of 10X calmodulin binding stock buffer, 8 μL of β-mercaptoethanol, 480 μL of protease inhibitor stock solution, and 10.32 mL of sterile distilled water.
12. Calmodulin wash 10X stock buffer: Prepare 10 mL of 10X calmodulin wash stock buffer containing 500 μL of 2 M Tris–HCl (pH 7.9), 2 mL of 5 M NaCl, 10 μL of 1 M $CaCl_2$, and 7.5 mL of sterile distilled water.
13. To prepare 10 mL of 1X calmodulin wash buffer, add 1 mL of 10X calmodulin wash stock buffer, 7 μL of β-mercaptoethanol, and 8.3 mL of sterile distilled water.
14. Calmodulin elution 1X stock buffer: Prepare 10 mL of 1X calmodulin elution stock buffer containing 50 μL of 2 M Tris-HCl (pH 7.9), 1 mL of 1 M ammonium hydrogen carbonate ($NH_4HCO_3$), 150 μL of 0.2 M EGTA, 7 μL of 14 M β-mercaptoethanol, and 8.8 mL of sterile distilled water.
15. $NH_4HCO_3$ (1 M): Dissolve 0.395 g of $NH_4HCO_3$ in 5 mL sterile distilled water. Make sure to use freshly prepared $NH_4HCO_3$ solution each time.

### 2.6. Protein Identification by Gel-Based Peptide Mass Fingerprinting Using Matrix-Assisted Laser Desorption and Ionization Time-of-Flight Mass Spectrometry

#### 2.6.1. Silver Staining the SDS-Polyacrylamide Gel

1. Fixer: 50% methanol and 10% acetic acid (AA) in 500 mL of sterile distilled water.
2. Sensitizer: Prepare 20 mg of fresh sodium-thiosulfate (Sigma cat#S-7143) in 1,000 mL of distilled water.
3. Silver nitrate solution: Dissolve 2 g of silver nitrate (Fischer Scientific cat#S181-100) in 1,000 mL of distilled water.
4. Developing solution: Add 1.4 mL of 37% formaldehyde and 30 g of sodium carbonate to 1,000 mL of distilled water. Avoid using glutaraldehyde as it might crosslink the proteins to the gel.
5. Stop solution: Dilute 5 mL of AA in 500 mL of distilled water.

*2.6.2. Gel-Based MALDI-TOF Mass Spectrometry*

1. ULTRAFlex II MALDI-TOF instrument (Bruker Daltonics, Billerica, MA) to run the samples and acquire spectral data.
2. MALDI target plate from Bruker (Bruker, cat#209519).
3. Bulk C18 reverse phase resin from Sigma (cat#H8261).
4. Knexus automation, a Windows-based program from Genomics Solutions Bioinformatics to perform automatic protein searches and to evaluate MALDI spectral results.
5. α-Cyano-4-hydroxycinnamic acid matrix solution (Fluka Buchs SG, Switzerland cat# 28480).
6. One millimolar HCl: Dilute 1 μL of 10 N HCl into 10 mL HPLC grade water. Solution is made fresh each day and stored on ice until use.
7. Trypsin stock solution: Dissolve 100 μg Boehringer Mannheim unmodified sequencing grade trypsin (Roche cat#1047 841) in 1 mL of 1 mM HCl. Store the trypsin stock solution at –80°C.
8. Digestion buffer: Add 9.12 mL of 100 mM $NH_4HCO_3$, 9.12 mL of HPLC grade water, 960 μL of 1% $CaCl_2$, 1 mL of trypsin stock solution. Prepare the solution fresh each day and store on ice until use.
9. One hundred millimolar $NH_4HCO_3$: Dissolve 0.79 g of $NH_4HCO_3$ in 100 mL of HPLC grade water.
10. Ten milliliter of 100 mM $NH_4HCO_3$ containing 10 mM DTT is prepared by adding 100 μL of 1 M DTT to 9.9 mL of 100 mM $NH_4HCO_3$. Solution is prepared fresh each day and stored in amber colored bottle at room temperature in dark until use.
11. Ten milliliter of 100 mM $NH_4HCO_3$ containing 55 mM iodoacetamide (Sigma cat#16125) is prepared by adding 0.103 g of iodoacetamide to 10 mL of 100 mM $NH_4HCO_3$. Solution is prepared fresh every day and stored in amber colored bottle at room temperature in dark until use.
12. Sixty-six percent acetonitrile (ACN), 1% AA: 66 mL HPLC grade ACN (Sigma cat#A998-4), 33 mL of HPLC grade water, 1 mL AA. Store the solution in a glass bottle for up to 1 month at room temperature.
13. Seventy-five percent ACN, 1% AA: 75 mL HPLC grade ACN, 24 mL HPLC grade water, 1 mL AA. Store the solution in a glass bottle for up to 1 month at room temperature.
14. Two percent ACN, 1% AA: 2 mL of HPLC grade ACN, 97 mL HPLC grade water, 1 mL AA. Store the solution in a glass bottle for up to 1 month at room temperature.
15. One percent $CaCl_2$: Dissolve 1 g of $CaCl_2$ in HPLC grade water to a final volume of 100 mL. Store the solution in a glass bottle for up to 1 month at room temperature.

16. Trifluoroacetic acid (TFA) (0.1%) stock solution: Dilute 0.25 µL of TFA in 24.75 µL of HPLC grade water.
17. Peptide calibration standard (Bruker cat#206195) solution: Dissolve the lyophilized peptide standards with 125 µL of 0.1% HPLC grade TFA.

*2.6.3. Gel-Free Liquid Chromatography–Tandem Mass Spectrometry*

1. LTQ tandem mass spectrometer (Finnigan Corp, San Jose, CA, USA) to run the samples, and XCalibur software to acquire tandem mass spectra and to control the instrument.
2. Digestion buffer: Prepare the digestion buffer by mixing 50 mM $NH_4HCO_3$ and 1 mM $CaCl_2$. Store the solution at –4°C prior to use.
3. Immobilized trypsin solution: Add 18.7 µL of digestion buffer, 1.8 µL PIERCE immobilized trypsin beads (PIERCE cat#20230), 0.9 µL immobilized trypsin beads (Applied Biosciences cat# 2-3127-00) and 0.06 µL of 1 M $CaCl_2$. Make sure that pH of the immobilized trypsin solution is ~8.0.
4. One hundred and fifty micrometer fused silica (Polymicro Technologies, Pheoenix, AZ, USA).
5. C18 reverse phase packing material (Zorbax eclipse XDB-C18 resin) from Agilent Technologies (Mississauga, ON, Canada).
6. Solvent A: 5% ACN, 0.5% AA, and 0.02% Heptafluorobutyric acid (HFBA; Sigma Aldrich cat#77247).
7. Solvent B: 100% can.
8. Proxeon nano HPLC pump (Proxeon Biosciences).

## 3. Methods

### 3.1. SPA Vector Construction and Gene-Specific SPA-Tagging in E. coli

*3.1.1. E. coli SPA Vector Construction*

1. All amplifications were performed using Bio-Rad PCR system (Bio-Rad, cat#170-8703). Following are the PCR cycling conditions: 94°C for 5 min; 30 cycles of 94°C for 1 min, 55°C for 1 min, 68°C for 2 min 10 s, followed by 68°C for 10 min.
2. The kanamycin-resistance ($Kan^R$) cassette was PCR-amplified from the plasmid pKD4 *(12)* with primers ECTAP-Forward (5′-CTAGATATCGTGTAGGCTGGAGCTGCTTC-3′) and ECTAP-Reverse (5′-CGCGGGCCCCATATGAATATC-CTCCTTAGTTC-3′).
3. The amplified PCR product was cut with restriction enzymes EcoRV and ApaI and ligated into the plasmid pBS1479 *(15)* containing the TAP-tag cassette. The resulting plasmid was

named as pJL72 and contains the TAP-tag and the "Kan^R^" marker for selection in *E. coli* (*see* **Note 1**).

4. The SPA-tag was amplified (using the above PCR conditions) from the plasmid pMZ13F *(14)* by PCR using the ECSPA-Forward (5′-CGCGGATCCATGGAAAAGAGAAGATGGAA-3′) and ECSPA-Reverse (5′-CTAGATATCTCCCCTCAGGTCACTACTTGT-3′) primers.
5. The amplified SPA-tag product was then cut with restriction enzymes BamHI and EcoRV and ligated into the plasmid pJL72 by replacing the TAP-tag with the SPA-tag between the sites 2,205 and 2,433. The resulting ~4.5 kb construct was designated as pJL148. In all cases, PCR-amplified products were purified using the Qiaquick PCR purification kit (Qiagen).

*3.1.2. Gene-Specific SPA-Tagging in E. coli*

1. The plasmid pJL148 construct enabled the cloning of the SPA-tag downstream of the target gene and production of the target protein as a C-terminal SPA fusion.
2. To increase the efficiency of recombination, we used the SPA template instead of the plasmid pJL148 as DNA template in PCR reactions. To generate the SPA template, the 27 bp SPA-Forward (5′ AGCTGGAGGATCCATGGAAAAGAGAAG 3′) and 27 bp SPA-Reverse (5′ GGCCCCATATGAATATCCTCCTTAGTT 3′) primers homologous to the 5′ and 3′ end of the SPA cassette with ~10 bp of the vector sequences, respectively, were designed to amplify the SPA-tag. The tag was amplified by PCR using the cycling conditions shown in **Subheading 3.1.1, step 1**.
3. The amplified PCR products were purified using the Qiaquick PCR purification kit (Qiagen).
4. The purified PCR products were run on a 1% DNA agarose gel at 100 V for 1 h. The 1.7 kb amplified SPA cassette was gel-extracted and purified using the Qiagen purification kit. The purified SPA-tag amplicon was suspended in 200 μL of sterile water or TE buffer (10 mM Tris–HCl, pH 7.5/1 mM EDTA) and quantified by spectrophotometry (Fisher Scientific, cat#S42669ND). The purified SPA-tag amplicon was stored at –20°C and served as a template for the gene-specific SPA-tagging.
5. In each case, we used a gene-specific forward primer containing 45 bp of nucleotide sequence immediately upstream of the target gene stop codon (variable region forward, VF) in frame with the tag-specific TF (5′-TCCATGGAAAAGAGAAG-3′) constant priming region, and a gene-specific reverse primer containing 45 bp of nucleotide sequence immediately downstream of the target gene stop codon

(variable region reverse, VR) in frame with the tag-specific TR (5′-CATATGAATATCCTCCTTAG-3′) constant priming site. The primer pairs VF-TF and VR-TR were used to amplify the SPA cassette and $Kan^R$ resistance gene flanked by gene-specific 45 bp sequences. These sequences are identical to regions flanking the target gene stop codon and are substrates for the λ-Red homologous recombination machinery.

6. The PCR product was purified using a Qiaquick PCR purification kit (Qiagen) to remove salts that might interfere with electroporation.
7. The purified PCR product was then introduced by electroporation into *E. coli* strain DY330, in which the λ-Red system was induced (*see* **Subheading 3.2.1** for details).
8. After transforming the linear PCR product, cells which had recombined the tag into the chromosome were selected by their resistance to kanamycin, and were screened for the expression of a tagged fusion protein by Western blot.
9. Strains expressing fusion proteins with C-terminal tags were then cultured on a larger scale to allow isolation of potential protein complexes by SPA purification.

### 3.2. Induction of λ-Red Recombination Functions, Preparation and Transformation of the Competent E. coli Cells

#### *3.2.1. Induction of λ-Red Recombination Functions and Preparation of competent E. Coli cells*

1. The Red expressing strain DY330 was grown overnight in 2 mL of LB medium at 32°C with shaking at 180 rpm.
2. Inoculate 1.4 mL of the overnight culture into 70 mL fresh LB in a 500 mL conical flask. The inoculum was grown at 32°C with shaking at 180 rpm until $OD_{600}$ reached 0.5 to 0.8.
3. Induction was performed by transferring the culture into a 250 mL conical flask. The flask was incubated in a water bath at 42°C by gently shaking at 200 rpm for 15 min.
4. Immediately after the 15 min induction, the flask was incubated in an ice water slurry bath for at least 30 min with gentle shaking.
5. The cooled 70 mL culture was poured into pre-chilled 50 mL polypropylene tubes and the culture was centrifuged at 3,993 × *g* for 6 min at 4°C.
6. The cell pellets were resuspended in 50 mL ice-cold sterile water and centrifuged once again at 3,993 × *g* for 6 min at 4°C. **Subheading 3.2.1**, **step 6** was repeated one more time.
7. The cell pellets were resuspended in 1 mL of cold water, transferred to a 1.5 mL Eppendorf tube, and centrifuged at maximum speed for 20 s at 4°C.
8. After this series of washing steps, the cell pellets were finally resuspended in 700 μL of ice-cold sterile water (*see* **Note 2**).

*3.2.2. Cell Transformation and Selection Procedure*

1. Mix 1 μL (~100 ng) of the purified linear donor DNA with 40 μL of the electro-competent cells in Eppendorf tubes. Incubate the cells on ice for 5 min. Transfer the mixture into a 0.2 cm precooled electroporation cuvette.
2. Place the cuvette in the cuvette holder and electroporate the cells with the DNA in a Bio-Rad GenePulser® II electroporator set at 2.5 kV, 25 μF with pulse controller of 200 Ω.
3. Immediately, add 800 μL of SOC medium to the cells in the cuvette and mix the contents by gently pipetting up and down. Transfer the cell contents from the cuvette into a 12 mL culture tube.
4. Incubate the transformed cells at 32°C with shaking at 190 rpm for 1–2 h prior to the selection of recombinants.
5. After 1–2 h of shaking, centrifuge at 3,000 × *g* for 3 min. Remove the excess SOC and mix the cells gently by pipetting up and down 3–4 times. One hundred microliters of the transformed cells are spread onto the LB agar plates containing 50 μg/mL kanamycin and are incubated overnight at 32°C to obtain $Kan^R$ colonies.

*3.2.3. Confirmation of Recombinants*

1. Two colonies were randomly picked from each transformed plate.
2. A single transformed colony was inoculated into 2 mL LB medium containing 50 μg/mL kanamycin and grown overnight at 32°C with shaking at 190 rpm.
3. One milliliter of the overnight culture was centrifuged for 30 s at 15,900 × *g*. The cell pellet was resuspended in 200 μL of 1X SDS sample buffer and boiled for 5 min.
4. Load 20 μL of the cell pellet resuspended in 1X SDS sample buffer onto a 12.5% SDS gel. After electrophoresis, immediately transfer the gel onto a nitrocellulose membrane and perform Western blotting.
5. If the Western blotting result indicates successful tagging of the target bait protein, add 1 mL of 50% glycerol to the leftover 1 mL of overnight culture, and store it in a cryogenic tube in –80°C.

***3.3. SDS-Polyacrylamide Gel Electrophoresis***

1. The following steps were performed using the Mini Protean 3 Cell (Bio-Rad cat# 165-3301) gel system. Make sure the glass plates, spacers, and combs are clean and free of dried gel fragments, grease, and dust.
2. For each gel, lay one small and one large glass plate, separated by spacers and an alignment card. Make sure to place the small glass plate on top of the spacers so that both sides and the bottom of the plates and the spacers are even.

3. Slide glass plates into the holder without tightening the screws. Make sure the glass plates are pushed all the way to the bottom before tightening the screws. If the assembly is correct, then the whole setup should snap into place when placed above the gray gasket.
4. Remove the alignment card. Now slide in the comb and mark a line at 2-3 cm from the bottom of the comb. Make sure that the comb thickness is the same as that of the spacers.
5. Carefully remove the comb. Pour the 12.5% polyacrylamide resolving gel up to the marked line. Overlay a thin layer of water-saturated *n*-butanol on the top of the gel. Allow the gel to polymerize for about 30 min.
6. Invert the gel to remove the *n*-butanol. Touch with filter paper to wick off residual liquid.
7. Pour the 7.5% polyacrylamide stacking gel to fill the remaining space between the glass plates. Insert the comb and allow the gel to polymerize for another 30 min.
8. Once the stacking gel has set, carefully remove the comb and use a 3-mL syringe fitted with a 22-gauge needle to wash the wells with running buffer.
9. Place the gel in the unit and add running buffer to the upper and lower chambers of the unit.
10. Load each well with 20 μL samples suspended in 1X SDS sample buffer. Make sure to include in two of the wells high and low range precision plus pre-stained protein molecular weight standards (Bio-Rad cat#161-0363). Complete the assembly of the gel unit and attach the power cords first to the apparatus, then to the power supply.
11. Turn on the power supply and run the gel at 150 V through the stacking gel and 200 V through the resolving gel. The gel is run until the blue dye front reaches the bottom.
12. After the electrophoresis, unclamp the gel plate assembly from the apparatus. Place the gel plate assembly on a paper towel. Use a thin spatula to carefully pry the upper glass plate away from the gel.
13. Immediately transfer the gel to the nitrocellulose membrane following the Western transfer protocol indicated below.

### 3.4. Western Blotting with the Chemiluminescent Method

#### 3.4.1. Western Blot Transfer

1. Fill the Bio-ice cooling unit with water and store it at –20°C until ready to use. These instructions assume the use of a Trans-Blot® Cell system.
2. Cut the nitrocellulose membrane slightly larger than the dimension of the gel.

3. Equilibrate the gel and soak the membrane, four pieces of Quick Draw™ blotting paper (Sigma Aldrich cat#P7796), and two scotch-brite® fiber pads in transfer buffer for 15 min. Make sure to wear gloves at all times when handling the membrane. Mark one side of the membrane for future reference.
4. Preparation of the gel sandwich is as follows:
   (a) Place the cassette with the gray side down on a clean surface.
   (b) Place one pre-wetted scotch-brite® fiber pad on the gray side of the cassette.
   (c) Put two pieces of wet blotting paper on the scotch-brite® fiber pad.
   (d) Place the equilibrated gel on top of the filter paper.
   (e) Place the pre-wetted nylon membrane on top of the gel.
   (f) Place the other two pieces of blotting paper over the membrane.
   (g) Be sure to remove any air bubbles trapped between the gel, membrane, and blotting paper layers. This is easily done by rolling a clean pipet over the sandwich. Complete the sandwich with the second Scotch-Brite® fiber pad.
5. Close the cassette firmly being careful not to move the gel and the filter paper sandwich. Lock the cassette with the white latch.
6. Insert the sandwich into the transfer apparatus with the membrane positioned between the gel and the appropriate electrode. Most polypeptides are eluted from SDS-polyacrylamide gels as anions and therefore the membrane should usually be placed between the gel and the anode.
7. Fill the transfer apparatus with transfer buffer. Pour the transfer buffer slowly to prevent bubble formation. Put the lid on the tank and activate the power supply. Transfers are accomplished at either 30 V overnight or 100 V for 1 h. Make sure to add a standard stir bar to help maintain even buffer temperature and ion distribution in the tank.
8. After the transfer is complete, unclamp the blot sandwich, remove the membrane, and allow it to air dry at room temperature. Make sure to mark the side of the membrane that is facing the gel. Mark the positions of the pre-stained markers, since they may fade away during detection.

#### *3.4.2. Confirmation of the Tagging in Transformants with M2 Antibody Against the FLAG Epitopes of the SPA-Tag*

1. After transferring the proteins to the nitrocellulose membrane, block non-specific binding sites on the membrane by incubating the membrane in blocking buffer for 1 h at room temperature.
2. Rinse the membrane briefly with wash buffer prior to adding the primary antibody. The primary anti-FLAG M2 antibody is usu-

ally diluted 1:5,000 in primary antibody buffer and incubated with the membrane for 1 h by shaking gently using a C2 platform rocking shaker (New Brunswick Scientific, Edison, NJ, USA).

3. Rinse the membrane with wash buffer once for 15 min, and then twice for 5 min.
4. Dilute 2 μL of HRP (horseradish peroxidase) - labeled secondary antibody in 40 mL of secondary antibody buffer and incubate with the membrane for 45 min at room temperature by shaking gently using a rocking shaker.
5. Rinse the membrane with wash buffer once for 15 min and then twice for 5 min.

*3.4.3. Image Processing*

1. Transfer the membrane to a shallow tray and incubate the membrane in the chemiluminescence reagent for 5 min by gently shaking (in the dark, if possible).
2. Remove excess chemiluminescence reagent by draining or blotting.
3. Cover the membrane with Saran wrap.
4. Expose the membrane to Kodak X-OMAT Blue Autoradiography film for 30 s. If necessary, expose the membrane for up to another 30 min.

*3.4.4. Stripping and Reprobing the Blots*

1. After the exposure of the film, wash the membrane for 20 min in TBS buffer.
2. Incubate the membrane for 30 min at 50°C in the stripping buffer.
3. Wash the membrane for 20 min in TBS buffer.
4. Incubate the membrane for 1 min in the Western lightning™ chemiluminescence reagent. Expose the film for at least 30 min to 1 h to make sure that the original signal is removed.
5. Wash the membrane again for 20 min in TBS buffer. The membrane is now ready for reuse.

### 3.5. Purification of SPA-Tagged Proteins

*3.5.1. Culturing SPA-Tagged E. coli Strains and Sonication*

1. Inoculate 100 μL of a SPA-tagged *E. coli* glycerol stock into 45 mL TB liquid medium supplemented with 5 mL of potassium salt solution and 25 μL of 100 mg/mL kanamycin solution in a 250 mL conical flask.
2. Grow the culture overnight at 32°C until late log phase. Note that the $OD_{600}$ of the overnight culture should be ~5 to 6.
3. Inoculate 50 mL overnight culture into 900 mL fresh TB supplemented with 100 mL of potassium salt and 25 μg/mL kanamycin in a 4 L flask. The culture is grown at 32°C with shaking at 250 rpm until the $OD_{600}$ reaches to ~4 to 5.
4. Transfer 1,000 mL *E. coli* cultures from the shaker to clean centrifugation bottles.

5. Centrifuge the *E. coli* cultures in a Beckman J6-HC centrifuge at 3,993 × *g* for 15 min.
6. Discard the supernatants and remove excess liquid by inverting the bottles on paper towels. Keep the centrifugation bottles on ice.
7. Add 25 mL of sonication working buffer to the centrifugation bottles and resuspend the *E. coli* cell pellets using a clean 25 mL pipet.
8. Transfer the resuspended cultures into 50 mL polypropylene Falcon tubes and snap freeze the Falcon tubes using liquid nitrogen. Frozen, resuspended cell pellets are stored at –80°C for future use.
9. Remove the Falcon tubes containing the frozen cell pellets from the freezer and place the Falcon tubes in cold water to minimize thawing time. Keep the samples at all times on ice when they are completely thawed.
10. Ensure that the notch of the flat-tip in the sonicator is fastened securely. Set the sonicator (Branson Ultrasonic Sonifier 450 analog, cat#23395) controls to Duty cycle: "50"; Timer: "Hold" and Output control: "7."
11. Transfer the cell samples to the stainless steel cup for sonication.
12. Place the stainless steel cup on ice in an appropriate sized box. Make sure to place the box on top of the sand at an appropriate height so that the flat tip is submerged into the liquid but not too close to the bottom of the stainless steel container. Try to avoid contact between the flat tip and the stainless steel container.
13. When ready, turn on the sonicator. Set timer to 5 min. Sonication is done for 3 min followed by 2 min of cooling to prevent overheating of the samples (*see* **Note 3**).
14. Pour the sonicated cell lysates into pre-chilled centrifugation tubes and place them on ice.
15. Centrifuge the lysates at 35,267 × *g* (16,000 rpm) for 30 min using a JA-17rotor (Beckman).
16. Remove the supernatants carefully from the centrifugation tubes, transfer them to 50 mL Falcon tubes, and snap freeze the samples with liquid nitrogen. The sonicated frozen cell extracts are stored at –80°C for future use.

#### 3.5.2. SPA-Tag Purification

##### Anti-Flag M2 Agarose Beads

1. Prior to use, 100 μL of anti-Flag M2 agarose beads are transferred into a column and the beads are washed twice with 1 mL of AFC buffer without DTT.
2. The Falcon tube containing the sonicated frozen cell extract is thawed by placing the tube in cold water.

3. The cell extracts are incubated with 3 μL of benzonase nuclease (Novagen cat#70746; 25 U) for 30 min at 4°C.
4. Add 500 μL of 10% non-ionic detergent Triton X-100 (final concentration of Triton X-100 should be 0.1%) and 150 μL of anti-flag M2 agarose beads (Sigma cat#A2220) to the Falcon tube containing the cell extracts. Briefly mix the tube contents by tilting the Falcon tube upside down, then rotate the Falcon tube for 3 h at 4°C using a LabQuake shaker (Barnstead/Thermolyne, cat#59558).
5. Centrifuge the tubes at 1,700 × *g* for 6 min. Carefully remove as much supernatant as possible, taking care not to disturb the loose bead pellet.
6. Resuspend the pellets in the remaining supernatant and transfer the beads into 0.8 × 4 cm Bio-Rad polypropylene prep columns (Bio-Rad cat#732-6008). Remove the bottom outlet plugs of the columns and allow the eluates to drain by gravity flow.
7. Wash the columns five times with 200 μL of 1X AFC working buffer and twice with 200 μL of TEV cleavage buffer.
8. Close the bottom outlet of the column. Cleavage is done in the same column by adding 200 μL of 1X TEV cleavage buffer and 5 μL (50 units) of TEV protease. Close the top of the column with a cap.
9. The column containing the beads is rotated overnight at 4°C.

Calmodulin-Sepharose Beads

1. Remove the top and bottom outlet plugs of the column after incubation with TEV protease and drain the eluates into fresh columns.
2. One hundred microliter of Calmodulin-Sepharose beads (Amersham Biosciences cat#17-0529-0), corresponding to 200 μL of bead suspension, is transferred into a column and washed twice with 10 mL of 1X calmodulin binding buffer.
3. Four hundred microliters of 1X TEV cleavage buffer and 1.2 μL of 1 M $CaCl_2$ are added to the eluate recovered after TEV cleavage.
4. The mixture is then transferred to the column containing the washed Calmodulin-Sepharose beads. After closing the column, rotate for 3 h at 4°C.
5. Remove the top and bottom plugs of the column and drain the eluate by gravity flow.
6. The beads are washed four times with 200 μL of 1X calmodulin binding buffer followed by two washes with 200 μL of calmodulin wash buffer.

7. The bound proteins are eluted in six fractions of 50 μL in a fresh effendorf tube using 1X calmodulin elution buffer.
8. Distribute the eluted fractions into two separate microcentrifuge tubes in volumes of 125 μLand 175 μL, respectively.
9. Dry the 175 μL eluted fractions to 50 μL using a Speedvac (Eppendorf Vacufuge cat#5301). Add half the volume of 3X SDS sample buffer to 50 μL of the dried eluate. The eluate and the sample buffer mixture are boiled for 5 min and load onto an SDS polyacrylamide gel, silver stained and the sliced peptide bands are analyzed by MALDI-TOF mass spectrometry.
10. In parallel with **step 9**, dry down the other 125 μL eluted fraction using a Speedvac. This dried sample is subjected to LC-MS for protein identification.

### 3.6. Protein Identification by Gel-Based Peptide Mass Fingerprinting Using MALDI-TOF Mass Spectrometry

The tagged and purified *E. coli* bait protein is separated by silver staining on an SDS-polyacrylamide gel, and peptide bands are cut out and analyzed by MALDI-TOF mass spectrometry. The mass spectrometry data is further analyzed to produce probable identifications of the proteins. Protein identification is finally linked back to the purified protein band. The various procedures, such as silver staining, in-gel trypsin digestion, extraction and purification of tryptic peptide fragments, spotting samples on MALDI target plates, acquisition of spectra, and protein identification are summarized below.

#### 3.6.1. Silver Staining the SDS-Polyacrylamide Gel

1. Place the gel in a clean staining solution containing fixer and agitate the gel on a shaker for 20 min.
2. Rinse the gel in 20% ethanol for 10 min.
3. Wash the gel twice with 500 mL double distilled water for 10 min. Note that thorough rinsing gives a low uniform background.
4. Discard the distilled water and agitate the gel in 500 mL sensitizer solution for 1 min.
5. Discard the sensitizer solution and rinse the gel twice in distilled water for 20 s.
6. Pour off the water and incubate the gel in 200 mL of 0.1% silver nitrate for 30 min.
7. Discard the silver nitrate and rinse the gel once with distilled water for 20 s to remove excess silver nitrate.
8. Wash the gel with fresh 50 to 75 mL developing solution for half a minute. Replace with fresh developing solution and agitate the gel slowly by hand constantly (*see* **Note 4**).
9. When desired intensity is achieved, discard the developing solution, and add 80 mL of stop solution to the gel. Incubate the gel with stop solution for a minimum of 20 min.

#### 3.6.2. Gel-Based MALDI-TOF Mass Spectrometry

##### Preparation of In-Gel Digests

1. Protein bands were excised from silver stained gels with a clean, sharp straight edge razor blade. The gel slices were excised as close as possible to the boundaries of the protein band and stored in –80°C freezers in 96-well polypropylene plates (Nunc cat#249946).
2. Gel slices stored in the –80°C freezer in a 96-well polypropylene plate are allowed to thaw for ~20 min. Remove the liquid that accumulates during thawing.
3. Shrink the gel slices with 200 µL of 100% ACN for ~10 min on an IKA Schuttler MTS 4 orbital shaker (VWR cat #82006-096) at 700 rpm (*see* Note 5). Remove all liquid.
4. Reduce the gel slices with 75 µL of 100 mM $NH_4HCO_3$ containing 10 mM DTT for 30 min in a 50°C heating block. The gel pieces should be covered with liquid. When hydrated, there should still be some fluid left.
5. Remove all liquid by centrifuging the plate ~500 rpm for 3–5 min and repeat **step 3**.
6. Alkylate the gel slices in the 96-well microtiter plate with 75 µL of 100 mM $NH_4HCO_3$ containing 55 mM iodoacetamide for 20 min in the dark.
7. Centrifuge the plate at low speed (~500 rpm) for 3–5 min. Remove all liquid. Repeat **step 3**.

##### In-Gel trypsin Digests

1. Hydrate the gel slices with 60 µL of digestion buffer containing trypsin for 30–45 min on ice. Gel pieces should be fully hydrated with trypsin solution on ice. Check after 15–20 min and add more digestion buffer containing trypsin, if necessary, to allow complete hydration.
2. Add 20 µL (if needed) of digestion buffer without trypsin and incubate samples overnight at 37°C (*see* **Note 6**).

##### Extraction of Tryptic Peptide Fragments

1. Transfer the extracted peptides into a clean 96-well polypropylene plate.
2. Add 100 µL of 100 mM $NH_4HCO_3$ to the gel slices and extract peptides by shaking at 700 rpm on an orbital shaker for 60 min at room temperature.
3. Briefly centrifuge, transfer the extracted peptides into the 96-well polypropylene plate, and add 2.5 µL of 100% AA to the extracted peptides.
4. Repeat **steps 2** and **3** one more time.

##### Purification of Tryptic Fragments

1. The following purification uses bulk C18 reverse phase resin from Sigma. Add 1.5 g of dry resin to a reservoir.
2. Wash two times with HPLC grade methanol and two times with HPLC grade 66% ACN, 1% AA prior to use. Add 75% ACN, 1% AA to prepare 5:1 resin slurry.
3. Add 2.5 µL of C18 slurry to the extracted peptides into the 96-well plate, the resin should float on top of the liquid. Shake

at moderate speed on an orbital shaker at 500 to 700 rpm for 45 min at room temperature.

4. Discard the supernatant by placing the pipet tips below the surface of the liquid to avoid aspirating any resin. Add 200 μL of 2% ACN, 1% AA. Shake briefly for 5 to 15 min at moderate speed of 500 to 700 rpm on an orbital shaker at room temperature.
5. Prepare in advance a 384-well Melt Blown Polypropylene (MBPP) Whatman filter plate (Whatman cat#7700-2117). Place the MBPP filter plate on top of a 384-well collection plate (Whatman cat#7701-3100). Wash the MBPP filter plate wells with 15 μL of 66% ACN, 1% AA. Centrifuge the MBPP filter plate for 1-2 min at 1,000 × *g*, and discard the filtrate collected in the collection plate.
6. Remove the supernatant from **step 4** by centrifuging the 96-well plate briefly for 5-15 min at 1,000 × *g* at room temperature.
7. Elute the peptides by adding 30 μL of 66% ACN, 1% AA. The resin should make a slurry and slowly pellet to the bottom. Shake the 96-well plate at high speed on an orbital shaker and incubate approximately for 5 to 10 min at room temperature. Ensure that all of the resin has entered into the slurry.
8. Transfer the liquid from the 96-well plate to the MBPP filter plate. Place 384-well collection plates (Whatman cat#7701-3100) under the MBPP filter plate and centrifuge for 3–5 min at 2,000 × *g*. The filtrate collected in 384-well collection plates is either spotted immediately onto the Bruker MALDI target plate or sealed with sealing foil and stored at –70°C until spotting onto MALDI targets.

Spotting Samples onto the MALDI Target Plates

1. Separate the Bruker MALDI target plate from its base.
2. Remove the previously spotted matrix spots from the MALDI target plate by washing the plate with 100% methanol. Then, gently rinse the plates with HPLC grade water. Wipe the plate with 100% methanol using Kimwipes (*see* **Note 7**).
3. Spot 1 μL of α-cyano-4-hydroxycinnamic acid matrix (Fluka Buchs SG, Switzerland cat#28480) solution and 1 μL of the purified trypsin sample onto the MALDI target plate. Make sure that there are no air bubbles and the tip does not touch the MALDI target plate. Allow samples to dry at room temperature.
4. For the bottom-most row, 1 μL of α-cyano-4-hydroxycinnamic acid matrix solution and 1 μL of peptide calibration standard (Bruker cat#206195) are spotted onto the MALDI target plate.

5. On completion of spotting, seal the protein plate (s) and store at –80°C. Wait until the spots on the MALDI target plate dry before acquiring the spectra.

#### Acquisition of Spectra from MALDI Target Plates

1. Double click on the flex control icon on an ULTRAFlex II MALDI-TOF instrument. When the log on information appears, Click OKAY. Open the FLEX control method: RP_pepmix.par.\
2. Wait for the method to upload.
3. Insert the spotted MALDI target plate into the source chamber and click on the green load button on the instrument.
4. Wait for the system to fully load the MALDI plate. Once the plate is loaded, the system status light on the FLEX control should turn green and say "READY." The sample status light should turn green and say "IN."
5. Acquire the spectra by clicking the start button and adjust the laser intensity such that the peptide peaks can be visualized.
6. Click on "Add" to the sum buffer.
7. Click on the "calibration tab" and choose peptide mix monoisotopic II from the drop down menu and a peptide reference list will appear.
8. Change the zooming factor to 0.5.
9. To manually calibrate the instrument click on each peptide in the reference list. The red vertical line indicates the corresponding peptide. Click on the monoisotopic peak (left of the highest peak) to visualize each of the peptides.
10. Click on accept fit result; now the instrument is calibrated.
11. To automatically acquire the samples, click on the AutoXecute tab. Load the .txt file by clicking on the select button. Make sure that the method to be run is "autoXtest." Check the box where it shows the autoX output.
12. Click on START automatic run and the system should start acquiring the spectra automatically.

#### MALDI-TOF Spectral Analysis and Protein Identification

1. MALDI spectral searches were performed using Knexus automation, a Windows-based program from Genomics Solutions Bioinformatics (Discovery Scientific, Inc., Vancouver, Canada) that automatically selects the spectral peaks and performs the searches.
2. After the installation of the Knexus program, the Fasta sequences of the *E. coli* genome were downloaded from the European Bioinformatics Institute database (http://www.ebi.ac.uk/) and were uploaded into the machine using the Knexus Database Installation Wizard.

3. Protein identification is done using the ProFound search engine, which matches the observed peaks against the database of theoretical peaks.
4. The program automates the running of ProFound on more than one spectrum using one set of conditions. In addition, a Java program was developed in-house, which automates the re-running of Knexus. It uses 72 varying parameter sets and evaluates the aggregate results to produce a set of identified proteins in each case. Based on these results, it calculates an aggregate score for each protein.
5. Using graphical interface software developed in-house, the user is able to specify where bands are located on the corresponding gel image. The entire area of the band is inputted, so that its total intensity is measured (*see* **Note 8**). When all the data have been entered into the system for a given band, an annotated gel image and a plot are produced in JPEG format. Examples of affinity-purified protein complexes analyzed by SDS-PAGE and silver staining followed by protein identification by trypsin digestion and MALDI spectral analysis are shown in **Fig. 2**.

#### *3.6.3. Gel-Free Liquid Chromatography–Tandem Mass Spectrometry*

##### Proteolysis and Sample Preparation for LC-MS

1. One hundred and twenty-five microliters of eluted fractions (described in **step 10** under the section Calmodulin-Sepharose Beads) were dried down using a Speedvac.
2. Dissolve the pellets in 20 μL of the immobilized trypsin solution.
3. Incubate the samples overnight at 30°C with rotation or agitation.
4. Stop digestion by adding 20 μL of the LC-MS solvent A and centrifuge the peptide mixtures for 5 min at maximum speed using an Eppendorf centrifuge.
5. Carefully recover 10–20 μL of the peptide mixtures into microcentrifuge tubes and immediately analyzed by LC-MS or store the samples at –20°C prior to use.

##### LC-MS Analysis and Protein Identification

1. Proteins are analyzed using single-dimension reverse phase chromatography coupled online to ion trap tandem mass spectrometry using standard conditions.
2. The micro-columns are packed with ~10 cm of 5 μm Zorbax eclipse XDB-$C_{18}$ resin and are interfaced to a custom electrospray ion source.
3. Place the packed column in line with the LC-MS instrument.
4. A Proxeon nano HPLC pump is used to deliver a stable tip flow rate of ~300 nL/min during the peptide separations.
5. Elution of the peptide is achieved using the following gradient that may be increased or decreased according to the sample

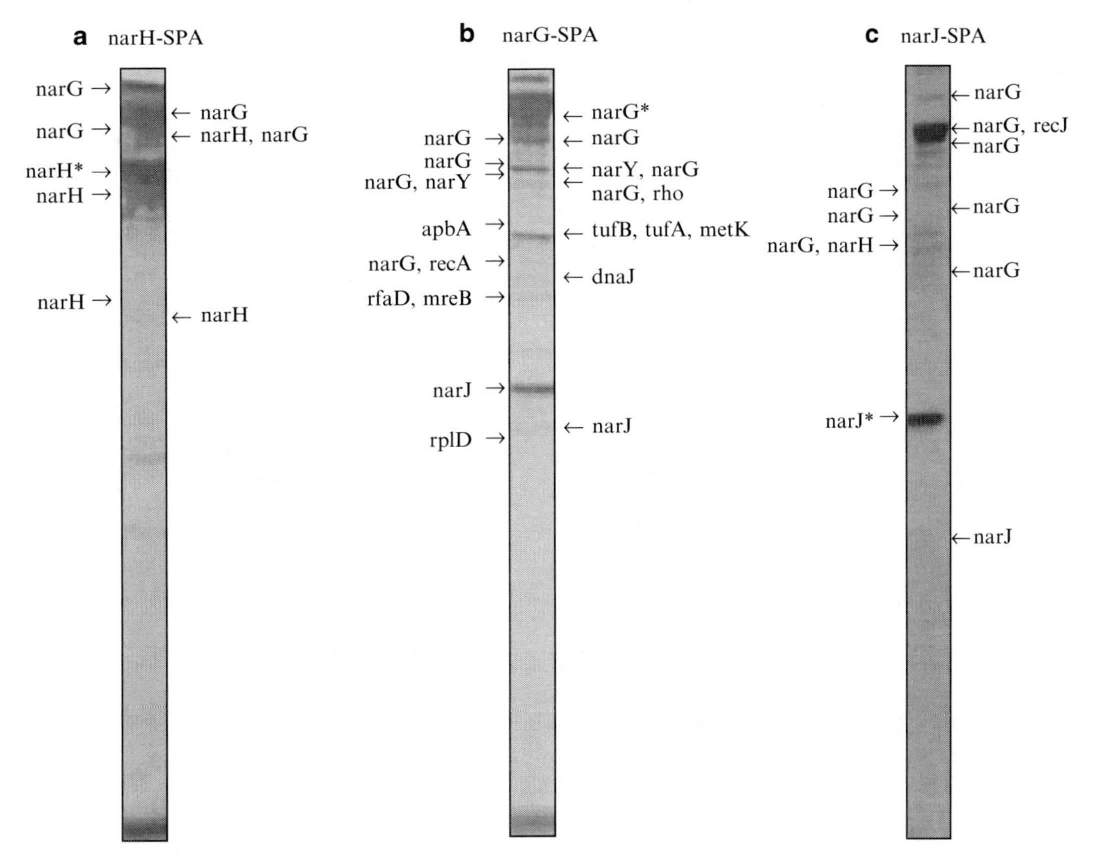


Fig. 2. Examples of silver-stained SDS polyacrylamide gels portraying NarHGJ (nitrate reductase) protein complexes recovered following affinity purification of *E. coli* SPA-tagged NarH, NarG, and NarJ, respectively. The SPA-tagged bait proteins are indicated at the top of each lane. Distinct gel bands (indicated by *arrowheads*) containing individual subunits of the purified complexes were identified by in-gel trypsin digestion followed by MALDI-TOF mass spectrometry. Note that the observed interactions in any given lane are validated by reciprocal SPA-tagging and purification of proteins shown in the other lanes. Asterisks indicate the recovery of the intact tagged bait protein from each purification. Protein degradation products are also routinely identified.

complexity: 15% of solvent B from 0 to 30 min, 40% of solvent B from 31 to 49 min, 80% of solvent B from 50 to 55 min, 80% of solvent B from 56 to 60 min, 100% of solvent A from 61 to 65 min, 45% solvent B from 66 to 70 min, 80% solvent B from 71 to 75 min, 80% solvent B from 76 to 80 min, 100% solvent A from 81 to 85 min, 100% solvent A from 86 to 105 min. The flow rate at tip of the needle is set to 300 nL/min for 105 min.

6. The mass spectrometer cycles run through successive series of 11 scans as the gradient progresses. The first scan in a series is a full mass scan and is followed by successive tandem mass scans of the most intense ions.
7. Proteins from the mixture are identified using the SEQUEST computer algorithm (*16*) and validated using the STATQUEST (*17*) probabilistic scoring program (*see* **Note 9**).

## 4. Notes

1. The plasmid pJL72 contains a version of the TAP cassette *(14)* that was placed adjacent to a selectable Kan$^R$ marker for use in *E. coli* and related bacteria.
2. Forty microliters of competent cells are sufficient for each standard electroporation reaction. For a greater number of reactions, larger numbers of cultures are prepared. It is very important to make sure that the electrocompetent cells are prepared fresh each time prior to transformation.
3. Avoid sonicating the samples continuously for 3 min as overheating of the instrument might lead to proteolysis of the proteins in the samples.
4. The developing solution must be made fresh each time. When the solution becomes cloudy, replace it with new developing solution. Make sure that all the gels are developed for about the same time.
5. The gel slices will become white and should feel gritty like grains of sand. Make sure to remove all the liquid when the gel slices are completely shrunk.
6. In each well ensure that there is enough liquid for the gel slices to be completely submerged.
7. If spots still remain on the MALDI targets, sonicate the plate for 10 min with 100% methanol followed by 5-min sonication with HPLC water and 100% methanol, respectively.
8. The band location algorithm tries to identify the complete area of each band. It is very important to look at the location that the computer has chosen for each band and make sure that it matches the gel image and accurately includes the full area of the band.
9. The SEQUEST database search algorithm was used to match the acquired spectra to peptide sequences encoded in the *E. coli* protein database downloaded from the European Bioinformatics Institute (http://www.ebi.ac.uk/) database. The probabilistic STATQUEST scoring program was programmed to evaluate and assign confidence scores to all putative matches. Proteins were considered as positive if detected with two or more high confidence peptide candidates, each passing with a minimum likelihood threshold cut-off of 90% or greater probability.

## Acknowledgments

The authors thank Wenhong Yang, and Xinghua Guo for tagging and purifying the *E. coli* bait proteins and Shamanta Chandran, Michael Davey, Peter Wong, and Constantine Christopoulos for assisting with mass spectrometry. This work was supported by funds from the Ontario Research and Development Challenge Fund and Genome Canada to A.E. and J.G.

### References

1. Ito, T., Chiba, T., Ozawa, R., Yoshida, M., Hattori, M., and Sakaki, Y. (2001) A comprehensive two-hybrid analysis to explore the yeast protein interactome. Proc Natl Acad Sci USA 98, 4569–4574.
2. Uetz, P., Giot, L., Cagney, G., Mansfield, T.A., Judson, R.S., Knight, J.R., Lockshon, D., Narayan, V., Srinivasan, M., Pochart, P., Qureshi-Emili, A., Li, Y., Godwin, B., Conover, D., Kalbfleisch, T., Vijayadamodar, G., Yang, M., Johnston, M., Fields, S., and Rothberg, J.M. (2000) A comprehensive analysis of protein-protein interactions in *Saccharomyces cerevisiae*. Nature 403, 623–627.
3. Gavin, A.C., Aloy, P., Grandi, P., Krause, R., Boesche, M., Marzioch, M., Rau, C., Jensen, L.J., Bastuck, S., Dumpelfeld, B., Edelmann, A., Heurtier, M.A., Hoffman, V., Hoefert, C., Klein, K., Hudak, M., Michon, A.M., Schelder, M., Schirle, M., Remor, M., Rudi, T., Hooper, S., Bauer, A., Bouwmeester, T., Casari, G., Drewes, G., Neubauer, G., Rick, J.M., Kuster, B., Bork, P., Russell, R.B., and Superti-Furga, G. (2006) Proteome survey reveals modularity of the yeast cell machinery. Nature 440, 631–636.
4. Krogan, N.J., Cagney, G., Yu, H., Zhong, G., Guo, X., Ignatchenko, A., Li, J., Pu, S., Datta, N., Tikuisis, A.P., Punna, T., Peregrin-Alvarez, J.M., Shales, M., Zhang, X., Davey, M., Robinson, M.D., Paccanaro, A., Bray, J.E., Sheung, A., Beattie, B., Richards, D.P., Canadien, V., Lalev, A., Mena, F., Wong, P., Starostine, A., Canete, M.M., Vlasblom, J., Wu, S., Orsi, C., Collins, S.R., Chandran, S., Haw, R., Rilstone, J.J., Gandi, K., Thompson, N.J., Musso, G., St Onge, P., Ghanny, S., Lam, M.H., Butland, G., Altaf-Ul, A.M., Kanaya, S., Shilatifard, A., O'Shea, E.,Weissman, J.S., Ingles, C.J., Hughes, T.R., Parkinson, J., Gerstein, M., Wodak, S.J., Emili, A., and Greenblatt, J.F. (2006) Global landscape of protein complexes in the yeast *Saccharomyces cerevisiae*. Nature 440, 637–643.
5. Veraksa, A., Bauer, A., and Tsakonas, S.A. (2005) Analyzing protein complexes in *Drosophila* with tandem affinity purification-mass spectrometry. Dev Dynam 232, 827–834.
6. Polanowska, J., Martin, J.S., Fisher, R., Scopa, T., Rae, I., and Boulton, S.J. (2004) Tandem immunoaffinity purification of protein complexes from *Caenorhabditis elegans*. Biotechniques 36, 778–780.
7. Tsai, A., and Carstens, R.P. (2006) An optimized protocol for protein purification in cultured mammalian cells using a tandem affinity purification approach. Nat Protoc 1, 2820–2827.
8. Knuesel, M., Wan, Y., Xiao, Z., Holinger, E., Lowe, N., Wang, W., and Liu, X. (2003) Identification of novel protein-protein interactions using a versatile mammalian tandem affinity purification expression system. Mol Cell Proteomics 2, 1225–1233.
9. Zeghouf, M., Li, J., Butland, G., Borkowska, A., Canadien, V., Richards, D., Beattie, B., Emili, A., and Greenblatt, J.F. (2004) Sequential Peptide Affinity (SPA) system for the identification of mammalian and bacterial protein complexes. J Proteome Res 3, 463–468.
10. Butland, G., Peregrı'n-Alvarez, J.M., Li, J., Yang, W., Yang, X., Canadien, V., Starostine, A., Richards, D., Beattie, B., Krogan, N., Davey, M., Parkinson, J., Greenblatt, J., and Emili, A. (2005) Interaction network containing conserved and essential protein complexes in *Escherichia coli*. Nature 433, 531–537.

11. Hu, P., Janga, S.C., Babu, M., Díaz-Mejía, J.J., Butland, G., Yang, W., Pogoutse, O., Guo, X., Phanse, S., Wong, P., Chandran, S., Christopoulos, C., Armavill, A.N., Nasseri, N.K., Musso, G., Ali, M., Nazemof, N., Eroukova, V., Golshani, A., Paccanaro, A., Greenblatt, J.F., Hagelsieb, G.M., and Emili, A. (2009) Global Functional Atlas of Escherichia coli Encompassing Previously Uncharacterized Proteins. PLoS Biol, In Press.

12. Hernan, R., Heuermann, K., and Brizzard, B. (2000) Multiple epitope tagging of expressed proteins for enhanced detection. Biotechniques 28, 789–793.

13. Yu, D., Ellis, H.M., Lee, E.C., Jenkins, N.A., Copeland, N.G., and Court, D.L. (2000) An efficient recombination system for chromosome engineering in *Escherichia coli.* PNAS 97, 5978–5983.

14. Rigaut, G., Shevchenko, A., Rutz, B., Wilm, M., Mann, M., and Seraphin, B. (1999) A generic protein purification method for protein complex characterization and proteome exploration. Nat Biotech 17, 1030–1032.

15. Datsenko, K.A. and Wanner, B.L. (2000) One-step inactivation of chromosomal genes in *Escherichia coli* K-12 using PCR products. Proc Natl Acad Sci USA 97, 6640–6645.

16. Eng, J.K., McCormack, A.L., and Yates, I.I.I.J.R. (1994) An approach to correlate tandem mass spectral data of peptides with amino acid sequences in a protein database. J Am Soc Mass Spectr 5, 976–989.

17. Kisliger, T., Rahman, K., Radulovic, D., Cox, B., Rossant, J., and Emili, A. (2003) PRISM, a generic large scale proteomic investigation strategy for mammals. Mol Cell Proteomics 2, 96–106.

# Chapter 23

# Bioinformatical Approaches to Detect and Analyze Protein Interactions


## Beate Krüger and Thomas Dandekar



### Summary

Protein-protein interactions are the building blocks of cellular networks and at the heart of cellular regulation. However, their experimental identification is still a challenge.

This chapter is concerned with the determination of protein-protein interactions by bioinformatical methods. These often can operate just on sequence information. Further required information is derived from public knowledge in literature databanks and biochemical databases as well as from the sequences themselves and iterative sequence comparisons. Further tools include domain analysis, structure prediction, and genome context methods. The results are predicted binary interactions and complete interaction networks.



**Key words:** Bioinformatics, Protein interaction network, Biological database, Domain analysis, Sequence analysis, Text-mining, Gene context, Regulation, Proteome, Prediction


## 1. Introduction

Protein–protein interactions are the building blocks of cellular networks and the heart of cellular regulation. This includes all sorts of pharmacological and biotechnological applications. Recent advances in experimental techniques allow large-scale collection of proteome and genome data; however, it is still a challenge to obtain accurate data on protein-protein interactions. This chapter is hence concerned with their determination by bioinformatic methods, which often can operate already just on sequence information (DNA or proteins sequence). We start from sources (genetic data and databases, literature), then look at individual proteins (domain analysis, sequence analysis), and


Jörg Reinders and Albert Sickmann (eds.), *Proteomics,* Methods in Molecular Biology, Vol. 564
DOI: 10.1007/978-1-60761-157-8_23, © Humana Press, a part of Springer Science+Business Media, LLC 2009

define binary interactions and protein interaction networks. A number of prediction examples and typical challenges complete the chapter.

## 2. Sources to Identify Protein Sequences and Interactions

### 2.1. Genetic Data and Databases

Scientists accumulated a lot of important biological data in the past years. Typically, such knowledge is now stored in detail in databases while reference books provide a more conceptual overview. The journal *Nucleic Acids Research (NAR)* *(1)* has an online overview on the current available databases, the Molecular Biology Database Collection. It contains links to many molecular biology databases that are freely available on the web. It is updated yearly including a special issue of NAR devoted to the topic with articles for each new or updated database. Currently (2006 onwards), it includes 968 databases.

#### 2.1.1. Primary Databases

Primary databases are those that collect data from scientists all over the world and contain sequences independent from their sources. The most important sequence databases are EMBL *(2)*, Genbank *(3)*, and DDBJ *(4)*, which are synchronized daily. They should be the first place to go when you are looking for sequences.

As the three databases often have redundancies in their data (e.g., the same sequence of interest was investigated by three different research teams and each sends in its own sequencing results to the database), there also exist nonredundant sequence databases. One of the best curated ones is the Swissprot database, which contains meticulously curated and annotated protein sequences with useful annotations and descriptions *(5)*.

#### 2.1.2. Secondary Databases

Databases that contain filtered or interpreted data from primary databases are termed secondary databases. They offer improved additional experiments and biochemically testable information for a given sequence. These include the following:

The RefSeq *(6)* collection: it provides a comprehensive, integrated, nonredundant set of sequences, including genomic DNA, transcript (RNA), and protein products, as well as a comprehensive functional annotation of some genome sequencing projects, including those of human and mouse.

The Biomolecular Object Network Databank (BOND) *(7)*: it is a new resource to perform cross-database searches of available sequences, interactions, complexes, and pathway information. BOND integrates a range of component databases including Genbank and BIND *(8)*, the Biomolecular Interaction Network Database.

The "Database of Interacting Proteins" (DIP, *[9]*): it documents experimentally determined protein-protein interactions. It currently catalogs roughly 5,900 proteins participating in about 10,000 distinct interactions contained in more than 80 organisms; the vast majority of interactions to date concern yeast, *Helicobacter pylori*, and man.

There are many more secondary databases containing special information in addition to the sequence. For instance, MEROPS *(10)* contains specific data on proteases and OMIM data *(11)* on human genetic diseases in addition to basic sequence information. Complete database lists and summaries are available online from the NAR web site *(1)*.

Meanwhile, there are also powerful searchtools to retrieve data from these various databases. All NCBI databases are, for example, searchable via Entrez (http://www.ncbi.nlm.nih.gov/sites/entrez), a graphical interface to search all NCBI databases concurrently. Quite powerful, at least in the commercial versions, is the sequence retrieval system (SRS at http://srs.ebi.ac.uk/srsbin/cgi-bin/wgetz?-page+quickSearch, *[12]*). Here, a number of databases are linked and connected to each other most efficiently using the Meta-language "Ikarus" to describe their data content and connections. A query manager helps to specifically refine query retrieval sets by adding, subtracting, and concatenating different queries with each other.

### 2.2. Text-Mining

Database resources contain structured information, as they are stored in standard formats in databases or files. In contrast, literature stores important information in a nonstructured way. Text-mining allows automatic detection of new, relevant, and correct information from textual data with the help of statistical and linguistic methods. Entrez/Pubmed *(13)* is an integrated, text-based search and retrieval system used at NCBI for the major databases, including PubMed, Nucleotide and Protein Sequences, Protein Structures, Complete Genomes, Taxonomy, and others. There are several more refined tools available for text-mining, e.g., Xplormed *(14)* (http://www.bork.embl-heidelberg.de/xplormed/).

## 3. Analyze Individual Proteins

The above-mentioned databases and text-mining techniques allow one to retrieve information and description on individual proteins and their function. Sequence and domain analysis allows next (1) to get a first function and domain description for new and uncharacterized proteins, (2) to rapidly identify domains that are directly involved in protein-protein interactions, and (3) the fine-tuning of interactions is modulated by protein modifications, which are not so easy to predict.

### 3.1. Sequence Analysis

Genes and proteins from model and other organisms are generally very well annotated and experimentally verified concerning their functions and interactions. When sequence alignment shows that experimentally well-characterized gene or protein sequences are sufficiently similar to nonannotated sequences, it is possible to transfer the functional characteristics from one to the other. The standard way a sequence alignment works is to compare two sequences and to find as much identical or similar positions (so-called matches) in the sequences as possible. To measure similarity searches, the alignments use amino acid substitution matrices, which contain a similarity score for each amino acid combination, based on careful analyzed sequence-structure-alignments (using, e.g., the software BlockMaker, Henikoff, 1991) such as the BLOSUM matrix (Henikhoff, 1992).

To identify interacting domains, local alignments are required. These find short corresponding sequences indicating similar domains within different proteins. A fast and good heuristic method for this is BLAST *(15)*. To interpret a result from such a sequence alignment search, the e-value (expected value) shows the possibility for a match by chance; "gaps" shows the number of gaps, "identities" is the number of perfect matches, and "positives" gives the number of conservative substitutions (found by the substitution matrix).

It can also be helpful to compare a set of sequences against each other and to get a consensus sequence, which represents the most likely amino acids for all positions within a sequence.

The best way to compare a set of sequences with each other is to use multiple alignments. It provides information about amino acid distribution on each position (whole sequence). A well-known heuristical tool for such a global multiple sequence alignment is ClustalW *(16)*.

The visualization of multiple alignment results can be done by consensus sequences, specific coloring of conserved positions, or more refined presentations such as sequence logos *(17)* or position specific matrices (PSSM) *(18)*.

Sequence logos show graphically by the amino acid's letter size the likelihood for a specific amino acid at that position in the form of a consensus sequence. A PSSM instead is a two-dimensional matrix with all amino acids on the $y$-axis and the consensus sequence at the $x$-axis. The numbers filling the matrix are scores to find a given amino acid on each position in the sequence.

### 3.2. Domain Analysis: Typical Domains

Different parts of a protein sequence fold into independent folding units or domains (50-150 amino acids long). A domain has a specific structure and usually carries also a specific function (e.g., catalytic domain, cofactor binding domain, regulatory domain). Thus, it is important to analyze all domains hidden in one sequence. Here, we give you some examples for domains typical for protein-protein interaction.

#### *3.2.1. Diaphanous FH3 Domain (PF06367)*

Formin homology (FH) proteins play a crucial role in the reorganization of the actin cytoskeleton, which mediates various functions of the cell cortex including motility, adhesion, and cytokinesis. Formins are multidomain proteins that interact with diverse signaling molecules and cytoskeletal proteins. The proline-rich FH1 domain mediates interactions with a variety of proteins, including the actin-binding protein profilin, SH3 (Src homology 3) domain proteins, and WW domain proteins. The FH2 domain is required to inhibit actin polymerization. The FH3 domain is less well conserved and is required for directing formins to the correct intracellular location, such as the mitotic spindle, or the projection tip during conjugation.

#### *3.2.2. WW Domain (PF00397)*

The WW domain is a short conserved region in a number of unrelated proteins, which folds as a stable, triple-stranded beta-sheet. This short domain of approximately 40 amino acids may be repeated up to four times in some proteins. The name WW or WWP derives from the presence of two signature tryptophan residues that are spaced 20–23 amino acids apart and are present in most WW domains known to date, as well as that of a conserved Pro. The WW domain binds to proteins with particular proline-motifs, [AP]-P-P-[AP]-Y, and/or phosphoserine-, phosphothreonine-containing motifs. It is frequently associated with other domains typical for proteins in signal transduction processes.

#### *3.2.3. SH3 Domain (PF00018)*

SH3 (Src homology 3) domains are often indicative of a protein involved in signal transduction related to cytoskeletal organization. They are found in a variety of intracellular- or membrane-associated proteins, proteins with enzymatic activity, in adapter proteins that lack catalytic sequences, and in cytoskeletal proteins.

The SH3 domain has a characteristic fold, which consists of five or six beta-strands arranged as two tightly packed anti-parallel beta sheets. The linker regions may contain short helices. The surface of the SH2-domain bears a flat, hydrophobic ligand-binding pocket, which consists of three shallow grooves defined by conservative aromatic residues in which the ligand adopts an extended left-handed helical arrangement. The function of the SH3 domain is not well understood, but they may mediate many diverse processes such as increasing local concentration of proteins, altering their subcellular location, and mediating the assembly of large multiprotein complexes.

#### *3.2.4. PDZ Domain (Also Known as DHR or GLGF) (PF00595)*

PDZ domains are found in diverse signaling proteins in bacteria, yeast, plants, insects, and vertebrates. PDZ domains can occur in one or multiple copies and are nearly always found in cytoplasmic proteins. They bind either the carboxyl-terminal sequences of proteins or internal peptide sequences. However, agonist-dependent activation of cell surface receptors is sometimes required to promote interaction with a PDZ protein.

PDZ domains consist of 80 to 90 amino acids comprising six beta-strands and two alpha-helices, A and B, compactly arranged in a globular structure. Peptide binding of the ligand takes place in an elongated surface groove as an antiparallel beta-strand interacts with the betaB strand and the B helix. The structure of PDZ domains allows binding to a free carboxylate group at the end of a peptide through a carboxylate-binding loop between the betaA and betaB strands.

#### 3.2.5. Protein Databases

These and further protein–protein interaction domains are also collected in databases. We silently introduced them earlier as we always mentioned the pfam identifies (PF00…) under which the above-mentioned domains are stored in the database pfam. Important domain databases are the following:

The **PFAM** database contains protein domain families. With the help of well-known domains and sequence comparisons, it is possible to detect similar function or relationships to unknown proteins *(19)*.

**SMART** is also a database for domain families. Here, you can find information about function, important amino acids, structure, and the phylogenetic development *(20)*.

The **PROSITE** database consists of a large collection of biologically meaningful signatures and motifs that are described as patterns or profiles. Each signature is linked to a documentation that gives useful biological information on the protein family, domain, or functional site identified by the signature *(21)*. With those patterns, it is easy to detect proteins with similar structures and functions.

The **InterPro** database summarizes data from PROSITE, PRINTS, Pfam, and ProDom and adds important literature references *(22)*.

### 3.3. Detecting Protein Modification

Detecting protein modifications is a challenge both for experimental techniques as for bioinformatics. However, these give a lot of additional pathophysiological information regarding the actual type of protein interaction. There are two major types:

(1) Genetic modifications, in particular, individual genetic variation of the encoded protein sequence for instance by single nucleotide polymorphisms (SNPs) in the DNA sequence of this individual compared with the whole population. In most cases, those changes do not have any effect on the resulting protein because the wobble took place in a noncoding area or does not cause a change in the amino acid. When the change did cause an amino acid change, the protein's function can be influenced in a positive or negative way. For swift bioinformatical detection, the database dbSNP collects nuclear polymorphisms *(23)*. Nonannotated SNPs can be detected by sequence comparison (tedious work; needs sufficient sequence data).

(2) Post-translation modification (PTM) of the protein sequence after it has been translated from the ribosome. This allows one to adapt the protein during time for different conditions, exemplified by protein phosphorylation for enzyme activity. Some proteins, for instance structure proteins such as collagen with amino acids hydroxylysin and hydroyprolin, undergo permanent PTM. Many of those changes are documented in the Swissprot database. In contrast, annotated protein modifications are difficult to detect. A large amount of data is needed for this; then, a number of methods exist, well known is the training of neuronal networks to detect protein modifications such as amidation or acetylation. For phosphorylation and glycosylation also, standard patterns (e.g., from PROSITE) exist to predict which amino acids are modified; however, the specificity of such standard patterns is low, yielding a high number of false-positives.

In general, a protein modification is important to understand the dynamics of protein-protein interactions. It leads to different binding behavior, modifies protein stability, enzyme activity, and protein–protein recognition. The latter point also allows one to predict such dynamic protein–protein interactions via the partner protein and specific involved domains such as PTB (PF08416; phosphotyrosine binding domain) or WW domain (PF00397; proline-rich peptide motifs) or different types of lectins for binding different types of glycosilations. In future, the combination of both bioinformatical strategies (prediction of modified amino acids and domain analysis) with powerful new techniques (e.g., detection of phosphorylation by MS) will allow large-scale improvement in this area.

## 4. Binary Interactions

We next build up interaction networks from the analysis of the data. Let us first consider binary interactions. These take place between a protein and a partner. Such a partner could be another protein, a DNA, or a ligand. A ligand is a molecule (mostly small) that binds to biological macromolecule (**Subheading 4.1**). Protein–protein interactions can also be directly modeled using structure information and docking techniques (**Subheading 4.2**). Typical domains and proteins for specific binary protein–protein interactions will be reviewed next (**Subheading 4.3**). Finally, we will briefly address complementary experimental techniques.

### 4.1. Predicting Protein–Ligand Interactions

Interaction predictions can be done using protein domains, information of interacting residues and their specific ligand, and training suitable hidden Markow models (HMMs) *(24)*. In the special case of the interaction prediction, an HMM contains aligned domain sequences and corresponding interaction profiles. Ligand-type specific interaction site information is stored in these profiles. Model parameters can be estimated via maximum likelihood combined with position-specific sequence weights and weighted pseudocount regularization. Such a model calculates site-associated probabilities for all possible states *(25)*. Many more interactions and ligands are predictable, less sophisticated, and accurate, but efficiently using databases such as PDB (for detailed structure-ligand interactions, surprisingly powerful including homology-based-predictions) and PFAM (taking the compiled large-scale sequence alignments and information from the database).

### 4.2. Predicting Protein Interactions: Docking

Protein–protein interactions can also be nicely explored using molecular modeling techniques or protein–protein docking. Pharmaceutical research employs docking techniques for a variety of purposes, most notably in the virtual screening of large databases of available chemicals to select likely drug candidates by their optimal binding to the modeled receptor pocket ("lock and key model," after Emil Fischer (1894) *(26)*, the substrate has to fit the enzyme like a key a lock). Two well-known software applications that include virtual screening for optimal ligands or binding are FlexX and Docks. FlexX is based on mathematical models for fields of force, which visualize the descriptors' physical energy. The lesser energy the binding costs, the more likely is the correct way the ligand will bind *(27)*. Docks *(28)* is a geometric method that estimates the inside structure of the binding pocket and compares it with the ligand's structure. The better the ligand's structure matches the pocket, the more likely is a good binding between ligand and binding pocket.

### 4.3. Examining Typical Domains and Proteins Involved in Binary Interactions

#### 4.3.1. Regulatory Interactions

Living cells need to receive and to react to external signals like changes in pressure. That process is also known as signal transduction. Such protein–protein interactions involve regulatory cascades, e.g., the kinase cascades involved in synthesis as well as degradation of glycogen or even complex cellular networks such as the control of cell cycle.

**Two-Component Systems** However, in its easiest form, the regulatory interaction consists of two protein components, best exemplified by two component (sensor) systems *(29)*. Signaling kinases have as substrates another protein (for metabolic substrates and networks, see later). In the example, a transmembrane-protein, also called sensor protein, reacts to changes in the environment by adding a phosphate group to the protein. In the second step, the phosphate group is transferred to a regulator protein. The activated regulator protein influences the gene expression.

**Transcription Factor Binding and Complexes** In this case, one partner is the DNA. Transcription factor binding sites and corresponding Transcription factors are stored in the TRANSFAC database *(30)*; detailed composite models of promotors and their binding partners are possible using the GENOMATIX software *(31)*. Furthermore, such an interaction can be predicted from the domain composition. To give an example, the secondary structure helix-turn-helix is also the general template for a structure motif of DNA-binding proteins (a short α-helix, the "recognition helix," which is linked via a flexible loop to a longer α-helix, the "positioning helix"). Like all transcription factors, helix-turn-helix proteins bind the DNA to activate or suppress the promoter activity.

#### *4.3.2. Enzyme Subunits*

These are in most cases binary interactions. Furthermore, such subunit interactions can be predicted, for example, by gene context methods and operon structure analysis. In some cases, homology models can also point to the direct interaction of (previously unrecognized) subunits *(32)*.

#### *4.3.3. Interactions with Membranes*

Protein binding to membranes plays a central role in many biological processes. Although there are many different types of membranes (e.g., cytoplasmic, Golgi, ER), which interact with many different proteins under very different circumstances, there are great similarities between all these interactions. Hence, understanding the biophysics of protein-membrane interaction is key for understanding such diverse phenomena as communication between the cell and its environment, trafficking of proteins in the cell, and viral infection.

One example is a G-protein coupled receptor (GPCR). Such receptors are within the cell membrane to transfer signals into the cell via GTP binding proteins. GPCR are responsible luminous-, smell-, and the taste-stimulus transfer. They play an important role in inflammatory processes, chemotaxis, endo- and exocytosis, and cell differentiation. Additionally, very often they are the target structure for hormones like adrenaline or glucagon.

Functions and binding partners are for many GPCRs still unknown. Even some viruses use a GPCR to enter the cell (e.g., HIV).

Based on the electrostatic characteristics of a protein and the membrane, in principle, it is possible to predict the way in which the protein binds to the membrane.

Bioinformatics tools for sequence and structure analysis can be combined with energy and electrostatics calculations to yield a fairly detailed account of a protein–membrane interaction.

Moreover, instructive crystal structures of protein–protein complexes and ligand-receptor complexes can be found in PDB database or the Cambridge Structural Database *(33, 34)*.

### 4.4. Experimental Techniques to Detect Binary Interactions

These include yeast two-hybrid systems as an in vitro method for protein–protein interaction detection in yeast. The basic principle *(35)* is to use a transcription factor that consists of two domains, a binding domain and a transcription activation domain.

The proteins of interest are fused to the activating and the binding domain, respectively. Provided that the two explorative proteins do interact with each other, the transcription factor is completed, and the gene expression takes place. A successful gene expression is proven by a translated and detectable protein. A number of other powerful techniques have appeared for screening for protein–protein interactions, such as the tap-tag technology *(36)*. Nevertheless, large-scale screening efforts vary surprisingly in the interactions they discover, as recently shown for yeast *(37–39)*. For any meaningful comparison and evaluation of such screens, detailed bioinformatical comparisons of complexes and members in protein complexes are critical *(39, 40)*.

## 5. Protein Interaction Networks

A single binary interaction is relatively easy to understand. But very often, it is not enough to understand the binary interactions, since interactions are embedded in a huge number of interactions and relationships. Just by using textual description, it is not possible to get an overview any more.

That is why development of new tools for target identification, drug design, or studies on causes for genetic diseases relies on a network perspective.

A basic network consists of edges and nodes. Nodes stand for proteins and edges for the different interactions between the proteins.

A pathway concerns a trajectory within the network, a sequence of events associated with specific signaling (e.g., MAP kinase pathway) or with specific metabolite conversions (e.g., glycolysis). One of the big surprises of bioinformatics was the discovery that the classical pathway concept (textbook pathways discovered by years of experimentation) can be generalized, formalized, and specific pathways be calculated (e.g., by elementary mode analysis) *(67)*. This showed that besides the classical pathways, many more pathways become not only apparent by such techniques but have important biological consequences (additional signaling pathways; further metabolic adaptation potential for an organism, etc.). Linear pathways can furthermore be compared in different organisms, for example, by pathway alignment *(68, 69)*. This allows one to identify organism's specific differences, alternative pathways, and detours, including potential pharmacological targets and potential to enhance biotechnological yield of desired products. Pathways can be divided into metabolic and regulatory pathways.

### *5.1. Metabolic Networks*

A metabolic pathway describes a set of reactions that are responsible for the transformation of a substance into another. The single reactions are wedded with the catalytic enzymes.

Important examples for metabolic pathways are citric acid cycle or glycolysis. All metabolic pathways together build a metabolic network. Metabolic databases such as KEGG *(41)*, WIT2, Panther, or Ecocyc help one to rapidly reconstruct metabolic pathways. This is also important if in proteomic experiments, several enzymes are found and a complete picture is desirable. Furthermore, software such as Metatool *(42)*, YANA *(43)*, and its successor YANAsquare *(44)* allow rapid setup of networks exploiting such databases, calculation of all hidden metabolic paths, and the distribution of flux values. Furthermore, the latter two also allow one to deduce metabolic networks (and protein–protein interactions for metabolic pathways) from few gene expression or metabolite data.

### *5.2. Regulatory Networks*

Regulatory pathways deal with the communication between cells and the organization control within the organism. Cell processes can be controlled by different signals.

All regulatory pathways together built the regulatory network. A database that visualizes regulatory pathways is the BIOCARTA database *(45)*. Several more tools for such regulatory pathways have appeared; a good introduction is the subway map of cancer (http://www.nature.com/nrc/journal/v2/n5/weinberg_poster/mobilization.html), a good database is the Gene Express System from Kolshanov *(46)* or the recent development of the REACTOME map *(47)*.

Dynamical and kinetic modeling of regulatory and metabolic networks is a current challenge. This includes software such as PLAS *(48)* (build on the S-system formalism from Voit *[49]*), metabolic control analysis *(49)*, and modeling with differential equations or even time series analysis *(50)*. Dynamical data and the proteins hidden behind gene expression changes can also be revealed by different gene expression analysis tools; here, we recommend the open source and free bioconductor suite based on the statistical package R *(51)*.

### 5.3. Prediction of Protein–Protein Network Partners

Next to networks that contain experimental or affirmed interaction data, there also exist databases that predict interactions based on their stored data.

One of those databases is the "Search Tool for the Retrieval of Interacting Genes/ Proteins"(STRING) *(53–55)*, which is a database with included web tool. It summarizes experimental and predicted gene and protein interactions and gives a likelihood for the interaction to be true (*see* **Fig. 1**). The search result is presented in a table that contains all possible interaction partners and in the form of a network.

For the prediction calculation, STRING uses interaction information mainly from four different sources, so called "*evidence types*:"

- Gene context prediction
- Own height throughput experiments
- Own coexpression

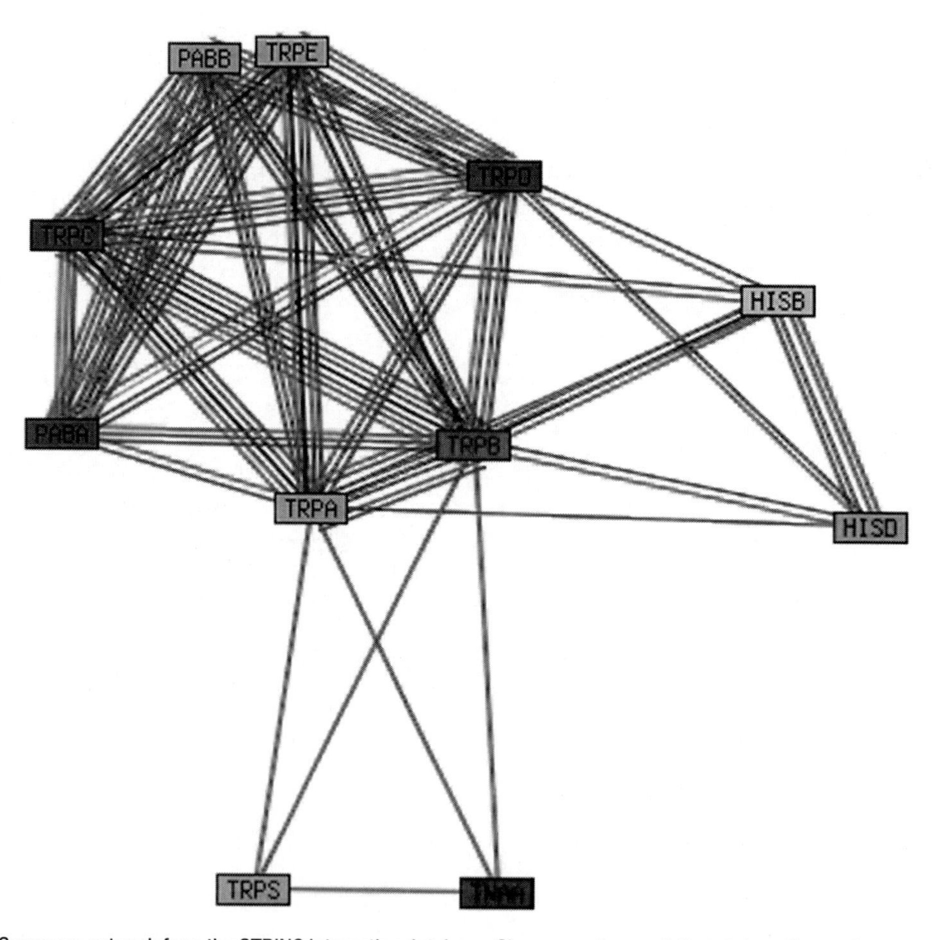


Fig. 1. Summary network from the STRING interaction database. Shown are the most likely interaction partners according to the database regarding the protein TrpB (*center*). If several types of evidence support a direct interaction, several lines connect the protein with the predicted interaction partner.

- Information from other databases
  (these include Bind *[8]*, Dip *[9]*, Pubmed *[13]*, Kegg *[41]*, Biocarta *[45]*, ArrayProspector *[56]*, Mips *[57]*, and Mint *[58]*)

Each evidence type receives an own row score, and all row scores are combined to one STRING score that represents the likelihood to be true for that specific interaction.

The STRING network presents all evidence types marked by different colors.

One big advantage of STRING is its size. The database currently contains 1,513,782 proteins in 373 species. With this large size, it is possible to transfer information from well-known model organisms to more unknown organisms. That means many more interactions are available in STRING than just the interaction sum of individual databases from single organisms.

#### 5.3.1. Genome Context

One evidence type in STRING uses genomic context methods, which is based on observations in the genome. Functionally related proteins are often coded by genes that are located near to each other in the genome or that have another relationship to each other in the genome.

They often experience a common selection pressure. Consequently, proteins with functional related genes often own similar behavior in different species. Genomic context can be divided into three different methods: gene-fusion, gene-neighborhood, and the phylogenetic profile.

#### 5.3.2. Gene-Fusion

During evolution, in some organisms, the former separated genes fuse to a common, bigger gene. Parallel to that observation in other organisms, the same genes keep separated (*Rosetta stone sequences*) *(59)*. That observation is a hint for a functional relationship between those two genes. The fusion's lifetime is only high when the coded proteins do have a functional relationship or are always used together. Represents for gene fusion are *TrpA* and *TrpB*, which are fused in *Escherichia coli.* **Figure 2** represents the different Genomic context methods. On the right side, the gene fusion is presented. For instance, the fusion between *TrpA* (*black*) and *TrpB* (*white*) is a hint that, also in the other listed species, a functional association can be found.

#### 5.3.3. Gene-Neighborhood

Very often, functional associated genes are located near to each other in the genome. By this, it is possible to regulate those genes together; especially when they have the same orientation, the genes can even be transcribed together. This allows the common presence of the coded proteins in the cell.

The described situation can especially be found in prokaryotes with their operon system *(60)*. An operon is a transcription unit that collects genes that are regulated in common.

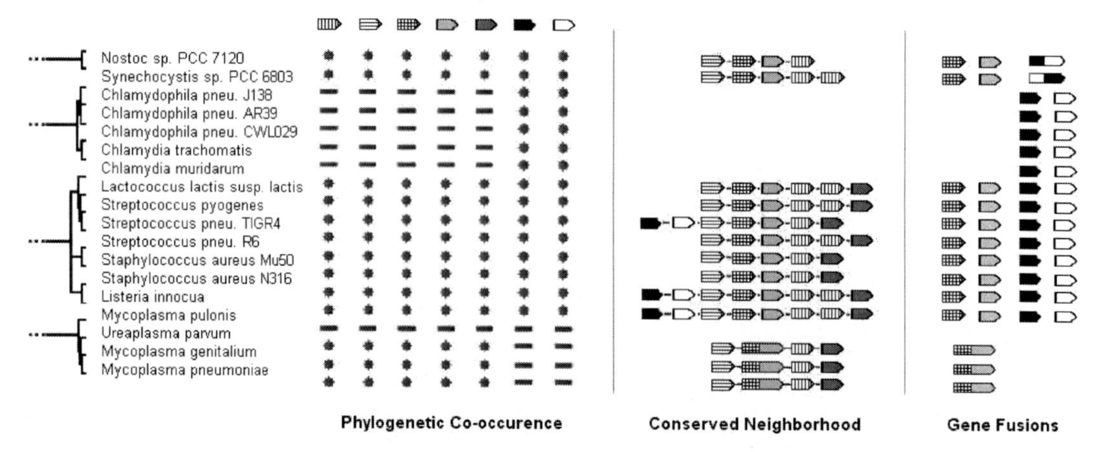


Fig. 2. Genomic Context Methods. Three different methods exploiting genome context conserved in many genomes for protein–protein interaction prediction are illustrated for the STRING database *(55)*. The prokaryotic genomes compared are given on the *left*. Methods (**a**) "Phylogenetic cooccurrence"–The first six protein genes as well as the last two genes are always together present (*filled dots*) or absent (*little bars*) in one genome. (**b**) "Conserved neighborhood"–Different shades indicate different genes, in the example for the central genes the same neighbors are found in most of the genomes compared, indicating interaction (physical or functional). (**c**) "Gene fusions" –The same reasoning applies for gene fusion events. These occur rare, but provide strong evidence. If two genes are fused in one or several genomes (*bottom three genomes*), this is a strong indication for interaction of their protein products.

The associated proteins in prokaryotes are also prediction pointers for the corresponding proteins in eukaryotes.

In the gene fusion method, genes have to be fused in at least one genome compared. A location near to each other in the genome is enough.

The middle part of **Fig. 2** illustrates gene neighborhood evidence. Genes (example: *Staphylococcus aureus) Mu50* Q99UB3 (fasciated), Q99UB4 (checkered), *TrpD* (light gray), and *TrpC* (striped) are present in all genomes, and additionally, they have the same orientation. For this reason, a functional relationship is likely for those genes.

*5.3.4. Phylogenetic Profile*

A phylogenetic profile is the most general form of genome context. The simple cooccurrence of genes is enough for the prediction independently from the location in the genome. Thus, the method is very sensitive to find associations, but it also underlies a huge background noise. For best accuracy, many genomes have to be compared and combined *(61)*.

The left part of **Fig. 2** visualizes the phylogenetic profile. Dots show the presence of gene and dashes represent the absence of gene. In that case, a functional association between *TrpA* (*black*) and *TrpB* (*white*) is very likely because they are always together present or absent. The same is true for the other five genes.

This is a tool that we partly developed and used in our own hands; however, a number of other tools allow assembly and prediction of protein–protein interaction networks also.

We mentioned already TRANSFAC and GENOMATIX regarding transcription factor networks and complexes. The BIOBASE system also allows rapid setup of networks. Gene context methods are also used by other tools *(62, 63)*. However, the basic philosophy should have become clear from our example tool.

## 6. Example Interactomes

In the following section, we will show experimental and predictive tools and databases on the basis of three different organisms and how it is possible to rapidly gain insight into protein–protein interactions involved, even if only few data are experimentally provided at start.

We use for our examples the techniques mentioned earlier, in particular protein-protein interaction databases (e.g. STRING) and the large-scale metabolic databases such as KEGG. This is complemented by sequence analysis, domain architecture studies before a rapid setup of the network is possible. Expert knowledge of the system and biological awareness are also critical.

### 6.1. Human Platelet's Interactome

Blood coagulation is a very complex process. One part of the process is the *NO/cGMP Signalcascade*(*see* **Fig. 3**).

**C**yclic **G**uanosin**m**ono**p**hosphate (cGMP) is the second messenger in platelets and takes part in the musculature's relaxation and minimization of platelet adhesion and aggregation.

To avoid blood vessels' obstruction, platelets normally exist only in a very low number. The platelet's reduction is possible via cGMP that is built by guanylyl-cycase. This enzyme is activated by nitrogen oxide (NO), which is built in the endothelium. But the cGMP concentration increase affected by NO is only transient.

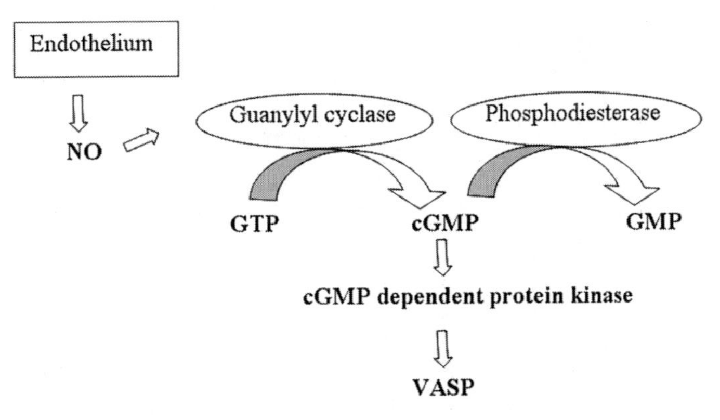


Fig. 3. cGMP function in human platelets. The cartoon sketches the steps from NO (nitric oxide) production to VASP (vasodilator-stimulated phosphoprotein) activation.

The platelet's inhibition is done via cGMP-depending protein kinase induced phosphorylization of the **Va**sodilator-stimulating **P**hosphoproteins (VASP). VASP is phosphorelated after NO treatment and is responsible for the actin-induced cell movements ***(64)***.

Even the described little part of the NO/cGMP signaling pathway includes many partly unknown interactions.

In the following section, we introduce the prediction tool STRING and show what can be found in experimental public databases.

#### *6.1.1. Guanylyl-Cyclase*

The guanylyl-cyclase is responsible for the GTP transformation in cGMP.

When you enter guanylyl-cyclase-activating protein in STRING, you will receive the result in **Fig. 4**.

On the left side of **Fig. 4**, you see the STRING network. The requested protein is marked in red. The edges' colors depend on the evidence type where the interaction was found with. On the right side, you see the interaction partners listed in a table together with a short description, the evidence type, and the evidence type's quality.

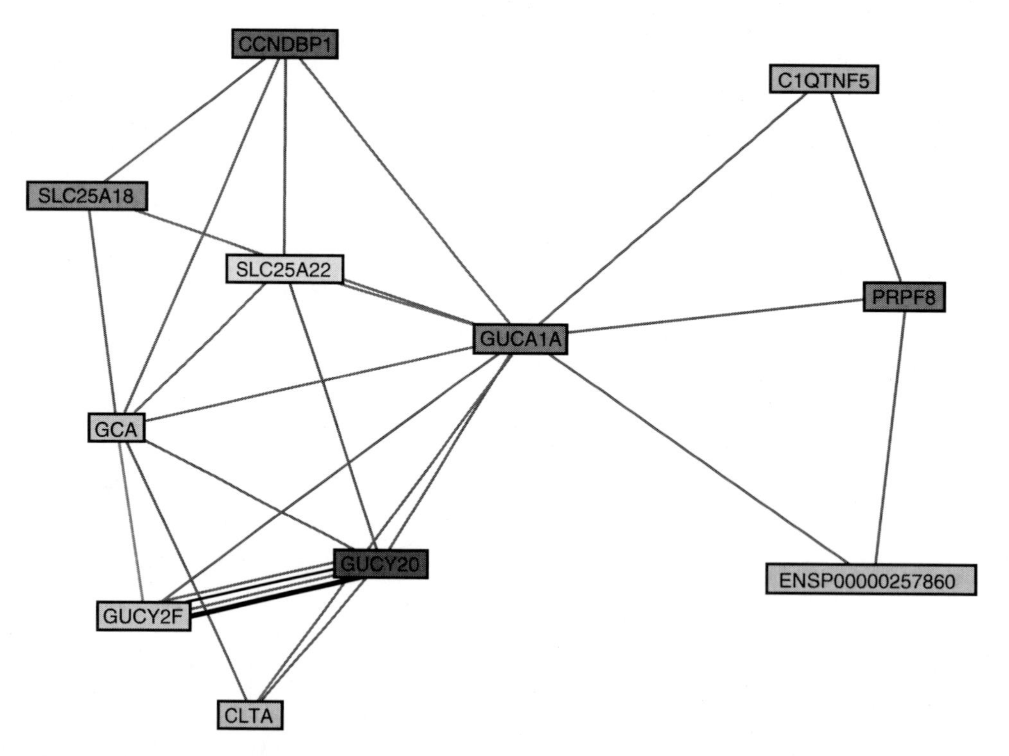


Fig. 4. Prediction of protein-protein interactions for guanylyl-cyclase activating protein 1 (GCAP1, *center*, it produces cGMP). The String database suggests a number of interaction partners (summary network, *left*), the evidence (*right*) is in all cases text mining with Bayesian confidence score from 0.996 down to 0.525.

In that case, the STRING results are mainly derived from the text-mining evidence type. For example, retinal guanylyl-cyclase precursor protein has a likelihood of $p = 0.986$. In that case, STRING found a known interaction partner. The comparison with KEGG database confirms the prediction. The cyclase precursor protein is an interaction partner of SLC5A22 ($p = 0.866$) and SLC25A18 ($p = 0.863$) in calcium metabolism, *see* **Fig. 5**.

The other five proteins predicted by STRING are widely unknown and to prove the prediction, experimental analysis would be needed.

Especially, complete C1q tumor nekrosefaktor precursor (C1QTNF5) with $p = 0.624$ could be interesting for more analysis.

1. **VASP**

   A second protein of the described signaling pathway is the until now relatively unknown VASP protein. In the STRING results, you see only one interaction partner ZYX with $p = 0.975$ found in another database (**Fig. 6**).

A very likely interaction partner for VASP is Vinculin ($p = 0.999$), found via experiments and literature. Only the Swissprot database protein description will give you a hint that the prediction is true.

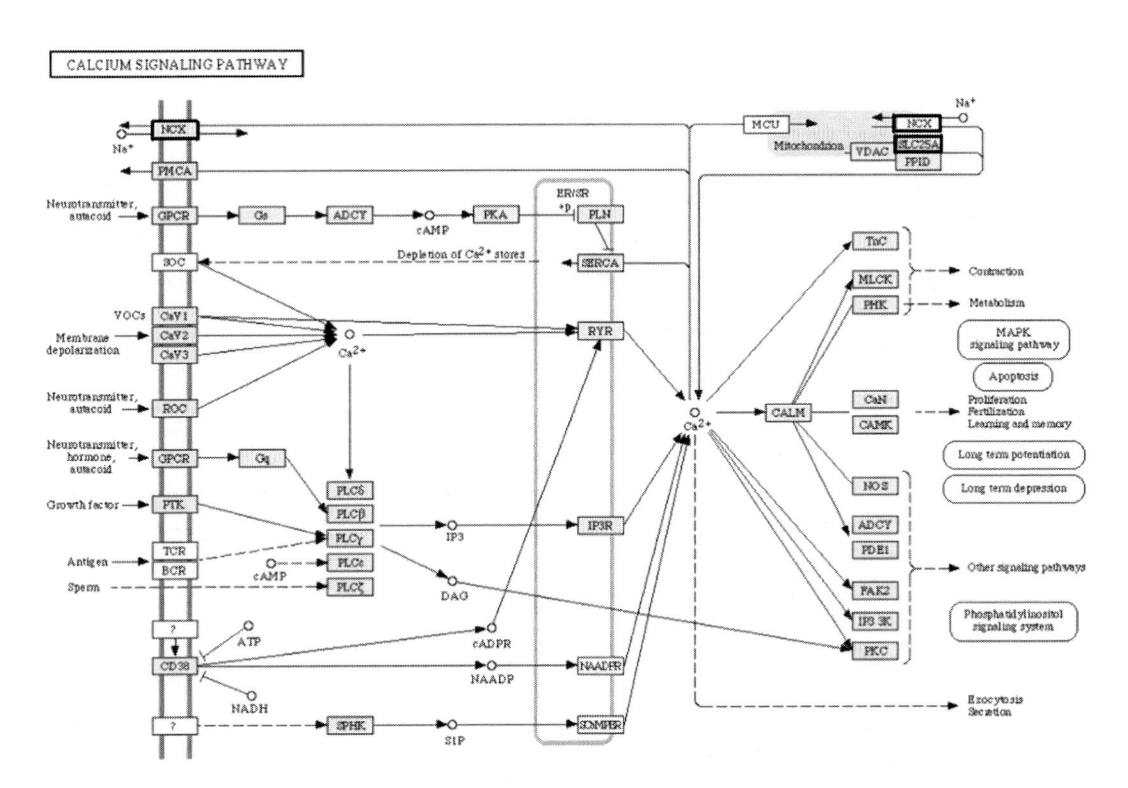


Fig. 5. Metabolism of calcium. Using the KEGG database, the enzymes found in man are shown in gray shading (*light green* according to the KEGG databank, map shown is hsa04020). The *right top corner* indicates in bold the example proteins SLC25A18 (SLC25A) and SLC5A22 (NCX).

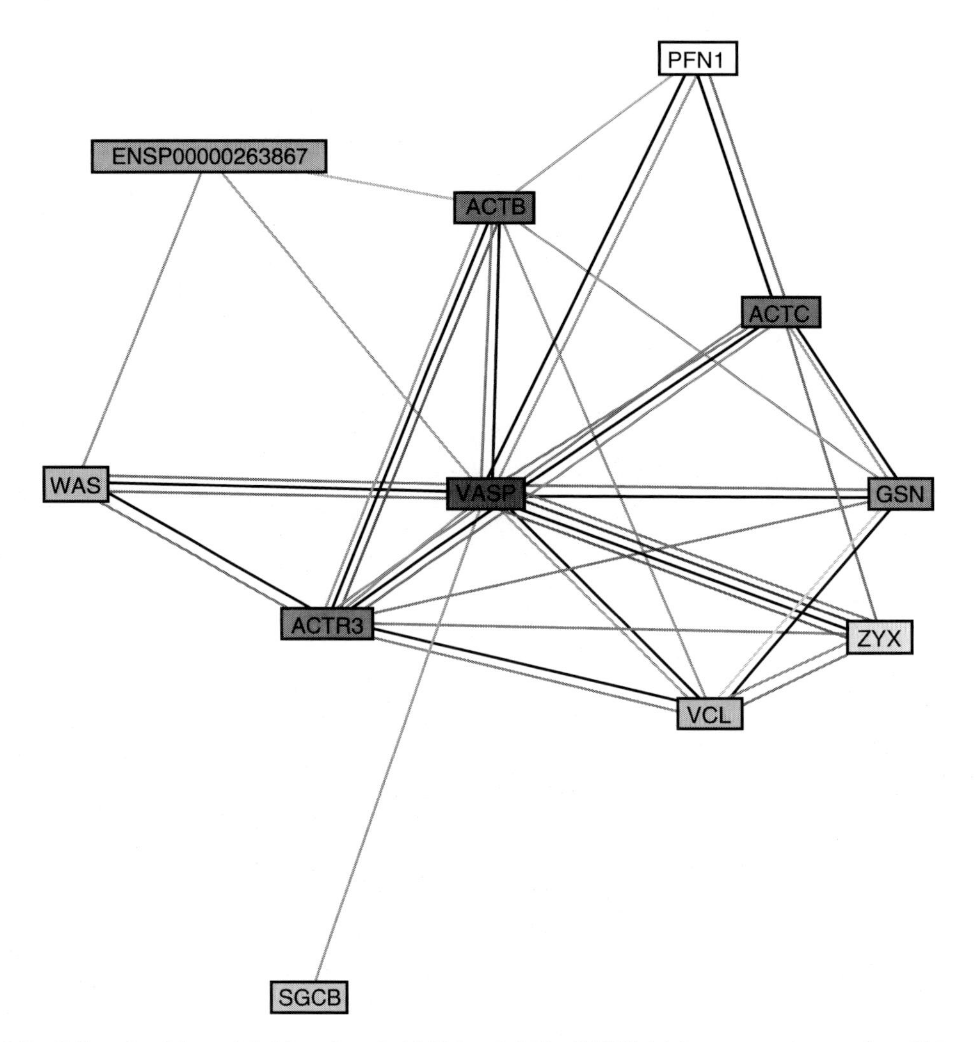


Fig. 6. Prediction of protein–protein interactions for VASP (*center*). The STRING database suggests a number of interaction partners (summary network, *left*). In this example, different evidence types such as experiments and text mining support different interactions (several connecting lines between the same protein partners).

Vinculin is involved in cell adhesion and actin-based microfilament deposition in plasma membrane.

A very interesting interaction partner suggested by STRING is the macrophage capping protein ($p = 0.930$) and the Wiskott-Aldrich syndrome protein ($p = 0.826$), both found via text-mining and experiments.

The macrophage capping protein is a calcium-sensitive protein, which reversibly inhibits actin filaments. It could play an important role in macrophage function. Both VASP and macrophage capping protein bind to actin. That is why there could be also interactions between these two proteins.

A similar situation can be found with the Wiskott-Aldrich syndrome protein. Wiskott-Aldrich syndrome is a congenital immunodeficiency caused by the lacking gene WASP. One of the disease's symptoms is a low number of platelets.

### *6.2. Listeria*

After the eukaryotic example, we turn to Listeria, gram-positive, and ubiquitous bacteria. Especially, three Listeria phenotypes are interesting for medicine:

– Listeria as commensal
– Listeria as pathogen in human
– Listeria as pathogen in animals

In the following section, we describe two examples from Listeria.

1. **PrfA (Transcription Factor)**

The first step is to search public interaction databases for affirmed interaction information about the protein or gene. In that case, neither the panther database nor the interdom database has information about PrfA.

However, the IntAct database has no information about PrfA in Listeria, but it contains information about PrfA (= rf1_ecoli) in *E.coli* (**Fig.** 7).

IntAct currently contains >65,000 binary and complex interactions imported from the literature and curated in collaboration with the Swissprot team, making intensive use of controlled vocabularies to ensure data consistency *(65)*.

More information about the found interaction partners for Prfa_Ecoli can be found in IntAct itself or in Swissprot. Interaction partners found by Interact are the following:

*gabt_ecoli:* Catabolite repression by glucose (repression relieved by GABA). Pathway: 4-aminobutyrate (GABA) degradation.

*nade:ecoli:* Catalyzes a key step in NAD biosynthesis, transforming deamido-NAD into NAD by a two-step reaction.

*rl4_ecoli:* One of the primary rRNA binding proteins, which initially binds near the 5 end of the 23S rRNA. It is important during the early stages of 50S assembly. It makes multiple contacts with different domains of the 23S rRNA in the assembled 50S subunit and ribosome.

*ch60_ecoli:* Prevents misfolding and promotes the refolding and proper assembly of unfolded polypeptides generated under stress conditions.

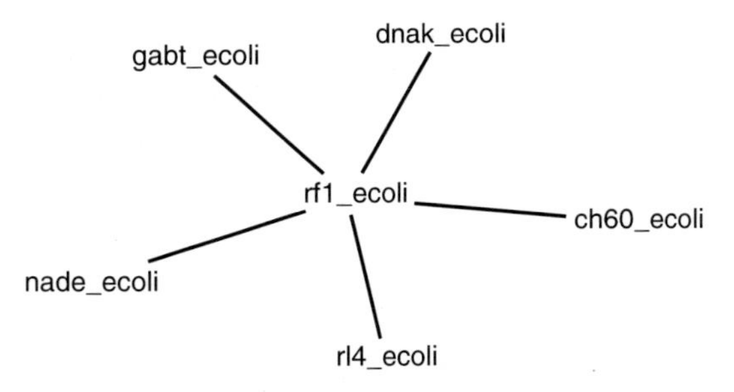


Fig. 7. Interactions predicted by the InterAct database. An InterAct interaction block is shown for the interaction partners of PrfA in *E. coli* (rf1_ecoli; a major regulator of virulence gene expression).

*dnak_ecoli:* Plays an essential role in the initiation of phage lambda DNA replication, where it acts in an ATP-dependent fashion with the dnaJ protein to release lambda O and P proteins from the preprimosomal complex. DnaK is also involved in chromosomal DNA replication, possibly through an analogous interaction with the dnaA protein,and also actively participates in the response to hyperosmotic shock.

The next step is to proof that a transfer from *E. coli* to Listeria is justified. That means we have to check whether the suggested interaction partner for *E. coli* has a homolog protein in Listeria. If a homolog cannot be found by name or in homolog databases like HomoloGene, we could search for similar sequences via BLAST comparisons. An example applying BLAST can be found in **Subheading 6.3**.

2. **Pentose Phosphate Cycle**

The pentose phosphate cycle is one possibility for the glucose catabolism. Its product can be used as basic modules for nucleotide biosynthesis. One important protein is the transketolase. It metabolizes D-xylulose 5-phosphate into fructose-6-phosphate (*see* Swissprot database) (**Fig. 8**).

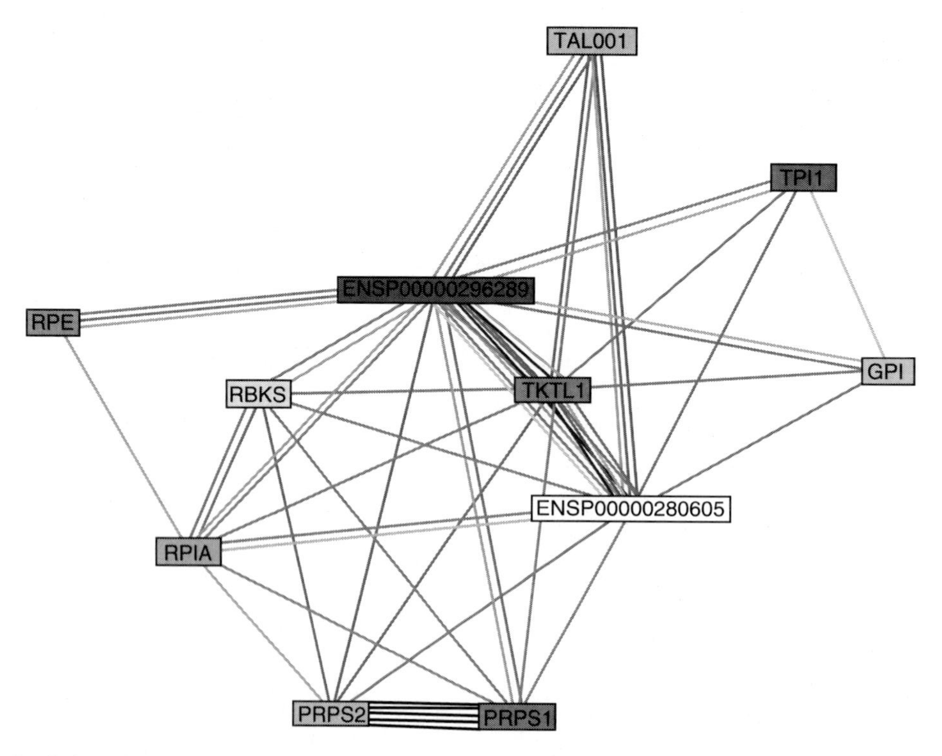


Fig. 8. Prediction of protein–protein interactions for transketolase (*top*; identifier ENSP00000296289). The STRING database suggests a number of interaction partners (summary network, *left*), the evidence (*right*) is from different interaction databases, and text-mining as well as further sources. *Lighter shades* show less strong evidence.

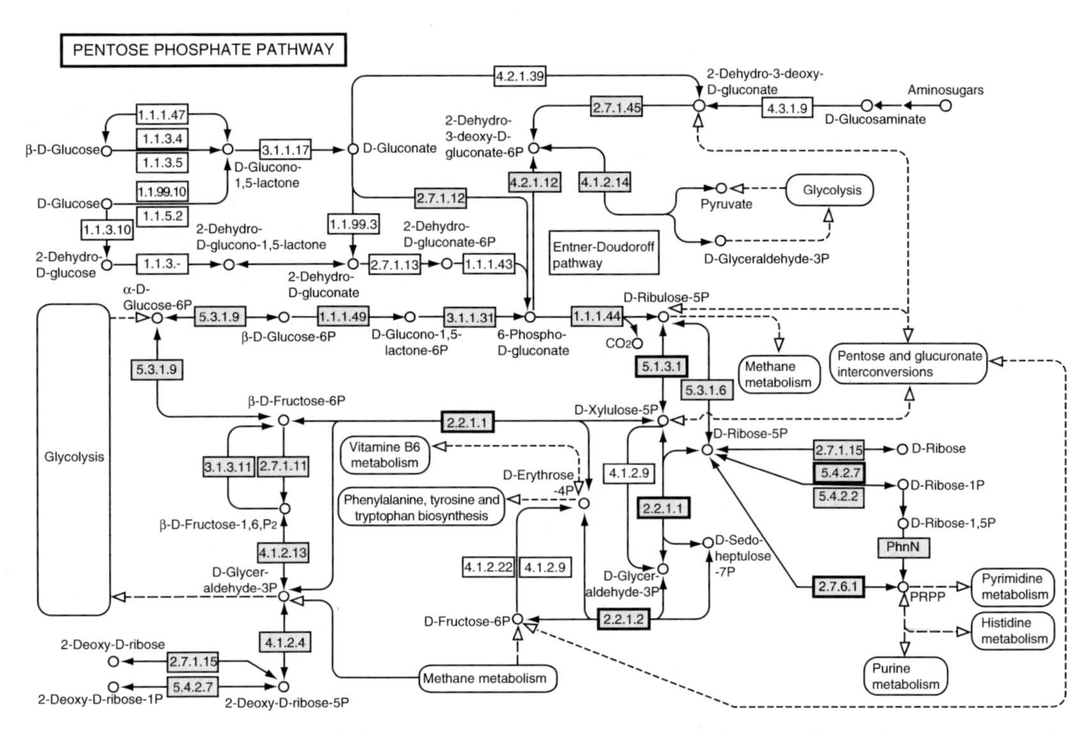


Fig. 9. Pentose phosphate pathway. Colored in gray are enzymes that appear in *L. monocytogenes* (we show here results from the KEGG database: lmo00030). To emphasize the interaction partners, we marked in this figure the transketolase (2.2.1.1) in bold. Further interaction partners are transaldolase (2.2.1.2), ribulose epimerase (5.1.3.1), ribokinase (5.4.2.7), pyrovat kinasae (2.7.6.1).

All STRING results in that case have a high likelihood over 90% to interact in Listeria.

Especially transaldolase, ribulose epimerase, and ribokinase get top scores and are with high probability real interaction partners. They can be found via STRING with the help of other databases and text-mining. Additionally, they can be found via KEGG in the pentose phosphate pathway map. We colored the concerning proteins in the map (**Fig. 9**).

### 6.3. Arabidopsis

The third example deals with the eukaryotic organism Arabidopsis, a small flowering plant related to cabbage and mustard. This genus is of great interest since it contains Thale Cress (*Arabidopsis thaliana*), one of the model organisms used for studying plant biology and the first plant to have its entire genome sequenced. Changes in the plant are easily observed, making it a very useful model.

Until now, we always considered examples where we already knew what kind of protein we had or at least where the protein was already annotated. Now, we will consider an example where we only have a sequence and do not know to what kind of protein it belongs to.

Here, we took the following amino acid sequence:

```
MRSSFLVSNPPFLPSLIPRYSSRKSIRRSRERFSFPESLRVSLLHGIRRNIEVAQGVQFD
GPIMDRDVNLDDDLVVQVCVTRTLPPALTLELGLESLKEAIDELKTNPPKSSSGVLRFQV
AVPPRAKALFWFCSQPTTSDVFPVFFLSKDTVEPSYKSLYVKEPHGVFGIGNAFAFVHSS
SVDSNGHSMIKTFLSDESAMVTAYGFPDIEFNKYSTVNSKDGSSYFFVPQIELDEHEEVS
ILAVTLAWNESLSYTVEQTISSYEKSIFQVSSHFCPNVEDHWFKHLKSSLAKLSVEEIHP
LEMEHMGFFTFSGRDQADVKELLNRETEVSNFLRDEANINAVWASAIIEECTRLGLTYFC
VAPGSRSSHLAIAAANHPLTTCLACFDERSLAFHAIGYAKGSLKPAVIITSSGTAVSNLL
PAVVEASEDFLPLLLLTADRPPELQGVGANQAINQINHFGSFVRFFFNLPPPTDLIPVRM
VLTTVDSALHWATGSACGPVHLNCPFRDPLDGSPTNWSSNCLNGLDMWMSNAEPFTKYFQ
VQSHKSDGVTTGQITEILQVIKEAKKGLLLIGAIHTEDEIWASLLLAKELMWPVVADVLS
GVRLRKLFKPFVEKLTHVFVDHLDHALFSDSVRNLIEFDVVIQVGSRITSKRVSQMLEKC
FPFAYILVDKHPCRHDPSHLVTHRVQSNIVQFANCVLKSRFPWRRSKLHGHLQALDGAIA
REMSFQISAESSLTEPYVAHMLSKALTSKSALFIGNSMPIRDVDMYGCSSENSSHVVDMM
LSAELPCQWIQVTGNRGASGIDGLLSSATGFAVGCKKRVVCVVGDISFLHDTNGLAILKQ
RIARKPMTILVINNRGGGIFRLLPIAKKTEPSVLNQYFYTAHDISIENLCLAHG
```

With the sequence, a BLAST search was done in which the sequence was compared with public databases, we pasted our sequence into EBI WuBlast and received the results in **Table 1**.

The first column contains the database id where the Blast search was done against. The next two columns comprehend a short description and the sequence's length. The further columns show the BAST results quality.

The "Score" column means a total score that includes not only the found positions but also mismatch and gaps. The better the alignment's quality, the higher is the score.

The "Identity" column contains the number of exact matches between the search sequence and the database sequence. The "Positive" columns include additionally the similar matches.

In the last column, the e-value should always be as small as possible, as it indicates the possibility for a chance similarity. Blast interpretation: results are sorted by e-value.

To carry on, we examined the *E. coli* protein Mend_ecoli. Although there are homologous proteins with a better score, we will take the model organism *E. coli* because very likely more data about that gene and its products exist.

Next, we considered the protein's environment and interaction partners. The closer the original sequences are to the model sequence, the more likely is the possible function and interaction partner transfer from one protein to another. In KEGG, we found the Mend_ecoli protein as part of the ubiquinone biosynthesis pathway (**Fig. 10**).

**Table 1**
**BLAST comparison identifies similar proteins. In this table, the EBI–WuBLAST result is shown, conducted for a protein with previously unknown function. While the three columns on the left contain information about sequence-related proteins found by the database search in the UNIPROT database using BLAST, the following columns on the right indicate the quality of the search (score, identical positions found, similar or "positive" positions found, expected value "E-value" to find such a similarity by chance: very low probability, all hits are highly significant)**

| DB ID | Source | Length | Scored | Identity % | Positive % | E-value |
|---|---|---|---|---|---|---|
| UNIPROT:Q9CAB2_ARATH | Putative menaquinone biosynthesis protein | 894 | 4332 | 94 | 94 | 0. |
| UNIPROTrA3A8E0_ORYSA | Hypothetical protein | 1742 | 1695 | 55 | 73 | 6.5e-209 |
| UNIPROT:A2X6H9_ORYSA | Hypothetical prateinr | 1871 | 1697 | 55 | 73 | 1.7e-208 |
| UNIPROT:Q3VQJ3_9CHLB | Menaquinone biosynthesis protein | 593 | 565 | 38 | 54 | 4.9e-84 |
| UNIPROT:Q3VX3S_PROAE | Menaquinone biosynthesis protein | 625 | 579 | 35 | 55 | 1.2e-79 |
| UNIPROT:Q3APX4_CHLCH | Menaquinone biosynthesis protein | 583 | 557 | 35 | 54 | 1.9e-76 |
| UNIPROT:Q8KBE8_CHLTE | 2-succinyl-6-hydroxy-2,4-cyclohexadiene-1-carboxylate synthase (EC 4.1.1.71) | 601 | 560 | 33 | 53 | 5.2e-75 |
| UNIPROT:Q3B612_PELLD | Menaquinone biosynthesis protein | 583 | 537 | 35 | 55 | 1.0e-70 |
| UNIPR0T:A1SRV2_PSYIN | 2-succinyl-6-hydroxy-2, 4-cyclohexadiene-1-carboxylic acid synthase/2- oxoglutarate decarboxylase (EC 4.1.1.71} | 581 | 424 | 37 | 54 | 2.3e-67 |
| UNIPROT:Q3GJG7_CHLPH | Menaquinone biosynthesis protein | 570 | 509 | 35 | 54 | 2.9e-66 |
| UNIPROT:Q6D7W4_ERWCT | Menaquinone biosynthesis protein | 560 | 413 | 40 | 56 | 2.0e-65 |
| UNIPROT:A1JKT8_,YERE8 | Menaquinone biosynthesis protein | 557 | 412 | 38 | 54 | 2.6e-65 |

(continued)

**Table 1 (continued)**

| DB ID | Source | Length | Scored | Identity % | Positive % | E-value |
|---|---|---|---|---|---|---|
| UNIPR0T:Q1 CHS9_YERPN | Menaquinone biosynthesis protein | 567 | 411 | 37 | 53 | 2.3e-64 |
| UNIPROT:Q74T53_YERPE | Menaquinone biosynthesis protein | 567 | 411 | 37 | 53 | 2.3e-64 |
| UNIPROT:Q1C6D7_YERPA | Menaquinone biosynthesis protein | 567 | 411 | 37 | 53 | 2.3e-64 |
| UNIPROT:Q7CJ76_YERPE | 2-oxoglutarate decarboxylase (Menaquinone biosynthesis protein) (EC 4.1.3.-) | 567 | 411 | 37 | 53 | 2.3e-64 |
| UNIPROT:Q4C7L2_CROWT | Menaquinone biosynthesis protein | 582 | 508 | 31 | 51 | 4.8e-64 |
| UNIPROT:MEND_ECOLI | Menaquinone biosynthesis protein menD [includes: 2-succinyl-6-hydroxy- 2,4-cyclohexadiene-1-carboxylate synthase; 2-oxoglutarate decarboxytase ] | 556 | 407 | 39 | 55 | 2.8e-63 |

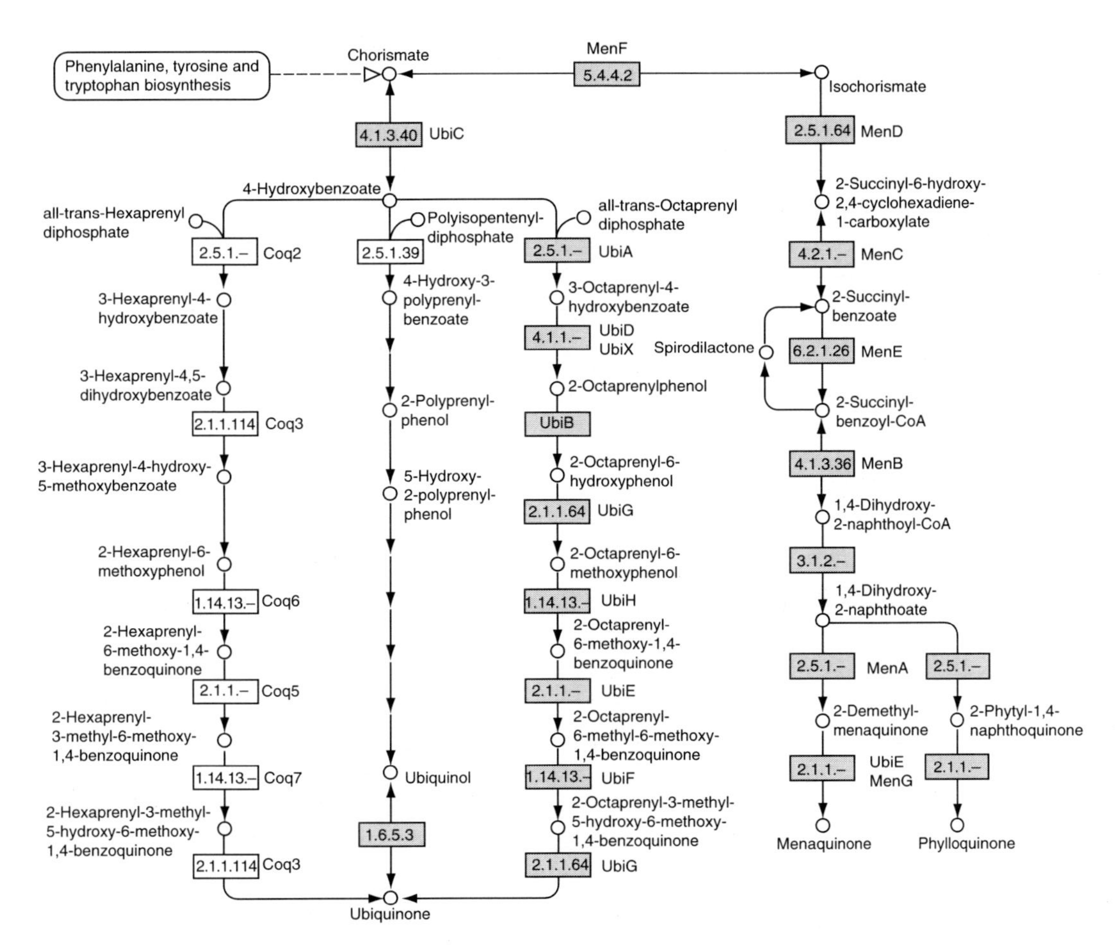


Fig. 10. The ubiquinone pathway in *E. coli* (00130). Results are shown from the KEGG database. The enzyme (EC number 2.5.1.64) encoded by the gene MenD Mend_ecoli is marked in bold.

Next, we examine the protein domain structure via the protein family domain database Pfam. The protein domain that can be found in Mend_ecoli is the Thiamine pyrophosphate (TPP) enzyme, N-terminal TPP binding domain.

A number of enzymes require TPP (vitamin B1) as a cofactor. Some of these enzymes are structurally related *(66)*.

The TPP domain interacts additionally with the following families:

| Domain Name | Description |
|---|---|
| TPP_enzyme_C | Thiamine Pyrophosphate enzyme, C-terminal TPP binding domain |
| TPP_enzyme_M | Thiamine Pyrophosphate enzyme, central domain |

That means proteins that contain one of the two mentioned domains are likely to interact with our examined protein.

The database Prosite contains many structure patterns. The pattern for our TPP domain is the following:

TPP_ENZYMES, PS00187; TPP enzymes signature

*Consensus pattern:* [LIVMF] – [GSA] – x(5) – P – x(4) – [LIVMFYW] – x – [LIVMF] – x – G – D – [GSA] – [GSAC]

In a further step, there could be a search with that pattern against the Swissprot database. These are of course only introductory examples and they focused on gene context methods and sequence comparison. Exploiting the other tools mentioned earlier would of course further broaden and extend these results.

## 7. Challenges in Protein-Protein Network Analysis

With an exponentially increasing amount of protein sequences available, their annotation becomes more and more important. For annotation, it is important to describe the function of the protein sequence and to identify the interaction partners. Moreover, interactions cannot be considered separately but have to be integrated into networks, as many proteins are involved in pathways and influence each other.

For annotation of new protein sequences and their interaction partners, we recommend the following workflow:

At first, it is always reasonable to have a look at experimental and other public databases like DIP, INTACT, KEGG, BIOCARTA, and others.

The second step is to use predictive databases and tools such as STRING to identify protein interactions on a network scale and programs such as FlexX to model individual interactions if more information is available.

To get more information on the protein sequence in question, you should do further analysis and interpretation. Important is information from other better known organisms having a homologous sequence or from proteins with similar structure. Homologies can be found in databases such as HomoloGene or you can identify them yourself using tools such as Blast, ClustalW, or more advanced approaches relying on PSSMs or HMMs.

When you find similar proteins, start looking in public databases for all experimental information available on these similar proteins to further confirm your results regarding assigned function or predicted interaction partners for your protein sequence.

### 7.1. Experimental Validation

Predictions cannot replace experimental data and experimental validation. However, predictive data can give important hints in which direction experiments should go. With the prediction of possible interaction partners, it is possible to reduce expensive experiments, as the best candidates can be selected and tested. This includes completely new, previously ignored interaction partners. Moreover, until now, physical binding data are only available for a few model organisms, in particular yeast. Applying predictions, it is possible to predict physical interactions automatically and for many organisms. Predictions mostly include data transfer between organisms; this applies to a large number of organisms and the description of their protein interactions. Biological functions of proteins and complexes can thus be better differentiated and experiments better planned.

### 7.2. Future Developments

Predictive methods normally depend on sufficient and available data. They become more powerful when more genes and proteins are annotated and more genomes are sequenced. Moreover, new and more powerful prediction methods are steadily coming up.

In future, prediction methods will gain even more importance to interpret the exponential increasing data and evidence types on proteins and protein–protein interactions. This chapter should stimulate own protein–protein interaction prediction efforts.

## Useful Weblinks

### Primary Databases

PubMed: http://www.ncbi.nlm.nih.gov/sites/entrez?db=pubmed
Swissprot: http://expasy.org/sprot/
EMBL: http://www.ebi.ac.uk/embl/

### Secondary Databases

RefSeq: http://www.ncbi.nlm.nih.gov/RefSeq/
MEROPS: http://merops.sanger.ac.uk/
OMIM: http://www.ncbi.nlm.nih.gov/sites/entrez?db=OMIM

### Interaction Databases

DIP: http://dip.doe-mbi.ucla.edu/
BOND: http://bond.unleashedinformatics.com
KEGG: http://www.genome.jp/kegg/
Biocarta: http://www.biocarta.com/
STRING: http://string.embl.de/

### Domain Databases

Pfam: http://www.sanger.ac.uk/Software/Pfam/
Smart: http://smart.embl-heidelberg.de/
Prosite: http://www.expasy.ch/prosite/
Interpro: http://www.ebi.ac.uk/interpro/

### Search Tools

NAR: http://www3.oup.co.uk/nar/database/c/
NCBI Entrez: http://www.ncbi.nlm.nih.gov/Entrez/
SRS: http://srs.ebi.ac.uk/srsbin/cgi-bin/wgetz?-page+quickSearch
Xplormed: http://www.bork.embl-heidelberg.de/xplormed/).

**Further Interesting Tools**

Blast: http://www.ebi.ac.uk/blast2/
ClustalW: http://www.ebi.ac.uk/clustalw/
BlockMaker: http://blocks.fhcrc.org/blocks/blockmkr/make_blocks.html
HMM: http://bioweb.pasteur.fr/seqanal/motif/hmmer-uk.html
Biobase: http://www.gene-regulation.com/pub/databases.html
Genomatix: http://www.genomatix.de/
Metatool: http://pinguin.biologie.uni-jena.de/bioinformatik/networks/
YANA: http://yana.bioapps.biozentrum.uni-wuerzburg.de/

## References


1. Galperin MY (2007) The Molecular Biology Database Collection: 2007 update. *Nucleic Acids Res* 35:D3–4.
2. Kulikova T, Akhtar R, Aldebert P, Althorpe N, Andersson M, Baldwin A, Bates K, Bhattacharyya S, Bower L, Browne P, Castro M, Cochrane G, Duggan K, Eberhardt R, Faruque N, Hoad G, Kanz C, Lee C, Leinonen R, Lin Q, Lombard V, Lopez R, Lorenc D, McWilliam H, Mukherjee G, Nardone F, Pastor MP, Plaister S, Sobhany S, Stoehr P, Vaughan R, Wu D, Zhu W, Apweiler R (2007) EMBL Nucleotide Sequence Database in 2006. *Nucleic Acids Res* 35:D16–20.
3. Benson DA, Karsch-Mizrachi I, Lipman DJ, Ostell J, Wheeler DL (2007) GenBank. *Nucleic Acids Res* 35:D21–5.
4. Tateno Y, Imanishi T, Miyazaki S, Fukami-Kobayashi K, Saitou N, Sugawara H, Gojobori T (2002) DNA Data Bank of Japan (DDBJ) for genome scale research in life science. *Nucleic Acids Res* 30(1):27–30.
5. Bairoch A, Apweiler R (1997) The SWISS-PROT protein sequence database: its relevance to human molecular medical research. *J Mol Med* 75(5):312–6.
6. Pruitt KD, Tatusova T, Maglott DR (2007) NCBI reference sequences (RefSeq): a curated non-redundant sequence database of genomes, transcripts and proteins. *Nucleic Acids Res* 35:D61–5.
7. Alfarano C, Andrade CE, Anthony K, Bahroos N, Bajec M, Bantoft K, Betel D, Bobechko B, Boutilier K, Burgess E, Buzadzija K, Cavero R, D'Abreo C, Donaldson I, Dorairajoo D, Dumontier MJ, Dumontier MR, Earles V, Farrall R, Feldman H, Garderman E, Gong Y, Gonzaga R, Grytsan V, Gryz E, Gu V, Haldorsen E, Halupa A, Haw R, Hrvojic A, Hurrell L, Isserlin R, Jack F, Juma F, Khan A, Kon T, Konopinsky S, Le V, Lee E, Ling S, Magidin M, Moniakis J, Montojo J, Moore S, Muskat B, Ng I, Paraiso JP, Parker B, Pintilie G, Pirone R, Salama JJ, Sgro S, Shan T, Shu Y, Siew J, Skinner D, Snyder K, Stasiuk R, Strumpf D, Tuekam B, Tao S, Wang Z, White M, Willis R, Wolting C, Wong S, Wrong A, Xin C, Yao R, Yates B, Zhang S, Zheng K, Pawson T, Ouellette BF, Hogue CW (2005) The Biomolecular Interaction Network Database and related tools 2005 update. *Nucleic Acids Res* 33:D418–24.
8. Bader GD, Betel D, Hogue CW (2000) BIND: the Biomolecular Interction Network Database. *Nucleic Acids Res* 28(1):235–42.
9. Xenarios I, Salwinski L, Duan XJ, Higney P, Kim SM, Eisenberg D (2002) DIP, the Database of Interacting Proteins: a research tool for studying cellular networks of protein interactions. *Nucleic Acids Res* 30(1):303–5.
10. Rawlings ND, Morton FR, Barrett AJ (2006) MEROPS: the peptidase database. *Nucleic Acids Res* 34:D270–2.
11. Amladi S (2003) Online Mendelian Inheritance in Man "OMIM". *Indian J Dermatol Venereol Leprol* 69(6):423–4.
12. Zdobnov EM, Lopez R, Apweiler R, Etzold T (2002) The EBI SRS server - recent developments. *Bioinformatics* 18:368–73.
13. Sood A, Ghosh AK (2006) Literature search using PubMed: an essential tool for practicing evidence-based medicine. *J Assoc Physicians India* 54:303–8.
14. Perez-Iratxeta C, Bork P, Andrade MA (2001) XplorMed: a tool for exploring MEDLINE abstracts. *Trends Biochem Sci* 26(9):573–5.
15. Altschul SF, Madden TL, Schaffer AA, Zhang J, Zhang Z, Miller W, Lipman DJ (1997) Gapped BLAST and PSI-BLAST: a new generation of protein database search programs. *Nucleic Acids Res* 25(17):3389–402.
16. Thompson JD, Higgins DG, Gibson TJ (1994) CLUSTAL W: improving the sensitivity

of progressive multiple sequence alignment through sequence weighting, position-specific gap penalties and weight matrix choice. *Nucleic Acids Res* 22:4673–80.

17. Schneider TD, Stephens RM (1990) Sequence logos: a new way to display consensus sequences. *Nucleic Acids Res* 18:6097–100.
18. Gribskov M, Homyak M, Edenfield J, Eisenberg D (1987) Profile scanning for three-dimensional structural patterns in protein sequences. *Comput Appl Biosci* 4(1):61–6.
19. Bateman A, Birney E, Durbin R, Eddy SR, Howe KL, Sonnhammer EL (2000) The Pfam protein families database. *Nucleic Acids Res* 28(1):263–6.
20. Letunic I, Copley RR, Pils B, Pinkert S, Schultz J, Bork P (2006) SMART 5: domains in the context of genomes and networks. *Nucleic Acids Res* 34(Database issue):D257–60.
21. Hulo N, Bairoch A, Bulliard V, Cerutti L, De Castro E, Langendijk-Genevaux PS, Pagni M, Sigrist CJ (2006) The PROSITE database. *Nucleic Acids Res* 34(Database issue):D227–30.
22. Mulder NJ, Apweiler R, Attwood TK, Bairoch A, Bateman A, Binns D, Bork P, Buillard V, Cerutti L, Copley R, Courcelle E, Das U, Daugherty L, Dibley M, Finn R, Fleischmann W, Gough J, Haft D, Hulo N, Hunter S, Kahn D, Kanapin A, Kejariwal A, Labarga A, Langendijk-Genevaux PS, Lonsdale D, Lopez R, Letunic I, Madera M, Maslen J, McAnulla C, McDowall J, Mistry J, Mitchell A, Nikolskaya AN, Orchard S, Orengo C, Petryszak R, Selengut JD, Sigrist CJ, Thomas PD, Valentin F, Wilson D, Wu CH, Yeats C (2007) New developments in the InterPro database. *Nucleic Acids Res* 35(Database issue):D224–8.
23. Wheeler DL, Church DM, Edgar R, Federhen S, Helmberg W, Madden TL, Pontius JU, Schuler GD, Schriml LM, Sequeira E, Suzek TO, Tatusova TA, Wagner L (2004) Database resources of the National Center for Biotechnology Information: update. *Nucleic Acids Res* 32: D35–40.
24. Durbin R, Eddy S, Krogh A, Mitchison G (1998). Biological Sequence Analysis. Cambridge University Press, Cambridge, UK
25. Friedrich T, Pils B, Dandekar T, Schultz J, Muller T (2006) Modelling interaction sites in protein domains with interaction profile hidden Markov models. *Bioinformatics* 22(23):2851–7.
26. Folkers G (1995) Lock and Key - A hundred years after, Emil Fisher Commemorate Symposium, *Pharmaceutica Acta Helvetiae* 69:175–269.
27. Steffen A, Kamper A, Lengauer T (2006) Flexible docking of ligands into synthetic receptors using a two-sided incremental construction algorithm. *J Chem Inf Model* 46(4):1695–703.
28. Kuntz ID, Meng EC, Shoichet BK (1994) Structure-based molecular design, *Acc Chem Res* 27:117–23.
29. Nixon BT, Ronson CW, Ausubel FM (1986) Two-component regulatory systems responsive to environmental stimuli share strongly conserved domains with nitrogen assimilation regulatory genes ntrB and ntrC. *Proc Natl Acad Sci* USA 83:7850–4.
30. Matys V, Fricke E, Geffers R, Gossling E, Haubrock M, Hehl R, Hornischer K, Karas D, Kel AE, Kel-Margoulis OV, Kloos DU, Land S, Lewicki-Potapov B, Michael H, Munch R, Reuter I, Rotert S, Saxel H, Scheer M, Thiele S, Wingender E (2003) TRANSFAC: transcriptional regulation from patterns to profiles. *Nucleic Acids Res* 31(1):374–8.
31. Klingenhoff A, Frech K, Werner T (2002) Regulatory modules shared within gene classes as well as across gene classes can be detected by the same in silico approach. *Silico Biol* 2:S17–26 Electronic publication: In Silico Biol. 1, 0020.
32. Gaudermann P, Vogl I, Zientz E, Silva FJ, Moya A, Gross R, Dandekar T (2006) Analysis of and function predictions for previously conserved hypothetical or putative proteins in Blochmannia floridanus. *BMC Microbiol* 9(6):1.
33. Berman H, Henrick K, Nakamura H, Markley JL (2007) The worldwide Protein Data Bank (wwPDB): ensuring a single, uniform archive of PDB data. *Nucleic Acids Res* 35(Database issue):D301–3.
34. Cambridge Structural Databse (CSD) http://www.ccdc.cam.ac.uk/products/csd/
35. Fields S, Song O (1989) A novel genetic system to detect protein-protein interactions. *Nature* 340:245–6. doi: 10.1038/340245a0.
36. Rigaut G, Shevchenko A, Rutz B, Wilm M, Mann M, Seraphin B (1999) A generic protein purification method for protein complex characterization and proteome exploration. *Nat Biotechnol* 17:1030–2.
37. Gavin AC, et al (2002).Functional organization of the yeast proteome by sytematic analysis of protein complexes. *Nature* 415:141–7.
38. von Mering Cet al. (2002) Comparative assessment of large scale datasets of protein–protein interactions. *Nature* 417:399–403.
39. Goll J, Uetz P (2006) The elusive yeast interactome. *Genome Biol* 7(6):223. Review.
40. Krause R, von Mering C, Bork P, Dandekar T (2004) Shared components of protein complexes – versatile building blocks or biochemical artefacts? *Bioessays* 26(12):1333–43.

41. Kanehisa M, Goto S, Kawashima S, Okuno Y, Hattori M (2004) The KEGG resource for deciphering the genome. *Nucleic Acids Res* 1:32 (Database issue):D277–80.
42. von Kamp A, Schuster S (2006) Metatool 5.0: fast and flexible elementary modes analysis. *Bioinformatics* 22(15):1930–1.
43. Schwarz R, Musch P, von Kamp A, Engels B, Schirmer H, Schuster S, Dandekar T (2005) YANA – a software tool for analyzing flux modes, gene-expression and enzyme activities. *BMC Bioinformatics* 6:135.
44. Roland Schwarz, Chunguang Liang, Christoph Kaleta, Mark Kuhnel, Eik Hoffmann, Sergei Kuznetsov, Michael Hecker, Garreth Griffith, Stefan Schuster, Thomas Dandekar (2007) Integrated network reconstruction, visualization and analysis using YANAsquare. *BMC Bioinformatics* 8:313 (10pp.)
45. BioCarta http://www.biocarta.com/genes/allPathways.asp.
46. Kolchanov NA, Ponomarenko MP, Kel AE, Kondrakhin Yu V, Frolov AS, Kolpakov FA, Goriachkovsky TN, Kel-Margulis OV, Ananko EA, Ignatieva EV, Podkolodnaia OA, Stepanenko IL, Merkulova TI, Babenko VN, Vorobiev DG, Lavryushev SV, Ponomarenko JV, Kochetove AV, Kolesov GN, Podkolodny NL, Milanesi L, Wingender E, Heinemeier T, Solovyev VV, Overton GC (1999) GeneExpress: a WWW-oriented integrator for databases and computer systems for studying the eukaryotic gene expression. *Biofizika* 44(5):837–41. ML 20089649.
47. Joshi-Tope G, Gillespie M, Vastrik I, D'Eustachio P, Schmidt E, de Bono B, Jassal B, Gopinath GR, Wu GR, Matthews L, Lewis S, Birney E, Stein L (2005) Reactome: a knowledgebase of biological pathways. *Nucleic Acids Res* 33(Database issue):D428–32.
48. Ferreira AEN, Ponces Freire AMJ, Voit EO (2003). A quantitative model of the generation of Ne-(carboxymethyl) lysine in the Maillard reaction between collagen and glucose. *Biochem J* 376(Pt1):109–121.
49. Savageau MA, Voit EO (1987) Recasting nonlinear differential equations as S-Systems: a canonical nonlinear form. *Math Biosci* 87(83):115.
50. Timmer J, Schwarz U, Voss HU, Wardinski I, Belloni T, Hasinger G, van der Klis M, Kurths J (2000) Linear and nonlinear time series analysis of the black hole candidate Cygnus X-1. *Phys Rev* E 61:1342–52.
51. Lottaz C, Spang R (2005) Stam – a Bioconductor compliant R package for structured analysis of microarray data. *BMC Bioinformatics* 6:211.
52. Fell DA (1992) Metabolic control analysis: a survey of its theoretical and experimental development. *Biochem J* 286 (Pt 2):313–30.
53. von Mering C , Jensen LJ, Kuhn M, Chaffron S, Doerks T, Krüger B, Snel B, Bork P (2006) STRING 7 – recent developments in the integration and prediction of protein interactions. *Nucleic Acids Res.* 2007 Jan 35 (Database issue):D358–362.
54. Mering C, Jensen LJ, Snel B, Hooper SD, Krupp M, Foglierini M, Joure N, Huynen MA, Bork P (2005) STRING: known and predicted protein-protein associations, integrated and transferred across organisms. *Nucleic Acids Res* 33(Database issue):D433–7.
55. von Mering C, Huynen M, Jaeggi D, Schmidt S, Bork P, Snel B (2003) STRING: a database of predicted functional associations between proteins. *Nucleic Acids Res* 31(1):258–61.
56. Jensen LJ, Lagarde J, von Mering C, Bork P (2004) ArrayProspector: a web resource of functional associations inferred from microarray expression data. *Nucleic Acids Res* 32 (Web Server Issue):W445–448.
57. Mewes HW, Frishman D, Mayer KF, Munsterkotter M, Noubibou O, Pagel P, Rattei T, Oesterheld M, Ruepp A, Stumpflen V.(2006) MIPS: analysis and annotation of proteins from whole genomes in 2005. *Nucleic Acids Res* 34(Database issue):D169–72.
58. Chatr-aryamontri A, Ceol A, Palazzi LM, Nardelli G, Schneider MV, Castagnoli L, Cesareni G.(2006) MINT: the Molecular INTeraction database. *Nucleic Acids Res* 35(Database issue):D572–4.
59. Galperin MY, Koonin EV (2000) Who's your neighbor? New computational approaches for functional genomics. *Nat Biotechnol* 18(6):609–13.
60. Zaslaver A, Mayo AE, Rosenberg R, Bashkin P, Sberro H, Tsalyuk M, Surette MG, Alon U (2004) Just-in-time transcription program in metabolic pathways. *Nat Genet* 36(5):486–91.
61. Huynen M, Snel B, Lathe W 3rd, Bork P (2000) Predicting protein function by genomic context: quantitative evaluation and qualitative inferences. *Genome Res* 10(8):1204–10.
62. Dietmann S, Aguilar D, Mader M, Oesterheld M, Ruepp A, Stuempflen V, Mewes HW (2006) Resources and tools for investigating biomolecular networks in mammals. *Curr Pharm Des* 12(29):3723–34.
63. Liu Y, Kuhlman B (2006) Rosetta Design server for protein design. *Nucleic Acids Res* 34(Web Server issue):W235–8.
64. Kuhnel K, Jarchau T, Wolf E, Schlichting I, Walter U, Wittinghofer A, Strelkov SV

(2004) The VASP tetramerization domain is a right-handed coiled coil based on a 15-residue repeat. *Proc Natl Acad Sci U S A* 101(49):17027–32.

65. Hermjakob H, Montecchi-Palazzi L, Lewington C, Mudali S, Kerrien S, Orchard S, Vingron M, Roechert B, Roepstorff P, Valencia A, Margalit H, Armstrong J, Bairoch A, Cesareni G, Sherman D, Apweiler R (2004) IntAct: an open source molecular interaction database. *Nucleic Acids Res* 32(1):D452–5.
66. Arjunan P, Umland T, Dyda F, Swaminathan S, Furey W, Sax M, Farrenkopf B, Gao Y, Zhang D, Jordan F (1996) Crystal structure of the thiamin diphosphate-dependent enzyme pyruvate decarboxylase from the yeast Saccharomyces cerevisiae at 2.3 A resolution. *J Mol Biol* 256(3):590–600.
67. Schuster, S., Fell, D. und Dandekar, T. (2000) A general definition of metabolic pathways useful for systematic organization and analysis of complex metabolic networks. Nature Biotechnology 18, 326–332.
68. Dandekar, T., Schuster, S., Snel, B., Huynen, M. und Bork, P. (1999) Pathway alignment: application to the comparative analysis of glycolytic enzymes. *Biochemical Journal* 343, 115–124.
69. Dandekar, T. und Schmidt, S. (2004) Metabolites and Pathway flexibility. In Silico Biology, 5 (No. 0012), pp. 1–13.

# INDEX

## A



## B



## C

## M



## N

## O



## P

## Q



## R



## S

## T

## U



## V



## W



## Z